THE STRUCTURE
AND FUNCTION OF MUSCLE

SECOND EDITION

VOLUME IV

Pharmacology and Disease

CONTRIBUTORS

E. B. BECKETT

T. M. BELL

F. D. BOSANQUET

G. H. BOURNE

P. M. DANIEL

E. J. FIELD

P. C. C. GARNHAM

M. N. GOLARZ DE BOURNE

JOHN T. HACKETT

RONALD A. HENSON

H. WARNER KLOEPFER

KARL MASON

H. B. PARRY

DAVID M. J. QUASTEL

JOHN N. WALTON

THE STRUCTURE

AND FUNCTION OF MUSCLE

Second Edition

VOLUME IV

Pharmacology and Disease

Edited by

Geoffrey H. Bourne

Yerkes Primate Research Center
Emory University
Atlanta, Georgia

ACADEMIC PRESS   *New York   San Francisco   London*   *1973*
A Subsidiary of Harcourt Brace Jovanovich, Publishers

ACADEMIC PRESS, INC.
111 Fifth Avenue, New York, New York 10003

United Kingdom Edition published by
ACADEMIC PRESS, INC. (LONDON) LTD.
24/28 Oval Road, London NW1

LIBRARY OF CONGRESS CATALOG CARD NUMBER: 72-154373

PRINTED IN THE UNITED STATES OF AMERICA

CONTENTS

1. Effect of Drugs on Smooth and Striated Muscles

David M. J. Quastel and John T. Hackett

2. Effects of Nutritional Deficiency on Muscle

Karl Mason

3. Virus Infections

T. M. Bell and E. J. Field

4. Parasitic Infections

P. C. C. Garnham

5. Histochemistry of Skeletal Muscle and Changes in Some Muscle Diseases

E. B. Beckett and G. H. Bourne
Revised by M. N. Golarz de Bourne and G. H. Bourne

6. Myopathy, The Pathological Changes in Intrinsic Diseases of Muscles

F. D. Bosanquet, P. M. Daniel, and H. B. Parry
Revised by G. H. Bourne and M. N. Golarz de Bourne

7. Clinical Aspects of Some Diseases of Muscle

Ronald A. Henson

8. Genetic Aspects of Muscular and Neuromuscular Diseases

H. Warner Kloepfer and John N. Walton

LIST OF CONTRIBUTORS

Numbers in parentheses indicate the pages on which the authors' contributions begin.

E. B. BECKETT, *Demyelinating Diseases, Newcastle General Hospital, Newcastle-Upon-Tyne, England* (289)

T. M. BELL, *Demyelinating Diseases, Newcastle General Hospital Newcastle-Upon-Tyne, England* (207)

F. D. BOSANQUET, *West End Hospital for Neurology and Neurosurgey and Maida Vale Hospital for Nervous Diseases, London, England* (359)

G. H. BOURNE, *Yerkes Primate Research Center, Emory University, Atlanta, Georgia* (289, 359)

P. M. DANIEL, *Department of Neuropathology, Institute of Psychiatry, The Maudsley Hospital, London, England* (359)

E. J. FIELD, *Demyelanating Diseases, Newcastle General Hospital Newcastle-Upon-Tyne, England* (208)

P. C. C. GARNHAM, *Imperial College Field Station, Ascot, Berkshire, England* (249)

M. N. GOLARZ DE BOURNE, *Yerkes Primate Research Center, Emory University, Atlanta, Georgia* (289, 359)

JOHN T. HACKETT, *Department of Pharmacology University of British Columbia, Vancouver, Canada* (1)

RONALD A. HENSON, *Neurological Department, London Hospital, White Chapel, London, England* (433)

H. WARNER KLOEPFER, *Department of Anatomy, Tulane University, New Orleans, Louisiana* (475)

KARL MASON, *National Institutes of Health, Bethesda, Maryland* (155)

H. B. PARRY, *Nuffield Institute for Medical Research, University of Oxford, Oxford, England* (359)

DAVID M. J. QUASTEL, *Department of Pharmacology, University of British Columbia, Vancouver, Canada* (1)

JOHN N. WALTON, *Regional Neurological Centre, General Hospital, Newcastle-Upon-Tyne, England* (475)

PREFACE

In the years that have elapsed since the first edition of this work was published in 1960, studies on muscle have advanced to such a degree that a second edition has long been overdue. Although the original three volumes have grown to four, we have covered only a fraction of the new developments that have taken place since that time. It is not surprising that these advances have not been uniform, and in this new edition not only have earlier chapters been updated but also areas in which there was only limited knowledge before have been added. Examples are the development of our knowledge of crustacean muscle (172 of 213 references in the reference list for this chapter are dated since the first edition appeared) and arthropod muscle (205 of 233 references are dated since the last edition). Obliquely striated muscle, described in 1869, had to wait until the electron microscope was focused on it in the 1960's before it began to yield the secrets of its structure, and 33 of 43 references dated after 1960 in this chapter show that the findings described are the result of recent research. There has also been a great increase in knowledge in some areas in which considerable advances had been made by the time the first edition appeared. As an example, in Dr. Hugh Huxley's chapter on Molecular Basis of Contraction in Cross-Striated Muscles (in Volume I) 76 of his 126 references are dated after 1960.

The first volume of this new edition deals primarily with structure and considers muscles from the macroscopic, embryonic, histological, and molecular points of view. The other volumes deal with further aspects of structure, with the physiology and biochemistry of muscle, and with some aspects of muscle disease.

We have been fortunate in that many of our original authors agreed to revise their chapters from the first edition, and it has also been our good fortune to find other distinguished authors to write the new chapters included in this second edition.

To all authors I must express my indebtedness for their hard work and patience, and to the staff of Academic Press I can only renew my confidence in their handling of this publication.

Geoffrey H. Bourne

PREFACE
TO THE FIRST EDITION

Muscle is unique among tissues in demonstrating to the eye even of the lay person the convertibility of chemical into kinetic energy.

The precise manner in which this is done is a problem, the solution of which has been pursued for many years by workers in many different disciplines; yet only in the last 15 or 20 years have the critical findings been obtained which have enabled us to build up some sort of general picture of the way in which this transformation of energy may take place. In some cases the studies which produced such rich results were carried out directly on muscle tissue. In others, collateral studies on other tissues were shown to have direct application to the study of muscular contraction.

Prior to 1930 our knowledge of muscle was largely restricted to the macroscopic appearance and distribution of various muscles in different animals, to their microscopical structure, to the classic studies of the electro- and other physiologists, and to some basic chemical and biochemical properties. Some of the latter studies go back a number of years and might perhaps be considered to have started with the classic researches of Fletcher and Hopkins in 1907, who demonstrated the accumulation of lactic acid in contracting frog muscle. This led very shortly afterward to the demonstration by Meyerhof that the lactic acid so formed is derived from glycogen under anaerobic conditions. The lactic acid formed is quantitatively related to the glycogen hydrolyzed. However, it took until nearly 1930 before it was established that the energy required for the contraction of a muscle was derived from the transformation of glycogen to lactic acid.

This was followed by the isolation of creatine phosphate and its establishment as an energy source for contraction. The isolation of ADP and ATP and their relation with creatine phosphate as expressed in the Lohmann reaction were studies carried out in the thirties. What might be described as a spectacular claim was made by Engelhart and

Lubimova, who in the 1940's said that the myosin of the muscle fiber had ATPase activity. The identification of actin and relationship of actin and myosin to muscular contraction, the advent of the electron microscope and its application with other physical techniques to the study of the general morphology and ultrastructure of the muscle fibers were events in the 1940's which greatly developed our knowledge of this complex and most mobile of tissues.

In the 1950's the technique of differential centrifugation extended the knowledge obtained during previous years of observation by muscle cytologists and electron microscopists to show the differential localization of metabolic activity in the muscle fiber. The Krebs cycle and the rest of the complex of aerobic metabolism was shown to be present in the sarcosomes—the muscle mitochondria.

This is only a minute fraction of the story of muscle in the last 50 years. Many types of discipline have contributed to it. The secret of the muscle fiber has been probed by biochemists, physiologists, histologists and cytologists, electron microscopists and biophysicists, pathologists, and clinicians. Pharmacologists have insulted skeletal, heart, and smooth muscle with a variety of drugs, *in vitro, in vivo,* and *in extenso;* nutritionists have peered at the muscle fiber after vitamin and other nutritional deficiencies; endocrinologists have eyed the metabolic process through hormonal glasses. Even the humble histochemist has had the temerity to apply his techniques to the muscle fiber and describe results which were interesting but not as yet very illuminating—but who knows where knowledge will lead. Such a ferment of interest (a statement probably felicitously applied to muscle) in this unique tissue has produced thousands of papers and many distinguished workers, many of whom we are honored to have as authors in this compendium.

Originally we thought, the publishers and I, to have *a book* on muscle which would contain a fairly comprehensive account of various aspects of modern research. As we began to consider the subjects to be treated it became obvious that two volumes would be required. This rapidly grew to three volumes, and even so we have dealt lightly or not at all with many important aspects of muscle research. Nevertheless, we feel that we have brought together a considerable wealth of material which was hitherto available only in widely scattered publications. As with all treatises of this type, there is some overlap, and it is perhaps unnecessary to mention that to a certain extent this is desirable. It is, however, necessary to point out that most of the overlap was planned, and that which was not planned was thought to be worthwhile and was thus not deleted.

We believe that a comprehensive work of this nature will find favor with all those who work with muscle, whatever their disciplines, and that although the division of subject matter is such that various categories of workers may need only to buy the volume which is especially apposite to their specialty, they will nevertheless feel a need to have the other volumes as well.

The Editor wishes to express his special appreciation of the willing collaboration of the international group of distinguished persons who made this treatise possible. To them and to the publishers his heartfelt thanks are due for their help, their patience, and their understanding.

Emory University, Atlanta, Georgia   GEOFFREY H. BOURNE
October 1, 1959

CONTENTS OF OTHER VOLUMES

Volume III: Physiology and Biochemistry

1

EFFECTS OF DRUGS ON SMOOTH AND STRIATED MUSCLE*

DAVID M. J. QUASTEL and JOHN T. HACKETT

* The work done in the authors' laboratory was supported by grants from the Muscular Dystrophy Association of Canada and the Medical Research Council.

I. Introduction

Of all areas of pharmacology, the study of action of drugs on muscle is at present one of the most active; a cursory glance at the titles in any current pharmacology journal will show that up to half the communications bear directly on smooth muscle pharmacology, and a large fraction of the remainder are highly relevant to the subject. It is a reflection of the major developments of the past few years that a large proportion of current reports serve mainly to support, amplify, and fill in a fairly coherent picture which has emerged in recent years of how drugs interact with muscle cells to modify their behavior. Nevertheless, it remains true that for not one drug can it be said that the mode of action is understood. Our ignorance stems from incompleteness of knowledge not only of the molecular structure of membranes and their receptors which interact with drugs, but also from lack of understanding of function at a relatively gross level. For example, the morphology of smooth muscle has only recently been investigated in sufficient detail to permit interpretation, in terms of an electrical model, of how action potentials and synaptic potentials may be conducted from one cell to another within the tissue (see Burnstock, 1970). The overall picture not only is incomplete in its details, but in many areas remains nebulous and even chaotic. If investigation continues at the present rate, the very near future will doubtless bring major changes not only in detail, but in concept, from what is presently hypothesized. For those areas where our understanding is minimal, only a comprehensive review could do justice to all the investigation which has been pursued, and such a review would of necessity become no more than a catalog of phenomenology. We intend, therefore, in this chapter to deal primarily with drug effects that can be interpreted in terms of our present understanding of muscle physiology. In particular, as a reflection of our own personal interest, wherever it is possible, attention will be focused on information elucidated by electrophysiological techniques, which provide a resolution of events in time and place still unmatched by other methods. The relative ease of applying these methods (and indeed, other techniques as well) to skeletal muscle as distinct from smooth has permitted extensive and detailed investigation of the physiological mechanisms involved in neuromuscular transmission and excitation–contraction coupling. As a result, it is often possible to describe drug actions in terms of interaction with specific steps in the activation of muscle contraction.

The pharmacology of smooth muscle relies a great deal upon analogies with the better studied and more thoroughly understood skeletal muscle system; often indirect evidence must serve to indicate basic similarities and differences. For this reason we will not pursue the usual practice of dealing separately with smooth and skeletal muscle. Indeed, smooth muscle is itself so diverse that from one type of smooth muscle to another there may be more differences than between skeletal muscle and smooth muscle in general. Moreover, we feel that little purpose can be served by the usual classification of drug action into "direct" and "indirect"; the semantic difficulties seem to us too great. Thus, one could define as "direct" only an action upon the contractile system itself, making the overwhelming majority of drugs "indirect" in their action. Alternatively, and more usefully, classification as "direct" might mean an action on the muscle itself as distinct from an action on, for example, the neuromuscular transmission process. What then of tubocurarine? Inasmuch as it acts by combination with receptors on the muscle membrane, it must be "direct" acting. But the effect of this combination, prevention of muscle activation by motor nerve stimulation, surely is an "indirect" effect by any reasonable definition.

The number and variety of drugs which affect muscle is enormous. Indeed, one would be highly surprised to learn of any drug that had no action at all. One is therefore compelled to make an arbitrary selection of what should be included as drugs affecting muscle. In this chapter we will restrict ourselves to the consideration of those agents whose effects in the whole animal are exerted largely on muscle and the actions of which are manifest upon muscle preparations *in vitro*. Hence, we exclude those drugs which modify muscle activity *in vivo* by effects upon the central nervous system, e.g., "muscle relaxants." Of course, we could almost equally well have excluded from consideration drugs whose main site of action is on nerve terminals, limiting discussion to agents which act subsynaptically and postsynaptically. In practice, however, when the effects of drugs on any kind of muscle are determined, it us usually a major problem to distinguish between pre- and postsynaptic loci of action. Indeed, a very large number of drugs act at both sites. Consideration only of postsynaptic actions would frequently lead to quite incorrect conclusions as to the actual effect of the drug *in vivo*. Unfortunately, this still leaves us with a field that could hardly be encompassed adequately by anything less than a multivolume encyclopedic review. Thus, much important work will perforce be omitted from consideration, and our references will often be illustrative, rather than complete.

Fortunately for our purpose, at this time little purpose would be served by even an attempt to do justice to all the extensive and important activity in the field of muscle pharmacology. There have recently been published several extensive reviews of the physiology and pharmacology of the skeletal neuromuscular junction, skeletal muscle, and smooth muscle. The recent book edited by Bülbring *et al.* (1970) contains a wealth of material dealing in detail with various aspects of smooth muscle physiology and pharmacology.

Figure 1 is a general scheme of the processes involved in neuromuscu-

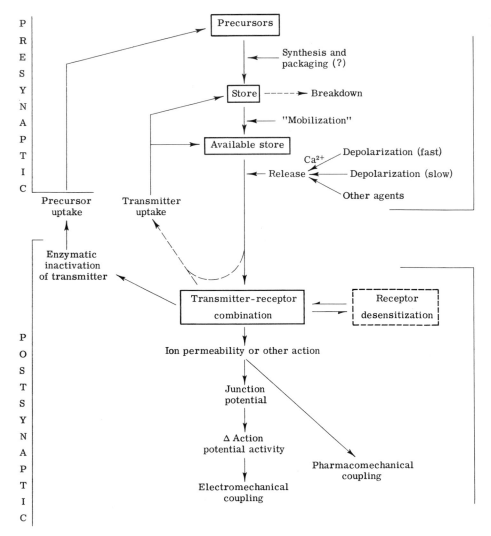

Fig. 1. General scheme of the processes involved in neuromuscular transmission.

lar transmission, derived on the basis of many studies on cholinergic and adrenergic systems. It provides a general framework in terms of which drug actions can be pigeonholed, even when a drug does not act either on the motor nerve terminal or upon the receptors mediating the normal transmission process. Such an agent can be presumed to act either via membrane receptors analogous to those mediating the action of the neurotransmitter substance, or by acting on steps in the activation of muscle contraction which are involved in all muscle responses, including the response to nerve stimulation.

II. Effects of Drugs on Nerve Terminals

A. General Considerations

With only a few exceptions, muscle activity is under nervous control. The control over skeletal muscle is virtually absolute; normally, all activation of muscle contraction is dictated by the central nervous system. Activity of smooth muscle, on the other hand, varies from completely autonomous (as shown by contraction of cells in tissue culture) to centrally modulated (vascular smooth muscle). Whatever the degree of nervous control, there seems no doubt that in all cases the action potential frequency of efferent neurons is translated to modification of muscle contraction by means of release of neurotransmitter by specialized parts of the neurons—i.e., *"boutons terminaux," "boutons-en-passage,"* "varicosities" (see Burnstock, 1970), which we shall refer to in general as "nerve terminals." Although electrotonic conduction of nerve impulses from nerve to muscle must remain a theoretical possibility, there is at present no electrophysiological evidence that this ever occurs, or histological evidence for junctions between nerve and muscle fibers which would permit such a mechanism to operate. In skeletal muscle and some examples of smooth muscle, there is close proximity between the nerve terminals—i.e., the structures which contain high densities of synaptic vesicles—and the neighboring muscle membrane, and the term "synapse" may be applied by extension from similar junctions in the central nervous system. In some smooth muscle, however, typical nerve terminals appear widely separated from the muscle fibers; there exists a large gap across which the transmitter must diffuse in order to act upon the muscle. For convenience, such an arrangement may also be considered as a synapse with pre- and postsynaptic elements separated spatially but not functionally.

It follows from the intimate functional relationship between nerve and muscle that actions of drugs on nerve terminals, either to inhibit or potentiate release of transmitter, will have potent effects upon muscle activity. From *a priori* considerations, it is possible to classify the presynaptic systems with which a drug might interact. Chemically mediated synaptic transmission, in order to operate, demands two basic mechanisms, namely, the transmitter substance must be released at appropriate times (i.e. as specified by neuronal action potentials) and the postsynaptic cells must possess specialized systems by which the transmitter can act. For release, there must exist (1) a mechanism for excitation–release coupling and (2) a system whereby transmitter is made available for release. At the very least, the latter entails machinery for synthesis of the transmitter and/or its active uptake by the nerve terminal. It seems generally true that there is also to be found a mechanism by which transmitter substance is stored within the nerve terminal, although this is not in principle an absolute requirement for chemical transmission. An abundance of evidence (see Douglas, 1968; Hubbard, 1970) indicates many similarities and perhaps identity of the release mechanisms for different transmitter substances. This may arise because the unit that is acted upon by the release system is not the individual molecule of transmitter, but a multimolecular package. For the skeletal neuromuscular junction this is now established beyond contention. The presynaptic nerve impulse acts by virtue of the depolarization associated with it to enhance the probability of liberation of quantal units of acetylcholine, whose spontaneous liberation is manifest as miniature endplate potentials (MEPPs) (see B. Katz, 1969). In the mammalian vas deferens, where the transmitter is noradrenaline rather than acetylcholine, the spontaneous junction potentials are also accelerated by presynaptic action potentials (Burnstock and Holman, 1962, 1966), and similar observations have been made at synapses as diverse as the junctions between the second order giant axon and the cell bodies of the third order giant axon in the stellate ganglion of the squid (Miledi, 1967), crayfish neuromuscular junction (Dudel and Kuffler, 1961), and the junction between Ia afferents and motoneurons in the mammalian spinal cord (Kuno, 1964). As yet it is unclear whether these similarities represent a fundamental identity in the systems subserving transmitter storage and secretion or whether there has been convergent evolution to a high efficiency mechanism whereby transmitter substance is stored and released in multimolecular packages. Nevertheless, it appears likely that one can distinguish between drug actions on the release mechanism, which should be similar for synapses with different transmitter substances, and drug effects which involve the specific transmitter substance. Another class of drugs is those affecting transmitter secretion indirectly, e.g., by modification of the nerve terminal action potential. Thus, we can distinguish three different kinds of pharmacology of presynaptic systems:

1. Drug effects on the transmitter release system *per se*.
2. Drug effects on nerve terminals which indirectly influence transmitter secretion.
3. Drug effects related to the specific transmitter substance—modification of transmitter synthesis, storage, or uptake.

B. Direct Effects on Secretion of Transmitter

The overwhelming bulk of research into the mechanisms by which neurotransmitter is secreted has been carried out using vertebrate nerve–skeletal muscle preparations. These have the enormous advantage that there is generally only one nerve terminal per postsynaptic cell (i.e., muscle fiber). Since the muscle fibers are not electrotonically coupled to one another, all the postsynaptic junctional activity that is recorded can be attributed to the effects of transmitter released by a single nerve terminal. Skeletal muscle fibers are large enough that they may be penetrated with microelectrodes, but at the same time their input impedance is sufficiently high that the depolarization associated with each packet of transmitter released from the nerve terminal can easily be distinguished from noise generated by the microelectrode tip. The investigations of Katz and his colleagues have established, beyond any reasonable doubt, that the MEPP represents the postsynaptic response to a spontaneously liberated single multimolecular package or "quantum" of acetylcholine (ACh) (reviewed by del Castillo and Katz, 1956; B. Katz, 1962, 1966, 1969; Martin, 1966; Hubbard *et al.*, 1969a; Hubbard, 1970). The endplate potential (EPP) is composed of quantal components identical to the MEPP and is the postsynaptic response to many packages of transmitter released almost synchronously following a presynaptic action potential. Thus, the number of MEPPs observed per unit time and number of quantal components of the EPP (quantum content) must always reflect presynaptic events. No alteration of postsynaptic response to ACh can modify the number of unit responses. The amplitude of MEPPs and of quantal components of the EPP are always identical (e.g., Elmqvist *et al.*, 1964; Elmqvist and Quastel, 1965a,b) and alteration of this amplitude reflects either change in the postsynaptic response to ACh or an alteration of the ACh content of each quantum. Acetylcholine content of the quantum can be altered by stimulating release of ACh in the presence of an inhibitor of ACh synthesis (Elmqvist and Quastel, 1965a), but it is only with depletion of the ACh content of the nerve terminal that there occurs a reduction in the amplitude of MEPPs (and of quantal components of the EPP), and this process takes some time to develop. Thus, it seems generally true that any effect of a drug rapidly to alter MEPP amplitude must reflect postsynaptic events, while any effect on frequency of MEPPs or quantum content of EPPs shows a presynaptic action. As a result it is possible unequivocally to separate pre- and postsynaptic actions of drugs, and, indeed, quantitate presynaptic actions to an accuracy limited only by the patience of the investigator, since it is numbers of quantal units that must be evaluated. As it turns out, there are many drugs which alter either the frequency of spontaneous MEPPs or quantum content of EPPs or both. For an appreciation of how these drugs act, it will be necessary to review briefly our present knowledge of the release mechanism, much of which is based upon very recent work. The major clues to an understanding of the system has come from consideration of the effects of divalent ions, calcium in particular, on spontaneous and evoked release of transmitter.

That calcium ions are essential to excitation–release coupling has been hypothesized for some time (see Douglas, 1968) and recently demonstrated clearly by B. Katz and Miledi (1967b). They showed that the release evoked by an action potential or local depolarization, after being reduced virtually to nothing by calcium-free solution, could be restored by local ionophoresis of calcium onto the nerve terminal immediately prior to the stimulus. The amplitude of endplate potentials is graded with external calcium and inhibited by magnesium in the manner expected if magnesium acts by competition with calcium for a binding site from which calcium is made available for the release process (Jenkinson, 1957). The gradation of quantum content with calcium concentration above the normal (e.g., Hubbard *et al.*, 1968b; Dodge and Rahamimoff, 1967) accounts for the decurarizing action of calcium observed by Feng (1936). Barium and strontium ions can evidently replace calcium in depolarization release coupling (Mines, 1911; Miledi, 1966; Elmqvist and Feldman, 1965a, 1966; Blioch *et al.*, 1968). At other synapses, divalent cations evidently act in the same way as at the skeletal neuromuscular junction, qualitatively if not quantitatively. For release of noradrenaline by nerve impulses, calcium (or barium or strontium) is required (Burn and Gibbons, 1965; Boullin, 1967; Kirpekar and Misu, 1967; Kirpekar and Wakade, 1968; Fujii and Novales, 1969), and the same is true for other synapses which are neither cholinergic nor adrenergic, e.g., frog spinal cord (B. Katz and Miledi, 1963) or squid giant synapse (Miledi and Slater, 1966). This calcium dependence is shared by a variety of other secretory systems (see Douglas, 1968; Rubin, 1970; Hubbard, 1970). Similar dependence upon calcium and inhibition by magnesium of a variety of secretory systems has recently been emphasized by Douglas (1968) as support for the hypothesis that all secretory systems share a stimulus–response coupling mechanism that is basically the same. Whether or not this be the case, the evidence is now overwhelming that the mechanism by which transmitter secretion is stimulated by presynaptic nerve impulses has the same ionic requirements at all synapses which have been studied.

Another important line of investigation has focused upon the response of the secretory mechanism to passive graded depolarization of the nerve terminal. This has been made possible by the availability of the puffer fish poison, tetrodotoxin, which selectively blocks the electroresponsive sodium current system which is necessary to action potential generation (Kao, 1966). It has become clear that the calcium dependence of release by nerve impulses reflects a calcium requirement of the coupling system by which release of tansmitter is graded with presynaptic depolarization, at the squid giant synapse (B. Katz and Miledi, 1967c; 1970a) and at the neuromuscular junction (B. Katz and Miledi, 1967b; Landau, 1969; Cooke and Quastel, 1972). B. Katz and Miledi (e.g., 1970a) have interpreted the role of calcium in transmitter release in terms of the "calcium hypothesis." Stated briefly, depolarization of nerve terminal membrane causes an increased permeability to calcium or, a positively charged complex, CaX, which then moves down its electrochemical gradient into the nerve terminal. In support of this hypothesis, there has been observed evidence for a positive inward current in nerve terminals, which is not blocked by tetrodotoxin (B. Katz and Miledi, 1968a,b, 1969b), and an excessive nerve terminal depolarization impedes transmitter release (B. Katz and Miledi, 1967c). The major difficulty of this hypothesis has been to explain the very different quantitative relationship between calcium concentration and transmitter release evoked by a nerve impulse on the one hand and on spontaneous release on the other. Although spontaneous release may be augmented by raised calcium (Hubbard, 1961) and greatly reduced by calcium depletion (Elmqvist and Feldman, 1965a), some release persists even at a concentration of free calcium ions of about 10^{-10} M, under which conditions there is no detectable response of the release system to presynaptic depolarization (Quastel *et al.*, 1971; see also Hubbard, 1970). Moreover, for the calcium-dependent fraction of spontaneous release, the relation between MEPP frequency and calcium concentration is far less steep than that between release evoked by a nerve impulse and calcium concentration. Indeed, at the mammalian neuromuscular junction, the relationship suggests that release of each quantum of transmitter requires the action of one calcium ion (in combined form) for spontaneous release but of three for evoked release, since release increases with nearly the third power of calcium concentration, in the appropriate range of calcium concentration (Hubbard *et al.*, 1968b). At the frog neuromuscular junction four calcium ions appear to be required (Dodge and Rahamimoff, 1967). Thus, there seems to be at least three distinct and separate release mechanisms: (1) spontaneous, calcium independent, (2) spontaneous, calcium dependent, and (3) depolarization-evoked, calcium dependent. To these three might be added two more: (4) potassium augmented spontaneous release and (5) the increased spontaneous release which follows conditioning presynaptic action potentials or a long-lasting passive depolarization, and is associated with facilitation. Potassium augmented release falls between spontaneous and impulse-evoked release in calcium dependence (Gage and Quastel, 1966), and facilitated MEPP frequency parallels spontaneous release in its calcium dependence, above 10^{-4} M calcium (e.g., Cooke and Quastel, 1973). The facilitation of EPPs following conditioning stimuli is well known to be more prominent when magnesium is raised or calcium reduced in the bathing medium (see Hubbard, 1970). It has been suggested that this facilitation can be attributed to persistence in the nerve terminal of some of the calcium that entered following the conditioning pulses (Rahamimoff, 1968). However, this model cannot account quantitatively for the facilitation of MEPP frequency. Indeed, unless it is supposed that four calcium ions can cooperate for release only just following a nerve action potential, the model predicts augmented MEPP frequencies that are linearly proportional to the quantum content of the EPPs.

Rather than several different release processes, each characterized by a certain calcium dependence, the data suggest that the relation between release and calcium concentration is a continuous variable that depends on the degree of nerve terminal depolarization. This would arise, for example, if each calcium ion or CaX molecule that enters the nerve terminal and comes into proximity to certain membrane sites (or hypothetical activator or membrane alteration produced as a result of depolarization in proportion to the amount of CaX on the membrane) acts independently to reduce an energy barrier that limits quantal discharge and hence multiplies the probability of release of a quantum of transmitter located nearby. If it is supposed that numbers of activating ions or molecules are randomly (i.e., Poisson) distributed, this model predicts the relation

$$f = ke^{\lambda(r-1)} = ke^{\alpha[CaX]}$$

where f is instantaneous frequency of quantal release, λ is the mean number of calcium ions (or activator) at a critical site, r is the multiplication of release probability by one calcium ion (or CaX, or activator), k is a constant, and α is a function of nerve terminal depolarization. Quantitatively, this model accounts for the observed calcium and magnesium dependence of EPPs and for the relatively flat relation between normal spontaneous release and calcium concentration. More important, it predicts by extension that agents might exist other than the combination of depolarization and membrane CaX that also can reduce the energy barrier (acting directly or by formation of activator), and with such an agent one should find (1) an exponential relation between MEPP frequency and concentration, and (2) an equal multiplication of transmitter release whether it be calcium-independent spontaneous, normal spontaneous, or evoked by depolarization. Such an agent is indeed to be found. As shown by Gage (1965), ethanol causes parallel multiplication of both spontaneous MEPP frequency and EPP quantum content. Correspondingly, it multiplies spontaneous release and release evoked by presynaptic depolarization to the same extent (Quastel *et al.*, 1971). Moreover, in the virtual absence of calcium (estimated free calcium 10^{-10} M) ethanol multiplies spontaneous release to the same extent as when calcium is present. The relation between release and ethanol concentration is exponential, MEPP frequency being about doubled for every 200 mM increment of concentration. When magnesium as well as calcium concentration is reduced to very low levels (by using EDTA) the response to ethanol is unaltered. It may further be noted that the effect of ethanol is without any apparent time lag, and fully reversible, at least up to 1.6 M. The multiplying effect of ethanol on release by presynaptic depolarization rules out any possibility that the action of ethanol is not exerted on the normal system of transmitter release, and it follows that this system does not in fact require calcium. The system which does require calcium is the one by which transmitter release is coupled to presynaptic depolarization.

In terms of this scheme, it is also relatively easy to fit in facilitation phenomena. Facilitation appears as a slow response of the release system to depolarization, distinct from the fast response which is the major component in impulse-evoked release in that it gradually builds up during nerve terminal depolarization and only slowly decays afterward, and is maximal at a calcium concentration of about 10^{-4} M (Cooke and Quastel, 1972). Like the effect of alcohol, it adds logarithmically to effects of any other activation of the release process; i.e., both spontaneous and evoked release are multiplied to much the same extent. To complicate the picture, there is yet a third effect of depolarization on the release process. This consists in an inactivation of the depolarization–release coupling system. During or following depolarization, there can be observed a flattening of the relation between release and presynaptic depolarization (Cooke and Quastel, 1972). Presumably, this represents an inactivation of membrane response to potential change similar to inactivation by depolarization of the sodium current system in nerve (Hodgkin and Huxley, 1952). If we accept the notion that it is calcium entering the terminal that acts on the release system, for which there is strong evidence against (Miledi and Slater, 1966; Hofmann, 1969) as well as for (B. Katz and Miledi, 1969a,b), then this inactivation would presumably be exerted on the system by which membrane calcium (or CaX) permeability increases in response to membrane depolarization.

Normal spontaneous transmitter release is now seen as merely the special case of release at the normal resting membrane potential of the nerve terminal. As at any other level of nerve terminal polarization, a true spontaneous probability of quantal liberation will be multiplied according to several factors, namely, (1) membrane CaX; (2) level of activation of the fast electroresponsive membrane system (perhaps calcium permeability), which represents a balance between activation and inactivation processes; (3) level of activation of the slow electroresponsive system, which will depend on a balance between a rate of growth and a rate of decay; (4) other factors such as osmotic pressure and, possibly, endogeneous substances acting like alcohol.

In terms of pharmacology, it is evident that drugs and ions might act in a variety of ways: (1) alteration of membrane CaX; (2) alteration of fast response of presynaptic membrane (calcium permeability?) to depolarization, (3) alteration of magnitude or time course of slow response to depolarization, and (4) direct, alcohol-like action. These different actions can be distinguished by study of drug effects on the response of MEPP frequency to presynaptic polarization and on spontaneous release in the absence of calcium (see Table I).

The multiplication of calcium-independent release of transmitter by ethanol is far from unique to this compound. Other alcohols act similarly, with potency increasing about threefold for each added methylene group. In addition, various general anesthetic agents, including urethane, pentobarbital, paraldehyde, chloral hydrate, chloroform, and ether, act in the same way (Quastel *et al.*, 1971; Quastel and Hackett, 1971). For many compounds, a large postsynaptic blocking action, which makes MEPPs small, forestalls an accurate assessment of effects on MEPP frequency, but for those drugs which could be tested, there is a close correlation between relative potency in increasing transmitter release and stabilization of erythrocyte membranes (as defined by P. M. Seeman,

TABLE I

EFFECTS OF POLARIZATION, IONS, AND SOME DRUGS THAT ACT
UPON TRANSMITTER RELEASE

Depolarization

Fast response — Step increase MEPP frequency (f) after synaptic delay

$$\log f = k + \phi(\Delta V)\,[\text{CaX}]$$

$\phi(\Delta V)$ linear between $f = 5/\text{sec}$ and $R = 1000/\text{sec}$ with slope S. [CaX]: total X $\approx 1:2$ at [Ca] = 0.15 mM.

Slow response — Asymptotic rise $\log f$ during pulse and fall of $\log f$ after pulse. $\tau \approx 20$ msec; requires $>10^{-10}$ M calcium; maximum at $\approx 10^{-4}$ M calcium

Inactivation of fast and slow(?) response — S reduced; effect persists after a depolarizing pulse

Indirect effects via sodium current system — Depolarization can trigger presynaptic action potential (immediate); maintained depolarization reduces or blocks presynaptic action potential

Hyperpolarization

Spontaneous release — Note $\phi(\Delta V)$ always remains appreciable; bursts of MEPPs develop progressively; requires no calcium

Anodal breakdown

Ions

K^+ — Immediate effect on $\phi(\Delta V)$: S increased; depolarizes (relatively slowly)

Ca^{2+}, Ba^{2+}, Sr^{2+}, Mg^{2+}, Mn^{2+}, Ni^{2+}, Zn^{2+} — See above: responses to depolarization: Ba^{2+} and Sr^{2+} substitute for Ca^{2+} on X, others compete with Ca^{2+}; slow decay of slow response to depolarization: Ca^{2+} and Mg^{2+} and others(?); inhibit effect of K^+ on $\phi(\Delta V)$: Ca^{2+}, not Mg^{2+}, others(?)

Na^+ — Extracellular: compete with Ca^{2+} on X (?); intracellular: increase Ca^{2+} entry (??); indirect: nerve action potential propagation

Drugs

Alcohols, chloral hydrate, urethane, paraldehyde, ether, chloroform, pentobarbital, chlorpromazine, raised osmotic pressure (π-impermeant solute), theophylline

$$\log f = k + \phi(\Delta V)\,[\text{CaX}] + k_1[\text{drug}_1] + k_2[\text{drug}_2] + \ldots + k_\pi \pi$$

Chlorpromazine and π reduce S; others(?) except ethanol which does not

1966). This correlation also applies to chlorpromazine, which is of special interest because this compound, unlike ethanol and other unchanged anesthetic molecules, acts to displace calcium from erythrocyte membranes (Kwant and Seeman, 1969; P. Seeman *et al.*, 1971). Perhaps related to this effect on calcium binding is the observation that chlorpromazine also acts to reduce the slope of the logarithm of MEPP frequency versus presynaptic depolarization (Quastel *et al.*, 1971). This action might also represent a reduction of the responsiveness of the nerve terminal membrane to depolarization. Raised osmotic pressure, well known to increase MEPP frequency (Fatt and Katz, 1952; Furshpan, 1956) even in the absence of calcium and independent of magnesium (Blioch *et al.*, 1968; Hubbard *et al.*, 1968a), tends to multiply depolarization-evoked release, in the same way as ethanol, but like chlorpromazine, also reduces the slope of the logarithm of release rate versus depolarization (Quastel *et al.*, 1971). Figure 2 summarizes the effects on the presynaptic transfer function of ethanol, chlorpromazine, osmotic pressure, and raised [Mg^{2+}] or lowered [Ca^{2+}]. It is based upon observation of MEPP frequencies up to about 1000/sec, and observations on EPPs, i.e., release evoked by a presynaptic action potential. It will be noted that a depression of depolarization–release coupling, e.g., by chlorpromazine, need not be associated with depression of EPP quantum content. Black widow spider venom, another agent that increased MEPP frequency independently of calcium, tends to increase rather than decrease quantum content of EPPs (Longenecker *et al.*, 1970), but this does not necessarily indicate lack of action on depolarization–release coupling.

One agent which has been shown to potentiate depolarization–release coupling is potassium, which also, of course, depolarizes nerve terminals. An increase of potassium concentration to 15 m*M* increases the slope of log MEPP frequency versus depolarization by about 25% (Cooke

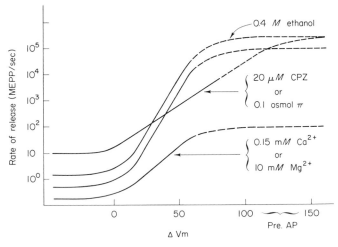

Fig. 2. Effects of ethanol, chlorpromazine (CPZ), osmotic pressure (π), and raised [Mg^{2+}] or lowered [Ca^{2+}] on the presynaptic transfer function.

and Quastel, 1969, 1972). This effect is blocked by raised calcium (8 mM), which accounts for the depressing effect of high calcium on "spontaneous" MEPP frequency in the presence of raised potassium (Gage and Quastel, 1966). The effect of potassium on depolarization–release coupling, of course, accounts for its action of increasing transmitter release in response to a presynaptic nerve action potential (Liley and North, 1953; A. Takeuchi and Takeuchi, 1961). At present, the physiological role played by this system linking extracellular potassium to transmitter release is obscure. Perhaps it acts at myoneural junctions to counteract a tendency for EPPs to become reduced because of depolarization (see below) if local potassium concentration becomes raised secondary to muscle activity. Whether any drugs act directly on this system is at present unknown, but it is likely that a variety of drugs, acting either pre- or postsynaptically, may tend to increase potassium concentration in the synaptic cleft and thereby to some extent act indirectly to increase transmitter release in response to presynaptic action potentials.

It is possible that other agents may act in the same way as potassium on depolarization–release coupling. Phenol, for example, we have found to have little or no effect on spontaneous MEPP frequency at a concentration which augments quantum content of EPPs (Otsuka and Nonomura, 1963).

There are a number of agents which affect transmitter release at the skeletal muscle junction that might well act either on the depolarization–release coupling system(s) as such or by altering calcium binding, i.e., CaX level. Thus, the antibiotics neomycin, streptomycin, and kanamycin can cause a neuromuscular blockage that is antagonized by calcium (Corrado and Ramos, 1960; Corrado *et al.*, 1959; Jindal and Deshpande, 1960). The block caused by neomycin has been studied using intracellular recording from endplates and was found to be similar to that produced by calcium deficiency or high concentrations of magnesium. An increase of calcium tended to reverse the effects (Elmqvist and Josefsson, 1962). In the absence of quantitative studies, either mechanism seems possible.

The neuromuscular blockade sometimes associated with bronchial carcinoma, the so-called myasthenic syndrome, is also characterized by a low quantum content of EPPs (Elmqvist and Lambert, 1968). Its relative insensitivity to raised calcium (Elmqvist, personal communication) suggests an interference with the coupling system per se. The persistence of facilitation indicates a sparing of the slow system of response to depolarization.

At the present time, there has been no investigation of possible effects of drugs on the slow response to depolarization or its counterpart, facilitation of transmitter secretion evoked by nerve impulses. That facilitation is not immune to modification is shown by the results of Rahamimoff (1968) and of Rosenthal (1969), which indicate that the time course of facilitation is prolonged by an increase of calcium or magnesium. It need hardly be pointed out that the possibility that facilitation might be based on entry of sodium into nerve terminals during nerve impulse generation is ruled out by the experiments which demonstrate that it occurs without nerve impulses and is then graded with presynaptic polarization (e.g., Cooke and Quastel, 1972).

After a large and/or long-maintained depolarizing pulse, the slow response could be found to persist for several minutes (Cooke and Quastel, 1972). It was therefore proposed that post-tetanic potentiation may simply represent a slowly decaying phase of the same process as reflected in facilitation. This conclusion is contrary to that of Gage and Hubbard (1966), who found that metabolic inhibitors selectively inhibited post-tetanic potentiation without affecting facilitation. However, the prolonged sequelae of presynaptic action potentials, in the presence of metabolic inhibitors acting to prevent active sodium transport, might antagonize or obscure an underlying continued facilitation.

Except for the cholinergic skeletal neuromuscular junction, there is at present very little information regarding the mechanisms by which transmitter release is controlled. As has already been pointed out, the characteristic effects of calcium and of magnesium at all synapses studied may merely reflect similarity of the mechanism of depolarization–release coupling, rather than identity of the actual release process. It might therefore be unwise to extrapolate experimental results for the skeletal muscle junction to other systems. The only adrenergic system that has been studied to any extent on lines similar to the skeletal junction is the mammalian vas deferens. Here, there can be recorded spontaneous and evoked junction potentials related to release of noradrenaline. The effect of osmotic pressure on the frequency of spontaneous junction falls far short of that at the skeletal junction. At the latter synapse, an increase of osmotic pressure by only 25% increases spontaneous release by as much as tenfold (Hubbard *et al.*, 1968a; Blioch *et al.*, 1968). In the vas deferens, it requires triple the normal osmotic pressure to double spontaneous release (Burnstock and Holman, 1962a). However, when nerve terminals are depolarized, the effect of osmotic pressure on release at the skeletal junction is much depressed (Quastel *et al.*, 1971). Thus, the adrenergic release system might differ from the skeletal cholinergic system only by where spontaneous release lies on the presynaptic transfer function, spontaneous release at this site corresponding to that related to considerable depolarization at the skeletal junction. If this is so, then any agent that depolarizes nerve terminals should increase release.

Another major quantitative difference between the adrenergic junction and the skeletal is the relative ineffectiveness of the excitation–release coupling process. Exact measures of quantum content at the junctions in vas deferens are impossible, because of variation in amplitude of the size of the spontaneous junction potential, which presumably correspond to quantal units of the evoked junction potential. However, it is apparent from the records of Burnstock and Holman (1962a) and of Kuriyama (1963) that it must be far lower in comparison to spontaneous release rate than in skeletal muscle and resembles release at the latter junction evoked in the presence of lowered calcium or raised magnesium. Unlike the latter, it cannot be raised very appreciably by increase of calcium. In terms of the model that has been put forward to account for the relation between release, calcium, and presynaptic depolarization (Cooke and Quastel, 1972), this could be based upon either a relative deficiency in the change of nerve terminal membrane (calcium permeability?) brought about by nerve terminal polarization, a deficiency in the amount of calcium-binding material on the terminal membrane or a high resting inactivation of the coupling system. The evoked junction potential in the vas deferens grows with repetitive nerve stimulation like the magnesium-inhibited EPP, but at low stimulation frequencies. Thus, facilitation is relatively prominent. However, whether this is by the same mechanism as that of the skeletal junction is not established. The direct effects, if any, of alcohol and anaesthetic agents on release by adrenergic terminals appear not to have been investigated. It should be noted that actions of such drugs simply on total noradrenaline release by nerve stimulation may be completely misleading as to the direct effects on the release system; actions on nerve transmission, nerve impulse generation, and nerve impulse propagation and amplitude in the varicosities could well predominate. Effects on spontaneous release are also difficult to interpret. For example, the finding that intraarterial thiopental has a vasoconstrictor effect on rabbit ear vessels that is potentiated by cocaine but abolished by reserpinization (Burn and Hobbs, 1959) clearly indicates a noradrenaline release by thiopental; but any number of hypothetical mechanisms might be proposed to account for this phenomenon. Unless the effect were shown to persist in the absence of calcium, one would hesitate even to suggest that it is basically similar to that of central depressants at the skeletal neuromuscular junction. Nevertheless, there seems no reason to doubt that noncholinergic nerve terminals should possess a pharmacology of the quantal secretion systems as rich as, although perhaps not the same as, the cholinergic.

C. Indirect Actions on Transmitter Secretion

In this section we will consider drug actions that are manifest in alteration of transmitter secretion, but in our opinion, probably do not represent direct actions on the secretory system(s). There is here the major problem of insufficient information, and the classification will at times appear arbitrary. For example, it now appears that the action of acetylcholine on transmitter secretion is unlikely to be secondary to depolarization of the nerve terminal (Pilar, 1969b), and the mechanism of action is totally obscure. Treating this phenomenon here rather than in the previous section is therefore based on no more than a guess.

1. NERVE IMPULSE GENERATION

The prime example of a potent, local, but entirely indirect action on transmitter secretion, is that of tetrodotoxin, the neurotoxin found in puffer fish, and of certain other toxins which act in the same manner, such as saxitoxin. The action of tetrodotoxin is limited to blockade of the sodium current system, which is fundamental to generation of the action potential in most excitable tissues (reviewed by Kao, 1966), the major exceptions being those in which the action potential depends on an electroresponsive increase in calcium rather than sodium permeability. Thus, tetrodotoxin blocks generation and propagation of the presynaptic action potential in both skeletal and smooth muscle. There is no apparent action on spontaneous junction potentials (e.g., Elmqvist and Feldman, 1966) and little if any action on the response of the secretory system to presynaptic depolarization (Landau, 1969). Tetrodotoxin and anything that acts similarly will act to block entirely the release of transmitter in response to nerve stimulation, without in the least indicating an action on the transmitter release system. Thus, the observation of Matthews and Quilliam (1964) that a variety of anesthetic agents reduce transmitter release by the orthodromically stimulated superior cervical ganglion might indicate no more than blockade of impulse conduction to the presynaptic nerve terminals. Even without complete block of presynaptic impulse generation, a reduction of presynaptic action potential amplitude, related to partial blockade of the sodium current system, that occurs with a variety of anesthetic agents (Hille, 1966; Blaustein, 1968), could account for these results. The direct effect of such agents, as judged by miniature endplate potential frequency, is to increase spontaneous release (Quastel *et al.*, 1971). For the same reasons, caution must be used in interpreting such phenomena as the reduction by morphine of acetylcholine release by nerve stimulation, in the intestine (Paton, 1957).

An example of local anesthetic action on transmitter release that is relatively selective for adrenergic terminals is that of bretylium, which is selectively concentrated by these terminals via the amine pump, explaining the reversal of its effect by, for example, amphetamine (see Boura and Green, 1965).

Tetraethylammonium (TEA) is probably an example of a drug that indirectly increases EPPs by an action on presynaptic nerve impulses. This agent is known to inhibit delayed rectification—i.e., the potassium current system of the excitable membrane, both in squid and vertebrate nerve (C. M. Armstrong and Binstock, 1965; Hille, 1967). It therefore prolongs the action potential and should thereby increase evoked release of transmitter. To this action may be ascribed its decurarizing action, which is based upon an increase in transmitter release (Stovner, 1958a). There is, however, a difficulty. At mammalian nodes of Ranvier, delayed rectification is much less prominent than at amphibian nodes and plays little part in terminating the action potential (Horaćkova *et al.*, 1968). Unless the nerve terminal is unlike the node in this respect, it seems difficult to account entirely for the effect on release in mammalian muscle in terms of the action on delayed rectification. However, at the frog neuromuscular junction, TEA does not affect the frequency of MEPPs, either spontaneous or evoked by raised potassium (Benoit and Mambrini, 1970), ruling out any direct action on the release system per se or depolarization–release coupling. Benoit and Mambrini (1970) also investigated the action of other agents that act to prolong the presynaptic spike, namely, uranyl, nickel, and zinc ions. These and TEA were found to delay the release of transmitter following a nerve impulse, as anticipated in view of the observation of B. Katz and Miledi (1967c) that prolongation of a large presynaptic depolarization delayed transmitter release. Only uranyl ions, however, acted similarly to TEA to increase EPP quantal content, and the effect of uranyl ions seemed unlikely to be related to prolongation of spike duration, since its facilitating action on release was as much or greater on spontaneous release as on the EEP. Nickel and zinc both depressed quantum content; nickel also depressed potassium-evoked release as did zinc, but only transiently. An interaction of these ions with the calcium-binding site, and competition with Ca in particular, could account for the failure of increased spike duration to be mirrored in augmented quantal content.

2. Inhibition of Sodium–Potassium Pump

Ouabain and other cardiac glycosides have been shown to have several effects on transmitter release at the frog myoneural junction. They cause a progressive increase in MEPP frequency, increased quantal content of isolated EPPs, eventual failure of impulse propagation, and spontaneous multiquantal release of ACh (Birks and Cohen, 1968a,b). As pointed out by Hubbard (1970), all but the increase of quantal content follow directly from nerve terminal depolarization (see Table I) if the spontaneous multiquantal release is actually related to generation of action potentials by the depolarized terminal. In contrast, with anoxia and metabolic inhibitors such as sodium azide, antimycin A, and dinitrophenol, which also would be expected to raise intracellular sodium, an acceleration of MEPP frequency is accompanied by a reduction in EPP amplitude (Kraatz and Trautwein, 1957; Gage and Hubbard, 1966; Beani et al., 1966; Elmqvist and Feldman, 1965b, Hubbard and Løyning, 1966). While Birks and Cohen argue that this reflects a more general action of these agents than that of the cardiac glycosides, Elmqvist and Feldman (1965b) have found an effect of ouabain more easily attributable to calcium mobilization than to a presumed action only on the sodium–potassium ATPase. In some of the earlier experiments with digoxin (Birks, 1963), it is not unlikely that the ethanol in which the drug was dissolved was quite sufficient to give rise to the increased EPP amplitude observed. It should also not be forgotten that efflux of potassium from both nerve terminals and muscle fibers tend to increase its concentration in the synaptic cleft, with a resulting tendency to increased EPP quantum content. It remains to be demonstrated that the major effects of cardiac glycosides are not mediated simply by nerve terminal depolarization, as appear to be the effects of the other metabolic inhibitors.

The major argument against this interpretation is that it has in fact been shown that an increased intracellular sodium (in squid axon) can augment the influx of calcium (Baker et al., 1969). Birks et al. (1968) have indeed interpreted the interaction of sodium and calcium on MEPP frequency in terms of competition of extracellular sodium with calcium for entry, together with augmentation of calcium entry by intracellular sodium. However, it should be recalled that Miledi and Slater (1966) found no effect on transmitter release of calcium injection into the squid nerve terminal, and a role for sodium entry in release of transmitter by a presynaptic action potential appears to be ruled out by B. Katz and Miledi's (1967c) finding that a passive depolarization of the squid nerve terminal in the presence of tetrodotoxin could be just as effective as an action potential in releasing transmitter. This implies that sodium entry during the presynaptic action potential does not influence the depolarization–release coupling system, a result which is contrary to the hypothesis of Birks *et al.* (1968).

Kirpekar *et al.* (1970) have recently studied the effects of metabolic inhibitors on release of noradrenaline from the perfused spleen of the cat. The results were analogous to what has been found for ACh release at the neuromuscular junction, namely, in increase of spontaneous release by ouabain, anoxia, and dinitrophenol, which extended, at least temporarily, also to release evoked by nerve stimulation or raised potassium. The experiments were performed on animals pretreated with phenoxybenzamine, which rules out effects secondary to inhibition of the amine uptake mechanism. It appears that depolarization of adrenergic terminals by inhibition of sodium transport can indeed evoke release of the transmitter, which is hardly surprising. However, the experiments confirm the similarity of cholinergic and adrenergic release mechanisms is that neither, apparently, requires metabolic energy.

3. ACETYLCHOLINE AND CHOLINOMIMETIC AGENTS

The action of ACh, its analogs, and cholinesterase inhibitors on nerve terminals has been the subject of considerable interest and speculation. Indeed, it has been suggested that ACh release by cholinergic nerve terminals is mediated by small amounts of ACh previously liberated by the same terminals (Koelle, 1962), apparently on no more cogent evidence than that the ACh analog carbachol can release ACh from perfused sympathetic ganglion (McKinstry *et al.*, 1963). It has recently been demonstrated unequivocally by Collier and Katz (1970) that the release by carbachol or ACh is not from the normal presynaptic store, i.e., depot ACh, but only from the so-called surplus ACh, which accumulates if a ganglion is perfused with an anticholinesterase. This finding is in keeping with the observation by Fatt and Katz (1952) that ACh does not increase MEPP frequency at the frog myoneural junction.

Ferry (1966) has considered in some detail the closely related hypothesis of Burn and Rand (1959), that noradrenaline release by adrenergic terminals is mediated by the prior release of ACh. On the whole, the evidence is unconvincing. Ehinger and his colleagues have recently examined the ACh content of cat iris before and after parasympathetic denervation. They concluded that if adrenergic terminals contain any ACh at all, it must be minimal, less than 6% of the ACh content of cholinergic neurons. Moreover, Ehinger *et al.* (1966) found no fall in choline acetylase in the cat iris with sympathetic denervation. Thus, the evidence suggests that noradrenergic nerve terminals contain no ACh to release, and the Burn and Rand hypothesis seems unlikely to be correct.

This is not to deny that ACh may have an important role in modulating release of transmitter by both cholinergic and adrenergic nerve terminals. There is cogent evidence that ACh does indeed depolarize nerve terminals as it does all nerve fibers unprotected by a myelin sheath (Gray and Diamond, 1957; J. M. Ritchie and Armett, 1963; discussed by Nachmansohn, 1964). The most direct evidence is that of Pilar (1969a,b), who, recording intracellularly from the giant nerve terminals in chick ciliary ganglion, found that ACh depolarized the terminals. Less direct, but very convincing, are the results of Hubbard *et al.* (1965) who found ACh to lower the threshold for electrical stimulation of motor nerve terminals via a focally placed microelectrode. Other things being equal, an alteration of threshold found by this technique can be taken to reflect alteration of presynaptic membrane potential. In these experiments, the stimulating electrode was close enough to the endplate to record extracellular MEPPs, and current spread to the first node of Ranvier would be very small. Indeed the only reason for their doubt that ACh depolarized the motor nerve terminals themselves was the lack of any acceleration of MEPP frequency. As illustrated in Fig. 2, the response of MEPP frequency to small depolarizations may be very small; only with depolarizations above about 10 mV does the relation between transmitter release and depolarization become steep. Whether this is also the case for adrenergic nerve terminals is not at present known. If the relation is steep near the resting potential then some noradrenaline release by ACh would be expected. However, Bell (1968) has found that tetrodotoxin blocks the response of a wide variety of tissues to nicotine as well as to sympathetic stimulation, without preventing their response to noradrenaline, indicating that the effects of nicotine were mediated by action potential generation rather than directly by depolarization of nerve terminals. It is in keeping with a depolarizing action of ACh that the effect of ACh on transmitter release by nerve impulses is to depress at the skeletal neuromuscular junction (Ciani and Edwards, 1963; Hubbard *et al.*, 1965) and in the chick ciliary ganglion (Pilar, 1969b). The ACh analog succinylcholine similarly depresses quantal content of EPPs (Edwards and Ikeda, 1962). Although decamethonium tends to increase quantal content of EPPs, this is apparently via an increase in the immediately available store of quanta, rather than an increase of release probability, which is depressed (Blaber, 1970). However, recent observations of Pilar (1969b) argue against the depression of release being related to depolarization. In the chick ciliary ganglion, the inhibition by ACh of quantal content of evoked synaptic potentials occurs after the membrane potential of the nerve terminal has returned to normal, i.e., after desensitization of the receptors responsible for the depolarizing response and after the presynaptic action potential has returned to control amplitude. Indeed, the experiments of Hubbard *et al.* (1965) demonstrate that there must be at least two kinds of ACh receptors at the mammalian motor nerve terminal; one blocked by d-tubocurarine, and therefore presumably nicotinic, and the other not blocked by *d*-tubocurarine, but in fact, acted upon by *d*-tubocurarine in the same manner as by ACh. It is of interest that *d*-tubocurarine can excite ACh-sensitive neurons in the thalamus (Andersen and Curtis, 1964) and cerebral cortex (Krnjević and Phillis, 1963a) and the characteristic bursty response of these cells to *d*-tubocurarine is suggestive of a muscarinic response mediated by a reduced potassium conductance (see below, p. 72). Such an action on nerve terminals would lead to a reduced threshold to focal extracellular stimulation, even without depolarization.

What of the apparent increase of available store by decamethonium (Blaber, 1970)? Here one can only speculate. It must be recalled that this "immediately available store" is defined in terms of rundown of EPP amplitude early in a tetanus and could as well represent a store of membrane material necessary for quantal release as an actual store of quanta of transmitter (Elmqvist and Quastel, 1965b; and see Hubbard, 1970). Experimentally, it appears also to be increased by hyperpolarization of the nerve terminal by externally applied current, and with raised potassium, and reduced by depolarizing current (Hubbard and Willis, 1962a,b, 1968; and see Hubbard, 1970). Hyperpolarization by extrinsic current will, of course, be associated with ingoing current across the terminal membrane; i.e., the membrane current will be in the same direction as when a depolarizing agent acts on the membrane of the nerve terminal but not on the axon nodes of Ranvier. Thus, if an apparent increased available store were related to inward membrane current, one would expect the same alteration for extrinsic hyperpolarization, raised potassium (which would reach the endplates of a superficial muscle fiber before the rest of the motor axon), and in all probability, depolarizing agents such as decamethonium, if they act on nerve terminals rather than nodes of Ranvier. By the same token, action potential activity could be expected to increase mobilization, a phenomenon which is indeed observed (Elmqvist and Quastel, 1965b). It seems not unlikely that this effect of intrinsic as distinct from extrinsic depolarization might sufficiently antagonize the effects of depolarization, to inactivate coupling and reduce action potential amplitude, that such presynaptic depolarization could give rise to a net increase of release evoked by presynaptic action potentials.

A muscarinic inhibitory action of ACh on noradrenaline release by adrenergic nerve terminals has been demonstrated by Lindmar *et al.* (1968) and by Haeusler *et al.* (1968). They showed that ACh, methacholine, or pilocarpine caused a reduction in the amount of noradrenaline released from the perfused rabbit or cat heart by nicotine. The inhibitory effects of these drugs on noradrenaline release were abolished by atropine. Although low doses of ACh may enhance responses to sympathetic nerve stimulation (e.g., Malik and Ling, 1969; Rand and Varma, 1970), the significance of this effect, which may represent no more than an accentuation of spontaneous release by depolarization of the nerve terminals, is open to question. Morphological studies have shown that in a variety of tissues, adrenergic and nonadrenergic (presumably cholinergic) nerve terminals lie in close proximity, and there are contacts of apparently synaptic character between them (Thoenen *et al.*, 1966). Functionally, one would expect at these sites high local concentrations of ACh, which would, by virtue of the depolarization it causes, tend to block propagation of impulses in the fine adrenergic terminal arborization. This, together with muscarinic inhibition of the release process, would represent an example of presynaptic inhibition.

Depolarization of nerve terminals, probably mediated by the nicotinic receptor, is also the most likely explanation for the observation that decamethonium and other ACh analogs, which classically depolarize postsynaptically, may cause blockade of conduction of presynaptic action potentials to motor nerve terminals, in skeletal neuromuscular preparations *in vitro* (Roberts and Thesleff, 1968; Galindo, 1970). Isolated nerve muscle preparations are notoriously susceptible to anoxia, which tends to block impulse conduction at branch points of motor axons (Krnjević and Miledi, 1958). Whether such an action occurs *in vivo* and contributes appreciably to the neuromuscular blocking action of these agents must be considered doubtful.

However, a reduction of fractional release of available transmitter release based upon either nerve terminal depolarization or upon the quite mysterious nondepolarization-dependent depression of depolarization–release coupling could well significantly modify the blocking action. It is a commonplace that during a partial neuromuscular block by succinylcholine and similar agents, in contrast to that of curare, the tetanic response to repetitive nerve stimulation is well maintained. The fall in muscle contraction during a tetanus is related to the rundown of EPP amplitude early in tetanus (e.g., Elmqvist and Quastel, 1965b), which is in turn related to depletion, as release is evoked, of the amount of transmitter available for release (or of hypothetical release material). An inhibition of fractional release of transmitter will correspondingly reduce the depletion, the rundown, and thus the extent to which transmission is sensitive to the frequency of nerve stimulation.

Depolarization of nerve terminals by ACh and by agents which act in the same way, including prostigmine (Hubbard *et al.* 1965), provides an explanation for the antidromic action potential activity which can be recorded from motor nerves following a maximal shock applied to the nerve (e.g., M. C. Brown and Matthews, 1960). This backfiring is greatly enhanced by anticholinesterases and occurs even when the muscle fibers are cut and depolarized (Barstad, 1962), precluding the possibility that it is due to ephaptic reexcitation of nerve terminals by the summed muscle action potential, although the latter can probably occur (D. P. C. Lloyd, 1942). This backfiring, with reexcitation of muscle fibers as the impulses are propagated down other branches of the axons affected, together with repetitive activation of muscle fibers by the prolonged EPP, explains the action of anticholinesterases and related agents to potentiate the muscle twitch response to maximum nerve stimulation. There is no good correlation between anticholinesterase potency and twitch potentiation (e.g., Blaber and Bowman, 1963); therefore the antidromic firing caused by these agents cannot be ascribed entirely to the back action on the nerve terminal of previously released ACh persisting because of cholinesterase inactivation. Evidently, the drugs act themselves on the nerve terminal, either to enhance ACh action or to depolarize the nerve terminals by themselves acting on the ACh receptor. Many anticholinesterases also act to depolarize postsynaptically like ACh (see below p. 68). Another possibility that must be considered is modification of presynaptic excitability by other mechanisms. Tetraethylammonium (TEA), for example, also causes antidromic activity in motor nerves and repetitive muscle activity, presumably because of delay of nerve terminal repolarization by inhibition of delayed rectification (Hagiwara and Saito, 1959). This action is not blocked by curare (Masland and Wigton, 1940).

The fact that twitch potentiation by anticholinesterases cannot be attributed to their postsynaptic action has apparently led some investigators to the conclusion that the only important effects of these agents are exerted presynaptically (reviewed by Riker and Okamoto, 1970). However, in view of the marked inhibition by curare of antidromic post-tetanic backfiring (Standaert, 1964), it seems impossible to attribute the decurarizing effect of anticholinesterases to their presynaptic action. In treatment of myasthenia gravis, on the other hand, effects on nerve terminals could play an important role. In combination with the postsynaptic action to preserve ACh released by the nerve terminal, a reduction in early tetanic rundown of EPPs, based upon a reduction of fractional release of available transmitter, would tend to prevent the rapidly developing weakness that is characteristic in this disease. Repetitive excitation of nerve terminals in response to a single impulse would also increase total muscle contraction.

As an agent which mimics ACh in the effect on motor nerve terminal excitability, it might be expected that tubocurarine would also act to reduce depolarization–release coupling and hence the fractional release of immediately available transmitter store evoked by a presynaptic action potential. The data available, however, indicate the contrary; i.e., tubocurarine's main action on release is probably to antagonize effects of endogenous ACh. Blaber (1970), using cut tenuissimus muscle of the cat and the method of Elmqvist and Quastel (1965b) for estimating fractinnal release, size of available store, and quantal mobilization rate, found tubocurarine to increase fractional release (about 60% by 7×10^{-7} M), with little effect on available store or mobilization, and therefore to increase quantal content. Such a finding, based upon reduced variance of EPPs, is unlikely to be secondary to sources of error implicit in the method. Hubbard *et al.* (1969b), working with cut rat diaphragm, found tubocurarine, at a rather larger effective dosage, to reduce net quantal content, on the basis of a large decrease of available store and mobilization, combined with a smaller increase in fractional release. Unfortunately, identical results could be expected merely on the basis of insufficient correction for nonlinearity of the postsynaptic response to ACh. This could arise, for example, if the equilibrium potential for transmitter were more negative than −15 mV, in the cut muscle. Similar considerations make it difficult to accept entirely the conclusions of Galindo (1970) that tubocurarine reduces the quantal content of EPPs in uncut, unblocked rat diaphragm. What may be most significant in interpreting these results is that in these experiments Galindo also found a reduction of MEPP frequency. Another important clue is that in myasthenia gravis, where MEPPs are reduced in amplitude but postsynaptic response to cholinomimetics is normal, quantal content of EPPs, and by extension, fractional release and available store, is the same as in curarized normal muscle (Elmqvist *et al.*, 1964). Thus, the apparent reduced fractional release and increased store in noncurarized muscle compared to curarized may be related to a back action on the nerve terminal either of ACh itself (compare the parallel action of decamethonium) and/or of raised potassium in the synaptic cleft, secondary to the postsynaptic action of ACh. The second mechanism, in particular, could account for alteration of MEPP frequency by tubocurarine. It will be objected, of course, that raised potassium increases rather than reduces fractional release of transmitter. However, it should be recalled that if fractional release were to increase progressively with each impulse early in a tetanus and mobilization were enhanced, there would appear a reduction of early tetanic rundown and hence an apparent reduction of fractional release and increase of apparent store.

Another point that is convenient to consider here is the mechanism of fasciculations produced in muscle by close arterial injection of ACh or analogs such as succinycholine. It has often been thought that this represents a direct stimulation of muscle fibers by ACh acting post-synaptically at endplates, although it is impossible to imagine how such an action could give rise to fasciculation as distinct from activation of single muscle fibers. Kato and Fujimori (1965) have demonstrated that this response is mainly reflex in nature, based upon increased motoneuron activity. Acetylcholine has long been known to increase the sensory discharge from mammalian muscle spindles, presumably by producing contracture of their intrafusal muscle fibers, an effect which is blocked by tubocurarine and enhanced by physostigmine (Hunt, 1952). It is of course possible that some muscle activity caused by intra-arterial ACh may be mediated by nerve terminal depolarization and consequent axon reflexes. The correctness of the view that direct post-synaptic effects at the endplates of extrafusal fibers play a negligible part in the response to intraarterial ACh has recently been confirmed by Okamoto et al. (1970), who found that after treatment of muscle with black widow spider venom, an agent which rapidly empties nerve terminals of ACh, there was very little contraction in response to ACh.

4. CATECHOLAMINES

It has long been recognized that stimulation of the synaptic nerve supply to the vasculature of fatigued frog skeletal muscle causes an increase in twitch tension (e.g., Orbeli, 1923), evidently due to the action of catecholamine diffusing from sympathetic nerve terminals to the nearby motor endplates. It is now evident that this effect is related to an action of catecholamine to increase quantal release of transmitter, both spontaneous and evoked (Krnjević and Miledi, 1958; Hidaka and Kuriyama, 1969b), as well as a postsynaptic effect to increase membrane resistance and hence response to ACh, in frog (Hutter and Loewenstein, 1955), fish (Hidaka and Kuriyama, 1969b), and rat (Kuba, 1970). Judging by the relative potency of noradrenaline, adrenaline, and isoprenaline, and of α and β blockers, the postsynaptic response is mediated by a β-adrenoceptor, while the presynaptic response is mediated by an α-adrenoceptor (Jenkinson *et al.*, 1968; Kuba, 1970; Hidaka and Kuriyama, 1969b). The increase in both MEPP frequency and evoked release is unlikely to be related to nerve terminal depolarization, not because depolarization necessarily reduces quantal content (see above—it may well act on available store in the same way as extrinsic hyperpolarization), but because the raised MEPP frequency persists in raised magnesium (Krnjević and Miledi, 1958). The effect is indeed highly reminiscent of that of ethanol and many other compounds (see Table I), which would suggest that a general activation of the release mechanism similar to that of ethanol may be produced as a result of presynaptic α-adrenoceptor interaction with catecholamine.

It has been suggested (Goldberg and Singer, 1969) that the catecholamine action is mediated by cyclic AMP. Since only beta mediated effects of catecholamines have been demonstrated to be associated with increase in cyclic AMP levels (see Sutherland *et al.*, 1968; Gilman and Rall, 1970), this proposition does not appear altogether well founded. Their major support for the hypothesis is that both dibutyryl cyclic AMP and theophylline can increase both MEPP frequency and, to a small extent, EPP quantal content. However, there was no potentiation by theophylline of the response to dibutyryl cyclic AMP. Theophylline also acts in the absence of calcium (Quastel and Hackett, 1971), perhaps in the same way as ethanol and many other organic compounds (see Table I). The same could as well be true for dibutyryl cyclic AMP; we have been able to find no reproducible effect of this compound.

In the gut, noradrenaline and adrenaline have been found to be potent inhibitors of ACh output evoked by field stimulation (Vizi, 1968; Paton and Vizi, 1969), and this action is antagonized by the α blocker phentolamine (Knoll and Vizi, 1970); isoprenaline is ineffective. It is unclear, however, to what extent the effect merely represents a postsynaptic inhibitory action on the cholinergic ganglion cells, rather than an effect on the transmitter release mechanism. By either mechanism, the effect would be expected to be less at high frequencies of stimulation, as has been observed (Paton and Vizi, 1969).

For bullfrog sympathetic ganglia, the evidence is clearer. According to Nishi (1970), adrenaline decreases the frequency of miniature EPSPs in raised potassium, and reduces quantal content of EPSPs without altering quantal size. The depressant action of adrenaline is antagonized by α-adrenergic blocking agents (phenoxybenzamine and dihydroergotamine). These observations explain previous data indicating inhibition by catecholamines of ganglionic transmission (see Nishi, 1970). The opposite effect of activation of presynaptic α-adrenoceptors in the ganglion from that at the neuromuscular activation is a striking reminder of how incorrect it may be to extrapolate from one synapse to another, even when both release the same transmitter substance and release mechanisms are similar in terms of calcium–magnesium interaction.

5. MISCELLANEOUS AGENTS

There are a number of agents which act at the skeletal neuromuscular junction to increase transmitter release (either spontaneous or nerve impulse evoked) by mechanisms which are at present poorly understood and may be multiple. Guanidine, for example, produces spontaneous twitching in muscle (Feng, 1938), which is related to the appearance of giant multiquantal MEPPs (Otsuka and Endo, 1960). It also, like TEA, increases the quantity of transmitter released per impulse and accelerates early tetanic rundown. Marois and Edwards (1969) have found that in frog sartorius muscle with transmission blocked by either d-tubocurarine or high magnesium–low calcium solution, quinidine elicits repetitive EPPs in response to single nerve stimuli, an effect which also occurred simply with low calcium (no magnesium). During the interval between the first and repetitive EPPs, MEPP frequency was raised, suggestive of maintained depolarization of nerve terminals between action potentials. Reduction in the usual inactivation of the sodium carrier system, which normally serves to terminate an action potential, could account for this effect, as well as, via prolongation of action potential duration, for increase of transmitter release by the action potential. It is difficult, however, to understand how multiquantal MEPPs could arise by this mechanism. Veratridine and germine monoacetate apparently act in a similar manner to guanidine, at least in terms of action on presynaptic action potential activity. Both cause twitch potentiation related to repetitive action potentials in nerve terminals (Okamoto and Baker, 1969; Detwiler and Standaert, 1969) and tend to relieve depression of transmission by tubocurarine or in myasthenia gravis. To what extent the decurarizing action is related to spike prolongation, or possibly, to effects on depolarization–release coupling, has not been determined.

Feldman *et al.* (1969) have reported the effects of amino acids on quantal release of transmitter at the rat neuromuscular junction. All amino acids tested, at 15 mM concentration, caused reduction of MEPP frequency raised by 15 or 20 mM potassium, but did not influence spontaneous MEPP frequency. Such an effect indicates either interference with depolarization–release coupling or, perhaps, inhibition of the specific effect of potassium to enhance depolarization–release coupling.

Caffeine (Elmqvist and Feldman, 1965a; Hofmann, 1969) and

theophylline (Goldberg and Singer, 1969) act to increase both spontaneous and nerve impulse-evoked transmitter release at the skeletal neuromuscular junction. However, the augmenting action of caffeine on
evoked release is different from that of theophylline, caffeine acting
to depress fractional release but increase quantal mobilization (Hofmann, 1969). The results of Elmqvist and Feldman (1965a) indicate
that displacement of calcium from a membrane store is probably the
major mechanism of caffeine action. From the model of release already
put forward (see above and Cooke and Quastel, 1972), a steady small
influx of calcium should lead to a parallel multiplication of spontaneous
and evoked release; that caffeine has the opposite action does not support
the idea that calcium entry is what normally activates the release mechanism. Calcium entry may instead serve as the stimulus to quantal mobilization. Caffeine also acts to increase MEPP frequency after strenuous
calcium depletion with EGTA (estimated free calcium, 10^{-10} M) and
this action is well maintained, rather than transient, indicating that it
is not mediated by mobilization of calcium from a residual store; the
same is true of theophylline (Quastel and Hackett, 1971). It is of interest
that theophylline acts simply to multiply release, like ethanol, and does
not increase the slope of the relation between log MEPP frequency
and ethanol concentrations. It is not inconceivable that theophylline
could act via inhibition of the diesterase which breaks down cyclic AMP
(Goldberg and Singer, 1969), the actual increase in release being mediated by the second messenger. However, the results argue strongly
against cyclic AMP being a mediator of the action of ethanol and other
agents, including depolarization, unless release is in general proportional
to a local cyclic AMP concentration, rather than related exponentially.
This would imply fantastic increases of local cyclic AMP concentration—up to 10^5-fold within half a millisecond after a presynaptic nerve
impulse—at some critical sites in the nerve terminal.

Among agents that increase both spontaneous and evoked release of
transmitter are phenol and catechols (Otsuka and Nonomura, 1963;
Kuba, 1969). The effect might well be identical to that of ethanol and
other similarly acting compounds (see Table I), the only difference
being a relatively high potency. But preliminary experiments using
mouse diaphragm (Hackett and Quastel, 1971) suggest a greater effect
on evoked than on spontaneous release. Such a result would indicate
some action to augment depolarization–release coupling. Batrachatoxin,
a steroidal alkaloid isolated from a South American frog, produces,
transiently, an enormous increase of MEPP frequency in rat diaphragm
muscles. This action probably can be attributed entirely to nerve
terminal depolarization, since batrachatoxin causes a profound depolar-

ization of muscle fibers (up to +15 mV). Activation of the sodium carrier system usually involved in action potential generation is indicated by antagonism of the effect by tetrodotoxin (Warnick and Albuquerque, 1970). Nonmaintenance of high MEPP frequencies, with excessive stimulus to release, appears to be a general phenomenon (see below, p. 32).

Barium is an ion known to substitute for calcium in depolarization–release coupling (see above). However, it causes more increase in spontaneous MEPP frequency that can be obtained by raising calcium concentration. Recently, Laskowski and Thies (1970) have found this effect to be transient, followed by a small reduction of spontaneous MEPP frequency, and to be associated with parallel depolarization of muscle fibers, suggesting that the presynaptic effect to increase MEPP frequency is also simply secondary to depolarization.

D. Effects on Synthesis, Storage, and Uptake

In recent years, there has been an upsurge of interest in drug effects on muscle which are secondary to actions on the synaptic machinery by which transmitter substance is stored, synthesized, or accumulated. Generally, drugs that act in this way are selective for nerve terminals of a particular type. For example, inhibition of ACh synthesis will affect cholinergic synapses, and agents that displace noradrenaline will act only at noradrenergic sites. In the preceding two sections we dealt with drug effects exerted on the secretory mechanism, either direct or secondary to actions on nerve terminals, which appear unlikely to depend upon which transmitter substance is handled by the terminal. However, from the standpoint of secretion processes, investigation of synapses other than the skeletal neuromuscular cholinergic junction has been minimal, and it is not unlikely that some of the pharmacology with which we have dealt may in fact be unique to this synapse. When it comes to storage mechanisms, it is also possible that some observations made at the neuromuscular junction are also ture for synapses in general. An example is the rundown in quantal content of EPPs at varied frequencies of nerve stimulation (Elmqvist and Quastel, 1965b). The characteristic decline in quantal content and its dependence upon frequency of stimulation were interpreted in terms of depletion of an immediately available store either of transmitter packages or of a membrane component required for release, together with repletion by mobilization from a backup store. The data also indicated a large increase in mobilization with increased frequency of stimulation, which was not simply secondary to depletion of the immediately available store. Noradrenergic transmission differs from normal skeletal cholinergic transmission in being characterized by growth of junction potentials during repetitive nerve stimulation (e.g., Burnstock *et al.*, 1964a), and it is therefore not possible to make the same analysis in terms of stores. Nevertheless, there seems no reason to doubt that at high frequencies of stimulation noradrenaline release, like ACh release, rapidly becomes mobilization limited and subsequently limited either by rate of formation of quantal packages or availability of unpackaged transmitter substance. At the skeletal junction, again, it has been observed that very high-level maintained stimulation of the release process is associated with a gradual falling off of the rate of transmitter release. With black widow spider venom or raised osmotic pressure, there is eventually almost complete loss of synaptic vesicles (Clark *et al.*, 1970), at a time when MEPP frequency has declined to almost zero. The total number of MEPPs released under these circumstances is about 500,000, in agreement with the total store of ACh expressed as full quanta, obtained by a different method of depleting the nerve terminals (Elmqvist and Quastel, 1965a). Elmqvist (1965) has reported a similar depletion phenomenon with increased potassium in the bathing medium. In isolated human muscle, MEPP frequency was sustained in 30 mM K^+, or in 40 or 50 mM K^+, provided magnesium was raised to limit maximum MEPP frequency. However, in 40 or 50 mM potassium with normal calcium and magnesium, the high MEPP frequencies did not continue, but declined exponentially to near zero. The total number of MEPPs that could be elicited was about 200,000, which is the same as the total presynaptic store of quanta determined in human muscle using ACh synthesis blockade (Elmqvist *et al.*, 1964).

These results suggest that it may be a general phenomenon that a maintained high-level stimulation of release also entails an inhibition of quantal formation in the nerve terminal. Nagasawa *et al.* (1970) have reported that release of neurohypophyseal hormone is by exocytosis; the vesicle that contains the hormone combines with the surface membrane and discharges its contents. If liberation of neurotransmitter is by the same process, any stimulus to transmitter release presumably acts by increasing the tendency of synaptic vesicles to combine with the membrane. To reform quanta would require endocytosis, i.e., a budding off of vesicles from the membrane. An increased tendency for vesicle combination with surface membrane might therefore be associated with a reduction of the rate at which new quanta are formed. With nerve stimulation, of course, no such complete depletion can be obtained. The difference, presumably, is that intermittent stimulation of the release process leaves time between pulses for reformation of quanta. Although sodium has been shown to be required for ACh synthesis (Quastel, 1962), it is unlikely that an increased intracellular sodium is what makes the difference between nerve stimulation and other methods of stimulating release. Warnick and Albuquerque (1970) have found that with batrachatoxin the enhanced MEPP frequency is transient, despite maintained depolarization, indicating that the same depletion phenomenon occurs with this stimulator of release. Since batrachatoxin apparently works by an increase in membrane sodium conductance, intracellular sodium in the presence of this toxin would be raised.

1. CHOLINERGIC SYSTEMS

The enzyme choline acetyltransferase—or choline acetylase (ChAc)—responsible for formation of acetylcholine from acetyl-CoA and choline (Nachmansohn and Machado, 1943; Korey *et al.*, 1951) is found in vertebrate brain and spinal cord, in high concentration in ventral roots, and in lesser concentration in mixed nerve trunks (Hebb, 1963). In rat diaphragm, most of the ChAc activity is concentrated in the part of the muscle containing the phrenic nerve and the majority of the nerve endings (Hebb *et al.*, 1964). Thus, the enzyme is present throughout the length of cholinergic axons. Turnover studies in perfused ganglia indicate, however, that transport of ACh down a nerve trunk could play only a minute role, if any, in sustaining ACh stores during stimulation. If synthesis in the nerve terminals were not active, only a few thousand nerve stimuli would suffice to deplete the ACh content (Birks and MacIntosh, 1961).

The effects of inhibition of ACh synthesis on neuromuscular transmission have been studied using the drug hemicholinium No. 3 (HC-3). Schueler (1955), who first described the hemicholinium series, was intrigued by the slowly developing neuromuscular blockade characteristic of this agent and its antagonism by choline, and suggested that it might act by inhibition of synthesis of ACh by competition with choline. MacIntosh *et al.* (1956) found that HC-3 inhibited synthesis of ACh only in intact nervous tissue—not in homogenates—and therefore proposed that HC-3 acts at the level of the nerve terminal membrane, where it competes with choline for an active transport system (Marchbanks, 1968).

At the neuromuscular junction, HC-3 is found to have two effects: (1) a postsynaptic blocking action, as exists with many other quaternary ammonium compounds (Thies and Brooks, 1961), and (2) a gradual reduction in size of MEPPs and quantal components of EPPs (in parallel) that can be attributed to blockade of ACh synthesis (Elmqvist and Quastel, 1965a). The alteration of quantum size is a strict function of how much the release system is stimulated after HC-3 has been added. If stimulation is continued sufficiently, MEPPs and EPPs become too small to be observed. By extrapolation, it was calculated that total extinction of MEPPs would occur when total release amounted to the equivalent of about 270,000 normal quanta (in the rat diaphragm), a quantity which presumably represents the normal total presynaptic store of ACh. It is of interest that there is no alteration of quantal content of EPPs or MEPP frequency concomitant with alteration of quantum size. From these phenomena, two important conclusions were drawn: (1) ACh, if normally packaged in the synaptic vesicles, must be mobile from vesicle to vesicle (otherwise complete synthesis blockade could result only in a mixture of full and empty quanta) and (2) the transmitter release system operates on the unit quanta or packages of ACh, independent of how much ACh is contained within them. The latter phenomenon is especially interesting in view of the recent demonstration by Jones and Kwanbunbumphen (1968) that the volumes of vesicles become reduced in parallel with MEPP amplitude after prolonged stimulation of the release system in the presence of HC-3. Elmqvist and Quastel (1965a) also observed that if stimulation of release were stopped during rundown of quantum size, with synthesis blockade maintained, the size of quanta rapidly recovered, partially, only to decline at a more rapid rate when release was reinitiated. This effect cannot be attributed to a residual synthesis, but implies some kind of transport of ACh to those quanta most favorably located for release.

More direct evidence on how ACh is handled in nerve terminals has emerged from studies of subfractions of brain tissue. The ChAc of brain synaptosomes has been found to behave as a soluble synaptoplasmic enzyme during subcellular fractionation (Whittaker, 1965; Fonnum, 1968; Potter *et al.*, 1968), implying cytoplasmic synthesis of ACh. The main storage of ACh is evidently in vesicles (see Whittaker, 1965). Uptake of synaptoplasmic ACh into vesicles, however, has not been demonstrated with isolated vesicles or synaptosomal preparations (Marchbanks, 1968; Whittaker, 1968). A. K. Ritchie and Goldberg (1970) have recently adduced evidence that ACh synthesis can be directly into vesicles as well as cytoplasmic. Presumably much of the nerve terminal ChAc is normally bound to vesicles (cf. Fonnum, 1967), and the ACh it forms can be transferred immediately into the vesicles. Vesicle-bound ChAc might indeed constitute a route whereby ACh can be transferred from one vesicle to another.

Experiments with fractional store labeling (with choline-^{14}C) have indicated that in both rat diaphragm (Potter, 1970) and in perfused ganglia (Collier and MacIntosh, 1969), there is preferential release of newly synthesized ACh, although most of the transmitter store does slowly contribute to release even when the stores are kept quite fully stocked. The preferential release of newly synthesized ACh might well reflect a mechanism whereby new vesicles are reformed from the presynaptic membrane. Although they would normally be refilled with ACh before being discharged, they could continue to be located relatively close to the membrane and hence more likely to be released than the bulk of the store. It has recently been observed (Cooke and Quastel, 1972) that there exists, in rat or mouse diaphragm, a population of small MEPPs that could represent incompletely filled vesicles. Triethylcholine, another quaternary ammonium compound, also blocks the synthesis of ACh in intact tissues (Bull and Hemsworth, 1963) and has a slow effect on quantal size at stimulated endplates identical to that of HC-3 (Elmqvist and Quastel, 1965a). Like HC-3, triethylcholine also depresses the postsynaptic end plate sensitivity of ACh (Bowman *et al.*, 1962; Thies and Brooks, 1961). However, higher concentrations of both drugs are necessary to block neuromuscular transmission by this mechanism than by blockade of ACh synthesis. WIN 4981 is another agent that acts by blockade of ACh synthesis (Gesler *et al.*, 1951). Studies of ACh metabolism in sympathetic ganglia (Matthews, 1966) have shown inhibition of ACh synthesis also by tetraethylammonium and troxonium. In contrast, tubocurarine, hexamethonium, and physostigmine in high concentrations did not affect maintained output or synthesis of ACh.

Since the inhibition of synthesis by HC-3 or triethylcholine becomes effective in blocking transmission only after nerve terminal stores have become appreciably depleted of transmitter, drugs with this action should first block the most activated cholinergic junctions in an animal. Animal experiments using tetanus toxin have borne out this prediction. The spasms of muscles affected by the action of the toxin were selectively reduced by triethylcholine before interference with neuromuscular transmission in other muscles (Bowman and Rand, 1961; Laurence and Webster, 1961). In principle, the idea of selectively blocking hyperactive synapses by partial inhibition of transmitter synthesis is an intriguing one and may have therapeutic applications. In terms of selectivity, it would be preferable to a false transmitter, which would act equally to block all synapses of the same type. Blockade of transmission by synthesis inhibition does not, of course, discriminate between muscarinic and nicotinic cholinergic junctions. Indeed, blockade by HC-3 and similarly acting agents such as WIN 4981 can provide circumstantial support for cholinergic transmission when the usual cholinergic blocking agents

fail to block a nerve-mediated response, e.g., in mammalian bladder
(Dhattiwala *et al.*, 1970). It remains to be demonstrated, however, that
they can act only at cholinergic synapses.

Metabolic inhibitors or lack of substrate might be expected to influence
ACh synthesis in either of two ways, inhibition of choline transport
or of the formation of acetylcoenzyme A. Either effect would be expected
to be manifest in the same way as the action of HC-3, a reduction
in MEPP size consequent upon release of the ACh stores. However,
no such effect has ever been described for anoxia, glucose deficiency,
or metabolic inhibitors (e.g., Hubbard and Løyning, 1966). It is possible
that under these conditions formation of material for the packaging
of ACh could be as much depressed as the ACh synthesis mechanism
itself. If this were the case, there would be no outstripping of ACh
synthesis by new quantal formation and no small quanta would result.

Myasthenia gravis is a disease of unknown etiology characterized by
rapid failure of muscle strength on exertion. Microelectrode study of
muscle biopsy specimens (Elmqvist *et al.*, 1964) have shown that the
lesion consists in a reduction of the amplitude of MEPPs and quantal
components of the EPP. Postsynaptic response to cholinomimetics is
quantitatively the same as in normal muscle, and MEPP (quantal) size
does not alter with stimulation. The presynaptic store of transmitter,
in terms of number of quanta, also is normal. Hence, it appears that
the disease affects the process by which ACh is packaged in vesicles;
each vesicle contains about one-fifth the usual quantity of transmitter.
However, this hypothesis has not yet been substantiated by direct
measurement of ACh content of myasthenic muscle; if the disease is
in fact presynaptic, the ACh content of nerve terminals, and ACh release
by any standard stimulus (e.g., raised potassium) should be one-fifth
that of normal muscle. Normal postsynaptic sensitivity to ACh is of
course not possible to determine experimentally. A normal response
could reflect equal and opposite changes of ACh receptor density and
area. It is only because of the *a priori* unlikelihood that there should
be two such alterations exactly compensating that a presynaptic locus
for the disease process is indicated. Treatment with anticholinesterases
remains rational therapy whatever the mechanism of the disease; it is
in either case not directed against the lesion per se.

Botulinum toxin acts to block cholinergic neuromuscular transmission
in smooth as well as skeletal muscle (e.g., Brooks, 1956; Ambache, 1951;
Ambache and Lessin, 1955). The mode of action at the skeletal junction
was studied by Brooks (1956). The effect is certainly mediated by inhibition of processes subserving transmitter release. EPPs become small
in amplitude, without any change in MEPP size. This reduction in

quantal content resembles the action of magnesium. However, botulinum toxin, unlike magnesium, reduces spontaneous MEPP frequency as much as or more than quantal content of EPPs. Thus, it is not depolarization–release coupling that is affected. Another unusual characteristic is the restoration of MEPP frequency and EPP altitude by tetanic nerve stimulation. This is far out of proportion to normal facilitation and suggests that the mode of action of the toxin may be to inhibit mobilization so that available store becomes depleted. As was noted above, mobilization is enhanced by nerve action potentials.

2. ADRENERGIC SYSTEMS

The pharmacology of noradrenergic nerve terminals, in contrast to cholinergic terminals, has recently become almost overwhelmingly rich. A plethora of drugs has been shown to affect uptake and storage of the transmitter or to depend upon these systems for their effect. The interactions of these agents has become a major new field in pharmacology with important applications in psychopharmacology and the treatment of hypertension. In this section, even more than elsewhere, our discussion will be much more brief and superficial than the subject deserves. A far more complete account can be found in the recent monograph by Iversen (1967).

A. STORAGE. It is now generally agreed that the major part of the catecholamines present in catecholamine-containing cells, such as the adrenal medullary cells and adrenergic neurons, is stored in specific particles, which appear as large (adrenal medulla) or relatively small granular vesicles. An electron-dense core is related to the presence of catecholamine (Wolfe *et al.*, 1962; Potter and Axelrod, 1963a,b; Hokfelt, 1968; also see Iversen, 1967), which is stored in association with protein and ATP (Schumann, 1958; Potter and Axelrod, 1963b; Austen *et al.*, 1967). Agranular vesicles are also found in noradrenergic terminal varicosities. These presumably represent vesicles that have discharged their contents (see Hubbard, 1970), since treatment with drugs that deplete stores or electrical stimulation in combination with inhibition of noradrenaline synthesis lead to granular vesicles becoming agranular, a process which can be reversed by administration of exogenous amine (de Iraldi and de Robertis, 1963; Bondareff and Gordon, 1966; Hokfelt, 1967; van Orden *et al.*, 1966). Simple incubation of vas deferens with noradrenaline *in vitro* raises the proportion of granular vesicles at the synapses from 70% to 90% (van Orden *et al.*, 1967). The suggestion by Burn (1963) that the agranular vesicles contain ACh thus appears unlikely. It may also be noted that the zinc iodide–osmium stain, which stains all the agranular vesicles of presumed cholinergic nerve terminals, does not stain vesicles in peripheral adrenergic synapses (Akert and Sandri, 1968; Akert *et al.*, 1968).

Granular vesicles isolated from bovine splenic nerves have been shown capable of accumulating noradrenaline against a concentration gradient through a mechanism that depends upon the presence of magnesium ions and ATP (von Euler and Lishajko, 1963a,b) and that is inhibited by reserpine (von Euler and Lishajko, 1963c). If sufficient noradrenaline is present in the medium, about 10^{-4} M, the result is a balance between uptake of the amine and its loss from the granules. In the absence of external amine, the granules become 50% depleted in less than 10 minutes (Stjärne, 1964). Thus, the maintenance of an overall noradrenaline content of the granules depends upon a dynamic equilibrium between the amine· in the granules and in the medium (and the presence of magnesium and ATP), and if reserpine is present, the granules lose their store of amine (von Euler *et al.*, 1963; Stjärne, 1964, reviewed by Potter, 1966), despite retardation of amine loss by reserpine (von Euler and Lishajko, 1963c). Amine granules from adrenal medulla similarly depend upon magnesium and ATP for maintenance of their catecholamine content (Kirshner, 1962; Carlsson *et al.*, 1963b). There is ample evidence that the monoamine depletion induced by reserpine *in vivo* (Carlsson *et al.*, 1957) is secondary to the blockade of amine uptake by the storage granules (see Carlsson, 1966); blockade by reserpine of the amine incorporation has been demonstrated *in vivo* (Bertler *et al.*, 1961) as well as *in vitro*.

The blockage of granule uptake by reserpine appears initially to be competitive; it can be presented by high amine concentrations in the cytoplam, induced, for example, by inhibitors of monoamine oxidase (Carlsson *et al.*, 1963b; Jonason *et al.*, 1963); later, however, it becomes effectively irreversible. The block can initially also be antagonized by other agents which appear to act in the same way as reserpine, but with shorter duration of action, e.g., tetrabenazine and prenylamine (Quinn *et al.*, 1959; Carlsson, 1966).

The spontaneous monoamine release that occurs as the result of reserpine treatment is primarily an intracellular event, oxidative deamination by monoamine oxidase taking place before most of the amine escapes from the cell (Carlsson *et al.*, 1957). Formation of *O*-methylated amines is reduced (Häggendal, 1963), presumably as a consequence of reduced release of the unaltered amine, catechol-*O*-methyl transferase not being located in the nerve terminal. If monoamine oxidase is inhibited, noradrenaline itself is released after treatment with reserpine, and the sympathomimetic response is much greater. However, the rate of depletion of noradrenaline from nerve terminals in many tissues has been shown to be dependent upon the nerve impulse frequency in the sympathetic nerves. Thus, the loss of noradrenaline is retarded by sympathetic decentralization, acute postganglionic denervation, or ganglionic blocking drugs (e.g., Hertting *et al.*, 1962) and is accelerated by sympathetic nerve stimulation (e.g., Malmfors, 1969). Sedvall (1964) found that after preganglionic denervation of the cat gastronemium muscle reserpine caused noradrenaline disappearance from the muscle in two phases. Some 85% of the amine was gone within 5 hours; the remainder disappeared only slowly over the next 19 hours. This 15% reserpine-resistant pool appeared to be very much smaller in normally innervated muscle and could be rapidly depleted by adrenergic nerve stimulation (Sedvall and Thorson, 1963). This result suggests that the fraction of nerve terminal store most available for release by presynaptic nerve impulses is relatively resistant to reserpine depletion.

Farnebo and Hamberger (1970) have recently studied the effect of reserpine on release by electrical stimulation of tritiated noradrenaline from the isolated rat iris. Stimulation increased release to the same extent as in nonreserpinized irides, the spontaneous release of tritium by reserpine being simply superimposed. Total content, of course, fell faster with stimulation. Thus, reserpine does not interfere with the normal release process, and the reserpine sensitive pool does contribute to release by nerve impulses. After depletion with reserpine, treatment with a monoamine oxidase inhibitor permits the reaccumulation of noradrenaline by the nerve terminal, but this noradrenaline is predominantly extragranular in location. A small portion, however, is taken up into a reserpine-resistant granular pool (Hamberger, 1967; Stitzel and Lundborg, 1967; Jonsson *et al.*, 1969). One might imagine that this reserpine-resistant granular pool corresponds to that of Sedvall (1964), but this is evidently not the case. The noradrenaline taken up by nerve terminals treated with reserpine and a monoamine oxidase inhibitor (nialamide) cannot be released by nerve stimulation (Farnebo and Hamberger, 1970). Thus, in order to be put in a form releasable by nerve impulses, cytoplasmic amine must first be taken up into the granules by the magnesium–ATP-dependent reserpine-sensitive mechanism, a conclusion also reached from *in vivo* studies (Malmfors, 1967; Häggendal and Malmfors, 1969). On the other hand, the precursor of noradrenaline, dopamine, taken up after reserpine and nialamide treatment can be released by nerve stimulation, probably as both noradrenaline and dopamine (Malmfors, 1967; Häggendal and Malmfors, 1969; Farnebo and Ham-

berger, 1970), indicating the persistence of a route for transfer of dopamine into a reserpine-resistant available store that noradrenaline itself
cannot follow. Once in the granules, it is presumably acted upon (to
form noradrenaline) by dopamine-β-hydroxylase, which is present in
the granules (Stjärne, 1966) and is, indeed, released together with
noradrenaline upon nerve stimulation (Gewirtz and Kopin, 1970). The
reserpine resistance of the uptake into this available store is probably
only relative. The major fraction of noradrenaline synthesis from
dopamine is blocked by reserpine, presumably because of the block
of dopamine uptake by the storage granules (Kirshner, 1962; Rutledge
and Weiner, 1967), and the resistant pool of Sedvall (1964) does eventually succumb to reserpine action. It is this resistant pool which presumably is the first to recover after reserpine treatment and which accounts
for the partial restoration of adrenergic transmission by exogenous
noradrenaline or precursors when the major pool remains depleted. Thus,
after a single dose of reserpine, a very small recovery of the storage
function appears about 48 hrs after administration of the drug, at a
time when tissue levels remain extremely low, and at the same time
there is a concomitant restoration of function (Andén *et al.*, 1964; Iversen
et al., 1965). This early reversibility does not appear to exist after exposure to very high concentrations of the drug (see Carlsson, 1966).
Similarly, there can be some temporary restoration of adrenergic transmission by exposure of reserpinized tissue to noradrenaline, or its
precursors—dopamine, L-dopa, *m*-tyrosine, or phenylalanine (e.g., Burn
and Rand, 1958, 1960; see Iversen, 1967). The restoration of response
to nerve stimulation is much less than the restoration of response to
tyramine (which, as will be seen below, appears to be via a different
noradrenaline release mechanism) but is much more persistent (Trendelenburg and Pfeffer, 1964). Iversen (1967) has pointed out that the
restoration of transmission after the administration of noradrenaline precursors appears rather rapidly, suggesting an unexpectedly rapid synthesis of noradrenaline. Recent studies on the release by sympathetic nerve
stimulation of noradrenaline derived from tritiated tyrosine indicate
that the noradrenaline liberated spontaneously is very largely recently
synthesized. As the frequency of nerve stimulation is increased the
endogenous store rather than newly synthesized noradrenaline is
drawn upon to a greater extent. These results agree with the idea
that the major store of transmitter functions as a reserve to be
drawn upon when the demand for release is high. Correspondingly,
the available pool tends normally to be turned over at a relatively high
rate. The granule fraction corresponding to this available pool is relatively resistant to reserpine (Sedvall, 1964) and as a result can provide

transmitter function when most of the vesicles are incapable of taking up and storing amines.

Burnstock and Holman (1962b) have studied the effect of chronic reserpine treatment on the junction potentials found in vas deferens. Spontaneous junction potentials were reduced, either in frequency or in amplitude, or both. The spontaneous junction potential size distribution curve normally merges into noise, and it is not possible to distinguish changes in amplitude from changes in frequency. This result is consistent with either a depletion of the number of quanta available for release or of the transmitter content of the quanta. Nerve stimulation-evoked junction potentials were reduced in amplitude, which is again consistent with either alternative. Most interesting was the finding that facilitation was slower to develop than in nonreserpinized preparations, suggesting that the store of transmitter immediately available for release was less depleted than the store required for maintenance of release.

Recently it has been established that the granule uptake mechanism can be affected by a large number of drugs more usually thought of in connection with membrane uptake system (see below) or postsynaptic events. Thus α-adrenoceptor blocking agents—including haloalkylamines (phenoxybenzamine, SY-28), dihydroergotamine, phentolamine, hydergin, azapetine, and chlorpromazine—and β-adrenoceptor blocking agents—including propranolol, dichlorisoproterenol, and pronethalol—all act like reserpine to slow spontaneous noradrenaline loss from granules and inhibit the magnesium–ATP-dependent granule uptake system (von Euler and Lishajko, 1966, 1968b). The dibenamine congener GD-131, which is a very weak α-blocker, inhibited the uptake system very little. It is of interest that in contrast to their antiadrenergic postsynaptic effect, the *l*- and *d*-forms of propranolol and of Aptin were equally potent in their effect on granule uptake (von Euler and Lishajko, 1968b). The β-blockers MJ 1999 and MJ 1998 had little reserpinelike effect. Thus, it would appear that the affinity of the granule receptor for α- and β-blocking agents differs from that of α or β postsynaptic receptors.

Various amines also have marked effects on the granule (von Euler and Lishajko, 1968a). In retarding uptake of noradrenaline, apparently by competition, dopamine, phenylethylamine, and metaraminol were found to be about equipotent; tyramine, amphetamine, and methamphetamine were less potent; and ephedrine and mephentermine were much weaker. Effects on the rate of release of noradrenaline from the granules did not correlate. Thus, dopamine had little effect, mephentermine and metaraminol strongly retarded release, while phenylethylamine, and amphetamines, and tyramine accelerated release, the last only to a very small extent. The noradrenaline-releasing effect of tyramine chiefly related to inhibition of noradrenaline reuptake. Tritiated tyramine was found to be taken up by the granules to a small extent approximately corresponding to the direct release of noradrenaline. Reserpine and phenoxybenzamine, as inhibitors of the uptake mechanism, blocked the noradrenaline releasing effect of tyramine.

Desmethylimipramine (DMI) is particularly known for its ability to block the uptake of noradrenaline at the axon membrane (see below). When tested on the release of noradrenaline from nerve granules, it was found to have a considerable inhibitory effect combined with a marked inhibition of labeled noradrenaline incorporation, i.e., a reserpine-like effect. In common with reserpine, it blocks the noradrenaline-releasing effect of tyramine (von Euler and Lishajko, 1968a). There is good evidence that these effects of DMI appear in intact tissue; i.e., DMI does in fact enter the nerve terminal cytoplasm and act on the granule uptake system. Thus, DMI, in contrast to amphetamine, acts to depress noradrenaline efflux from sympathetic nerve endings in rat heart induced by metaraminol to a greater extent than it inhibits metaraminol influx (Leitz, 1970). Reid et al. (1969) found that DMI did not prevent the uptake into adrenergic neurons of large doses of exogenous noradrenaline, but prevented the exogenous amine from exchanging with endogenous stores. DMI has also been found to inhibit the formation of labeled noradrenaline from tyrosine without influencing the formation of labeled dopamine (Nybäck et al., 1968), and to inhibit the synthesis of octopamine from tyramine (Steinberg and Smith, 1970), again indicating a block of transfer of amine to the granules where dopamine-β-oxidase is located.

It might be expected that efficient inhibitors of granular uptake, like reserpine, should tend to cause partial emptying or depletion of noradrenaline stores. This has been shown to occur for phenoxybenzamine (Schapiro, 1958) and chlorpromazine (Johnson, 1964), but appears not to be the rule for the series of blocking agents studied by von Euler and Lishajko (1968b). In a large dose, propranolol, one of the most active agents, was found to cause a moderate lowering of noradrenaline in tissues (von Euler and Lishajko, 1968b). Thus, it does not appear that the action of adrenergic blocking agents, in usual doses, owes much to any effect on transmitter stores.

B. AXONAL UPTAKE (UPTAKE₁). It was first suggested by Burn (1932) that exogenous catecholamines might be taken up into storage sites in nervous tissue. That this is indeed the case has been demonstrated by a number of investigators (see Iversen, 1967). The importance of this mechanism in the disposition of circulatory catecholamine was first shown by Axelrod *et al.* If allowance is made for regional differences in blood flow to various tissues there is a good correlation between the amount of noradrenaline-^{3}H taken up and the endogenous noradrenaline content, i.e., the density of noradrenergic nerve terminals (Wurtman *et al.*, 1964; Kopin *et al.*, 1965). After surgical sympathectomy the ability of tissues to take up catecholamine is severely reduced (Hertting *et al.*, 1961; Hertting and Schiefthaler, 1963). In immuno-sympathectomized animals, uptake of noradrenaline-^{3}H is also markedly suppressed. The persistence of a small uptake after sympathectomy of course indicated that there can be some accumulation of noradrenaline by extraneuronal tissue, but also raised the question as to whether these results genuinely indicated that the main uptake is neuronal. Radio-autographic evidence has shown clearly that exogenous noradrenaline-^{3}H is indeed concentrated in postganglionic sympathetic nerve terminals (Wolfe *et al.*, 1962; Wolfe and Potter, 1963).

It is clear that neuronal uptake of noradrenaline cannot be accounted for in terms of the uptake system of the intracellular granular vesicles. Whereas granules lose their amine content if exposed to solutions containing low concentrations of amine, the concentration ratio between nerve terminals and medium is probably of the order of 10,000:1 or greater (Iversen, 1967). Noradrenaline-^{3}H has been found to be accumulated by heart tissue, of which sympathetic terminals constitute only a minute fraction, to a level thirty or forty times that of the perfusing medium (Iversen, 1963; Lindmar and Muscholl, 1964). The uptake mechanism, moreover, is saturable at remarkably low external concentrations of amine. The Michaelis constant for L-noradrenaline is about 0.27×10^{-6} M (Iversen, 1963), thus the affinity of noradrenaline for the membrane uptake system is at least 1000-fold that for uptake by isolated granules ($K_m \sim 10^3$ M). The stereo and structural specifity of the membrane uptake system also differs markedly from that of granules (see Iversen, 1967). Moreover, reserpine has no effect on the initial entry of noradrenaline into intact tissue (Kopin *et al.*, 1962; Iversen *et al.*, 1965). Indeed, two distinct phases could be recognized in the uptake of noradrenaline-^{3}H by tissue—a fast one, with a half time of less than 5 min, and a second, much slower, with a half time of approximately 20 min (Iversen, 1963). In animals pretreated with reserpine, the second phase was absent, suggesting that the normal slow phase represents slow accumulation of amine in the intraneural storage granules after rapid uptake into the axoplasm (Iversen *et al.*, 1965). Correspondingly, Hamberger *et al.* (1967) have found, using fluorescence microscopy, that after treatment with reserpine and administration of noradrenaline or α-methylnoradrenaline the accumulated amine was diffusely distributed throughout the sympathetic preterminal axons, rather than concentrated in the terminal varicosities.

The amine uptake mechanism of noradrenergic axon terminals has been studied in considerable detail. It shares with a number of other active transport processes a dependence upon extracellular sodium (Bogdanski and Brodie, 1966; Iversen and Kravitz, 1966; Horst *et al.*, 1968) and a susceptibility to inhibition by ouabain and other metabolic inhibitors (Dengler, 1964; Green and Miller, 1966). Calcium, but not magnesium, also appears to be necessary to the uptake process (Gillis and Paton, 1967).

There are a number of drugs that act to inhibit the axonal uptake of noradrenaline and other amines that can be transported by the same system. Cocaine, for example, is relatively selective in this regard. It clearly inhibits the uptake system without altering tissue content of noradrenaline (Whitby *et al.*, 1960; Muscholl, 1961). Its action is apparently competitive (e.g., Iversen, 1963). Iversen (1965a, 1967) has listed the relative activity of a number of drugs that are powerful inhibitors of the noradrenaline uptake system—DMI and related tricyclic antidepressants, quanethidine, bretylium, and α- and β-blockers. Corresponding to the ability of these drugs to block the amine uptake system, there can often be observed distinct potentiation of the action of catecholamine. Thus, the α-blocker phenoxybenzamine potentiates the actions of adrenaline and noradrenaline on rabbit atria (Stafford, 1963).

The relative affinities of different amines for the uptake site was studied by Burgen and Iversen (1965). In general, phenolic hydroxyl groups enhanced the affinity, as did α-methylation (the D-enantiomer having more activity than the L-), and β-hydroxylation, N-substitution, and O-methylation reduced the affinity. The phenylethylamine structure could be replaced with saturated five- or six-membered ring structures without producing a marked decrease in the affinity for the uptake site. Even indoleamine (5-HT) had an appreciable affinity. These experiments did not distinguish between drugs themselves transported by the uptake system and those not transported. Subsequently, several of the amines that were shown to be potent inhibitors of noradrenaline transport were found also to act as substrates for the uptake process. For example, metaraminol, the compound with highest affinity according to inhibition of noradrenaline-^{3}H uptake, was found to be avidly taken up by the axonal uptake system even after reserpinization (Carlsson and Waldeck, 1965, 1966). α-methyltyramine was found to be accumulated in isolated rat heart by a process which could be inhibited by low concentrations of noradrenaline or uptake-blocking drugs (Iversen,

1967), and the uptake of α-methylnoradrenaline by tissues is also well established (e.g., Malmfors, 1955a). However, α-methyldopamine, although a potent inhibitor of noradrenaline uptake, appears not to be taken up by tissues obtained from reserpinized animals (see Muscholl, 1966a). At least some of the drugs that block uptake are also transported. For example, quanethidine, bethanidine, and debrisoquin have been shown to be actively accumulated by rat heart slices; the uptake of bethanidine-^{3}H was shown to be inhibited competitively by quanethidine, debrisoquin, bretylium, tyramine, metaraminol, and desipramine (DMI) (Mitchell and Oates, 1970).

Physiologically, of course, the function of the axonal uptake system is to recapture noradrenaline released by nerve terminals. Its importance in this regard is emphasized by the recent demonstration that the discharge of noradrenaline, which can be found on stimulating sympathetic nerves in the presence of phenoxybenzamine to block uptake, can be ten times higher than the maximum rate of *de novo* synthesis of noradrenaline (Hedqvist and Stjärne, 1969). It is also of considerable interest that the uptake of amino acid precursors of noradrenaline, or methylated derivatives, appears not to be via the amine pump. Cocaine does not block this process (Gulati *et al.*, 1970). The amino acid transport system, like the amine pump, does require sodium ions.

C. Extraneuronal Uptake (Uptake$_2$). Although this subject properly belongs in the section concerning postsynaptic pharmacology, it will, for convenience, be dealt with here. It was pointed out above that sympathectomy does not completely eliminate the ability of tissues to take up noradrenaline. Histochemical studies using the fluorescence technique of Falck and Hillarp have also shown that certain cells—notably smooth muscle, the endothelium of brain capillaries, rat salivary gland cells, and cardiac muscle—have the ability to accumulate noradrenaline intracellularly (Gillespie and Hamilton, 1966; Hamberger, 1967; Hamberger *et al.*, 1967; Ehinger and Sporrong, 1968). This system has recently been studied in the spleen by Gillespie *et al.* (1970). Most tissue, with the notable exceptions of lymphoid tissue, red pulp, and phagocytic cells, showed this uptake, after perfusion with noradrenaline in high concentrations. It was unaffected by chronic postganglionic denervation or reserpine. Cooling the tissue to 15°C or less, phenoxybenzamine, or normetanephrine prevented the accumulation of noradrenaline by arterial smooth muscle but not by connective tissue or stellate cells. Thus, there appear two kinds of extraneuronal uptake. The uptake by arterial smooth muscle cells is probably to be identified with the second system of amine uptake (uptake$_2$) described by Iversen (1965b), for which the threshold concentration of noradrenaline is higher than the saturating concentration for uptake$_1$. This uptake system seems to lack stereospecificity. Like uptake$_1$, it is blocked by α-adrenergic blockers. However, it is also blocked by normetanephrine in low concentrations which do not block uptake$_1$, and metaraminol, a potent blocker of uptake$_1$ is quite ineffective on uptake$_2$ (Burgen and Iversen, 1965; see also Gillespie *et al.*, 1970). Isopropylnoradrenaline is an excellent substrate for uptake$_2$ but is not taken up by uptake$_1$ (Callingham and Burgen, 1966).

It now appears that uptake$_2$ is in fact not a threshold phenomenon but operates even at low levels of extraneuronal noradrenaline, serving to transport the amine to sites of its metabolic degradation, mainly by catechol-*O*-methyltransferase. Drugs that are known to block uptake$_2$ have been found to block the extraneuronal formation of metabolites from noradrenaline applied at low concentration (Eisenfeld *et al.*, 1966, 1967). In the presence of inhibitors of both monoamine oxidase and catechol-*O*-methyltransferase, uptake$_2$ can be demonstrated to occur even with very low noradrenaline concentrations as a nonmetaraminol-sensitive uptake of noradrenaline (Lightman and Iversen, 1969). As a postsynaptic amine binding mechanism, it is tempting to imagine that uptake$_2$ might in fact be identical with the adrenergic receptor(s). There is convincing evidence, however, that this is not the case. For example, higher concentrations of phenoxybenzamine are required to block uptake$_2$ than to block the α-receptors, and in blocking uptake$_2$, phenoxybenzamine acts competitively and its action is easily reversed (Lightman and Iversen, 1969; Gillespie *et al.*, 1970; Eisenfeld *et al.*, 1967). Moreover, trabecular smooth muscle in the spleen possesses α-receptors, but not the uptake mechanism (Gillespie *et al.*, 1970). Burgen and Iversen (1965), on the basis of structure–activity relationships, were forced to the conclusion that α- and β-receptors were not related to either uptake$_1$ or uptake$_2$.

The question naturally arises, which normally is the main mechanism responsible for terminating noradrenaline transmitter action on smooth muscle cells, uptake$_1$ or uptake$_2$? One line of evidence suggests that it may be neither; that in the spleen, at least, the noradrenaline that is released tends first to be bound to the α-receptors. In the perfused spleen, output of noradrenaline (overflow) evoked by nerve stimulation is characteristically increased by phenoxybenzamine (G. L. Brown and Gillespie, 1957), which is known to block both uptake$_1$ and uptake$_2$, but only slightly increased by cocaine, which blocks only uptake$_1$. Phentolamine, which is a potent α-adrenergic blocker, has little effect on noradrenaline uptake either in spleen or heart, and yet increases overflow (Hertting *et al.*, 1961; Kirpekar and Cervoni, 1963; Blakely *et al.*, 1963), suggesting that the overflow is related to α-adrenoceptor blockade. Kirpekar and Wakade (1970), testing the relative potencies of three related haloalkylamines (phenoxybenzamine, SY28, and GD-131), have found that effectiveness in increasing overflow of noradrenaline on nerve stimulation correlated with α-blocking activity rather than uptake blocking action; GD-131 in doses capable of blocking uptake of infused noradrenaline did not increase overflow. Kalsner and Nickerson (1969b) found that GD-131, like other blockers of uptake$_2$, blocks the enzymic inactivation of exogenous noradrenaline.

However, even though these results indicate that noradrenaline may be effectively bound to receptor, at least temporarily, it does not appear that this binding necessarily terminates transmitter action. Thus, in rat aortic strips, GD-131 very effectively potentiates and prolongs the action of added noradrenaline (Kalsner and Nickerson, 1968b). In the isolated trachea of the guinea pig, there is a good correlation between inhibition of uptake of isoprenaline (by uptake$_2$) and potentiation of the response to isoprenaline (which is β-receptor mediated) by a number of agents including phentolamine and phenoxybenzamine, both α-blockers (Foster, 1969).

D. INDIRECT ACTING AMINES. Barger and Dale (1910) introduced the term "sympathomimetic amine" to characterize the action of a wide variety of amines structurally related to adrenaline and noradrenaline that evoked physiological responses similar to those consequent upon stimulation of sympathetic nerves. It was subsequently discovered that the actions of two of these amines, tyramine and ephedrine, were reduced by sympathetic postganglionic denervation or by cocaine, although responses to adrenaline were enhanced (Tainter and Chang, 1927; Burn and Tainter, 1931; Burn, 1932; Bülbring and Burn, 1938). After denervation, drugs such as phenylethylamine and amphetamine behaved like tyramine, whereas noradrenaline behaved like adrenaline (Lockett, 1950). It was Burn (1932) who first suggested that the effects of tyramine and ephedrine might not be exerted directly on adrenergic receptors but might depend on the integrity of adrenergic nerve endings.

Classification of the quantitative differences in the actions of individual amines was made by Fleckenstein and Burn (1953), who tested a large series of amines on the normal and chronically denervated nictitating membrane of the cat. They could distinguish three groups: (1) catecholamines, whose actions were markedly enhanced by denervation, (2) amines (including ephedrine) containing a β-hydroxyl group but not more than one phenolic hydroxyl group, which were slightly less effective after denervation, and (3) amines (including tyramine and amphet-amine) lacking a β-hydroxyl group and having no more than one phenolic hydroxyl group, which caused little or no contraction of the nictitating membrane after its denervation. Later, it was observed that the same differentiation could be made on the innervated nictitating membrane according to whether or not cocaine inhibited or potentiated the response, and it was suggested that, while catecholamines act directly at the receptors, the indirect sympathomimetics act by releasing nor-adrenaline from the nerve endings, cocaine acting to block this action (Fleckenstein and Bass, 1953; Fleckenstein and Stöckle, 1955). Amines in group 2 evidently acted by a combination of both direct and indirect actions. Similar studies were conducted by Innes and Kosterlitz (1954). Soon after Carlsson *et al.* (1957) had discovered that reserpine depleted tissue noradrenaline and abolished the pressor effect of tyramine, several workers restudied the classification of sympathomimetic amines. Burn and Rand (1958) observed that pretreatment with reserpine decreased the actions of indirectly acting amines, while enhancing the effects of catecholamines. The responses of different organs to the indirectly acting amines could partly be restored by infusions of noradrenaline. This was taken as evidence that noradrenaline stores previously depleted by reserpine could be replenished from circulating noradrenaline. Testing of a wide range of sympathomimetic amines in reserpinized preparations on a variety of organs confirmed the older classification (Innes and Krayer, 1958; Holtz *et al.*, 1960; Kuschinsky *et al.*, 1960; Maxwell *et al.*, 1959; Schmitt and Schmitt, 1960; Schmidt and Fleming, 1963). Trendelenburg *et al.* (1962a,b) were able to determine the relative pro-portion of direct to indirect action for each individual amine, by using short-term reserpine depletion, which is not associated with the super-sensitivity to direct action which occurs after long-term depletion. The classification they obtained corresponds well to that of Fleckenstein and Burn (1953). However, it was now evident that a clear-cut distinction between direct, mixed, and indirect acting agents is not possible. The amines possess a continuous spectrum of actions ranging from almost pure direct to almost pure indirect; the overall action of each amine will depend on the relative potency of its two types of action, which may vary in different tissues. The classification shown in Table II is based on Table 1 of the review by Trendelenburg (1963).

It now seems to be generally accepted that the indirectly acting amines exert their sympathomimetic effects via release of noradrenaline from the adrenergic nerve terminals, to which they gain access via the amine pump (uptake$_1$). This explains why cocaine blocks their action and why their effects disappear after denervation or depletion by reserpine.

TABLE II

CLASSIFICATION OF AMINES

Direct acting	Mixed acting	Indirect acting
Noradrenaline	Norsynephrine	p-Tyramine
Adrenaline	Phenylethanolamine	β-Phenylethylamine
Isoprenaline	Phenylpropanolamine	Amphetamine
α-Methylnoradrenaline	Metaraminol	Methamphetamine
Dihydroxyephedrine	Ephedrine	Mephentermine
Dopamine	m-Tyramine	
Phenylephrine		
Synephrine		

Direct evidence of the displacement of noradrenaline from tissues by tyramine and other indirectly acting amines has been found (Stjärne, 1961; Lindmar and Muscholl, 1961; Chidsey *et al.*, 1962). However, Lindmar and Muscholl (1961) observed that the noradrenaline output produced by tyramine from a perfused heart was only about one-seventh that produced by a dose of a nicotinic (ganglion stimulant) drug which gave rise to an equivalent cardioaccelerator effect. Thus, it is apparent that tyramine acts not only to release noradrenaline, but also to potentiate the effect of the transmitter that is released, presumably by blockade of uptake$_1$. By the same token, restoration of responses to noradrenaline in preparations depleted of noradrenaline by reserpine may at least in part be attributed to potentiation of noradrenaline effect by blockade of reuptake, rather than to noradrenaline release per se (Muscholl, 1966b). There is, in fact, no clear idea at present of why the indirect acting amines should cause the release of noradrenaline. Simple blockade of the axonal uptake system seems not to account for this; cocaine, for example, is a potent blocker of uptake$_1$ but does not cause release of noradrenaline. Quantal release of the transmitter secondary to presynaptic depolarization is easy to rule out. Tyramine release does not depend upon calcium in the medium and is, indeed, inhibited by high calcium (Lindmar *et al.*, 1967; Farmer and Campbell, 1967). Indeed, it seems unlikely that the release of noradrenaline by indirect acting amines is quantal at all, since it can be shown to occur when the normal quantal packaging system is blocked by reserpine, provided nerve terminals are noradrenaline replenished as a result of monoamine oxidase inhibition or added noradrenaline, although under the same conditions noradrenaline cannot be released by nerve stimulation (de Schaepdryver *et al.*, 1963; Carlsson *et al.*, 1966; Farnebo and Hamberger, 1970).

It was already pointed out that indirect acting amines can displace noradrenaline from storage granules (von Euler and Lishajko, 1968a), but any such release should simply be into the cytoplasm of the axon terminals. Thus, it seems that in order to explain the release of noradrenaline by tyramine and similar agents, one must postulate either that the noradrenaline in nerve terminals but not in granules must be in some way bound so that these amines can act to displace it, or more likely, that they in some way act to increase the axonal membrane permeability to the free transmitter, allowing it to leak out from the terminals.

E. Cocaine. The paradoxical blockade by cocaine of the sympathomimetic actions of indirect acting amines was one of the important clues leading to current understanding of how these agents act and of the mechanism by which noradrenergic axons take up noradrenaline. However, the question has persisted, is cocaine potentiation of noradrenaline induced responses altogether secondary to inhibition of neuronal uptake? A variety of evidence has been adduced both for and against this hypothesis. Recently, this problem appears to have been settled. Trendelenburg *et al.* (1970) have found that with methoxamine, an amine not taken up by the amine pump in the cat's nictitating membrane, there is no potentiation of its sympathomimetic action by cocaine. In rabbit aortic strips, on the other hand, there is potentiation by cocaine of the response to methoxamine (Kalsner and Nickerson, 1969a), and there is cocaine-induced hypersensitivity to a variety of amines that cannot be attributed to blockade of axonal uptake (Maxwell *et al.*, 1966). Similarly, in the cat spleen the action of isoprenaline (α-mediated) is potentiated by cocaine, even though there is no cocaine-sensitive isoprenaline uptake system (Davidson and Innes, 1970). Thus, there are organ and/or species differences. At some adrenergic synapses, the sole potentiating action of cocaine is related to blockade of uptake, at others there is an added postsynaptic action.

In keeping with the presynaptic mechanism of cocaine potentiation, it seems to be a general phenomenon that agents that inhibit uptake also increase sensitivity to noradrenaline, provided their postsynaptic blocking activity is not so great as to mask this effect. Agents having both these effects include imipramine and desmethylimipramine (DMI), methylphenidate and quanethidine, amines such as tyramine and antihistamines such as tripelennamine (see Trendelenburg, 1966a). These results clearly imply that presynaptic uptake (uptake$_1$) represents a major mechanism for removal of noradrenaline from the extracellular space, thereby shortening the duration of transmitter action. However, other evidence (see above) clearly implicates uptake$_2$ as a mechanism that rapidly takes up released noradrenaline and implies significant binding by receptors themselves. If, indeed, uptake$_1$ represented the primary mechanism for removal of transmitter, one would expect to observe an increase and prolongation of adrenergic synaptic potentials recorded, for example, from vas deferens. In fact, although cocaine causes a substantial increase in the sensitivity of the vas deferens to added noradrenaline, it has little effect on the response to sympathetic nerve stimulation (Bentley, 1966), and only slightly, if at all, prolongs the time course of spontaneous and evoked junction potentials (Holman, 1967). Thus, the major system for disposal of transmitter probably depends upon the relative importance of subsynaptic response to noradrenaline and response by receptors some distance removed from the site of release, and therefore varies from one tissue to another. Where the major physiological action of the transmitter is exerted on receptors close to the nerve terminals, diffusion or postsynaptic binding probably is most important in terminating its action. Exogenous noradrenaline, of course, will tend to be mopped up by the uptake systems, limiting its penetration to receptor sites unless uptake is blocked, and cocaine under these circumstances will have a major potentiating action. The same should apply where the major action of endogenous noradrenaline is exerted on receptors relatively far removed from the sites of release. Evidently the relative roles of uptake$_1$ and uptake$_2$ will also differ from tissue to tissue; trabecular muscle in the spleen in contrast to arterial smooth muscle appears to be devoid of uptake$_2$ (Gillespie *et al.*, 1970).

F. BRETYLIUM AND GUANETHIDINE—ADRENERGIC NEURON-BLOCKING DRUGS. There are a number of drugs, currently important in the treatment of hypertensive diseases, that act to prevent the normal release of noradrenaline from adrenergic nerves in response to sympathetic stimulation (see Boura and Green, 1965). These drugs include bretylium, bethanidine, debrisoquine, xylocholine (TM 10), and guanethidine, all compounds that share a basically similar structure—a highly basic unit (e.g., quaternary amine) linked by a one- or two-carbon chain to a ring. With bretylium, bethanidine, and debrisoquine, blockade is not associated with any appreciable depletion of catecholamines (e.g., Brodie and Kuntzman, 1960), and with guanethidine, depletion occurs much later than the block of noradrenaline release (Cass and Spriggs, 1961). However, these drugs do cause the release of noradrenaline, like tyramine, resulting in a sympathomimetic response (Boura and Green, 1959; Abbs, 1966) and are accumulated in sympathetic terminal arborizations (Boura *et al.*, 1960) Bisson and Muscholl, 1962; Schanker and Morrison, 1965), evidently largely by the axonal amine uptake system, since accumulation is inhibited by such agents as desipramine, metaraminol, and tyramine (Mitchell and Oates, 1970). The characteristic antagonism of these agents by amphetamines (R. Wilson and Long, 1960; Day and Rand, 1962; Day, 1962) is presumably related to blockade of their uptake. Chronic blockade by guanethidine, which is associated with depletion, is but poorly antagonized by amphetamine (Follenfant and Robson, 1970).

These results have been interpreted in two different ways. Abbs (1966) and Carlsson (1966) suggested that the released catecholamine might be coming from a small part of the presynaptic store that is essential for the proper functioning of the adrenergic nerves; i.e., that these agents selectively deplete the immediately available transmitter store. Boura and Green (1965), on the other hand, have argued that these drugs work essentially as local anesthetics that are selectively accumulated in the sympathetic nerves by the membrane uptake system. Abbs and Robertson (1970) have recently demonstrated that bretylium does cause depletion of a supernatant fraction of a homogenate including sympathetic nerve terminals, and this depletion correlates with blockade of adrenergic nerve function. However, the possibility remains that this correlation is fortuitous, merely reflecting the influx of bretylium into the nerve terminal. The critical evidence appears to be that of Blinks (1966), who found that the positive inotropic effects produced by field stimulation within the atrial wall were not prevented by even very high concentrations of bretylium or xylocholine. Thus, bretylium or xylocholine does not appear to prevent the release of noradrenaline evoked by local stimulation of the nerve endings, a result which definitively contradicts the selective depletion hypothesis. It should be pointed out that accumulation of a local anesthetic by the nerves would be expected to produce a nerve block in the region of preterminal branching, rather than in the terminals themselves (see G. Campbell, 1970).

Whether this explanation also applies to the action of guanethidine is more doubtful. Rand and Wilson (1967) have shown that in a group of guanidine compounds there was no direct relationship between local anesthetic activity of individual drugs and their potency as adrenergic neuron-blocking agents. J. Wilson (1970) has examined the relative activity of guanethidine and three related compounds with about equal local anesthetic potency. One of these showed no amphetamine-sensitive uptake and virtually no adrenergic neuron-blocking activity, and another acted similarly to guanethidine in both respects, which fits the local anesthetic theory. However, one of the compounds (3-cychohexyl-amino-n-propylguanidine) displayed amphetamine-sensitive uptake to an even greater extent than did guanethidine without corresponding adrenergic neuron-blocking activity. Thus, it seems likely that for guanethidine, which is indeed a much more potent depleter of noradrenaline stores than bretylium and xylocholine, the selective depletion theory is more likely to be true, rather than the local anesthetic theory.

Further support for the local anesthetic theory for the action of bretylium is the finding that it does not accumulate in storage granules (Costa *et al.*, 1964) and is not released by the action of ACh from the perfused cat spleen, although noradrenaline and false transmitters that are stored in synaptic vesicles are released (Fischer *et al.*, 1966). The depleting action of bretylium is presumably via the same mysterious mechanism by which indirect acting amines exert their effect. It is much more marked on amines such as metaraminol, octopamine, and amphetamine, which tend to accumulate extragranularly, than on noradrenline (Krauss *et al.*, 1970).

Evidently quite unrelated to the adrenergic neuron-blocking activity of bretylium and bethanidine is their ability partially to restore responses to tyramine in reserpinized animals (Malmfors and Abrams, 1970). This effect appears to be related to monoamine oxidase inhibition by these compounds (Clarke, 1970). It is of interest that in pithed reserpinized animals the ability of bretylium to restore the response to tyramine was reduced, suggesting that the nerve impulse flow down sympathetic nerves may modify either the rate of uptake of bretylium or the rate of catecholamine synthesis in the nerve endings.

Burnstock and Holman (1964) have reported the effects of bretylium on evoked and spontaneous junction potentials in the vas deferens. Bretylium was found to cause a rapid depression of the spontaneous potentials, even before nerve transmission was blocked. After 30 or 40 min exposure, when the nerve response was abolished, there was a conspicuous but transient increase in spontaneous discharge. Thus, blockade of the response to nerve stimulation could not have been due to selective depletion of available transmitter store, which must be the source for the spontaneous junction potentials. Why frequency of spontaneous discharge should rise, however, seems obscure. One guess is that it may represent an alcohol-like effect (see Table I) produced by a local anesthetic agent that has been concentrated in the nerve terminals. The initial depressing effect of bretylium on the spontaneous potentials may well have been related to a postsynaptic local anesthetic action, similar to that of procaine (Burnstock and Holman, 1964). It should be emphasized that in the vas deferens it is difficult to distinguish alteration of spontaneous junction potential frequency from alteration of amplitude.

G. Synthesis—False Transmitters. The presently accepted series of reactions by which noradrenaline is normally synthesized *in vivo* was first suggested by Blaschko (1939). L-Tyrosine is hydroxylated by tyrosine hydroxylase to form L-dopa, which is in turn decarboxylated by aromatic L-amino acid decarboxylase, to form L-dopamine, which is acted upon by dopamine-β-hydroxylase to form noradrenaline (see Iversen, 1967). The rate-limiting step is evidently the conversion of tyrosine to dopa (Levitt *et al.*, 1965). Nevertheless, inhibition of decarboxylase or of β-hydroxylase has been shown to delay dopamine or noradrenaline synthesis if it is increased beyond the physiological rate by the administration of an excess of substrate (Dengler and Reichel, 1957) or by infusing dopamine into animals whose noradrenaline stores were previously depleted by metaraminol (Nikodijevic *et al.*, 1963).

Nagatsu *et al.* (1964) isolated tyrosine hydroxylase and found several *in vitro* competitive inhibitors of this enzyme. These include several analogs of tyrosine. (e.g., α-methyl-p-tyrosine) and several catechols including dopacetamides (Carlsson *et al.*, 1963a,b). *In vivo*, the administration of α-methyl-p-tyrosine was found to cause severe depletion of noradrenaline stores in heart, spleen, and brain stem, with no impairment of tissue-binding capacity for noradrenaline (Spector *et al.*, 1965). There was corresponding miosis and relaxation of the cat nictitating membrane, but no severe hypotensive effects, suggesting either that the very low level of noradrenaline synthesis that continued was sufficient to maintain blood pressure or that α-methylnoradrenaline acting as a false transmitter might have been formed from α-methyl-p-tyrosine (Maître, 1965; Spector, 1966). Van Orden *et al.* have recently found that relatively large amounts of α-methyl noradrenaline can indeed be formed from α-methyl-p-tyrosine, especially in the vas deferens, which therefore appears, histochemically, to be resistant to depletion by blockade of transmitter synthesis.

Histochemical observations using the fluorescence method have shown that after inhibition of synthesis using dopacetamides (H22/54 or H33/07) the rate of depletion of noradrenaline in adrenergic terminals in the iris is increased by sympathetic nerve stimulation (Malmfors, 1965a,b). The decrease of specific fluorescence was greatest in the terminal varicosities. Thus, the effects of transmitter blockade in adrenergic systems seems to be similar to that studied in cholinergic systems. Blockade of synthesis has become a useful method for studying *in vivo* rates of catecholamine turnover (Loizou, 1971).

Since the mechanisms for uptake and storage of noradrenaline are not entirely selective, it is hardly surprising that a number of compounds have been found to replace noradrenaline in nerve terminals and act as false transmitters. Metaraminol, for example, replaces noradrenaline in storage sites (Andén, 1964) and is released by sympathetic nerve stimulation (Crout *et al.*, 1964). α-Methyl-*m*-tyrosine is taken up by nerve terminals and then decarboxylated and β-hydroxylated to form metaraminol; the depletion of tissue noradrenaline is accompanied by an accumulation of metaraminol (Gessa *et al.*, 1962; Andén, 1964). Metaraminol apparently functions as a fairly effective transmitter. Several authors have reported that adrenergic transmission is unaffected by α-methyl-*m*-tyrosine (Andén, 1964; Carlsson, 1964), but Haefely *et al.* (1965b) found significant reduction of the effects of sympathetic nerve stimulation in the cat nictitating membrane and spleen. Other substances that can be taken up by nerve endings and released in response to nerve stimulation include adrenaline (Rosell *et al.*, 1964), and β-hydroxylated amines such as octopamine, α-methyloctopamine, phenylethanolamine, and norephedrine, but not non-β-hydroxylated amines such as tyramine, α-methyltyramine, phenylethylamine, and amphetamine (Fischer *et al.*, 1965). Kopin *et al.* (1965) have shown that after prolonged administration of monoamine oxidase inhibitors octopamine does in fact accumulate in sympathetically innervated tissues, where it presumably acts as an ineffective false transmitter. This could account for the sympathetic blockade and hypotension associated with prolonged treatment with monoamine oxidase inhibitors.

The synthetic substance 6-hydroxydopamine is also taken up by nerve terminals, where it displaces noradrenaline from the granular stores, resulting in an increase in the granulation of the intraaxonal vesicles (Furness *et al.*, 1970). At the same time, it causes marked damage to the axons; within a few hours there is loss of neuromuscular transmission, and eventually there is complete degeneration and disappearance (2 weeks) of the distal part of adrenergic fibers (Tranzer and Thoenen, 1968; Malmfors and Sachs, 1968). Thus, there is effected a chemical sympathectomy. The action is blocked by desmethylimipramine—indicating that the integrity of the axonal amine pump is necessary for the intracellular accumulation of the amine—and it is accelerated by reserpine—suggesting that uptake into the granular store delays the rise in axoplasmic concentration to a critical concentration. Even earlier than the first histological signs of axon damage, there is an increase in the frequency of spontaneous junction potentials in the vas deferens, presumably related to presynaptic membrane depolarization secondary to membrane damage (Furness *et al.*, 1970). The mechanism by which 6-hydroxydopamine damages the nerve terminals is quite obscure.

In 1954, Sourkes found that α-methyldopa is a competitive inhibitor of the decarboxylation of dopa to dopamine *in vitro*, and subsequently this agent was found to be a potent antihypertensive agent (Oates *et al.*, 1960; see Muscholl, 1966b). It later became evident that after administration of α-methyldopa, there is formation of α-methylnoradrenaline, which functions as an effective, although false, transmitter (Carlsson and Lindqvist, 1962; Day and Rand, 1963, 1964; Muscholl and Maître, 1963). Thus, the mechanism of the hypotensive activity became difficult to understand. One explanation has been that the effect is mediated centrally, α-methylnoradrenaline or α-methyldopamine there acting as an ineffective false transmitter, or the drug may act directly (Henning and von Zwieten, 1968; Ingenito *et al.*, 1970). Ayitey-Smith and Varma (1970) have recently shown that the drug acts to reduce blood pressure in immunosympathectomized animals in which quanethidine was ineffective, as well as in control hypertensive and normal rats. The hypotensive effect of methyldopa was not antagonized by either α- or β-adrenoceptor or ganglion-blocking agents or by the inhibition of dopamine β-oxidase. This result leaves open the possibility of action in the central nervous system, mediated via the adrenal medulla.

III. Postsynaptic Drug Actions

A. General Considerations

At synapses generally, the transmitter substance released by the nerve terminal acts to cause a change in the physiological activity of the postsynaptic cell via combination or interaction of the transmitter with a postsynaptic receptor. The concept of a chemoreceptor was introduced by Paul Ehrlich (1913) to account for the specificity of action of drugs on different cell types. It remains an operational concept. The receptor for a drug exists, by definition, to account for the action of the drug. As yet, no one has succeeded in isolating a receptor substance or characterizing it chemically, and it remains possible that a receptor consists in a combination of substances in or on the membrane arranged to form a certain configuration, rather than a distinct molecular species.

In addition to the postsynaptic receptor, synaptic transmission demands that there also exist machinery by which the transmitter–receptor combination acts to alter the physiological activity of the cell. As will be discussed below, the mechanism is usually but not always an alteration of the permeability of the postsynaptic membrane to certain species of ions. This permeability change usually, though again not always, leads to flow of ions down their electrochemical gradients, which in turn entails an alteration of the membrane potential. In muscle where contractile activity is graded with membrane potential, the alteration of membrane potential will be reflected directly in alteration of muscle tension. More generally, the alteration of potential triggers or inhibits action potential generation, contractile activity being modified relatively indirectly. Alternatively, the ionic flux, of calcium in particular, may itself lead to muscle contraction.

Besides this basic machinery, it seems to be a general rule that there exists at synapses a mechanism by which transmitter substance is removed from its site of action, either by enzymic hydrolysis (e.g., acetylcholinesterase at cholinergic synapses) or by active uptake of the substance by the nerve terminals which previously released it (e.g., the amine pump at noradrenergic synapse). It remains possible that the true function of the uptake mechanism is not to terminate transmitter action so much as to conserve the transmitter substance.

It follows from the general system by which transmitter substances act that there are numerous possibilities for interaction by drugs. Thus, a drug may act like the natural transmitter, interacting with the same receptor sites, having the same postsynaptic action, and even being hydrolyzed or taken up like the transmitter itself. Such action is of course demonstrated by the natural transmitter applied artificially. However, it is not in general true that the actual transmitter substance will mimic exactly the action of the same substance when it was released by nerve terminals. This difference, of course, stems from the high local concentration and extremely rapid application to the specialized subsynaptic areas, which is characteristic of endogenous as opposed to exogenous transmitter. Instead, exogenous transmitter substance applied indiscriminately to a whole tissue and for a long period of time may only partially mimic the action of natural transmitter, and by occupying the receptor sites normally available for the endogenous substance, occlude the normal neuromuscular transmission process. Thus, study of the action of natural transmitters or analogs which closely mimic their actions of necessity entails considerable care in how the substance is applied. The ionophoretic method of drug application is the only one capable of imitating at all closely that which normally obtains at a synapse, but even with this technique it should be understood that gross quantitative differences may appear when interactions with other drugs are to be studied.

The second class of selectively acting drugs includes those which apparently combine with the receptor for the natural transmitter, or antagonist, but when they do so, the combination does not lead to the membrane change associated with the combination of receptor with agonist. If the combination of drug and receptor is reversible, leaving receptor again in a state to combine with agonist with the combination acting in the usual way, then the drug is a competitive antagonist and its effects on dose–response curves will obey the Michaelis–Menton kinetics which describe competitive inhibition of enzyme activity (Gaddum, 1937; Ariëns, 1964; Waud, 1968; Werman, 1969). On the other hand, if drug–receptor combination entails changes in the receptor that are irreversible, the kinetics become those of noncompetitive inhibition, even if the drug combines at the same sites on the receptor as the agonist. The frequent appearance of partial activity of the competitor–receptor combination, in which case the competitor is also a partial agonist, also leads to considerable complications in applying kinetic theory.

In theory, drugs might also act to facilitate or inhibit the combination of agonist and receptor by allosteric action on the receptor. Although no unequivocal evidence of such interaction has yet been described for synaptic receptors, it is not unlikely that many "competitive" inhibitors actually act by such a mechanism (Ariëns, 1964).

Drugs can and do interfere with postsynaptic or muscle activity at other levels than that of the transmitter–receptor combination. Examples are numerous of rather nonselective inhibitory activity. That of general anesthetics, for example, which in all likelihood is related not to any action on the receptor itself but rather inhibition of the membrane alterations normally subsequent to agonist–receptor combination. At present, there is no method by which agonist–receptor combination can be assessed except in terms of the physiological response of the tissue, i.e., an end effect several stages removed from the interaction of receptor and agonist. Often, with these techniques, saturation of tissue response tends to occur before saturation of receptor and by uncritical application of kinetic analysis almost any blocking drug may appear to be acting by competition with the agonist (Werman, 1969). Where the agonist works by producing permeability or ionic conductance change, this problem may be avoided if the actual conductance change is measured. However, there remains no way to distinguish between noncompetitive actions mediated at the receptor level from actions on the mechanisms by which conductance changes are brought about. Further down the sequence are other points where drugs may act. Thus, the postsynaptic action potential generating mechanism may be modified either to increase or reduce excitability of the cell. This can occur either with drugs that interfere directly with the sodium and potassium current systems or by actions on the passive membrane properties such as to alter threshold for impulse generation. The system that couples contractile activity of muscle to membrane events is also susceptible to drug action.

In skeletal muscle, it is now relatively easy to analyze drug actions in terms of site and mechanism. The problems are, with few exceptions, purely technical. To go into the variety of methods that can be employed would be out of place here. Many can be found in the recent monograph of Hubbard *et al.* (1969a). Basically, they rest upon the relative ease of measuring the electrical characteristics of the muscle membrane and the availability of drugs that act selectively at various stages in the normal excitation process. Tetrodotoxin, for example, blocks action potential generation in the muscle, as well as in nerve, without affecting the response of the endplate to ACh. Tubocurarine acts to depress the endplate response to ACh without appreciable effect on electrical properties. Anticholinesterases can be fairly specific in prolonging the time course of action of released ACh. The arrangement of one synapse per muscle fiber permits unequivocal differentiation of drug effects on amplitude and frequency of miniature synaptic potentials. The ease of investigation has perhaps sometimes led to overinterpretation of results. For example, the phenomenon of ACh desensitization, which is found at the skeletal neuromuscular junction, has been assumed to be analogous to the selective tachyphylaxis which can be found with various drugs in smooth muscle and therefore has been attributed to alteration in the receptor for ACh (B. Katz and Thesleff, 1957b). In the absence of any other agonist that acts at the endplate, evidence for selectivity and therefore involvement of the receptor is quite lacking (see Waud, 1968).

In smooth muscle, on the other hand, the anatomical organization is much more complicated than in skeletal muscle (see Burnstock, 1970). Indeed, the use of the terms "synapse" and "synaptic" can often be justified only by analogy. "Presynaptic nerve terminals," or "varicosities" often do not lie in close proximity to the surface of muscle fibers, and under these circumstances the transmitter released must diffuse considerable distances and act on "postsynaptic" receptors distributed over a wide area. Moreover, smooth muscle cells are generally linked by electrotonic junctions, so that electrical activity cannot be restricted to any one cell, but spreads throughout the "syncitium." Voltage clamping therefore becomes nearly impossible. Thus, problems which are relatively simple in skeletal muscle, such as determination of any membrane permeability changes associated with the action of transmitter, becomes an exercise of great difficulty in smooth muscle. Another complication is that in some forms of smooth muscle, gut in particular, drugs may act postsynaptically on neural elements as well as on muscle fibers directly. It has been one of the major problems of pharmacology in recent years merely to separate direct action from such indirect action, often by use of tetrodotoxin which selectively inhibits the action potentials of the nerves. The lack of action of tetrodotoxin on smooth muscle action potentials, however, has made it extremely difficult to sort out the extent to which excitatory drug effects may be mediated via enhanced spike activity rather than more direct actions.

Smooth muscle, of course, is notorious for its susceptibility to a variety of compounds, including some that occur normally in the body and whose physiological function may well include modification of smooth muscle activity—i.e., histamine, 5-hydroxytryptamine (5-HT), prostaglandins, angiotensin, kinins, adenine nucleotides, vasopressin, and oxytocin, as well as ACh and catecholamines. Indeed, there is now compelling evidence that adenosine triphosphate or a related nucleotide may be a transmitter substance for some postganglionic "parasympathetic" neurons in the gut, and evidence also exists for roles of histamine and 5-HT as transmitters in smooth muscle. All these compounds appear to act via receptors quite distinct from those involved in cholinergic or adrenergic transmission, and for a number of them specific antagonists exist. Hormonal modification of smooth muscle activity, for example that of the uterus by steroid hormones, represents another type of pharmacological action which is unfortunately beyond the scope of this chapter.

It should not be forgotten that the observed action of a drug on a muscle can depend very greatly on quite minor variations in structural organization. For example, the action of ACh at nicotinic skeletal muscle junctions is probably everywhere the same. However, a normal rat diaphragm *in vitro* cannot be made to contract appreciably by addition of ACh to the bath. In a denervated diaphragm, the receptor area responsive to ACh extends to the whole membrane surface. ACh now causes a much greater depolarization, as total conductance change is much larger, and the muscle will contract vigorously (Axelsson and Thesleff, 1959). The slow skeletal muscle fibers of the frog, particularly abundant in the rectus abdominis muscle, possess numerous junctions along their whole course (Kuffler and Vaughan Williams, 1953a,b) and an abundance of well distributed ACh receptors. Thus, like denervated "twitch muscle," they respond to ACh with a contracture. The intrafusal fibers of mammalian spindles also are multi-innervated; i.e., one intrafusal fiber possesses several motor endplates (Barker, 1948), and the fibers are capable of slow graded contractions (Kuffler *et al.,* 1951) in response to either nerve stimulation or applied ACh or an analog such as succinylcholine. The same is true of slow mammalian extraocular muscle (Hess and Pilar, 1963; Eakins and Katz, 1966). In smooth muscle, contractile response to excitatory agonists is a general phenomenon, suggesting that receptors are well distributed. It seems generally true, however, that structures densely innervated tend to be less sensitive to the action of transmitter applied exogenously than structures whose innervation is sparse. For example, no pharmacologist would choose vas deferens for assay of noradrenaline.

This section will deal with the effects on smooth and skeletal muscle of agents which are either known or putative transmitters or whose physiological function may be to modify muscle activity, and agents whose main effects are exerted by interaction with receptors for these substances. Our emphasis will be on the mechanism by which drugs act, rather than on the multiplicity of responses that can be obtained from various types of muscle. The general question of pharmacomechanical coupling (i.e., direct effects of drugs to influence mechanical activity without interposition of membrane potential change or action potential activity) will be dealt with in Section IV, as will effects of miscellaneous agents that also have more or less direct actions on muscle contraction.

B. Acetylcholine, Cholinomimetics, and Anticholinesterases

1. SKELETAL MUSCLE

The most studied and presumably best understood drug effects on muscle are those of the natural transmitter substance ACh on the skeletal muscle endplate and the actions of agents that mimic or modify the effect of ACh on skeletal muscle. Indeed, the experiments of Claude Bernard (1856) on the mechanism of curare action provided the first convincing evidence of a specific mechanism for neuromuscular transmission as distinct from impulse propagation in the nerve or muscle.

It is now established that the action of transmitter on the muscle membrane is to increase membrane conductance to sodium, potassium, and calcium ions (A. Takeuchi and Takeuchi, 1959, 1960; N. Takeuchi, 1963). Previously, Fatt and Katz (1951) had shown that the EPP was generated by an inward current flow localized to the endplate region of the muscle fiber, that this current flow was associated with an increased membrane conductance, and that the ion permeability change could not be solely to sodium, but also had to include potassium, and perhaps chloride. With the voltage clamp method, Takeuchi and Takeuchi found that there was no chloride conductance, and the change in conductance to sodium and potassium was nearly the same ($\Delta gNa/\Delta gK = 1.29$). That calcium permeability also increases was dramatically illustrated by B. Katz and Miledi (1967d, 1969b); MEPPs of close to normal amplitude could be recorded in the presence of solution in which all the sodium was substituted for by calcium. The participation of calcium is especially interesting since it indicates the possibility of direct pharmacomechanical coupling (A. V. Somlyo and Somlyo, 1968), even in skeletal twitch muscle, where it is absolutely clear that the normal mechanism by which muscle contraction is triggered by ACh is via the relatively indirect sequence: EPP→action potential→calcium mobilization. Indeed, it is not difficult to observe pharmacomechanical coupling in normal striated muscle. In the presence of ethanol (~ 1 M), MEPPs in a rat diaphragm are large, prolonged, and at about twentyfold the normal frequency. Under these circumstances, muscle contractures localized to endplate areas can clearly be observed with the microscope.

Recently, the important question has been raised as to whether the conductance changes for sodium and potassium ions are indeed simultaneous, as is demanded if the ACh–receptor combination acts to induce formation of a single membrane channel. If they were not simultaneous, it would raise the possibility of specific drug actions on either of the conductance channels and therefore, if the same applied to other synapses, the possibility of converting excitation to inhibition and vice versa. Such separation was suggested by Maeno (1966), and it was found by Gage and Armstrong (1968), who studied miniature end-plate currents (MEPCs) with a point voltage clamp, that the time course of the MEPC was different depending on whether postsynaptic membrane potential was set to exclude the current of one or the other ion. Kordaš (1969) found that end-plate currents (EPCs) at or near the equilibrium potential for the transmitter—i.e., at a membrane potential where outward potassium current balances inward sodium current through the channel(s) set up by ACh—were not diphasic, as would be required if, in fact, sodium and potassium channels had a different time course. He explained the results of Maeno (1966) and Gage and Armstrong (1968) on the basis of a variation with membrane potential of the time course of a common channel. It may be objected that dispersion in time of the quantal components of the EPC (B. Katz and Miledi, 1965b) would tend to prevent the appearance of the diphasic EPC. Maeno *et al.* (1971) have recently found such diphasic EPCs in the presence of procaine which seems to act selectively to prolong the sodium conductance change.

Evidence has also been obtained recently indicating that the ACh receptors on skeletal muscle may differ from one another according to their location. In frog sartorius muscle, the transmitter equilibrium potential, which is a measure of relative sodium and potassium conductance, is much more negative (-41 mV) for the extrajunctional receptors, which are not normally exposed to ACh released by the nerve terminal, than for the subsynaptic receptors (−15 mV) (Feltz and Mallart, 1970). The ACh receptors which are found all over the surface of the denervated muscle fiber are apparently similar to the normal extrajunctional receptors (Mallart and Feltz, 1968). Certainly, the enormously expanded receptor area of denervated muscle should also lead to exaggeration of any small depolarizing activity, normally too small to be observed, and such an effect might explain the anomolous depolarization and mechanical contractions produced by tubocurarine in denervated mammalian muscle (McIntyre *et al.,* 1945), but it is difficult to exclude the possibility of genuine difference in the receptor. The contracture of denervated muscle in response to catecholamines is not blocked by *d*-tubocurarine, which does block the response to ACh, and therefore cannot be ascribed to interaction with ACh receptors (Bowman and Raper, 1965). Instead, it appears that there is development of α-adrenoceptor in the chronically denervated muscle (Paterson, 1963), in addition to the usual β-adrenoceptors (see Bowman and Nott, 1969).

The importance cannot be exaggerated of the fact that the postsynaptic membrane change resulting from the ACh-cholinoceptor combination is a conductance change. From the simplified model of Martin (1955), for a time long compared to the time constant of the membrane,

$$v = \frac{g}{G + g} (V_{tr} - V_m) \tag{1}$$

where v is the potential change, g is the conductance change caused by transmitter action, G is the input conductance of the cell, V_m is the resting membrane potential of the cell, and V_{tr} the transmembrane potential at which the conductance change would not alter membrane potential. Rearranging,

$$\frac{g}{G} = \frac{v}{V_{tr} - V_m - v} \tag{2}$$

The expression on the right hand side of Eq. (2) also gives a value that is linearly proportional to g/G in the case of a conductance change that is brief compared to the membrane time constant, but the proportionality factor varies with the resting membrane time constant (unpublished calculations, Leung and Quastel). It follows from these equations that the depolarizing action of any ACh-like agonist will be altered not only by alteration of the conductance change caused by ACh (i.e., either an effect on the receptor or subsequent processes), but also by any change of the resting potential or of the membrane resistance. The resistance of the resting membrane is a function mainly of its permeability to potassium and chloride ions. Substitution of chloride by less permeable anions such as nitrate, sulfate, or iodide, therefore, increases the amplitude of the EPP and MEPP (Oomura and Tomita, 1961; Hofmann *et al.*, 1962).

It has recently been proposed by B. Katz and Miledi (1970b) that the response of the muscle membrane to ACh is made up by accumulation of minute unit responses, each related presumably, to the opening of one (sodium and potassium) membrane channel, perhaps in response to one ACh–receptor combination. As predicted by the appropriate model, depolarization by ACh was accompanied by an increase in the noise recorded by an intracellular electrode. The steady level of depolarization (ΔV) resulting from the micro endplate potentials of amplitude a, occurring at a rate, n per second, is given approximately by the formula, $\Delta V = na\tau$, where τ is the membrane time constant. For an MEPP or EPP where all the ACh induced conductance is virtually simultaneous, the amplitude will be close to na, where n is the number of unit responses. Hence, any alteration of membrane resistance and therefore time constant will change to a much greater extent the steady depolarization induced by added ACh, or an active analog, than it will the amplitude of an MEPP. In keeping with this, we have found that butanol, which prolongs the time course of MEPPs, increases the depolarization resulting from application of carbachol at a concentration that actually reduces MEPP amplitude to less than half the normal size.

In general, it can be stated that only the voltage clamp method can give any exact measure of drug effects causing or modifying a conductance change. Failing this, the action of drugs on the amplitude and time course of MEPPs can be of value, providing passive membrane properties are also measured. However, measurements of depolarization, or higher order effects one or more stages further removed from the conductance change, such as muscle contraction, whether graded or twitch in character, can be interpreted only with the greatest difficulty. Werman (1969) has pointed out that measurement of a response that does not begin until a threshold depolarization is reached cannot give the limiting slope of the relation between the logarithm of response and the logarithm of drug concentration, which is a measure of the number of molecules involved per receptor. This probably accounts for the general assumption that synaptic events involve only one molecule of transmitter per receptor. His replot of the data of Jenkinson (1960), of ACh-induced depolarization of the frog endplate clearly shows this limiting slope to be greater than one, and close to two, indicating either that two molecules of ACh are involved in activating each ACh receptor or that the unit conductance change is normally produced by more than one ACh–receptor combination.

It is of interest that the postsynaptic action of ACh at the postjunctional membrane seems to be directly sensitive to the concentration of divalent and polyvalent ions, though not dependent on their presence. Thus, in the frog sartorius the reduction by carbachol in effective membrane resistance (i.e. input resistance) of muscle fibers completely depolarized by high potassium, was a function of $[Ca^{2+}]$, $[Mg^{2+}]$, and $[La^{3+}]$ (Lambert and Parsons, 1970). The optimum $[Ca^{2+}]$ was about 2 mM and the optimum $[Mg^{2+}]$ about 8 mM; in the absence of either, the resistance charge was of the order of half the maximum. Lanthanum in increasing concentrations progressively increased the resistance change produced by carbachol.

A. Desensitization. There are a number of quaternary ammonium compounds that appear to act in the same way as ACh to depolarize skeletal muscle fibers by interaction with ACh receptors. Typical examples are choline, carbachol, succinylcholine, decamethonium, and tetramethylammonium. Nicotine also acts in the same manner. The compounds also probably act on the nerve terminal to depolarize it, and this may account, at least in part, for their neuromuscular blocking activity. This blocking action is also shared by exogenous ACh; indeed, the only difference in effect between ACh and these compounds seems to be related to high susceptibility to acetylcholinesterase of the natural transmitter.

The endplate depolarization caused by such an agent may initially result in the generation of action potentials, isolated or in trains, depending upon the extent of the depolarization, the rapidity with which it is brought about, and the electrical properties of the fiber membrane. As the depolarization persists, however, the muscle fiber action potential generating system is depressed, presumably because of inactivation of the sodium current system responsible for generation of the action potential. Thus, EPPs become unable to elicit propagated impulses in the muscle fiber and there is a block of transmission.

With prolonged exposure of the endplate to the depolarizing agent, the muscle membrane slowly becomes repolarized, effectively to its normal resting potential, and at the same time the endplate becomes refractory to added ACh or any other of the depolarizing agents (Thesleff, 1955; Nastuk, 1967). This phenomenon is generally known as desensitization. With prolonged bath application, concentrations of a drug that are too low to cause any appreciable initial depolarization can lead to quite profound depression of the response to a higher concentration of depolarizing agent. For example, in both normal and myasthenic human intercostal muscle, the depolarization caused by carbachol (10^{-5} gm/ml) was reduced from 17 mV to 5 mV by prior administration of only 10^{-6} gm/ml decamethonium (C_{10}) or 10^{-7} gm/ml carbachol. There was no difference between the responses of normal and myasthenic muscle, either in the direct effect of carbachol or in desensitization (Elmqvist *et al.*, 1964). After a prolonged bath application of a depolarizing agent, recovery of receptor sensitivity is a very slow process, taking of the order of 15 minutes or more (Thesleff, 1955). B. Katz and Thesleff (1957b) studied the time course of desensitization using ionophoresis of drugs to the endplate membrane together with intracellular recording of the membrane potential. Their results showed that the rate of development and degree of desensitization increased with the concentration of the depolarizer; 50% desensitization of the receptors could occur within a few seconds of drug application. The recovery half time upon withdrawal of the drug was 3–5 sec and was independent of the drug concentration. The following scheme was presented:

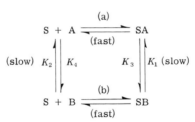

Here S is the steady agonist concentration, SA is the effective agonist–receptor compound, and SB is the refractory compound. Thus, by this hypothesis the receptor can exist in two states: active (A) and inactive (B), with the agonist normally combining most with the inactive form. This rather complicated model was forced by the observation that the time course of recovery from acute desensitization could be faster than the time course of development of desensitization. Although this scheme explains well the relationship between the rate of desensitization and the recovery of initial sensitivity, as well as the dependence of the rate of desensitization on the amount of ACh applied, it fails (Magazanik and Vyskočil, 1970) to explain several facts, namely, (1) that many cholinomimetics which differ in their affinity for the receptor, (i.e., "a") when applied in amounts causing the same initial depolarization, desensitize to almost the same extent, (2) that there is no effect of curare on the rate of desensitization, and (3) the very long time course of recovery after prolonged bath application (Thesleff, 1955). To account for the last finding it would be necessary to postulate yet another inactive state of the receptor with a very slow time course of formation and recovery to the normal state. The other findings are more difficult to reconcile with the Katz and Thesleff scheme. Moreover, the rate of desensitization is a function of the cation composition of the bathing solution to a greater extent than is the depolarizing response to ACh (e.g. Manthey, 1970). Magazanik and Vyskočil have found that multivalent cations which increase the rate of development of desensitization fit in the following sequence:

$$Mg^{2+} \ll Ca^{2+} \leq Ba^{2+} < Sr^{2+} \ll Al^{3+} < La^{3+}$$

This sequence is similar to that found for the effects of these ions on other membrane properties, e.g., stabilization of permeability of lobster axon membrane (Blaustein and Goldman, 1968). Moreover, the rate of desensitization is also a function of the membrane potential; it is slowed by depolarization and accelerated by hyperpolarization. These authors, therefore, have proposed that desensitization is not exerted at the receptor level, but consists in an altered ability of the receptor–agonist combination to increase ion permeability. Of course, these results could also be accommodated by the previous model in terms of action of multivalent ions and membrane potential on the rate constants. Indeed, the models are not exclusive if one defines the receptor operationally as comprising both the molecular configuration with which agonist interacts and the mechanism by which ionic channels in the membrane are opened. At present, there seems to be no method of measuring separately the activity of these two components of the receptor.

B. ANTICHOLINESTERASES. At the endplate, hydrolysis of ACh by the enzyme acetylcholinesterase appears to be an important mechanism in terminating transmitter action. When cholinesterase is inactivated, there is prolongation of the time course of the EPP (Fatt and Katz, 1951). Acetylcholinesterase in skeletal muscle is localized almost exclusively to the postsynaptic membrane and there only in apposition to the nerve terminals (Couteaux, 1958; Davis and Koelle, 1967). It is generally accepted (Koelle, 1970) that there are two sites on the enzyme that are essential to its activity: the anionic site, which binds electrostatically the cationic head of the ACh molecule, and the esteratic site, where the ester bond is activated. In general, anticholinesterase agents can be divided into three classes. Simple quaternary compounds, such as tetraethylammonium ions, and somewhat more complex quaternary structures, such as edrophonium, compete with substrate at the anionic binding site. Physostigmine, neostigmine, and related compounds that possess a carbamyl ester linkage or urethane structure in addition to a quaternary or tertiary ammonium group probably form a complex in which the inhibitor is attached at both the anionic and esteratic sites; its hydroylsis then proceeds, extremely slowly, in a manner similar to that of ACh. The reaction between acetylcholinesterase and organo-phosphorus inhibitors, such as diisopropylfluorophosphate (DFP) at the esteratic site, is such as to form a relatively stable phosphorylated enzyme.

In view of the fact that all classes of anticholinesterases bind to the enzyme in essentially the same way as does ACh, it is perhaps not surprising that these agents can also, to varying extents, react with the ACh receptor.

Riker and Wescoe (1946) found that neostigmine directly depolarizes postsynaptically at the neuromuscular junction, causing excitation inde-pendently from cholinesterase inhibition. Similar action by all cho-linesterase inhibitors, together with the accompanying desensitization, provides an explanation for the observation that all such agents, in suffi-cient concentration, have a curarizing action and block neuromuscular transmission. Often, this is attributed to a build-up of ACh at the end-plate, but this explanation seems very unlikely. Only when the dose of a cholinesterase inhibitor is relatively low and its application has been of short duration can antagonism of its blockade by tubocurarine be demon-strated. More generally the blocking effect of cholinesterase inhibitors tends to add to that of tubocurarine (Axelsson *et al.*, 1957). Moreover, tetraethylammonium ions, which act to augment transmitter release, can to a considerable extent restore transmission after partial blockade by DFP (Stovner, 1958a,b). If the DFP blockade were due to accumulation of released ACh, the opposite result would be expected. The most plausi-ble explanation of neuromuscular blockade by anticholinesterase agents is that these agents all cause desensitization.

It is important not to confuse the two different effects of anticholines-terases on neuromuscular transmission—potentiation of the twitch in noncurarized preparations and anticurare action. Twitch potentiation in all probability usually represents an admixture of several effects. The anticholinesterase action causes a prolongation of the EPP with a result-ing tendency for double or more repeated firing of the muscle action potential in response to a single nerve volley. The direct postsynaptic depolarizing action may add to this tendency. In addition, after a nerve volley, MEPPs are increased in frequency, and when these are enlarged and prolonged, they occasionally summate to trigger muscle action po-tentials. A presynaptic depolarizing action of anticholinesterases can also lead to repetitive firing of a fraction of the nerve terminals, propagated by axon reflexes to many of the other nerve terminals. Such a presynaptic action probably accounts in large part for the potent twitch-potentiating action of agents such as 3-hydroxyphenyltrimethylammonium and 3-hydroxyphenyltriethylammonium, which are only weak anticholinesterases. It may also lead to some decurarizing action; it was found by Hubbard *et al.* (1965) that curarization did not block the action of prostigmine to reduce the threshold for nerve terminal excitation, nor its effect to intensify and prolong the period of supernormal excitability after a presynaptic action potential. Nevertheless, there seems no reason to doubt that the major decurarizing action of anticholinesterases is exerted via the postsynaptic inactivation of cholinesterase, with the resulting increase in amplitude and duration of EPPs. It should not be forgotten that in a muscle curarized to the level of a partial block, most EPPs are only just below or above threshold for muscle action potential generation. A quite small increase in amplitude can therefore account for a large increase in muscle contraction. In addition, an increase in duration, by allowing piling of EPPs during tetanic trains, can permit subthreshold EPPs eventually to trigger action potentials. The same considerations also apply in myasthenia gravis, where there is evidently a deficiency of the ACh content of each transmitter quantum (see above, page 36).

2. Smooth Muscle

A. Acetylcholine as Parasympathetic Transmitter. Since the time of the early experiments of Dale and of Loewi (see review by Dale, 1937), it has generally been considered that ACh is the mediator that is released by postganglionic parasympathetic nerve terminals. As pointed out by Gershon (1970), the case for ACh has been based mainly on four points: (1) the similarity or identity of action between ACh and the effects of parasympathetic nerve stimulation; (2) the parallel antagonism by atropine of neural effects and those of exogenous ACh; (3) the potentiation of both effects by eserine; and (4) the demonstration of release of ACh, or a substance indistinguishable from it by biological assay, upon parasympathetic nerve stimulation, with eserine present.

Taken together, these four points make a strong case. They do not, however, rule out another possibility, i.e., that exogenous ACh might exert its action presynaptically to release unknown transmitter substance(s) whose action is blocked by atropine and potentiated by eserine. The ACh released by nerve stimulation could be from nonneural sources, released in conjunction with the postsynaptic action of the true transmitter(s).

It is only recently that cogent evidence has been provided indicating that exogenous ACh acts directly on smooth muscle. In an apparently nerve-free preparation of guinea pig ileum, which does not release ACh and is inexcitable by electrical stimulation of brief duration, the sensitivity to ACh is not decreased relative to other drugs (Paton and Zar, 1965, 1968). Moreover, eserine produces no contracture in the denervated longitudinal muscle strip. In the innervated preparation, even in the presence of a sufficiently high concentration of tetrodotoxin to block conduction of neuronal action potentials, eserine does produce a contracture provided it is applied in sufficient concentration (10^{-6} gm/ml) (Gershon, 1967a). It is of interest that the contracture produced by a lower concentration of eserine (10^{-7} gm/ml) is inhibited by ganglionic blocking agents such as hexamethonium (Erdmann and Heye, 1958) and is abolished by tetrodotoxin (Gershon, 1967a). The potentiation of parasympathetic nerve action by eserine and the parasympathomimetic effects of eserine (usually ascribed to accumulation of postganglionically released ACh), both of which phenomena are generally quoted as evidence for a role of ACh in mediating parasympathetic neuromuscular transmission, are evidently mainly related to effects at the level of the ganglionic synapse.

Although these results clearly indicate a transmitter role for ACh, they cannot exclude the possibility that there may be noncholinergic parasympathetic postganglionic neurons. It is well known that the action of parasympathetic nerves is usually more resistant to atropine than is the action of exogenous ACh, and parasympathetic effects vary considerably in their sensitivity to atropine (Henderson and Roepke, 1934; Dale, 1937; Ambache, 1955). In particular, parasympathetic excitation of the urinary bladder is extremely resistant to atropine (Henderson and Roepke, 1934). Since parasympathetic ganglion cells are usually located in or close to the target organ, and ganglionic transmission can be blocked by sufficient atropine (Bainbridge and Brown, 1960; A. M. Brown, 1967), only a sensitivity to atropine at low concentrations can be taken as evidence for cholinergic transmission. In the bladder there is ACh release by postganglionic parasympathetic nerve stimulation (Carpenter and Rand, 1965), and bladder contraction is potentiated by eserine (Carpenter, 1963). However, there exists a nerve-mediated excitatory effect of transmural stimulation of the bladder which is insensitive to atropine at a concentration a thousand times that blocking all response to exogenous ACh and not potentiated by eserine (Ambache and Zar, 1970). The easiest way to reconcile these results is to suppose that ACh may be the mediator at only a fraction of the parasympathetic nerve endings in the bladder, another transmitter being released at the others. The atropine-resistant twitch persists after desensitization to adenosine triphosphate (see below) and is not blocked by an antihistamine, which blocks the action of histamine on the bladder muscle, nor by methysergide, which blocks 5-HT effects, nor by α- nor β-adrenergic blockers. Amino acids are inactive (Ambache and Zar, 1970). Thus, the transmitter is unknown, unless, of course, it actually is ACh acting on receptors unaffected by either nicotinic or muscarinic blocking agents. It may be of significance that transmission to the bladder, in contrast to skeletal neuromuscular transmission, is more susceptible to type D botulinum toxin than to type A toxin, although this may be characteristic of parasympathetic nerve endings. After treatment with botulinum toxin, transmission was found to be inhibited only slightly less than ACh release in response to parasympathetic nerve stimulation (Carpenter, 1967). There exists an even better candidate for a noncholinergic postganglionic excitatory parasympathetic pathway. Kottegoda (1969) has found that contractions of the circular smooth muscle of the gut in response to repetitive electrical stimulation were not blocked by inhibitors of cholinergic transmission such as hyoscine, morphine, hemicholinium, or type D botulinum toxin. Similar results were obtained by Ambache and Freeman (1968), who also excluded 5-HT, histamine, or prostaglandin E.

B. MECHANISM OF ACETYLCHOLINE ACTION. The action of ACh on most smooth muscle, with the notable exception of resistance vessels, is to depolarize. Spontaneous activity is initiated, or, if present, action potential frequency is increased (see Kuriyama, 1970). In spontaneously active smooth muscle, the increased spike frequency may be due to an action of ACh on specific pacemaker areas or conversion of nonpacemaker into pacemaker cells by its depolarizing action. In the rat ureter, impulses normally originate only from the renal end, but local application of ACh to any region of the tissue can cause this area to become a pacemaker (Burnstock *et al.*, 1963). The spontaneous slow oscillations of potential found in some smooth muscle are also influenced by ACh. For example, in longitudinal muscle of the cat, ACh produces slow potentials with trains of spikes on their crests (Burnstock and Prosser, 1960). In guinea pig jejunum, both atropine and tetrodotoxin can reduce the slow potentials, suggesting that normal spontaneous membrane activity is influenced by release of ACh from nerve terminals (Kuriyama *et al.*, 1967a,b). However, in the cat stomach, the same type of slow potential does not appear to be caused by nervous activity; neither atropine, nor tetrodotoxin have any effect on it (Papasova *et al.*, 1968).

Inasmuch as it depolarizes, the action of ACh on smooth muscle resembles its action on skeletal muscle. However, there are many points of difference. As is well known, the depolarization in smooth muscle is mediated by a muscarinic rather than nicotinic receptor. ACh action can be imitated by muscarine and is antagonized by atropine; nicotine and tubocurarine act only at the level of the parasympathetic ganglion cells. It is perhaps because of the different receptor that the cholinergic excitatory junction potential evoked by postganglionic nerve stimulation and recorded intracellularly from esophagus (Ohashi and Ohga, 1967), taenia coli (M. R. Bennett, 1966b), or colon (Gillespie, 1962b, 1964), exhibits an extremely long latency (80–220 msec), rise time (150–400 msec), and total duration (up to 1 sec). Similar long latencies and durations have been recorded for muscarinic junctions in cardiac muscle (del Castillo and Katz, 1955) and in salivary glands (Lundberg, 1958; Creed and Wilson, 1969). Creed and Wilson (1969) were able to conclude that at least 100 msec of the latency of the secretory potential in the cat submaxillary gland occurred after ACh reached the cell membrane and was therefore related to delay in ACh–receptor combination or in activity of the combination.

Another site where muscarinic excitatory postsynaptic potentials (EPSPs) occur is sympathetic ganglia. The late negative (LN) wave, which has a long latency and duration (several seconds), and persists after curarization and is blocked by atropine (Eccles and Libet, 1961), represents a slow EPSP. It is also blocked by dibenamine, which does not inhibit the fast nicotinic EPSP (Eccles and Libet, 1961) but does block the response of smooth muscle to ACh (e.g., Burgen and Spero, 1968). A similar sequence of rapid nicotinic action and slow delayed muscarinic action is seen in Renshaw cells in the spinal cord either after appropriate nerve stimulation or ionophoresis of ACh (Curtis and Ryall, 1966). The excitation of cortical pyramidal tract cells by ACh is also characteristically muscarinic in nature, being inhibited by atropine while dihydro-β-erythroidine is ineffective (Krnjević and Phillis, 1963b). The time course of ACh action, with an onset that may be delayed for several seconds and a prolonged duration, is reminiscent of that on smooth muscle and in ganglia. In both cortical neurons and ganglia, in contrast to smooth muscle where the technical difficulties are formidable, it has been possible to study the potential change caused by ACh in a manner similar to that previously employed at the skeletal neuromuscular junction. Krnjević (1969) found that the depolarizing action of ACh on cortical neurons is associated with a decrease in membrane conductance, rather than the increase that would be expected if the ionic mechanism of ACh action were the same as at the skeletal neuromuscular junction, i.e., an increase of permeability to both sodium and potassium. He therefore proposed that ACh acts at this site by reducing potassium conductance. Such an action would lead to an increase of excitability out of proportion to the concurrent depolarization, which is indeed observed. His evidence, however, was open to the criticism that if ACh acted on dendrites rather than the somata of the cells studied, an increased membrane conductance might not be apparent. Recently, compelling evidence in favor of Krnjević's suggestion has been reported by Weight and Votava (1970), who investigated the effects of postsynaptic membrane polarization on the slow EPSP in sympathetic ganglia. The enhancement of the EPSP by extrinsic postsynaptic depolarization and reversal by hyperpolarization are compatible only with a transmitter action to reduce resting potassium conductance.

Thus, it seems to be a general phenomenon that muscarinic receptors mediate only a slow transmitter action and, at the one site where it has been possible to determine the mechanism of excitatory muscarinic action, it is via a reduction of potassium conductance rather than an increase in sodium and potassium conductance. Nevertheless, the bulk of the evidence available regarding the excitatory action of ACh on smooth muscle is rather more compatible with the view that the depolarization by ACh is secondary to increases in sodium and potassium conductances similar to that at the skeletal neuromuscular junction.

Bülbring and Kuriyama (1963b) observed that in smooth muscle cells of taenia coli, depolarization by ACh was graded with the membrane potential prevailing at the time of application. When the membrane potential was low, either because of spontaneous depolarization or raised potassium, the effect of ACh was diminished. The grading of effect with membrane potential was compatible with an ACh equilibrium potential of -20 to -26 mV (M. R. Bennett, 1966a). In raised potassium, the apparent equilibrium potential was less, as would be expected if ACh action were to increase both sodium and potassium permeabilities. It would not be impossible to reconcile these findings with the alternative mechanism, that of reduced potassium conductance by ACh. For example, it was seen previously that net depolarization caused by ACh application to skeletal muscle depends greatly on membrane time constant and therefore membrane resistance. If membrane resistance in smooth muscle declined greatly with membrane potential, this reduction could cause a grading of ACh depolarization with resting potential that did not reflect a deviation from ACh equilibrium potential. Indeed the fact that ACh may actually hyperpolarize in solution containing 40 mM potassium (Burnstock, 1958; Bülbring and Kuriyama, 1963b) cannot be explained on the basis of a parallel sodium and potassium permeability increase, but would fit a reduced potassium conductance if the equilibrium potential for chloride were more negative than the resting potential. The enhancement of ACh depolarization by raised sodium or calcium and the ineffectiveness of ACh in the absence of calcium (Bülbring and

Kuriyama, 1963b) would fit either hypothesis, while the lack of action of ACh in the absence of potassium is equally unexpected in terms of both possible mechanisms. The strongest evidence in favor of the idea that ACh acts in smooth muscle in a manner similar to that at the endplate is the finding of Hidaka and Kuriyama (1969b) that iontophoretic application of ACh increased membrane conductance in conjunction with membrane depolarization. However, it is not impossible that ACh may act in more than one way, as will be discussed below.

The data on ion fluxes in smooth muscle, although often quoted in support of the idea of a muscarinic ACh ionic mechanism similar to the nicotinic, in fact provides little support and in some respects raises serious doubts. Durbin and Jenkinson (1961) studied smooth muscle depolarized by raised potassium to avoid flux alterations secondary to depolarization and found carbachol to increase uptake and efflux of ^{36}Cl and ^{82}Br as well as ^{42}K, and also changes in ^{24}Na and ^{45}Ca fluxes suggesting nonselective increase in permeability to all these ions. Increase of calcium efflux by ACh and cholinomimetic agents, without change of influx, has been found in nondepolarized muscle (Schatzman, 1961, 1964). Quantitatively, ^{42}K efflux from taenia coli or longitudinal muscle of the ileum can be increased one hundredfold by ACh (Burgen and Spero, 1968). If sodium permeability change were about the same, a concentration of ACh producing about 0.1% of the maximal change in efflux would be expected to produce a large depolarization and considerable activation of muscle contraction. Such is indeed the case with ACh, for which the ratio of the dose giving half maximal efflux activation to that for half maximal contraction activation is about 1000:1. However, with carbachol, this ratio is much lower, about 330:1, and with other muscarinic agonists the ratio is much lower still. For example, with muscarine the ratio was found to be 14, with oxotremorine 9, with trimethylammonium 3.2, and with pilocarpine 2.2 (Burgen and Spero, 1968). Thus, with agonists other than ACh, there was no activation of contraction at doses where the potassium efflux and presumptive increase of permeability to potassium and the other ions was many times greater than that which would be expected to cause maximal activation of contraction if sodium permeability increase paralleled potassium permeability. It is impossible to escape the conclusion that there is no quantitative relation between ^{42}K efflux rate and the excitatory action of ACh and cholinomimetics on smooth muscle. At first sight, the data suggest that there are two distinct muscarinic receptors, one mediating the excitatory action and the other the efflux response, but this is unlikely in view of the fact that the reversible ACh antagonists atropine, lachesine, tricyclamol, methylscopolamine, and benzilyldimethylbutanol appeared to have the same affinity constants for both receptors, and the irreversible alkylating antagonist benzilylcholine mustard (Gill and Rang, 1966) caused the same shift in dose response curves for both responses (Burgen and Spero, 1968). Moreover, raised calcium or magnesium, lowered sodium, or high osmotic pressure have the effect of largely annulling the differences in structure–activity relationships of the two responses. These results have been interpreted as suggesting that both responses are due to the activation of the same kind of receptor, the sequelae determining the contractile response being different from those determining the efflux response (Burgen and Spero, 1970). Nevertheless, it seems clear that the contractile response to low doses of ACh or carbachol, at optimal calcium or magnesium, cannot be related to the same membrane action as the ionic flux response, and the mechanism of the former action remains obscure. Conversely, it is difficult to imagine that in conjunction with the efflux response that reflects increased potassium permeability, there is an increase in sodium permeability of the same magnitude, as at the skeletal muscle junction. If there were, maximal depolarization would have to occur at concentrations of agonist far less than one-fifth to one-half of that causing a maximal efflux response. A relatively very small increase in sodium and/or calcium permeability could account for the contractile response correlated with the ^{42}K efflux response, but this provides no explanation for the differential actions of various agonists. One possibility, of course, is that the relative sodium (and/or calcium) and potassium permeability increases produced by activating the receptor depends upon the agonist acting upon the receptor; the recent finding that sodium and potassium conductance changes produced by ACh at the skeletal junction can be altered selectively makes this hypothesis less unattractive than it would otherwise be. A simpler interpretation is that there are two distinct actions of ACh, mediated by receptors that are dissimilar in their affinities for ACh analogs, but possess the same accessory group at which competitive blocking agents interact (see below, page 106). At one kind of receptor, the agonist leads to increase of potassium permeability, probably with some associated increase of sodium and/or calcium permeability, while the major effect normally determining contractile response is mediated by other receptors activated at lower ACh concentrations. Here, a reduction in potassium conductance could just as well give rise to increased action potential activity and consequent contractile response as an increase in sodium conductance.

C. Vascular Smooth Muscle. The predominant peripheral effect of ACh and cholinomimetic agents is vasodilatation; in some species there appear to be cholinergic vasodilator nerve fibers (see Uvnäs, 1966; Mellander and Johansson, 1968; A. P. Somlyo and Somlyo, 1968, 1970). ACh can also constrict. In some preparations, there is evidently a considerable adrenergic component to this response. Thus, in dogs, the renal vasoconstrictor effect of ACh can be eliminated with adrenergic blocking agents (see A. P. Somlyo and Somlyo, 1970). However, in certain types of isolated vascular preparations (e.g., rabbit aortic strips), cholinergic agents have a direct excitatory (i.e., vasoconstrictor action) (Furchgott, 1955) in addition to an inhibitor effect exerted by lower concentrations of ACh than those producing vasoconstriction (Jelliffe, 1962). In canine saphenous vein, only a vasoconstrictor action of ACh is demonstrable (Clement *et al.*, 1969). In the umbilical artery, there are both dilator and constrictor effects of ACh, both blocked by atropine (Gokhale *et al.*, 1966).

Su and Bevan (1967) have found depolarization by ACh of the intimal layer of pulmonary artery, in which noradrenaline is without excitatory effect. In the portal vein, electrical recording (Funaki and Bohr, 1964; Nakajima and Horn, 1967) has shown a diphasic action of ACh, which may be typical for preparations where ACh has both inhibitory and excitatory action, namely, an initial period of hyperpolarization followed by a secondary depolarization. Paradoxically, the initial period of hyperpolarization was not associated with block of spontaneous electrical activity and increased amplitude of action potentials. In rat mesenteric arterioles, on the other hand, which show no excitatory effect of ACh, ACh caused abolition of action potentials and a decrease of slow wave amplitude (Steedman, 1966).

With regard to mechanism of ACh action, virtually nothing is known. It is tempting to assume that the inhibitory ACh action is by the same mechanism as that by which ACh acts to inhibit the myocardium, i.e., an increase in potassium permeability (Trautwein and Dudel, 1958; Hutter, 1961). However, it has recently been stated that ACh has a relaxant action on depolarized rabbit aortic strips, i.e., an effect that could not be ascribed to hyperpolarization via raised potassium conductance (Somlyo and Vinall, unpublished, cited in A. P. Somlyo and Somlyo, 1970).

D. Anticholinesterases. The potentiation of both neural effects and those of exogenous ACh by eserine or other anticholinesterases has long been considered as supporting evidence for the hypothesis that ACh is the postganglionic parasympathetic transmitter. In addition, anticholinesterases increase peristalsis and enhance spontaneous activity of intestinal smooth muscle (Salerno and Coon, 1949; Shelley, 1955; Harry, 1963) and also produce contractures (E. D. Adrian *et al.*, 1947). The contracture is clearly related to the neural release of ACh and not due to a direct action on muscarinic receptors, being absent after treatment with botulinum toxin (Ambache and Lessin, 1955), and it is blocked by atropine (F. B. Hughes, 1955; Erdmann and Heye, 1958). The contracture is graded with the degree of cholinesterase inhibition (Shelley, 1955) and is seen only in the innervated, and not the denervated, longitudinal muscle strip of guinea pig ileum (Paton and Zar, 1965, 1968). Similarly, tracheobronchiolar muscle is contracted by eserine (Dixon and Brodie, 1903) and by a variety of organophosphorus anticholinesterases (e.g., Douglas, 1951). The contracture is prevented by hemicholinium (Carlyle, 1963). One anticholinesterase, Mipafox (diisopropyl phosphodiamidic fluoride), although it potentiates the action of acetylcholinesterase, does not itself cause contractures on either bronchial or intestinal muscle (Carlyle, 1963; Harry, 1963), and this exception would militate against the usual interpretation that the contracture is due to an accumulation of ACh released at postganglionic endings. However, Mipafox evidently has an added action to inhibit release of ACh and is in any case a weak inhibitor of cholinesterase (Zar, 1966).

There is now strong evidence that ganglionic activity is involved in the contractures. Gershon (1967a) found that the contracture that follows the administration of 10^{-7} gm/ml of eserine is abolished by tetrodotoxin. Previously, Erdmann and Heye (1958) found that ganglionic blocking agents such as hexamethonium inhibited the contracture. With an eserine concentration of 10^{-6} gm/ml, Gershon (1967a) found the reappearance of contracture sensitive to atropine but not to tetrodotoxin. This result was interpreted as indicating some spontaneous release of ACh from postganglionic terminals, which will accumulate sufficiently to control intestinal smooth muscle when cholinesterase is completely inhibited. The response to the lower dose, tetrodotoxin absent, evidently reflects neuronal activity, but the question remains whether this neuronal activity was normally present or was, in fact, at least partially induced by the cholinesterase, which seems more likely. It is also possible that at least a portion of the action of anticholinesterases may be secondary to stimulation of ACh release from nerve terminals of both pre- and postganglionic neurons.

C. Catecholamines

1. Actions on Smooth Muscle

The actions of adrenaline and noradrenaline on smooth muscle can be classified as either excitatory or inhibitory. In general, vascular smooth muscle is excited, as is the smooth muscle of bladder and vas deferens (e.g., Bacq and Monnier, 1935), while in the gastrointestinal tract adrenaline is inhibitory. However, in the muscularis mucosa of pig esophagus, adrenaline causes depolarization and the initiation or increase of frequency of action potentials (Burnstock, 1960), and in the dog stomach, an initial inhibitory effect of adrenaline is followed by depolarization and contraction (Ichikawa and Bozler, 1955). The action of adrenaline on the uterus varies with the species and hormonal state of the animal. Thus, in rat uterus, the major action is inhibitory, to relax the smooth muscle; in cat, adrenaline relaxes the estrogen-dominated nonpregnant uterus but contracts the pregnant uterus. The rabbit or human uterus is made to contract by adrenaline (reviewed by Burnstock *et al.*, 1963). Excitation is generally associated with activation of α-receptors, an inhibition with activation of β-receptors (Ahlqvist, 1948, 1962, 1966).

A. MECHANISM OF EXCITATORY ACTION. It is generally considered that the mechanism of the excitatory action of noradrenaline at α-adrenergic synapses is similar to that of ACh at the skeletal neuromuscular junction, i.e., an increase in conductance for sodium, potassium and presumably calcium ions. The only strong evidence for this hypothesis is that of Dorward and Holman (1967; and see Holman, 1970) who were able to record inverted spontaneous junction potentials in mouse vas deferens smooth muscle cells depolarized by injected current (to $+20$ mV) in the presence of concentrations of potassium sulfate that reduced the resting membrane potential to about 0 mV. This experiment indicates clearly an increase rather than a decrease in potassium conductance during transmitter action. Since the junction potentials are depolarizing at normal resting potential, they must be associated with a concurrent increase in Na^+ and, or Ca^{2+} conductance. It would therefore be anticipated that the junction potentials would be made larger by hyperpolarization of the smooth muscle cells, as are the corresponding potentials at the skeletal neuromuscular junction. In fact, this experiment has failed. In neither mouse nor guinea pig vas deferens has it been possible to demonstrate any increase in amplitude of the junction potentials (which are adrenergic) with hyperpolarization of muscle fibers by injected current (Dorward and Holman, 1967; Holman, 1967; M. R. Bennett, 1967). However, these negative results might be accounted for by the electrical coupling between cells in this preparation. Hyperpolarization at one point might have little influence on the membrane potential at the points where the junction potentials are generated. Holman (1970) has reported that in some cells of the mouse vas deferens there was a demonstrable increase in input conductance during the early part of an evoked junction potential. This would constitute strong evidence in favor of the hypothesis, were it not for the possibility that the change of conductance might be related to the development of a local response which was superimposed on the junction potential rather than to transmitter action (Holman, 1970).

The excitatory effect of adrenaline and noradrenaline or arteries and veins evidently resembles that of visceral smooth muscle to ACh (see Burnstock and Holman, 1966). Small arteries of the rat *in vivo* respond to adrenaline (10^{-9} gm/ml) by a marked increase in the frequency of firing of action potentials. Direct application of noradrenaline to the adventitial layer of the vessel effectively mimics the action of noradrenaline released upon nerve stimulation (Steedman, 1966). Anterior mesenteric and portal veins of various species also show an increase in the frequency of action potentials, in response to concentrations of noradrenaline as low as 10^{-10} gm/ml (e.g., Cuthbert and Sutter, 1965; Cuthbert, 1967; Funaki, 1967). This increase is related to an increase in the number of spikes per burst, to an increase in the frequency of bursts, or to both, and the electrical effect is associated with an increase in amplitude and/or frequency of phasic contractions. With higher concentrations, noradrenaline causes a continuous discharge of action potentials and eventually maintained depolarization (Nakajima and Horn, 1967). These responses are presumably mediated by α-adrenergic receptors, since they can be blocked by phenoxybenzamine and other α-receptor antagonists (Holman *et al.*, 1967).

Although the electrical responses of vascular smooth muscle to noradrenaline are clear, it remains debatable whether these responses can account entirely for the contractile responses that are observed. Su *et al.* (1964) obtained tension responses of pulmonary artery to sympathomimetic drugs without change of electrical activity. Cuthbert and Sutter (1965), Axelsson and Högberg (1967), and Johansson *et al.* (1967) have shown records in which the contractile responses of veins are obviously partially dissociated from the alteration of electrical activity. Statistical analysis of the relation between electrical and mechanical activity supports the impression that there is an increase in contractile activity relative to electrical activity with time after administration of noradrenaline (appendix by Gudmundsson, in Axelsson, 1970). However, the objection remains that the recording of potentials of smooth muscle by the sucrose gap method could be considerably distorted by any alterations in membrane resistance. Unfortunately, the same would apply to intracellular recordings. It is difficult to see how this problem can be resolved.

How vascular smooth muscle is normally excited by endogenous nor-adrenaline also remains a problem. It is now well established (see Speden, 1970) that the noradrenergic innervation of most arteries and arterioles is confined to the adventitia of the vessels; nerve terminals lie just outside the media and do not penetrate it. Adrenergic nerve fibers are present in the media only in some arteries of the iris (Ehinger, 1966) and pineal gland (Owman, 1965) and some thick walled cu-taneous veins (Ehinger *et al.*, 1967). Blood vessels in the central nervous system are almost devoid of adrenergic innervation (Carlsson *et al.*, 1962; Malmfors, 1965a; Ehinger, 1966).

Thus, in most arteries, most muscle cells are at a considerable distance from nerve terminals. Even the outermost muscle fibers are separated from nerve terminals by a gap which is rarely less than 1000 Å (Verity and Bevan, 1967). The long diffusion distances involved have led to doubt whether the transmitter ever can reach the innermost muscle cells of the larger arteries in sufficient concentration to activate the muscle (Thaemert, 1963; Fuxe and Sedvall, 1964). Indeed, Folkow (1964a,b) has suggested that there is a functional separation of the smooth muscle cells into two layers: an inner one which is spontaneously active and responsible for basal vascular tone, and an outer nonspon-taneously active layer of a multiunit type of muscle (Bozler, 1948) under nervous control. Moreover, not all the transmitter released may be available for diffusion to the muscle cells. Drugs that block the uptake of noradrenaline by nerve terminals are effective in potentiating the response of blood vessels to adrenergic nerve stimulation, indicating that such uptake normally limits the amount of noradrenaline reaching muscle receptors. For example, either cocaine or chronic denervation largely abolished the relative insensitivity of a rabbit ear artery to ex-traluminal as compared to intraluminal noradrenaline (de la Lande *et al.*, 1967). Cocaine potentiation of the constrictor response to nerve stimulation was less than the potentiation of the response to extraluminal noradrenaline, suggesting that nerve stimulation may cause the imme-diate rerelease of newly taken up noradrenaline, as occurs in the cat nictitating membrane (Trendelenburg, 1966b).

Excitation by electrotonic spread of depolarization from cell to cell via low resistance pathways is clearly an alternative to activation by diffusion of transmitter. Although membrane fusion and areas of close apposition of cell membranes are not always frequent (Verity and Bevan, 1967; Cliff, 1967; Rhodin, 1967), the very success of external methods of electrical recording from strips of large arteries, such as the rabbit aorta (Shibata and Briggs, 1966) and sheep, dog, and bovine carotid arteries (Peterson, 1936; Keatinge, 1964; Cuthbert, 1967), indicates that electrical connections, (i.e., low resistance pathways) must exist between the individual muscle cells. Correspondingly, rhythmic spontaneous conducted contractions can occur in isolated arteries under certain circumstances, i.e., chronic denervation or pretreatment with quanethidine or reserpine. α-Adrenergic blockade fails to modify this activity, indicating its nonneural generation (Johansson and Bohr, 1966). The anterior mesenteric vein of the rabbit develops spontaneous activity particularly readily (Cuthbert and Sutter, 1964). Given the existence of electrotonic junctions, local depolarization by noradrenaline in the innervated areas could give rise to action potentials propagated to fibers not directly acted upon by the transmitter or to passive depolarization of these cells, with or without consequent action potential generation.

Even if the normal mechanism of excitation of the noninnervated smooth muscle is via diffusion of noradrenaline, the question remains, by what mechanism are the cells excited? There are three likely alternatives, by no means exclusive: (1) depolarization leading to action potential generation and thus secondarily to calcium influx; (2) depolarization leading to calcium influx; and (3) pharmacomechanical coupling, i.e., influx of calcium directly caused by calcium permeability increase secondary to transmitter action. The relative roles of these mechanisms is yet to be determined. Evidence for the last alternative will be considered in Section IV.

B. MECHANISMS OF INHIBITORY ACTIONS. The relaxation of intestinal smooth muscle by adrenaline is associated with block of spontaneous action potential activity and hyperpolarization which depends upon the level of the membrane potential when the drug was administered (e.g., Bülbring and Kuriyama, 1963c). The rise in membrane potential is evidently the result of two distinct actions: (1) inhibition of muscle fibers directly, by increase of membrane permeability to potassium, and (2) disexcitation, i.e., reduction of excitatory activity by inhibition of intramural ganglion cells. It is now clear that both adrenaline and noradrenaline act via α-adrenoceptors to increase potassium and chloride conductances. Both were found to increase ^{42}K efflux and ^{42}K uptake in depolarized taenia coli (Jenkinson and Morton, 1965, 1967a,b,c; Bülbring *et al.*, 1966), although under normal conditions the effect was not readily detectable, perhaps because the hyperpolarization of the membrane reduces the outward movement of potassium. More recently, experiments using the sucrose gap method have shown that the hyperpolarization by catecholamines is associated with increased chloride conductance as well as increased potassium conductance; the effects of altering potassium and chloride concentrations and of changing membrane polarity indicate that these conductance changes are responsible for the altered membrane potential (Bülbring and Tomita, 1968a, 1969a,b,c). These effects on membrane conductance do not, however, account entirely for the suppressing effect of catecholamines on spontaneous spike activity. There is in addition an action that is not associated with membrane hyperpolarization and that is evidently mediated by β-adrenoceptors, since it can be blocked by propranolol. For example, in the presence of phentolamine, which blocks α-adrenoceptors, there is no hyperpolarizing effect of catecholamines nor increase of membrane conductance, but the spike inhibition by adrenaline, noradrenaline, and isoprenaline remains. This can be blocked by propranolol (Bülbring and Tomita, 1968a). The action on spontaneous action potentials is exerted not on the spike mechanism, but on the threshold for action potential generation. As in the presence of raised Ca^{2+}, a larger and faster action potential can be evoked in the presence of adrenaline if a sufficiently strong stimulus is applied (Brading et al., 1969b). Bülbring and Tomita (1969c) have suggested that the β action may consist in a reduction or spontaneous removal of calcium from certain membrane binding sites, this removal being normally responsible for the pacemaker potential via reduction of potassium conductance and consequent depolarization.

There is strong evidence that the β-mediated positive inotropic effects of catecholamines on the heart, as well as most of the metabolic effects of the catecholamines, in liver, muscle, and adipose tissue are mediated by cyclic adenosine 3',5'-monophosphate (cyclic AMP) (Sutherland and Rall, 1960; Sutherland et al., 1965; Sutherland and Robison, 1966). There is now considerable evidence that cyclic AMP is also involved in the β-receptor-mediated inhibitory action of catecholamines on intestinal smooth muscle. In the case of taenia coli, there is not only evidence that cyclic AMP levels do increase following exposure to adrenaline, but some degree of correlation between the physiological effect and the increase in cyclic AMP (Bueding et al., 1966). Bowman and Hall (1970) have recently reported that the dibutyryl analog of cyclic AMP, which penetrates cell membranes more readily than cyclic AMP and is more resistant to breakdown by phosphodiesterase (Butcher et al., 1965), produces in the rabbit intestine an inhibition of slow onset which closely resembles that produced by β-adrenoceptor agonists. Theophylline and imidazole, which respectively inhibit and activate the phosphodiesterase responsible for destroying cyclic AMP (Butcher and Sutherland, 1962), were also tested. Theophylline itself inhibited and also potentiated inhibition by β-adrenoceptor agonists. Imidazole had the opposite effects. Although these drugs also had nonspecific effects, the data strongly support the hypothesis that β-adrenoceptor-mediated inhibition in the intestine is related to activation of adenylcyclase and intracellular formation of cyclic AMP.

Until recently, it was believed that adrenergic nerves in the intestine terminated directly in the muscle layers and that the adrenergic inhibitory function was caused by a direct action of noradrenaline in the smooth muscle cells. Studies with fluorescence microscopy have now demonstrated that what few adrenergic nerve terminals exist in the muscle layers probably belong to the blood vessels. Most adrenergic nerve terminals in the intestine invest around the intramural ganglion cells (Norberg, 1964, 1967; Jacobowitz, 1965; Hollands and Vanov, 1965). Correspondingly, it has been observed that uptake of tritiated noradrenaline by the small intestine is confined to the neural elements in the myenteric plexus (Marks *et al.*, 1962), and the inhibitory effects of stimulation of sympathetic nerve supply of the gastrointestinal tract are not accompanied by any distinct inhibitory junction potentials (Gillespie, 1962a; M. R. Bennett *et al.*, 1966a). Thus, the neurogenic inhibition of intestinal motility is probably exerted by an indirect effect on the myenteric plexus; some direct inhibition may be produced by diffusion of transmitter from the myenteric plexus to adjoining muscle, or by catecholamines released from the adrenal medulla (Gershon, 1967b). An action of catecholamines on ganglion cells was suggested by McDougal and West (1954) on the basis of the finding that the contraction of longitudinal muscle produced by nicotine (i.e., via excitation of ganglion cells) was reduced by doses of the catecholamines which had little or no effect on the contractions produced by histamine or ACh (i.e., via a direct action on the smooth muscle). This inhibition could be abolished by α-adrenoceptor blocking agents, indicating that the action of catecholamines on the ganglion cells is α-mediated, a conclusion reinforced by further evidence that α-receptors alone are concerned in the adrenergic inhibition of the peristaltic reflex of the isolated intestine of both guinea pigs and rabbits (see Lee, 1970). Indeed, there is good evidence that α-receptors are confined to the ganglion cells in longitudinal muscle of guinea pig ileum (Kosterlitz and Watt, 1965; A. B. Wilson, 1964), though not in taenia coli or circular muscle (Lee, 1970).

In contrast to the behavior of taenia coli, vascular smooth muscle cells show alteration of membrane potential in response to the inhibitory action of catecholamines mediated by β-adrenoceptors. In rabbit portal vein, isoprenaline reduced and even abolished spontaneous electrical activity, and in preparations where there was pacemaker-type depolarization between bursts of action potentials, isoprenaline caused hyperpolarization (Holman *et al.*, 1968). This effect was blocked by propranolol. In the rat portal vein, on the other hand, reduction of action potential activity and tension development was associated with depolarization (Johansson *et al.*, 1967), perhaps because of some α-receptor activation. The rabbit main pulmonary artery also responds with a hyperpolarization. In general, the hyperpolarizing action of isoprenaline is lost in the presence of raised potassium, although some relaxing effect may still remain (Johansson *et al.*, 1967), a finding most easily explained in terms of a β-receptor-mediated action of isoprenaline, due at least in part to increase in potassium and/or chloride conductance (Holman, 1969). In raised potassium, membrane potassium conductance would be expected to be high, and a further increase in potassium conductance due to isoprenaline would have little effect. A. V. Somlyo *et al.* (1970) have recently reported evidence supporting the hypothesis that this β-receptor-mediated effect, like others, is secondary to the production and action of cyclic AMP. This hypothesis predicted three results: (1) theophylline, as an inhibitor of phosphodiesterase, should potentiate the response to isoprenaline, (2) cyclic AMP or dibutyrylcyclic AMP should cause inhibition and hyperpolarization, and (3) this hyperpolarization should not occur in raised potassium.

Experimentally, theophylline alone had little action on the potential of pulmonary artery strips, but it markedly increased the response to a concentration of isoprenaline which in itself had little hyperpolarizing action. Dibutyrylcyclic AMP caused significant hyperpolarization, but only in the presence of theophylline, and in raised potassium (10 mM) this action was lacking. The relatively high concentrations of cyclic nucleotide required for the effect suggested that the site of hyperpolarizing action is not the outer surface of the cell membrane (A. V. Somlyo *et al.*, 1970). These results serve to emphasize that what might otherwise appear to be quite divergent hypotheses regarding the mode of action of a drug, one involving membrane conductance charge and the other a metabolic action, need not be exclusive.

2. Actions on Skeletal Muscle

Although there is no adrenergic innervation of skeletal muscle, it has long been known that catecholamines have considerable effects on skeletal muscle activity. Presumably, these effects may be of some importance *in vivo*, brought about either by noradrenaline diffusing from release sites in the vasculature or by circulating adrenaline from the adrenal gland. Presynaptic actions of catecholamine modifying neuromuscular transmission have already been discussed; there exist also postsynaptic actions, i.e., direct effects on the skeletal muscle fibers.

Fast-contracting (white) skeletal muscle responds to adrenaline with an increase in twitch tension associated with an increase in the total duration of the twitch (Goffart and Ritchie, 1952; Bowman and Zaimis, 1958); thus, there is an increase in fusion when incomplete tetanic contractions are recorded. Maximal tetanic contraction can be depressed (Goffart and Ritchie, 1952), but this may well be merely a consequence of vasoconstriction (Bowman and Zaimis, 1958). When muscle contraction is depressed by excess potassium chloride, the potentiating action of catecholamines is especially well marked (Bowman and Raper, 1964).

Slow-contracting muscle on the other hand (e.g., cat soleus) responds to adrenaline with a decrease in the tension and duration of maximal twitches. Thus the tension and degree of fusion of incomplete tetanic contractions are markedly reduced (Bowman and Zaimis, 1958). The same effect is produced in slow motor units in human muscles (Marsden and Meadows, 1970). This may represent the mechanism by which adrenaline and isoprenaline enhance physiological tremor.

It is notable that when a fast muscle (cat flexor digitorum longus) was converted to a slow muscle by cross innervation with soleus motoneurons, the response to catecholamines remained that of a normally innervated fast-contracting muscle. Denervation, which slows the contraction speed of fast muscle, also does not change the characteristic response (Raper, 1965, quoted in Bowman and Nott, 1969). However, soleus muscle that has been made fast by innervation with flexor digitorum longus motoneurons responds to catecholamines as does a normal fast muscle (Bowman and Raper, 1962). Chronically denervated muscles respond to small doses of catecholamines by an increase in the frequency of spontaneous action potentials (i.e., of fibrillary potentials), an effect associated with an increase in membrane potential; large doses produce sufficient increase in membrane potential to inhibit spike generation (Bowman and Raper, 1965).

With regard to mechanism, the actions of adrenaline appear to be mediated by β-receptors; isoprenaline is consistently more effective than adrenaline, and β-receptor blocking drugs such as propranolol, pronethalol, and dichlorisoproterenol are effective in blocking the action (e.g., Bowman and Raper, 1967). There is little, if any, effect of adrenaline on muscle membrane potential (Krnjević and Miledi, 1958); Dockry *et al.* (1966) found that isoprenaline acted to stimulate sodium–potassium transport in soleus muscles loaded with sodium. This action would tend to raise membrane potential to some extent. It has been suggested that the β-mediated actions of catecholamines on muscle contraction may be mediated by formation of cyclic AMP, but evidence for this seems still to be lacking (Bowman and Nott, 1969).

The chronically denervated rat diaphragm also possesses α-receptors by which sympathomimetic amines can cause contracture *in vitro* (Paterson, 1963). The superior rectus muscle of the cat, which in many ways responds to added drugs like a chronically denervated muscle, is also contracted by catecholamines (Eakins and Katz, 1967). These α-mediated effects are presumably secondary to depolarization caused by an increased ionic permeability similar to that caused by ACh.

D. Histamine

Since the first studies of the pharmacology of histamine (Dale and Laidlaw, 1910, 1911), there has been extensive and detailed investigation of the actions of this agent on a variety of smooth muscle preparations. For reviews, the reader is referred to Kahlson and Rosengren (1965), Parrot and Thouvenot (1966) and to recent symposia (Symposium on Histamine, 1965, 1967). It may suffice here to call attention to the fact that histamine acts to constrict most large blood vessels (excepting cerebral) but to dilate small arterioles and venules. It excites the smooth muscle of bronchioles and the uterus (except in the rat, where it causes constriction), but 'has little effect on the smooth muscle of bladder and iris. Smooth muscle of the gastrointestinal tract is moderately sensitive. In guinea pig taenia coli, histamine causes depolarization which is associated with an increase in the frequency of action potentials, an effect similar to that of ACh (Bülbring, 1957; Bülbring and Burnstock, 1960) and, as with ACh, ^{42}K efflux is increased from intestine (Born and Bülbring, 1956). The major difference in the action of histamine is that its depolarization is slower than that of ACh and the degree of tachyphylaxis is greater (Bülbring and Burnstock, 1960).

Undoubtedly, a component of the response of gastrointestinal muscle to histamine is secondary to the stimulation of neural elements in the myenteric plexus (Ambache, 1946; Paton and Vane, 1963). Histamine is well known to stimulate cutaneous nerve endings (causing itching and pain). It also stimulates (via depolarization) chromaffin cells of the adrenal medulla (Douglas *et al.*, 1967) and neurons in sympathetic ganglia (Trendelenburg, 1967), especially the adrenergic (Aiken and Reit, 1969). However, the major effect of histamine is probably directly on the smooth muscle cells. A dose of atropine, which abolishes ACh-induced contracture of intestinal smooth muscle, has little effect on the response to histamine (Schild, 1947). Moreover, histamine continues to cause a contracture after depolarization by excess potassium (Evans *et al.*, 1958), a maneuver that excludes either mediation by neural elements or effects on muscle membrane potential or action potential activity. Thus, it seems likely that the action of histamine on smooth muscle membrane in gut resembles that of ACh and involves an increase of permeability to calcium, as well as to potassium, and presumably sodium.

Histamine antagonists, such as mepyramine, serve to demonstrate the specificity of the histamine receptor. Just as atropine blocks the response of smooth muscle to ACh without influencing to any extent the response to histamine, these substances can abolish the action of histamine at doses without effect on the response to ACh. Rocha e Silva (1966) has discussed in some detail the possible structure of the histamine receptor.

The question of whether histamine is a neurotransmitter is still highly controversial. The only strong evidence in favor of this view relates to the sympathetically mediated reflex dilatation of peripheral vascular beds that follows a rapid increase in systemic blood pressure (Beck and Brody, 1961; Beck, 1964; review by G. Campbell, 1970). The reflex does not involve the cholinergic sympathetic vasodilation, since it is not specifically reduced by atropine treatment (Lindgren and Uvnäs, 1954), and it has generally been thought that the vasodilatation represents merely a diminution in vasoconstrictor activity. However, the reflex is antagonized by antihistamines (Beck, 1965; Brody et al., 1967; Tuttle, 1966). Moreover, there is good evidence for histamine release coincident with the reflex vasodilatation, a phenomenon difficult to attribute to enhanced perfusion of normally poorly perfused areas (as suggested by Glick et al., 1968) in view of Tuttle's (1967) finding that after preloading with the histamine precursor histidine-^{14}C, the bulk of the increased output of label was in the form of histamine-^{14}C, rather than in unaltered histidine. To those who find the evidence weakest (Glick et al., 1968), the most difficult problem to answer was the action of antihistamines in blocking the vasodilatation; the answer was that these drugs might act like cocaine by inhibiting neural uptake of noradrenaline. Indeed, both cocaine and the antihistamine pyribenzamine reduce the reflex vasodilatation and enhance the direct vasoconstrictor action of noradrenaline. If the vasodilatation were passive, cocaine and similarly acting agents would delay it, and therefore decrease the amplitude of the relatively brief reflex dilator response. According to this argument, the blockade of the reflex vasodilation by antihistamines should be accompanied by blockade of the histamine output that normally accompanies it. Conversely, a continued increased output of histamine in these circumstances would indicate that the vasodilatation is mediated by histamine. Brody (1968) has in fact found that the antihistamine pyribenzamine, while blocking the vasodilatation, does not block the concomitant histamine release. Cocaine, on the other hand, does decrease the output of histamine. These results are strongly in favor of a mediator role for histamine. Whether histamine is in fact the transmitter substance of the vasodilator neurons must remain open to doubt. It remains conceivable that histamine release is a secondary event, although necessary to rather than consequent upon the vasodilatation.

E. 5-Hydroxytryptamine

An enormous body of information has been acquired on the occurrence and effects of 5-hydroxytryptamine (5-HT) in vertebrates and invertebrates (reviewed by Garattini and Valzelli, 1965; Erspamer, 1966; Welsh, 1968; Born, 1970). There is some evidence that this substance may function as a neurotransmitter in the central nervous system (reviewed by Phillis, 1970).

Most preparations of vertebrate smooth muscle *in vitro* respond to 5-HT by contraction. In mammals, 5-HT contracts smooth muscle of blood vessels, the gastrointestinal tract, the urogenital tract, and the bronchial tree. Sensitivities vary; at present, the most widely used organ for the bioassay of 5-HT is the strip preparation of the rat stomach (Vane, 1957), which responds to concentration as low as 10^{-10} M. In the stomach and intestine, a number of sympathomimetic amines, including amphetamine and phenylephrine exert excitatory effects mediated by 5-HT (Vane, 1960; Innes and Kohli, 1969).

The action of 5-HT on smooth muscle results from effects on neural elements as well as directly on the muscle fibers. Only with the smooth muscle in the amniotic membrane of the chick embryo, which contains no nervous tissue at all, can it be certain that the contractile response to 5-HT is mediated solely by receptors on or in the muscle cells (Evans and Schild, 1953). Fortunately, neural and muscle sites differ in their sensitivities to 5-HT antagonists. Thus, lysergic acid diethylamide (LSD), a very effective antagonist of 5-HT in the rat uterus (where presumably all or most of 5-HT action is direct), can reduce by no more than half the effect of 5-HT on guinea pig ileum (Gaddum and Hameed, 1954). Complete blockade can be achieved by the further addition of atropine or morphine, which are also highly potent antagonists to 5-HT on guinea pig ileum (Rocha e Silva *et al.*, 1953) but are in themselves not capable of completely blocking the action of 5-HT (Kosterlitz and Robinson, 1955). To explain these observations, Gaddum and Picarelli (1957) proposed that in guinea pig ileum 5-HT acts on two different sites: (1) the D-receptor, located on the smooth muscle cells—this receptor can be blocked by LSD, dihydroergotamine, or phenoxybenzamine (Dibenzyline)—and (2) the M-receptor, located in nervous tissue—effects mediated by these receptors could be blocked by agents which inhibit cholinergic nerve transmission, i.e., atropine and morphine.

More recent results have confirmed this analysis. Brownlee and Johnson (1965) have demonstrated that 5-HT induces release of ACh from the guinea pig ileum. In circular muscle of the guinea pig ileum, contractions caused by 5-HT as well as nicotine could be abolished by botulinum toxin, hemicholinium (HC-3), and atropine or hyoscine (Harry, 1963). When guinea pig ileum is prepared in such a way as to remove Auerbach's plexus, maximal contractions produced by 5-HT are reduced by as much as 90% (Paton and Zar, 1968). Moreover, agents have been found that selectively block the action of 5-HT on neural receptors as distinct from muscle receptors. The biguanides (Fastier *et al.*, 1959; Gyermek, 1964a) and quaternary derivatives of 5-HT (Gyermek, 1964b) first stimulate and then block the neural receptors.

The action of 5-HT on neural elements is further distinguished by a rapid tachyphylaxis. When guinea pig ileum was continuously exposed to a relatively high concentration of 5-HT, initial contraction was followed after 10–15 min by relaxation. The preparation then responded to other drugs, but to neither 5-HT nor tryptamine (Gaddum, 1953). There was no such tachyphylaxis in rat uterus. However, the action of 5-HT on the circular muscle, which is also indirect, shows much less tachyphylaxis (Harry, 1963).

These results, indicating 5-HT receptors on neural elements, presumably the myenteric ganglion cells, raise the possibility that 5-HT might be a neurotransmitter at synapses on these cells. In support of this hypothesis is the finding by a number of investigators that 5-HT is released from stomach or intestine in response to stimulation of nerves, electrically or with drugs. For example, Burks and Long (1966a,b, 1967a,b) found 5-HT release in response to parasympathetic nerve stimulation, ACh, angiotension, barium chloride, and nicotine. These are all spasmogenic agents, however, and it seems not unlikely that the 5-HT release was predominantly from enterochromaffin cells in the mucosa as a consequence of mechanical activity, rather than from tryptaminergic nerve terminals, a conclusion strengthened by the finding that atropine inhibited nicotine-induced 5-HT release. Atropine, of course, also prevents nicotine-induced contraction of the ileum. Burks and Long (1967b) also found that the spasmogenic activity of morphine and related agents was reduced by tachyphylaxis to 5-HT and by depletion of 5-HT by reserpine. These results certainly suggest that the contractile response to morphine is 5-HT mediated, but from here to a role of 5-HT as a neurotransmitter is a rather large step. Much more convincing evidence for a transmitter role for 5-HT on myenteric ganglion cells comes from studies on mouse and guinea pig stomach.

Release of 5-HT from the stomach, following vagal or transmural nerve stimulation, was attributed by Paton and Vane (1963) and A. Bennett et al. (1966) to release from mucosa as a consequence of mechanical activity. However, Bülbring and Gershon (1967) found release from the mouse stomach even after asphyxiation and in the presence of hyoscine, which blocked nearly all the contractile response. After blocking nicotinic receptors in the myenteric plexus, by pentolinium or desensitization by nicotine, both vagal stimulation and 5-HT caused a relaxation, the response to 5-HT being blocked by tetrodotoxin and hence neurally mediated.

The inhibitory responses to vagal stimulation and 5-HT were found to be reduced or blocked by the biguanides and by bufoteninium iodide, a quaternary derivative of 5-HT, agents which inhibit neurally mediated effects of 5-HT in gut, and the inhibitory response to vagal stimulation could be abolished by 5-HT desensitization, which was apparently specific to 5-HT. These drugs did not block the inhibitory response to a nicotinic stimulator (DMPP), and their effects cannot therefore be ascribed to any nonspecific inhibition of the ganglion cells in the inhibitory pathway.

Thus, the results of Bülbring and Gershon (1967) amply satisfy the basic criteria for identification of a transmitter substance: (1) release on nerve stimulation, (2) imitation of transmitter action, (3) parallel and selective blockade of the neurally evoked response and the response to the putative transmitter. Further support for the hypothesis of a tryptaminergic link in the vagal inhibitory pathway came from experiments with reserpine (Bülbring and Gershon, 1967), although the stores of 5-HT in the stomach are known to be particularly resistant to reserpine depletion (A. Bennett et al., 1966). In reserpinized animals, the inhibitory vagal responses were found to be weaker than usual and more susceptible to fatigue, and the relaxant response to 5-HT larger. After exposure to 5-HT, responses to vagal stimulation were greater and the fatigue less, as would be expected if nerve terminal stores were replenished from exogenous 5-HT.

Evidence for 5-HT localization in the myenteric plexus, however, remains ambiguous. According to most investigators the fluorescent nerve terminals show the green color characteristic of catecholamines, rather than the yellow of 5-HT. The exception is the report of Tafuri and Raick (1964), who found a yellow fluorescence in the preganglionic terminals surrounding ganglion cells in guinea pig gut wall. Radioauto-graphic studies have shown that after injection of mice with 5-hydroxy-tryptophan the label is to be found in the region of preganglionic nerve terminal plexuses about myenteric ganglion cells (Gershon *et al.*, 1965; Gershon and Ross, 1966a,b), but Taxi and Droz (1966) have found the label also to be present in what are generally considered to be adrenergic axons in the rat vas deferens.

1. Blood Vessels

Erspamer (1966) has reviewed in some detail the actions of 5-HT on vascular smooth muscle. Originally, the action of this drug was con-sidered to be entirely vasoconstrictor and therefore hypertensive (Page, 1954; Furchgott, 1955); however, vasodilator actions also exist (Gordon *et al.*, 1958, 1959; McCubben *et al.*, 1962). On large arterial segments, adrenaline and 5-HT excite synergistically (Gordon *et al.*, 1958, 1959; de la Lande *et al.*, 1966), while in small arterioles, 5-HT acts to oppose the vasoconstrictor action of noradrenaline and adrenaline (Gordon *et al.*, 1958, 1959). According to de la Lande and his colleagues, 5-HT sensitizes the rabbit ear artery not only to noradrenaline, adrenaline, and sympathetic nerve stimulation, but also to histamine and angiotensin.

2. Autonomic Ganglia

It has already been pointed out that a large part of the action of 5-HT on intestinal smooth muscle can be attributed to excitatory actions on the intramural ganglion cells (M-receptors). Similarly, 5-HT is well known to stimulate neurons in autonomic ganglia, and this action is blocked by neurotropic 5-HT antagonists (Gyermek and Bindler, 1962a,b). Doses of 5-HT in themselves too low to stimulate, potentiate the response to submaximal ganglionic stimulation (Trendelenburg, 1956a, 1967) and to ACh (Trendelenburg, 1956b). The direct stimulat-ing effect was blocked by nicotine but not by hexamethonium. In the superior cervical ganglion of the rat, Jéquier (1965) found 5-HT to cause an initial brief depolarization, but after several hours incubation with solution containing 5-HT or 5-hydroxytryptophan, the effects of ACh and preganglionic stimulation were markedly depressed.

The physiological significance of these actions of 5-HT on ganglion cells must remain open to doubt. Sympathetic ganglia neither contain appreciable amounts of the amine (Gaddum and Paasonen, 1955) nor is there any enhancement of 5-HT release from the perfused ganglion by preganglionic stimulation (Gertner *et al.*, 1959) in the presence of iproniazid, a monoamine oxidase inhibitor without which no 5-HT release could be detected.

Recent studies on the cat stellate ganglion (Aiken and Reit, 1969) have revealed significant differences in the sensitivity to various agents of the two ganglion cell populations, adrenergic and cholinergic. Thus, the cholinergic neurons are much more sensitive to 5-HT and less sensitive to other agents (e.g., ACh and histamine) than the adrenergic neurons. Such a finding reemphasizes the fact that there must be distinct 5-HT receptors on the ganglion cells. Whether these represent anything more than a developmental accident is unknown.

Perhaps closely related to its action of ganglion cells is the action of 5-HT on chromaffin cells of the adrenal gland to cause depolarization and consequent (?) secretion of catecholamine (Douglas *et al.*, 1967). As with other drugs, the secretory response depends on the presence of calcium and is probably mediated by entry of calcium into the cells (Douglas, 1968), but it seems hardly necessary to suppose that this is related to a direct action to increase calcium permeability rather than an increase in calcium influx secondary to depolarization.

3. MECHANISM OF ACTION OF 5-HT

Most detailed information regarding how 5-HT acts at the cellular level comes from studies of molluscan neurons and smooth muscle, where it may well be a physiological transmitter (Gerschenfeld and Stefani, 1966). The excitatory action of 5-HT is restricted to certain neurons in the molluscan central nervous system, and here causes depolarization associated with increased membrane conductance, attributable to a rise in sodium permeability. Surprisingly, atropine blocks the excitant actions of 5-HT as well as ACh, but there is a desensitization or tachyphylaxis to 5-HT that is not associated with any alteration of ACh sensitivity, and LSD also selectively blocks the 5-HT response (Gerschenfeld and Stefani, 1966). These results are consistent with the generally held belief that 5-HT receptors, like those for ACh at the skeletal neuromuscular junction, are located on the membrane surface. However, Twarog (1960, 1967) has obtained evidence that in some molluscan smooth muscle this is unlikely to be true. When the anterior byssus retractor muscle of the lamellibranch *Mytilus edulis* is stimulated by ACh or electrical stimulation, tension is maintained after stimulation has ceased. This maintained tension, known as catch, does not involve a continuation of the active state of the contractile mechanism (Twarog, 1954) and is not accompanied by breakdown of high-energy phosphates (Nauss and Davies, 1966). 5-HT abolishes this catch phenomenon without altering membrane potential, which is indeed also not altered during catch (Twarog, 1960). Twarog (1968) has speculated that 5-HT inside the muscle fiber acts to increase the binding of intracellular calcium to a relaxing system; an action of 5-HT to increase calcium binding to certain lipids has been demonstrated by Woolley and Campbell (1960).

F. Adenine Nucleotides

It is now well established that there exists a noncholinergic, non-adrenergic parasympathetic inhibition of gastrointestinal smooth muscle mediated by intramural nervous elements (see G. Campbell, 1970). It has long been known that under certain circumstances relaxation of the stomach can be caused by stimulation of the vagus; this response is best seen when atropine has been applied to eliminate the cholinergically mediated excitatory response, or by stimulation of various structures in the brain (e.g., Semba *et al.*, 1964). Paton and Vane (1963) demonstrated that the inhibitory response of the stomach to transmural stimulation was not prevented by treatment with adrenergic-blocking drugs, although these drugs did block the inhibitory response to stimulation of the perivascular sympathetic nerve supply to the stomach. Similar relaxations resistant to adrenergic blockade have since been obtained from small and large intestine of many mammals, including man. (G. Campbell, 1970). Electrophysiological studies of guinea pig taenia coli have shown that transmural stimulation causes a hyperpolarization of the smooth muscle membrane, with a latent period of about 150 msec and lasting about 1 sec. The response is unaffected by treatment with quanethidine or bretylium and is certainly neurogenic (Burnstock *et al.*, 1964b; Bennett *et al.*, 1966b; Bülbring and Tomita, 1966, 1967). The hyperpolarizing response is evidently related to an increase in potassium conductance of the subsynaptic muscle membrane, analogous to the action of ACh on the heart (Bennett *et al.*, 1966b).

The vagal preganglionic pathway for this inhibitory system has already been discussed, in connection with 5-HT. From the results of Bülbring and Gershon (1967), it appears that 5-HT may be the transmitter acting on the inhibitory ganglion cells for a fraction of the vagal fibers; a partly cholinergic transmission is indicated by the partial blockade of vagal relaxation, which can be achieved with the nicotinic-blocking drug pentolinium.

Burnstock *et al.* (1970) have recently reported evidence indicating that adenosine triphosphate (ATP) or a related nucleotide is the transmitter substance released by the nonadrenergic inhibitory neurons in the gut. Their evidence is similar to that by which ACh and noradrenaline have been recognized as transmitters and is as strong as was the evidence for these agents until a few years ago. The first criterion, presence of the transmitter candidate in the nerve terminals associated with the inhibitory transmission, is certain to be true for ATP, in view of its wide distribution in all tissues. More specifically, axon profiles in the toad lung and in the gut of both toad and mammals have been found to contain large granular vesicles unlike those of cholinergic or adrenergic axons (Rogers and Burnstock, 1966; M. R. Bennett and Rogers, 1967), but the contents of these vesicles have not yet been identified. The second criterion for ATP or a related compound has been satisfied. Burnstock and his collaborators have shown that stimulation of the vagal nonadrenergic inhibitory innervation of the stomachs of both toads and guinea pigs is associated with an increase in the venous efflux of compounds which are likely to be breakdown products of adenine nucleotides. In contrast, stimulation of the cholinergic gastric sympathetic innervation did not increase the resting efflux of nucleosides; the stimulated release of nucleosides appeared to be a specific concomitant of activity in the nonadrenergic inhibitory fibers.

The third criterion, that the putative transmitter mimic the activity of nerve stimulation, has also been satisfied. ATP and ADP were the most potent of a series of purine and pyrimidine derivatives tested for inhibitory activity on the isolated guinea pig taenia coli and in twelve other gut segments previously shown to contain nonadrenergic inhibitory nerves; ATP caused inhibition not blocked by tetrodotoxin. It was already shown that ATP causes hyperpolarization of taenia coli cells (Imai and Takeda, 1967a; Axelsson and Holmberg, 1969).

Perhaps the most convincing evidence for the parallel action of ATP and the inhibitory transmitter comes from study using the rabbit ileum, where the response to ATP was tachyphylactic. When tachyphylaxis to ATP had been produced, there was a consistent antagonism of responses to nonadrenergic inhibitory nerve stimulation, but not of responses to adrenergic nerve stimulation. Such a finding is hard to reconcile with any hypothesis other than an action of ATP on the same receptors as mediate the inhibitory response to nerve stimulation. Quinidine was also found to antagonize the response to ATP and to nonadrenergic nerve stimulation, although this effect was to be found only with concentrations greater than those preventing the effects of applied noradrenaline. The parallel action on both responses also supports the hypothesis that ATP (or AMP/ADP) may be the inhibitory transmitter substance.

The effect of ATP on gut segments was not solely inhibitory. In higher concentrations than those causing inhibition, ATP was found to cause an excitatory response in guinea pig ileum, rabbit stomach, and mouse colon, and in rat and toad stomach an excitatory response followed the inhibitory (Burnstock *et al.*, 1970). Such findings do not disturb the hypothesis that ATP is the inhibitory transmitter (ACh and noradrenaline have both excitatory and inhibitory transmitter functions) and in fact raise the question whether ATP may at certain sites act as an excitatory transmitter rather than an inhibitory transmitter. However, most other suggestions for a transmitter role of ATP have it as an inhibitory transmitter.

J. Hughes and Vane (1967) observed a relaxation of rabbit portal veins by low-frequency transmural stimulation, the same sort of stimulation as evokes the nonadrenergic inhibitory response in gut. This effect was evidently neurogenic, since it was blocked by local anaesthetics and by tetrodotoxin. It was, however, blocked by neither α- nor β-adrenoceptor blocking agents, nor by anticholinergic, antihistiamine, or 5-HT blocking agents. These authors in fact suggested adenine nucleotides as the possible inhibitory agent.

Another neurogenic inhibitory response that is insensitive to agents that block the action of the known transmitters is the regional vasodilatation that occurs with antidromic stimulation of sensory nerves, e.g., the local flare which follows local skin irritation. Holton (1959) has presented evidence that ATP is released from mammalian afferent dendritic terminals in concentrations sufficient to cause regional vasodilatation. The possibility of a vasodilator function of adenine nucleotides in skeletal muscle has been thoroughly reviewed by Haddy and Scott (1968), who considered this possibility unlikely because of the rapid inactivation of adenosine to inosine. This argument would not apply, however, if an adenine nucleotide functioned as a neurotransmitter rather than a local hormone, and it has recently been demonstrated that there is indeed release of ATP from exercising frog nerve–skeletal muscle preparations (I. A. Boyd and Forrester, 1968).

It has been recognized for some time that adenine nucleotides have both excitatory and inhibitory actions on vascular smooth muscle (see A. P. Somlyo and Somlyo, 1970). In the rabbit pulmonary bed, for example, vasoconstrictor effects of ATP are pronounced (Lunde *et al.*, 1969), although the vasodilator effect of ATP predominates in the 5-HT-constricted canine pulmonary circulation (Rudolph *et al.*, 1959). There is, however, little evidence to indicate whether in fact ATP or a related compound functions also as an excitatory transmitter.

G. Polypeptides and Prostaglandins

There exist a number of substances which are found in the body, exert powerful effects on smooth muscle, and appear likely to function as hormone modulators of muscle function rather than as transmitters of excitation or inhibition from nerve to muscle. Of course, general or local hormonal action may also be a function of any or all of the substances already discussed as likely but as yet unproved transmitter agents—histamine, 5-HT, and adenine nucleotides. Similarly, it would not be overly surprising if in fact any of the present arbitrary collection of autacoids and hormones turns out to serve a more important role in the control and integration of smooth muscle activity than is presently supposed. In this section, we shall consider first the actions on muscle of angiotensin, the neurohypophyeal peptides, and plasma kinins; and second the actions of prostaglandins, lipid substances whose actions on smooth muscle presumably reflect important biological functions; just what these functions are remains quite unclear.

1. ANGIOTENSIN

Page and Bumpus (1961) and Peart (1965) have dealt extensively and authoritatively with advances in the field of the renin–angiotensin system, and A. P. Somlyo and Somlyo (1970) have recently considered in some detail the actions of angiotensin on vascular smooth muscle. Angiotensin is formed by hydrolysis from a plasma globulin, angiotensinogen, by the action of the enzyme renin, which is secreted by the kidney. It acts to constrict arteries and veins, the main effect being on peripheral resistance vessels. For example, the central artery of the rabbit ear is less sensitive to angiotensin than the more distal arteries (de la Lande and Waterson, 1968). Angiotensin action is associated with considerable tachyphylaxis, which varies depending on the particular vascular smooth muscle tested. This tachyphylaxis is reversed *in vivo* by the action of the enzyme angiotensinase (Khairallah *et al.*, 1966). During normal pregnancy, which is associated with an increased tolerance to angiotensin, this enzyme is found in increased quantity (see A. P. Somlyo and Somlyo, 1970).

The effect of angiotensin on vascular strips is evidently direct and not secondary to the release of noradrenaline from nerve endings. Reserpinized animals remain sensitive to angiotensin, and cocaine in sufficient concentration to block the effects of tyramine does not block the effects of angiotensin (Türker and Karahüseyinoglu, 1968). Moreover, cross tachyphylaxis does not occur in rat aortic strips between tyramine and angiotensin (Palaič and Khairallah, 1968), and completely nerve-free placental vessels are constricted by angiotensin (Ward and Gautieri, 1968). Studies in man of the effects of α-adrenergic receptor blockade and sympathetic denervation have indicated that catecholamines mediate little or none of the pressor actions of angiotensin (Scroop and Whelan, 1968). Nevertheless, some of the constrictor actions of angiotensin are probably neurally mediated, perhaps via reflexes. Thus, the effect of intravenous (but not intraarterial) angiotensin on human hand blood flow is inhibited by quanethidine or phenoxybenzamine (Henning and Johnsson, 1967). In cats, angiotensin causes the release of adrenal medullary catecholamines (Feldberg and Lewis, 1965), an effect which correlates with the pressor activity of angiotensin (Staszewska-Barczak and Konopka-Rogatko, 1967) and which is related to a direct depolarizing action on the chromaffin cells (Douglas *et al.*, 1967). It also causes the release of aldosterone from adrenal cortical cells (Gross, 1968). In relatively high concentrations, it apparently acts on the amine pump of adrenergic nerve terminals, inhibiting the uptake of noradrenaline by rat arteries *in vitro* (Palaič and Khairallah, 1967), but it seems unlikely that such a cocaine-like effect is significant to its action (see A. P. Somlyo and Somlyo, 1970).

Angiotensin also acts to contract most nonvascular smooth muscle preparations *in vitro*, an effect which seems most likely to be secondary to excitation of nervous elements rather than direct action on the smooth muscle cells. For example, Khairallah and Page (1963) found that the effect of angiotensin on the guinea pig ileum is enhanced by anticholinesterases and depressed by atropine and other agents that depress cholinergic transmission. These results are most easily interpreted in terms of an action on the cholinergic intramural ganglion cells. In the cat stellate ganglion, however, Aiken and Reit (1969) found that the ganglion cell stimulating action of angiotensin is confined to the adrenergic cells; there was no stimulation of the cholinergic neurons mediating sweat secretion in the foot pads. It is possible, of course, that the cholinergic ganglion cells in gut differ in their sensitivity from those in sympathetic ganglia.

With regard to its mode of action at the cellular level, little is known.

According to A. P. Somlyo and Somlyo (1970) angiotensin causes depolarization of vascular smooth muscle, and angiotensin tachyphylaxis
is associated with repolarization. Such results suggest an excitatory action
secondary to membrane depolarization resulting from a nonspecific increase in conductance to cations, presumably including calcium, since
the contractile response to angiotensin persists in strips of arteries depolarized by high-potassium solutions and then partially relaxed by amyl
nitrate (Keatinge, 1966). However, Shibata and Briggs (1966) have
reported that the rabbit aortic strip can be partially contracted by angiotensin without any accompanying potential change, and Keatinge (1966)
found that in sheep carotid arterial strips the depolarization produced
by angiotensin ceased and repolarization started at a time when the
contraction was still developing. Parallel recordings of contraction and
membrane potential from smooth muscle are difficult to interpret, since
one can never be sure of homogeneity of cell response, but these observations must suggest that at least a portion of the action of angiotensin
may not be secondary to a simple combined sodium–potassium–calcium
permeability increase.

The action of angiotensin on neurons in sympathetic ganglia is also
not clear. In substimulant doses, angiotensin acts greatly to potentiate
transmission in the superior cervical ganglion, and in some preparations,
this enhancement was found to last for several hours (Panisset *et al.*,
1966). Very low doses tend to inhibit transmission, an effect abolished
by dihydroergotamine, a finding which could implicate noradrenaline
release in mediating this effect (Haefely *et al.*, 1965a; Panisset *et al.*,
1966). It is of considerable interest that the tachyphylaxis of ganglion
cells produced by histamine extends to angiotensin (but not to bradykinin) and, correspondingly, mepyramine, a histamine antagonist, decreases ganglionic responses to angiotensin. However, angiotensin
tachyphylaxis does not extend to histamine. On the basis of these findings,
Lewis and Reit (1966) have suggested that the response to angiotensin
depends upon activation of the histamine receptor as well as the specific
angiotensin receptor. The facilitatory action of angiotensin on transmission at concentrations below those necessary for direct stimulation is
most easily accounted for in terms of facilitation of ACh release from
cholinergic nerve terminals; direct evidence for this is lacking.

2. VASOPRESSIN AND OXYTOCIN

The actions of the neurohypophyseal hormones on smooth muscle
have been reviewed in some detail by A. P. Somlyo and Somlyo (1970).

Not only is there considerable species variation in responses of various tissues to these agents, but there is also enormous variation in response according to the hormonal state of the animal. Modification of the peptide structure alters activity. Indeed, it has been possible to make analogs of vasopressin with a high ratio of pressor activity to antidiuretic activity (Berde *et al.*, 1964). Interpretation of much of the literature on oxytocin has been confused by the use of chlorbutanol as a preservative in a number of commercial preparations. This compound is a vasodilator and smooth muscle relaxant and accounts for the entire vasodepressor effect of commercial synthetic oxytocin in cats, for example (R. L. Katz, 1964). The sympathetic nervous system also influences the mammalian vasodilator activity of oxytocin. Lumbar sympathectomy, reserpinization, or treatment with dihydroergotamine, an α-adrenoceptor blocker, eliminates this action and converts it to vasoconstrictor in dog limb preparations. Stimulation of the sympathetic trunk (Haigh *et al.*, 1965) or the infusion of adrenaline after sympathectomy (S. Lloyd and Pickford, 1967) reestablishes the vasodilator action. On the other hand, the depression of blood pressure in birds by oxytocin is not modified by sympathectomy or reserpinization (S. Lloyd and Pickford, 1961; Wooley and Waring, 1958). In the spleen, vasopressin appears to potentiate the effect of catecholamines released during nerve stimulation (P. J. Bernard *et al.*, 1968). It evidently does not affect catecholamine storage, release, or uptake (Hertting and Suko, 1966).

The mechanism(s) by which oxytocin and vasopressin exert their effects on uterine, vascular, and intestinal muscle is perhaps even less well understood than for other agents that excite or inhibit smooth muscle. The mediation of effects by specific and probably similar receptors is indicated by the ability of methylated derivatives of neurohypophyseal peptides to act as inhibitors of oxytocin (Bisset and Clark, 1968) and rat pressor (Krejcí *et al.*, 1967; Law and du Vigneaud, 1960) activity and also to inhibit the action of neurohypophyseal peptides on smooth muscle *in vitro* (Vávra, Krejcí, and Kupková, cited in Rudinger, 1968).

A peculiarity of these hormones is their marked potentiation by magnesium or manganese. Schild (1969) recently studied extensively the potentiation of responses of depolarized myometrium to neurohypophyseal peptides by cobalt, manganese, nickel, zinc, magnesium, and iron. The results suggested that the peptides and their receptors combine with these metals to form ternary coordination complexes which mediate the hormonal effect, the intrinsic activity of the cation–receptor–peptide complex being greater than that of the binary receptor–peptide complex. This accounts for the observation that in magnesium-free solution there is a decreased maximal response of depolarized myometrium and avian pulmonary artery to vasopressin and of canine iliac arteries to oxytocin (Schild, 1969; A. P. Somlyo and Somlyo, 1970). The action of magnesium is manifest on inhibitory as well as the excitatory actions of neurohypophyseal peptides, as is the blocking actions of the peptide inhibitors, and Somlyo and Somlyo (1970) therefore suggested a common receptor mediating both excitatory and inhibitory activity.

Although oxytocin can produce contraction in uterine smooth muscle completely depolarized by potassium (Evans *et al.*, 1958; Schild, 1969), it is likely that under physiological conditions oxytocin acts through a change in membrane potential, involving an increase in sodium permeability. When the membrane potential is near threshold for the discharge of action potentials, the depolarizing action can produce spike discharges (Marshall, 1964). The effect of oxytocin is markedly enhanced by increase of sodium concentration, depressed in low sodium, and ineffective at below 10% normal sodium (Marshall, 1963, 1964). However, oxytocin effect is also diminished in low calcium (Berger and Marshall, 1961; Marshall and Csapo, 1961; Marshall, 1968), and it seems likely that oxytocin also mobilizes calcium ions as carriers of inward charge. Kao (1967) has suggested that oxytocin alters the relation between membrane potential and inactivation of a membrane sodium carrier system involved in action potential generation in such a way that oxytocin reduces inactivation and promotes action potential generation.

3. Plasma Kinins—Bradykinin and Kallidin

These two peptides are both formed from a common precursor or precursors (see Pierce, 1968) in the plasma globulin fraction (kininogen) by the action of trypsin and plasmin, or by a group of endogenous proteolytic enzymes (kallikreins) which occur in urine, saliva, blood, etc., or by proteolytic enzymes in certain snake venoms. Structurally, bradykinin is a nonapeptide of known composition (Boissonnas *et al.*, 1960), and kallidin is identical but for an additional N-terminal lysine residue (Webster and Pierce, 1963). These are but two of a number of peptides, which also includes eledoisin, physalaemin, and substance P, that have similar pharmacological properties. For the whole group, the term "kinin" has now been adopted (see Schacter, 1968). A physiological role in inflammation seems likely (see reviews by Schacter, 1964; Erdös, 1966; Melmon and Cline, 1967; Symposium, 1968).

High levels of bradykinin have been found in umbilical cord blood at delivery. Since bradykinin has marked constrictor effects on the umbilical vessels but a dilator effect on the pulmonary artery, it has been suggested to function in fetal to neonatal adjustments of the circulation (A. G. M. Campbell *et al.*, 1968; Melmon *et al.*, 1968).

The plasma kinins act in extremely low doses on peripheral vascular smooth muscle to cause vasodilation. On extravascular smooth muscle, they excite at some sites (e.g., rat uterus, guinea pig ileum, bronchial smooth muscle) and at other sites inhibit (e.g., rat duodenum). These actions are not blocked by cholinergic or adrenergic blocking agents or denervation (reviewed by A. P. Somlyo and Somlyo, 1970). They also stimulate sensory nerve endings, causing pain (D. Armstrong *et al.*, 1957), and insufficient doses, stimulate adrenal medullary secretion associated with depolarization of the chromaffin cells (Douglas *et al.*, 1967). In autonomic ganglia, bradykinin acts to stimulate adrenergic but not cholinergic ganglion cells (Aiken and Reit, 1969).

With regard to mechanism of action little is known. Keatinge (1966) found that contraction of sheep carotid arterial strips, induced by bradykinin, was only at first associated with depolarization. With bradykinin, as with angiotensin, repolarization started at a time when the contraction was still developing. There was also tachyphylaxis to bradykinin, as to angiotensin, but no cross tachyphylaxis between the two peptides, indicating distinct receptors. Bradykinin also resembled angiotensin in its ability to contract arterial strips depolarized by high potassium, and then partially relaxed by amyl nitrite, without any detectable alteration of membrane potential. These results not only indicate a direct action on muscle contraction, perhaps via calcium mobilization, but imply a dissociation of this action from that causing depolarization; i.e., it seems unlikely that the pharmacomechanical coupling could be simply a consequence of a small calcium conductance associated with sodium and presumably potassium conductance.

4. PROSTAGLANDINS

Originally discovered independently, in 1933 and 1934, by Goldblatt and by von Euler (see von Euler and Eliasson, 1967) as vasodepressor and smooth muscle contracting material in human seminal fluid and extracts of prostate, these lipid substances have recently attracted a great deal of attention. Their chemistry, natural distribution, and pharmacological actions have been summarized recently in a number of books and reviews (Pickles, 1967; Bergström *et al.*, 1968; Ramwell and Shaw, 1968, 1970). It is now evident that these agents account for the activity of such tissue extracts as "Darmstoff" (Suzuki and Vogt, 1965), the "irins" (Ambache *et al.*, 1966), and the vasodepressor material of renal medulla (Daniels *et al.*, 1967). Some fourteen types have been identified, and they are widely distributed in mammalian tissues (Samuelsson, 1965; Horton, 1965). Their actions are many and vary with tissue, species, and particular type of prostaglandin. Polyphloretin phosphate, a polymeric phosphorylated polyanionic derivative of phloridzin, has recently been found to be an antagonist to many of the actions of prostaglandins on smooth muscle (Eakins *et al.*, 1970).

The only function for prostaglandins that seems generally accepted is a hormonal one; produced by the male, they act to influence the tone of the female reproductive tract when absorbed from the vagina (Sandberg *et al.*, 1963; Eliasson and Posse, 1965; Horton *et al.*, 1965).

Prostaglandins are present in the central nervous system (Holmes and Horton, 1968), and a release of prostaglandin-like substances from many regions of the brain and spinal cord has been described. Moreover, iontophoretically applied prostaglandins of the same type have selectively excitatory or inhibitory actions on neurons in the brain stem and cerebral cortex (see Phillis, 1970). They have been shown to depolarize dorsal root nerve terminals as well as motoneurons in the isolated toad spinal cord (Phillis and Tebēcis, 1968).

It therefore seems likely that the compounds may have some function related to transmission at synapses, influencing the release of transmitter or the excitability of the postsynaptic membrane (Phillis, 1970). Release of prostaglandins as a result of nerve stimulation has been shown to be secondary to the postsynaptic action of transmitter rather than directly from the excited nerve terminals (reviewed by G. Campbell, 1970). Thus, sympathetic nerve stimulation causes release of prostaglandins from the spleen, and to an extent exceeding the amount extractable, indicating that the prostaglandin was formed during nerve stimulation. However, injections of catecholamines into the spleen also cause prostaglandin release, and the α-adrenergic blocking drugs phentolamine and phenoxybenzamine, which do not prevent the release of catecholamines but prevent activation of postsynaptic receptors, do block the release of prostaglandins (Davies *et al.*, 1966, 1968; Ferreira and Vane, 1967).

Similarly, but in this case for a cholinergic system, excitation of the gastrointestinal tract by vagus nerve stimulation is associated with the release of prostaglandins, but this effect is blocked by hyoscine and spontaneous release is accelerated by ACh (Coceani *et al.*, 1967; A. Bennett *et al.*, 1967; Bartels *et al.*, 1968).

On the other hand, release of prostaglandins from the rat diaphragm secondary to stimulation of the phrenic nerve is not blocked by *d*-tubocurarine (Ramwell *et al.*, 1965; Ramwell and Shaw, 1967). However, it is probable that this release is from vascular smooth muscle or adipose tissue following the excitatory action of catecholamines released from sympathetic adrenergic nerve fibers in the phrenic nerve (G. Campbell, 1970).

The functional role of the prostaglandins released as a consequence of transmitter action is still quite unclear. One possibility is that in some instances they act as local tissue hormones in a negative feedback mechanism, moderating the excitatory influence of the innervation (G. Campbell, 1970). Thus, the prostaglandins (E_1 and E_2), which have been claimed to be released into the rat stomach lumen during vagus nerve stimulation, are potent inhibitors of gastric secretion (Robert *et al.*, 1967), and the amount of prostaglandin released from rat epididymal fat pads following sympathetic nerve stimulation would be adequate to inhibit completely the free fatty acid-mobilizing effect of nerve stimulation (Berti and Usardi, 1964). However, prostaglandins cause excitation rather than inhibition of most intestinal smooth muscles, exceptions being inhibition by prostaglandin E_2 of the contraction of the circular muscle in human stomach (A. Bennett *et al.*, 1968) and guinea pig ileum (Kottegoda, 1970). Both inhibitory and excitatory effects of catecholamines on a number of smooth muscle preparations are antagonized by prostaglandins, but usually after a period of potentiation (Clegg, 1966). On the other hand, in spleen, prostaglandin E_2 is secreted, but causes no alteration in the response to nerve stimulation or catecholamines (Davies and Withrington, 1968).

There has been little analysis of the mechanism of prostaglandin action on muscle. As might be expected from the excitatory actions of these agents on some central neurons, there is a neurotropic component of the action on intestinal smooth muscle; in guinea pig ileum and rabbit jejunum, atropine can block the stimulant action of a dose of prostaglandin E_1 that gives a response about 50% of the maximum, but not the effects of higher doses. On isolated rat fundus, on the other hand, the response to prostaglandin E_1 was not affected by atropine or hexamethonium (Coceani and Wolfe, 1966), and the action is presumably entirely musculotropic. How either neurotropic or musculotropic actions are exerted is still quite unknown. Butcher (1970) has recently reviewed the evidence that prostaglandins exert these actions in a variety of tissues via increased formation of cyclic AMP. For muscle actions, the evidence is still very scanty.

H. Receptor Blocking Agents

There are a number of drugs that display little or no action of their own on muscle but serve to block, more or less selectively, the action

of agonists such as ACh, catecholamines, histamine, 5-HT, and neurohypophyseal hormones. Often, these drugs are more selective than the
agonists they oppose. Atropine, for example, inhibits the action of ACh
on muscarinic receptors at far lower doses than it antagonizes the action
of ACh on nicotinic receptors. The two kinds of adrenergic receptors,
α and β, are better defined in terms of the drugs that block their activity
than in terms of the relative potency of different catecholamines as
originally proposed by Ahlqvist (1948). Similarly, different classes of
drugs oppose the neurotropic and musculotropic actions of 5-HT (see
above).

As an example of the selectivity of receptor blocking agents, Speden
(1969) has recently found that a component of the vasoconstrictor
response of arteries to sympathetic nerve stimulation cannot be
blocked by the α-adrenergic blocking agents phentolamine, phenoxybenzamine, and dibenamine, although the response is certainly mediated
by noradrenaline, since it is blocked by reserpinization and by bretylium
(reversed by amphetamine) in the usual way and potentiated by cocaine. Thus, there evidently exist α-receptors that cannot be blocked
by α-receptor blocking agents. More familiar is the classification of
β-adrenoceptors into β_1 and β_2 on the basis of the relatively potent inhibition of β-adrenergic activity in the heart and adipose tissue by ICI-50172,
which has little effect on receptors in blood vessels or trachea (Barrett *et
al.*, 1968), while butoxamine (Levy, 1966; Burns *et al.*, 1967) and H
35/25 (Levy and Wilkenfeld, 1969) show selective affinity for β_2-recep-
tors, i.e., those in blood vessels, trachea, and skeletal muscle (Bowman
and Nott, 1970). On the other hand, there may be a high degree of
nonselectivity of blocking agents in terms of cross sensitivity of different
types of receptors. Dibenamine (dibenzyline), for example, discriminates
between α- and β-adrenoceptors, acting on β-receptors to a negligible
extent, if any, but also blocks receptors for ACh (muscarinic), histamine,
and 5-HT (e.g., Nickerson, 1949; Furchgott, 1954; H. Boyd *et al.*, 1963).

1. COMPETITIVE BLOCKADE

Receptor blocking agents acting at any one receptor can be subclassi-
fied according to the mode of interaction with the receptors. Generally,
two kinds of interaction are recognized, competitive and noncompetitive.
This classification is based primarily on how the agents act to alter
the dose–response curve obtained by measuring responses of an
assay tissue to increasing concentrations of the agonist drug. Analysis
is in terms of receptor theory, of which full consideration is beyond
the scope of this discussion, and for which the reader is referred to

reviews by Waud (1968) and Werman (1969) and the book by Ariëns (1964). However, taking a simplistic view, that receptor binds only one molecule of agonist or inhibitor and using the usual occupancy rather than rate theory, one obtains for competitive inhibition the formula,

$$\frac{[AR]}{R_t} = \frac{1}{1 + \dfrac{K_a}{[A]}\left(1 + \dfrac{[I]}{K_i}\right)} \tag{3}$$

where $[AR]/R_t$ is the proportion of receptors occupied by agonist, K_a is the dissociation constant of the agonist–receptor combination, $[A]$ is the concentration of agonist, $[I]$ is the concentration of inhibitor, and K_i to the dissociation constant of the inhibitor–receptor complex. If tissue response is proportional to the proportion of receptor occupied by agonist, then the Lineweaver–Burk plot, i.e., a plot of (response)$^{-1}$ versus $[A]^{-1}$, gives straight lines which intercept at $[A]^{-1} = 0$. With noncompetitive inhibition, on the other hand, addition of inhibitor leads to an apparent reduction of R_t, and the Lineweaver–Burk plot gives lines which intercept the abscissa at $[A]^{-1} = -[K_i]^{-1}$. Qualitatively, any demonstration that the maximal response of the tissue is depressed by an agent must be considered proof of noncompetitive inhibition, or antagonism, and excludes simple competition as a mechanism. Unfortunately, the converse does not apply. Straight-line Lineweaver–Burk plots that intercept at $[A]^{-1} = 0$ do not prove competitive inhibition unless it can be demonstrated that response is indeed linearly proportional to the occupancy of receptors by agonist. For example, it can be shown that even if conductance change were proportional to $[AR]/R_t$, consideration of depolarization of striated muscle by ACh, rather than conductance change, would give Lineweaver–Burk plots characteristic of competitive inhibition, even for non-competitive inhibitors (e.g., Hubbard, *et al.*, 1969a; Werman, 1969). In order to circumvent the problem of, in general, an unknown relation between response and the proportion of receptors occupied by agonist, a useful method is that of equal responses. From Eq. (3), equal responses of the tissue, at various concentrations of inhibitor, should represent equal arguments of the function on the right hand side, i.e.,

$$\frac{K_a}{[A]}\left(1 + \frac{[I]}{K_i}\right) = C$$

rearranging,

$$[A] = \frac{K_a}{C}\left(1 + \frac{[I]}{K_i}\right) \tag{4}$$

where C is a constant. Hence, graphs of [A] versus [I] for arbitrary levels of tissue response should give straight lines with an intercept at [A] = 0, [I] = $-K_2$. Moreover, if [A]′ is the concentration in the presence of inhibitor that gives a response equal to that produced by [A] in the absence of inhibitor, then

$$\frac{[A]'}{[A]} = 1 + \frac{[I]}{K_i} \tag{5}$$

This relation was applied by Jenkinson (1960) to demonstrate that the action of tubocurarine to block ACh action at the frog endplate was compatible with competition and to obtain the affinity constant of the receptor for d-tubocurarine. From the relation in Eq. (5), one can proceed further. Rearranging and taking logarithms,

$$\log\left(\frac{[A]'}{[A]} - 1\right) = \log [I] - \log K_i \tag{6}$$

A graph of log {([A]′/[A]) −1} versus log [I] is known as an isobole and gives a relatively sensitive measure of whether in fact the kinetics of the inhibitor interaction are competitive, especially if inhibitor concentrations extend over several orders of magnitude (Ariëns, 1964). However, although an inhibitory drug may fully meet these criteria for a competitive mode of action, whether it does in fact act competitively generally remains impossible to determine. Thus, it now seems clear that maximal responses of tissue to agonists may occur with only a minute fraction of receptors occupied. Deviation of an isobole from linearity for a noncompetitive inhibition of the receptor (as distinct from other noncompetitive actions on muscle contraction) would not be expected until one was working at concentrations of inhibitor and agonist so high as to call into question the possibility of quite nonspecific actions of both.

Ariëns (1964, 1967) has pointed out that for many of the familiar agonist–competitive antagonist pairs, such as α-adrenergic and α-adrenergic blocking drugs, cholinergic and anticholinergic drugs, histaminergic and antihistaminic drugs, and 5-HT and 5-HT antagonists, there is little obvious relationship between the structure of the agonists and their respective antagonists. This situation contrasts with the obvious structural resemblance between vitamins and antivitamins and between enzyme substrates and competitive antagonists thereof. Moreover, the drug antagonists show much less selectivity than antivitamins or antimetabolites. He has therefore suggested that the competitive antagonists may bind to sites close to but not identical with the receptors proper. The multipotent action of chlorpromazine, which blocks simultaneously α-adrenergic, cholinergic, histaminic, and 5-HT receptors, is then explained on the basis of structural similarity of accessory areas of the different receptors and a relative lack of those groupings on the inhibitor, which by interacting with receptors by the same process as binds the normal agonist, tend to confer specificity. The fact that interaction of the blocking agents is clearly competitive in terms of kinetic interaction, i.e., isoboles are linear with a slope of 45° (see Ariëns, 1964, 1967; Burgen and Spero, 1968), does not militate against this hypothesis; it is only necessary to suppose that combination of receptor with agonist entails displacement of the antagonist. The term "competitive" continues to be appropriate, provided one takes a broad view of the receptor as an entity, and considering the practical difficulties involved in distinguishing between the two possible modes of competitive antagonism, it may hardly be worthwhile attempting the distinction. It would seem, however, that the term "surmountable," which avoids any conclusion as to the exact mode of action, might be better applied to describe the effect of a blocking drug that follows competitive kinetics.

There are, of course, many examples where selective antagonism of an agonist seems very likely to be on a true competitive basis. For these drugs there is not only a fit to kinetics consistent with competition, but a marked structural resemblance between the antagonist and the agonist. Examples are the peptide analogs of neurohypophyseal hormones (for references, see A. P. Somlyo and Somlyo, 1970), quaternary derivatives of 5-HT (Gyermek, 1964b); β-adrenergic blocking agents such as dichlorisoproterenol, pronethalol, and propranolol (see Ariëns, 1967; Blinks, 1967); and nicotinic blocking agents such as hexamethonium, and indeed, d-tubocurarine, where the separation of the two quaternary nitrogens is the same as the optimum for bis-onium compounds such as decamethonium and succinylcholine, which are agonists rather than competitive blockers. The structural resemblance of these antagonists to the agonists is to close that it comes as no surprise that, for example, the quaternary derivatives of 5-HT stimulate 5-HT receptors before blocking (Gyermek, 1964b), and in denervated skeletal muscle, where ACh receptors are evidently slightly different from normal subsynaptic receptors, tubocurarine can cause a contracture (see p. 63). Similarly, in its effect on nerve terminals, tubocurarine can oppose ACh, but also acts itself in the same manner as ACh (Hubbard *et al.*, 1965). The action of bis-onium ions on nicotinic ACh receptors exemplifies another difficulty. Not only is the distinction between agonist and antagonist an extremely fine one, as exemplified by the demonstrable ACh-mimicking action of d-tubocurarine, but a pure ACh-mimicking action is manifest in a neuromuscular blockade due to receptor desensitization that is only slightly different from a competitive blockade. There is no reason why a desensitization type of blockade should not be synergistic with a competitive one. Indeed, tubocurarine acts only to intensify a blockade related to receptor desensitization (Thesleff, 1958). In terms of kinetic theory, of course, competitive blocking action is to be expected of agents which have affinity for receptor, but relatively little or no intrinsic activity, and one expects a continuous gradation of compounds between those showing predominantly agonist activity and those with predominantly competitive blocking activity. Where agonist activity is also associated with receptor desensitization, both types of action will tend to block. It is therefore likely that the neuromuscular blockade produced by a large number of tertiary and quaternary ammonium compounds may be by a mixture of competitive and desensitizing actions. Antihistamines, for example, have a blocking action which is synergistic with tubocurarine but resistant to prostigmine (Choksey and Jindal, 1965); the resistance to anticholinesterase is characteristic of depolarizing blockers and therefore suggest an admixture of desensitizing action as well as curare-like action. Similar mixed blocking actions might well occur at other receptors where there is prominent tachyphylaxis, perhaps similar in nature to the desensitization to ACh that has been observed at the skeletal junction, such as the neuroreceptors for 5-HT and some of the receptors for adenine nucleotides. It should also be pointed out that there are numerous examples of competitive blocking agents having agonist actions on receptors other than those in which the blocking action is most manifest. This is true especially of α-adrenergic competitive blockers. Ergot alkaloids, for example are well known also as oxytocic agents and are indeed general stimulators of smooth muscle. They also have potent actions on the central nervous system uncorrelated with α-adrenergic blocking activity (see review by Nickerson and Hollenberg, 1967). Tolazoline and phentolamine are competitive α-adrenergic blockers that also have sympathomimetic activity, (largely by stimulation of β-receptors), parasympathomimatic activity (including atropine sensitive stimulation of the gastrointestinal tract), and histamine-like activity (including stimulation of gastric secretion and peripheral vasodilatation) (see Nickerson and Hollenberg, 1967). Compounds very similar chemically are employed as antihistamines. The benzodioxans, including piperoxan, constitute another class of α-adrenergic blockers. These also stimulate many kinds of smooth muscle, including bronchi, intestine, uterus, and both coronary and peripheral blood vessels, leading to an increase of blood pressure unless the blood pressure is raised by circulating catecholamine (see Nickerson, 1949).

The best studied and perhaps best established example of competitive receptor blockade is that produced by *d*-tubocurarine at the skeletal neuromuscular junction. There can be no doubt that tubocurarine does act to block ACh action on receptors. Nevertheless, the question is often raised whether this action accounts entirely for its neuromuscular blocking activity. The answer seems to be that it does. When added to a skeletal muscle in increasing concentrations, tubocurarine progressively reduces in parallel the amplitude of extracellularly recorded EPPs and potentials evoked by ionophoresed ACh (e.g., del Castillo and Katz, 1957; Beránek and Vyskočil, 1967). MEPPs and quantal components of the EPP are also depressed in parallel with the EPP when quantal release is inhibited by raised magnesium (Martin, 1955; Beránek and Vyskočil, 1967). Moreover, neuromuscular blockade based upon deficiency of ACh content of quanta or postsynaptic blockade of ACh action begins when there is an inhibition by about 80%; i.e., when MEPPs are about 20% the usual amplitude (Elmqvist *et al.*, 1964; Elmqvist and Quastel, 1965b). In human muscle *in vitro*, a depression of this magnitude is produced by a bath concentration of about 4×10^{-7} gm/ml tubocurarine (Elmqvist and Quastel, 1965b). This corresponds to about half the usual dose used in surgery. Thus, whatever effects tubocurarine may have presynaptically, its neuromuscular blocking action is quantitatively accounted for by its postsynaptic action. Other agents that apparently act in the same manner are many. They include gallamine, β-erythroidine, the toxiferines, tetraethylammonium, the hemicholiniums, ganglionic blockers such as mecamylamine, conventional antihistaminic agents, and tris. However, detailed studies of the mechanism of action of most of these agents are still lacking, and it is possible that local anaesthetic activity plays a part in their neuromuscular blocking action. The structural features that are required for a curare-like action have been considered in some detail by Cavallito (1967). In general, the nondepolarizing blocking agents differ from the depolarizing agents (e.g., succinylcholine, decamethonium) in having a relatively rigid and bulky structure. For this reason, Bovet (1951) introduced the terms pachycurare for tubocurarine and similar agents, and leptocurare, for the depolarizing blocking agents.

2. NONCOMPETITIVE BLOCKADE

There are several forms of noncompetitive drug antagonism that can be observed. The first consists simply in physiological antagonism at the cellular level. Thus, in most intestinal smooth muscle, ACh and catecholamines have antagonistic actions. Similarly, α-adrenoceptor-mediated excitation of blood vessels may be antagonized by β-adrenoceptor-mediated inhibition. Rather more subtle is the antagonism to β-blocking activity (Somani and Lum, 1965; Lucchesi *et al.*, 1967). *d*-tubocurarine manifest by decurarizing agents such as anticholinesterases. The anticholinesterase leads to competitive antagonism by ACh of the receptor blockade by curare because of the increased amount and perseverance of the endogenous ACh resulting from inhibition of its hydrolysis.

Less trivial is the noncompetitive antagonism which is generally ascribed to a "local anaesthetic" action. It is a rare agent that does not, in sufficient concentration, have some action to block electrical activity of nerve or muscle, but it is far from established that the nonspecific antagonisms observed in fact represent only one kind of interaction with electrogenic systems in the membrane. Characteristically, this kind of action displays classic noncompetitive kinetics; depression of the response to agonist is associated with depression of the maximal response that can be elicited by agonist. Since in all known systems maximal activation of response can be elicited by activation of only a very small fraction of the total available receptors, these kinetics imply that a rate-limiting step subsequent to receptor activation is involved. Examples are numerous of this type of blockade; two will suffice. One is the blocking action of atropine at the skeletal neuromuscular junction, which is demonstrably noncompetitive (Kirschner and Stone, 1951). In denervated muscle, the action of tubocurarine to block ACh-induced depolarization is significantly reduced (Beránek and Vyskočil, 1967), and tubocurarine can cause a contracture, indicating a difference between the new receptor formed consequent to denervation from those normally present in the subsynaptic membrane. Nevertheless, the blocking action of atropine is unaffected (Beránek and Vyskočil, 1967). A second example is the α-adrenoceptor blocking action of pronethalol, a competitive β-adrenergic blocker. This blockade is associated with a clear depression af the maximal response of the rat vas deferens to noradrenaline (Ariëns, 1967). Pronethalol also blocks post-tetanic twitch potentiation in the soleus muscle (Standaert and Roberts, 1967), indicating an action on repetitive action potential discharge in either the muscle fiber or the nerve terminal or both, and is consistent with a local anaesthetic action. In the heart, antiarrhythmic effects of β-adrenergic blocking agents, including dichlorisoproterenol, pronethalol, and propranolol, are evidently related largely, if not entirely, to a quinidine-like action rather than

3. Alkylating Agents

The most important variety of noncompetitive blockade is the relatively selective and irreversible blockade of receptors that can be obtained by agents, related to nitrogen mustards, which act by alkylation of some grouping in or near specific receptors, rendering the receptors inactive (Harvey and Nickerson, 1954). In a sense, these agents might be considered competitive. During the first stage of blockade by halo-alkylamines such as dibenamine or phenoxybenzamine, the interaction with the receptor is apparently mediated by the same relatively weak forces involved in the attachment of most agonists and classic competitive antagonists (Nickerson, 1957). The presence of a specific agonist or competitive antagonist during this stage can protect the receptor from haloalkylamine blockade (Nickerson and Gump, 1949; Furchgott, 1954). This receptor protection is quite specific. Although the alkylating agents are rather nonselective, acting on α-receptors for catecholamines and receptors for histamine, 5-HT, and ACh, the presence of a specific agonist protects only the receptors to that agonist. For example, nor-adrenaline prevents blockade only of α-adrenergic receptors, and apparently not the receptors for ACh or histamine. This phenomenon of specific receptor protection is interpreted as indicating that the site of receptor alkylation by the haloalkylamine is close to, if not identical to, the site of agonist intereaction. Undoubtedly the specificity of the attachment by the alkylating agent can be modified by specific groupings on the alkylating agent. Benzilylcholine mustard, for example, appears to be more selective than dibenamine for the alkylation of muscarinic ACh receptors (Gill and Rang, 1966). However, the possibility that it is accessory sites rather than receptor itself that is alkylated gains considerable support from the finding that "irreversible" haloalkylamine blockade of α-adrenoceptors can be partially reversed by trypsin. This is associated with removal of the haloalkylamine, which remains bound to peptide fragments (Graham and Katib, 1966; Mottram and Graham, 1970; Mottram, 1970).

4. Spare Receptors

Early kinetic studies using haloalkylamine blockade of α-adrenoceptors revealed a phenomenon quite unexpected at the time. Although the blockade was effectively irreversible, dose–response curves to agonist were shifted in a parallel manner and the maximal response of tissues to agonist was unaffected until the dose of the blocking drug was sufficient to require many times the original agonist dose for an equivalent response. It is now considered that this kind of shift of dose–response curve should be characteristic of a noncompetitive inhibition of the receptor, as distinct from subsequent steps in muscle activation, provided full activation of tissue response can occur with activation of only a small fraction of the receptors present on the membrane surface (Stephenson, 1956; Nickerson, 1956). For the fraction of receptors that are redundant, in the sense that full activation can take place without them, the completely misleading terms "spare receptors" or "receptor reserve" have been applied. In terms of the hypothesis, it is because of these spare or reverse receptors that responses are normally elicited by relatively low quantities of the agonist. There is no reason to believe that spare receptors are not just as often activated during physiological events as any others. The evident explanation for apparent spare receptors is simply that the response measured is not linearly related to the fraction of receptors occupied or activated by the agonist. Generally, the response that is measured is a second order or higher order event. For example, one might obtain dose–response curves relating the contraction of ileum in response to ACh. The contractile response may be the result of three distinct sequelae (as far as is known) of increased membrane permeability to certain ionic species: (1) depolarization, (2) action potential generation, (3) direct mobilization of calcium ions. With the exception of the last response, which may well be small under usual conditions, none of these effects could be expected to be linearly related to the first order effect of ACh, i.e., an increased membrane conductance. Werman (1969) has therefore argued that ordinary kinetics can indeed be applied correctly, with the assumption of linearity between receptor activation and response, provided the response measured represented the first order effect of the agonist, e.g., conductance change in the case of ACh action. If the spare receptor exists for the conductance change, however, this is not the case. Burgen and Spero (1968) measured the alteration in response to ACh and related agonists of ^{42}K or ^{86}Rb efflux from gut. Since high potassium altered this response little, if at all, this should have been a direct measure of the membrane permeability change. They found that the alkylating agent benzilylcholine mustard shifted log dose–response curves in parallel to the right and did not depress the maximum response until the carbachol dose had to be multiplied three hundred times to obtain the control response. In terms of spare receptor theory, this indicates a high receptor reserve for the first order response; activation of only one receptor out of three hundred would normally result in full activation of the membrane permeability change. However, dibenamine showed no receptor reserve at all for the efflux response, as distinct from the contractile response. Evidently there are two possible explanations. Either dibenamine has an added effect to inhibit the efflux response, so that receptor reserve is not seen with this drug, or, conversely, there is no receptor reserve and the apparent reserve observed with benzilylcholine mustard reflects an action of this substance not on the actual receptors where ACh inter-acts but on accessory groups where alkylation causes allosteric effects on the receptor, reducing its affinity for the transmitter. Since it is diffi-cult to imagine how any of three hundred receptors in an array could suffice to open just one membrane channel, which is the clear implication of the spare receptor interpretation of this result, the second explanation is obviously easier to accept. By extension, the accuracy of measurement of spare receptors by the use of alkylating agents must remain open to doubt.

5. CHEMICAL MODIFICATION OF RECEPTORS

There has recently been considerable interest in characterizing phar-macological receptors in chemical terms (see reviews by Ehrenpreis *et al.*, 1969; Burgen, 1970). Much of this work appears to us highly speculative and based upon assumptions that are difficult to justify. Recent investigation of the nicotinic ACh receptor, however, has led to an apparently well founded model.

Albuquerque and his colleagues have studied the effects of enzymes on the ACh sensitivity of chronically denervated mammalian muscle. A variety of proteolytic enzymes were found to have no effect on ACh sensitivity, although they inactivated cholinesterase (Albuquerque *et al.*, 1968). The receptors were also unaffected by lysolecithin, phospho-lipase A, or phospholipase C, although these agents had distinct actions on muscle membrane resistance and action potential activity (Albuquerque and Thesleff, 1968). However, a sulfhydryl blocker (*p*-chloromercuribenzoate) and a disulfide bond-reducing agent (dithio-erythreitol) both reduced the ACh sensitivity of the muscle membrane, the effect being only slowly reversed by washing. These findings suggested that protein is part of the receptor structure.

Karlin (1969) has studied more extensively the reduction and reoxida-tion of the nicotinic ACh receptor of electroplax of the electric eel. With reduction of the receptor by dithiothreitol, there occurs an altera-tion in the relative potency of ACh analogs. In the reduced state, the receptor binds ACh and carbamylcholine less well, the binding is trans-lated less efficiently into a permeability change, and the apparent co-operation between sites decreases. The response of the reduced receptor to decamethonium, on the other hand, is greatly increased, and hexa-methonium becomes a depolarizing agent rather than a competitive blocking agent. These effects can be reversed by reoxidation, with restoration of normal responses. If the reduced receptor is alkylated by various maleimide derivatives, however, reoxidation no longer can occur. From the interactions of the various compounds, Karlin (1969) has been able to suggest a model. He sees the receptor as a protein with a disulfide bond in the vicinity of the active site, where "an activator such as ACh bridges between a negative subsite and a hydrophobic subsite, causing an altered conformation around the negative subsite and a decrease of a few angstroms in the distance between the two subsites." Although speculative and necessarily incomplete, the model seems to account well for the alterations in sensitivity produced by chemical manipulation.

IV. Drug Effects on Muscle Contraction

A. *Excitation–Contraction Coupling*

It is now generally accepted that the calcium ion is the final activator of the contractile system of muscle (see reviews by Hasselbach, 1964; Sandow, 1965; Weber, 1966). In skeletal muscle, the mechanism of its action has now been clarified (see review by Ebashi and Endo, 1968). The effect is exerted on the calcium-receptive protein troponin, which together with tropomyosin is located along the thin filament. In the absence of free calcium ion, the troponin molecule inhibits the interaction of myosin and actin, and this inhibition is removed by calcium ions. Relaxation depends upon the sarcoplasmic reticulum (relaxing factor), which contains specific sites that bind calcium, provided ATP is present, and may transport the initially bound calcium into the lumen of the reticulum.

The trigger for muscle contraction is evidently the release of calcium from the sarcoplasmic reticulum, initiated by spread of membrane depolarization through the internal membrane T system (A. F. Huxley and Taylor, 1958), which is directly open to the extracellular space (H. E. Huxley, 1964). When this continuity is interrupted, as occurs after glycerol treatment, depolarization of the surface membrane is no longer associated with muscle contraction (Howell, 1969). Polarization of the internal membrane system appears to be a prerequisite for its sensitivity to electrical stimulation (Natori, 1965). Recently, Costantin (1970) reported that the radial spread of depolarization in the T system is augmented by an increase of sodium conductance in the T tubule membrane which is sensitive to tetrodotoxin. However, there is no all-or-nothing response when a steady depolarization is applied to the mouth of a single T tubule (A. F. Huxley and Taylor, 1958), indicating that the sodium current system in the tubule does not become regenerative. Since most of the chloride conductance of the muscle fiber resides in the surface membrane, rather than the membrane of the T system (Hodgkin and Horowicz, 1960a; Eisenberg and Gage, 1969), this observation probably cannot explain the remarkable potentiation of twitch responses and contractures that is produced by nitrate and other anions (Hodgkin and Horowicz, 1960c; Foulks and Perry, 1966). Ebashi and Endo (1968) have suggested that this potentiation may be related to an effect exerted after entering the endoplasmic reticulum, perhaps inhibition of the calcium uptake process, since calcium binding by the isolated sarcoplasmic reticulum was shown to be inhibited by these anions (see Ebashi and Endo, 1968).

It has been a question for some time whether the release of activator (i.e., calcium) secondary to depolarization of the T system is entirely a passive event simply graded with depolarization. Hodgkin and Horowicz (1960b), studying potassium-induced muscle contractions observed that in the region of −50 mV there was an exceedingly steep relation between muscle tension and membrane potential, suggesting a regenerative process in the activation mechanism. Moreover, R. H. Adrian *et al.* (1969), employing a voltage clamp technique, observed that addition of long and short depolarizing pulses gave results expected of a system that becomes regenerative at about −50 mV. Recently, Endo *et al.* (1970) demonstrated that in skinned frog muscle fibers, calcium itself causes the release of calcium from the endoplasmic reticulum; this is not secondary to any effect of calcium on the uptake process.

Thus, the sequence of excitation–contraction coupling in striated twitch muscle is as follows. The nerve action potential is followed by release of transmitter (ACh) that interacts with the postsynaptic membrane to cause a depolarization, which if large enough, triggers a muscle action potential that is propagated longitudinally to either end of the muscle fiber. This action potential is analogous to that of nerve, based upon increase of sodium ion conductance in response to membrane depolarization (Nastuk and Hodgkin, 1950). Current also spreads transversely down the sarcotubular system, augmented by local activation of the sodium current system, and the resulting depolarization causes release of calcium into the myoplasm. The resulting free calcium in the myoplasm acts on troponin to derepress and hence cause activation of the contractile elements. The calcium is subsequently taken up by the sarcoplasmic reticulum; there is evidence that this reuptake may be to some extent a delayed effect of depolarization of the T tubular membrane on the sarcoplasmic reticulum (see Ebashi and Endo, 1968). The uptake system does not immediately restore the initial state. If a depolarization is maintained, calcium ions taken up by the sarcoplasmic reticulum is not released by further depolarization, but becomes ready for release in response to depolarization if the membrane is repolarized. Recovery from inactivation is a process taking many seconds, and is smaller and slower the smaller the repolarization.

For smooth muscle, the evidence is far less complete, but it is generally assumed that the activation of muscle contraction is basically analogous to that of striated muscle, the only differences being those that arise from anatomical differences. The same intracellular machinery, i.e., actin, myosin, and a longitudinal sarcotubular system with a high ATPase activity, has been shown to be present. It is notable that the sarcotubular system is very poorly developed in smooth muscle when compared to the elaborate structure found in skeletal and cardiac muscles (see Burnstock, 1970). Considerable evidence has accumulated that calcium ions play the same central role in contraction of smooth muscle as they do in striated muscle (Waugh, 1962a,b; Briggs, 1962; Hinke *et al.*, 1964; Schild, 1964). One major difference from smooth muscle is the strong dependence of smooth muscle contraction on the external calcium ion concentration (see Schild, 1964), which, together with the relative deficiency of the sarcoplasmic reticulum, suggests that the calcium ions that induce contractions come mainly from the external solution or from superficial membrane binding sites (see Goodford, 1970). However, some calcium probably comes from protected sites. Imai and Takeda (1967b) found guinea pig taenia coli to contract a little in response to isotonic potassium solution with an external calcium concentration as low as 10^{-8} M. Slow skeletal fibers of the frog are, in this regard, similar to smooth muscle (Lüttgau, 1963; Imai and Takeda, 1967b).

One major difference between smooth and striated muscle is the character of the action potential. In smooth muscle, the action potential resembles that of crustacean muscle, where the action potential, or spike, is related to entry of calcium rather than sodium ions (Fatt and Ginsborg, 1958; Hagiwara and Naka, 1964; Hagiwara and Nakajima, 1966; Ozeki *et al.*, 1966) in that (1) it is not blocked by tetrodotoxin (taenia: Kuriyama *et al.*, 1966; ureter: Washizu, 1966; artery: Keatinge, 1968a). (2) it is blocked by manganese in a low concentration that does not influence sodium spikes (taenia: Nonomura *et al.*, 1966; Brading *et al.*, 1969b; ureter: Kuriyama *et al.*, 1967a) and (3) barium can substitute for calcium in generation of the action potential (taenia: Hotta and Tsukui, 1968; Bülbring and Tomita, 1968b, 1969c). Another important point of difference is the action of procaine, which in crustacean muscle augments rather than inhibits spike activity (Fatt and Katz, 1953; Hagiwara and Nakajima, 1966), but in nerve and skeletal muscle inhibits the action potential by depressing the sodium current system (Shanes *et al.*, 1959; Inoue and Frank, 1962; Narahashi *et al.*, 1967). Washizu (1968) found that procaine acted to enhance action potential activity in both taenia coli and ureter.

This evidence, of course, does not in itself indicate that the normal charge carrier in the smooth muscle action potential is calcium as in the crustacean muscle. Indeed, Redfern *et al.* (1970) recently found that in chronically denervated rat skeletal muscle, the action potential cannot be blocked with tetrodotoxin, although it is sensitive to extracellular sodium concentration in the usual way. However, there is good evidence that smooth muscle action potentials are indeed calcium spikes. In taenia or ureter, action potentials have been observed to persist in the absence of sodium, provided calcium is present, and provided sodium chloride is replaced by sucrose rather than tris chloride (Bülbring and Kuriyama, 1963a; Brading *et al.*, 1969b; Kobayashi, 1965; Kuriyama and Tomita, 1970). Even with replacement of sodium by tris, Keatinge (1968a) found that arterial muscle could generate spikes, provided calcium was present and the muscle somewhat depolarized. Thus, in the absence of sodium, calcium can carry the inward current associated with the action potential. Since the overshoot and the rate of rise of the spike is actually increased in low sodium, it seems likely that the spike current is indeed normally carried by calcium rather than sodium. The plateau phase of the action potential in the cat ureter appears to be an exception; it evidently depends upon sodium as well as calcium entry, although it is unaffected by tetrodotoxin (Kobayashi and Irisawa, 1964; Kobayashi, 1965).

There is now evidence that the immediate source for the calcium that enters during spike activity is not the extracellular solution but a binding site at the outside of the membrane that continues to be replenished from a storage site inside the cell by a calcium transport mechanism (Bülbring and Tomita, 1970). This accounts for the fact that the rate of abolition of the spike by removing calcium is much slower than the loss of calcium by diffusion from the extracellular space (Brading and Jones, 1969), and that the effects of calcium depletion are highly temperature-sensitive (Brading *et al.*, 1969a). The rate of disappearance of the spike in calcium-free solution containing 0.1 mM EGTA is influenced by magnesium and sodium ions in the manner expected if these ions compete with calcium for membrane sites (Bülbring and Tomita, 1970). Hagiwara and Takahashi (1967) demonstrated that in the barnacle muscle fiber, which produces a calcium spike, the overshoot is related to the amount of calcium, or a divalent ion that can substitute for calcium, adsorbed at the membrane.

The fact that the inward current of the action potential is carried by calcium ions in smooth muscle gives the action potential a more direct role in excitation–contraction coupling than in skeletal muscle. Spike activity in response to membrane depolarization is not only a device that augments spread of excitation to cells and areas of cells not directly excited, but also presumably provides much of the calcium required for activation of the contractile elements. It is now a major problem in pharmacology to determine how much of the excitatory effects of drugs on smooth muscle is mediated by each of the three mechanisms by which drug action can cause mobilization of calcium to the myoplasm from the extracellular fluid or membrane binding sites: (1) direct increase of membrane calcium permeability or calcium mobilization from a membrane store (pharmacomechanical coupling), (2) membrane depolarization leading to calcium mobilization, and (3) calcium influx related to action potential and slow wave activity. Moreover, there are drugs that apparently act even more directly on the mechanisms by which calcium is handled inside the cell. Direct actions of drugs on the actin–myosin contractile system is another possibility. Generally speaking, the analysis of drug effects in terms of the mechanism of excitation–contraction coupling is far from complete. In the following sections we shall call attention to what little is now understood.

B. Pharmacomechanical Coupling

There are three independent lines of evidence for the hypothesis that drugs that excite smooth muscle owe a large part of their activity to direct pharmacomechanical coupling, i.e., to an action not mediated by an alteration of membrane potential or of membrane electrical activity. The first of these, the dissociation that can sometimes be observed between electrical activity and contractile response, has already been discussed (see p. 79). The second is the finding that the maximal contractile responses of the same tissue to different supramaximal chemical stimuli may be markedly unequal; for example, angiotensin, vasopressin, and adrenaline acting on a strip of vascular muscle (A. V. Somlyo et al., 1965), and noradrenaline can induce a maximal contraction greater than that produced by potassium, with less depolarization (A. V. Somlyo and Somlyo, 1968). Third, and most cogently, a variety of drugs that contract smooth muscle do so even when the muscle is completely depolarized by substitution of potassium for sodium ions in the bathing medium. This last fact was first discovered by Evans and Schild (1957) and investigated more thoroughly by Evans et al. (1958), who studied responses of depolarized chick amnion, rat uterus, strips of cat ileum longitudinal muscle, guinea pig and rabbit ileum, and the retractor anterior byssus muscle of *Mytilus* to a variety of stimulant drugs, including ACh, histamine, 5-HT, and oxytocin; to inhibitory drugs such as isoprenaline; and to drug antagonists such as atropine and mepyramine. Provided the tissues were cooled to about 20°C, the initial contraction resulting from potassium depolarization was not maintained. The preparations then responded to various agonists in typical fashion; these responses were blocked by the appropriate antagonists, the isoprenaline had the same activity as on normally polarized preparations. These observations were extended to vascular smooth muscle by Waugh (1962a,b), who found calcium-dependent contraction of depolarized arterial muscle by adrenaline (also see A. P. Somlyo and Somlyo, 1970). It should be pointed out that not all relaxant effects of drugs on smooth muscle exist when the muscle is depolarized by raised potassium. For example, the relaxing effect of phenylephrine in taenia coli, which is exerted purely via α-adrenoceptors, is absent in depolarized preparations, although isoprenaline, which acts via β-adrenoceptors, remains active (Jenkinson and Morton, 1967c). As has already been discussed, α-adrenergic inhibition in intestinal muscle is considered to be mediated by an increase of potassium conductance, causing hyperpolarization. In a potassium-depolarized preparation, this mechanism could not operate. The response to activation of β-adrenoceptors, on the other hand, is evidently mediated by formation of cyclic AMP. How this substance causes inhibition is quite unknown; the persistence of isoprenaline action in potassium-depolarized preparations indicates that the mechanism does not depend upon the membrane potential.

In smooth muscle that is completely depolarized by ouabain, the picture is somewhat different than with potassium depolarization (Matthews and Sutter, 1967). After an initial contracture, presumably due to depolarization, the muscle relaxed spontaneously even at 37°C. In taenia coli, ACh, histamine, and 5-HT were then without effect, although added 50 mM potassium chloride produced a contracture. In the rabbit anterior mesenteric vein treated in the same way, both noradrenaline and 5-HT elicited a contractile response. From these results several conclusions may be drawn. First, it appears that potassium may cause muscle contraction by a mechanism distinct from the depolarization it causes. The effect calls to mind the fact that potassium, unlike depolarization, acts to potentiate the secretory response of the motor nerve terminal to depolarization, perhaps by increasing calcium influx (Cooke and Quastel, 1969, 1972). Second, ouabain prevents the response in taenia to agonists, although depolarization by potassium does not. One possible explanation is that the calcium pool which is mobilized by these agents may become depleted if an active calcium transport system which normally keeps this pool full is blocked by ouabain. It has already been pointed out that there is evidence that the calcium pool in intestinal muscle which subserves action potential activity apparently depends upon an active transport system.

In the vein, on the other hand, noradrenaline and 5-HT continue to act when the muscle has been depolarized by ouabain. Presumably the calcium store mobilized by these agents in the vein does not depend upon the integrity of a ouabain sensitive transport system; indeed, the response to noradrenaline was augmented by ouabain (Matthews and Sutter, 1967). It should not therefore be concluded that the calcium involved in responses of vascular smooth muscle noradrenaline is the same as that involved in contraction induced by potassium. Indeed, the very persistence of a noradrenaline response after relaxation in the presence of isotonic potassium makes this unlikely. It is now generally accepted that potassium and noradrenaline differ in the way in which they utilize calcium to elicit contractions of vascular smooth muscle. In calcium-free solution, responses to noradrenaline persist for extended periods, whereas responses to potassium decline rapidly (Waugh, 1962b; Hinke *et al.*, 1964; Hudgins and Weiss, 1968). Moreover, after extended calcium depletion, responses to noradrenaline were found to reappear abruptly at a calcium concentration of about 0.5 mM, whereas responses to potassium appeared to return in parallel with the calcium concentration over a considerable range. These observations indicate that the contracture elicited by noradrenaline is secondary to calcium influx from a relatively tightly bound and saturable pool, whereas the potassium contracture utilizes calcium from a more labile and probably more superficial pool.

Although there can be no doubt of the existence of direct pharmacomechanical coupling, the question remains as to what extent it is involved in responses of normally polarized smooth muscle to various agonists. The strongest evidence that it is important comes from the studies of A. V. Somlyo and Somlyo (1968), who found that the relative maximal responses to supramaximal doses of various agonists were the same in potassium depolarized as in normally polarized vascular smooth muscle. One cannot escape the conclusion that the maximal response of the normally polarized muscle to the agonists reflects the same mechanisms as in depolarized muscle, i.e., pharmacomechanical coupling. However, whether this applies to submaximal doses of the agonists remains obscure. As A. V. Somlyo and Somlyo (1968) point out, agonists that induce membrane depolarization must bring into play all three of the systems by which calcium may be made to move into the myoplasm. These three modalities (graded electrical, action potential, and pharmacomechanical) will mutually reinforce one another. The relative importance of the three systems for physiological activation by transmitter substance will depend upon the muscle, its density of innervation, the distribution of receptors, and the level of polarization of the muscle fibers. A small amount of transmitter may well act by causing a moderate depolarization which in itself mobilizes little calcium but facilitates action potential generation. It will also facilitate spread of action potentials into cells that would otherwise be quiescent and thus increase spontaneous integrated activity. With larger amounts of the transmitter, the greater depolarization and increased membrane conductance would tend to inhibit action potential generation and spread (because of increased outward leak of inward current generated by the spike mechanism), but this would be more than compensated by the more direct mobilization of calcium in response to the membrane depolarization. Lastly, with saturation of the previous mechanism, direct calcium mobilization by transmitter action will play the major role in causing muscle contraction.

C. General and Local Anaesthetic Agents

Anaesthetic agents are well known to depress the direct excitability of skeletal muscle and in many instances to inhibit neuromuscular transmission (e.g., Overton, 1901). Thesleff (1956) investigated the actions on isolated frog sartorius muscle of pentobarbital, urethan, chloral hydrate, chloralose, paraldehyde, and tribromethanol. These agents all acted to raise the threshold for direct electrical stimulation of the muscle fiber, except pentobarbital, which in low concentration reduced threshold, perhaps because it also depolarized the fibers. Since this depolarization was associated with an increased membrane resistance, it appeared likely that this drug causes a reduction in potassium permeability. Of the other agents, tribromethanol and urethan reduced membrane resistance to a small extent; chloralose and peraldehyde increased membrane resistance. There was little if any effect on resting potential. Yamaguchi (1961) also studied the action of general anaesthetics (urethan, ether, and chloroform) on frog striated muscle, finding that in low doses these agents actually increase excitability to direct stimulation. With intermediate concentrations, excitability was depressed, while with high doses locally applied, there was abrupt and irreversible blockade of action potential conduction. Moreover, high doses of the anaesthetics cause actual muscle contracture. This is followed by a profound reduction of membrane potential. It is of interest that the ability of the anaesthetics to produce contracture was maintained in muscle completely depolarized by isotonic potassium chloride solution.

General anaesthetic agents also tend to block neuromuscular transmission by depression of endplate response to ACh, i.e., MEPP amplitude. This was observed for pentobarbital (Thesleff, 1956), for ether (Karis *et al.*, 1966), and for halothane (Gissen *et al.*, 1966). We have found it to be a general feature of general anaesthetics (e.g., chloroform, paraldehyde, tribrom- and trichlorethanol, chlorpromazine). Exceptions are methanol, ethanol, and propanol, which increase MEPP size (Gage, 1965; Inoue and Frank, 1967). Longer-chain alcohols (butanol, pentanol, etc.) we have found to depress MEPP size, although they prolong the duration. The mechanism of the amplitude potentiation by ethanol is not known. Gage found it to increase membrane resistance in rat diaphragm, and since the amplitude potentiation persisted after cholinesterase inhibition by neostigmine, he attributed the increase of MEPP size to increased membrane resistance. However, Inoue and Frank (1967) found that ethanol reduced membrane resistance of frog muscle but still increased MEPP amplitude. At low concentrations of ethanol (less than 4%), MEPP amplitude was increased without any indication of prolongation, making an anticholinesterase action unlikely. The difference in behavior of the membrane resistance in frog and mammalian muscle is most easily accounted for by supposing a differential effect of ethanol on potassium and chloride conductance, with potassium conductance being reduced and chloride conductance increased, and a small difference in the relative resting permeabilities of the different muscles to the two ions. Harris and Ochs (1966) found that amytal caused a large decrease in potassium permeability in frog muscle, as did also mepyramine and the local anaesthetics cocaine and procaine. It should not be forgotten that these agents all act in the same way to stabilize erythrocyte membranes against hemolysis, apparently by simple accumulation in the lipid phase of the membrane, (Seeman, 1966).

In intestinal smooth muscle (guinea pig ileum), the effects of ethanol were studied by Hurwitz *et al.* (1962). Ethanol was found to inhibit the sustained contraction of potassium-depolarized preparations, acting as a noncompetitive antagonist to calcium, which tended to reverse this inhibition. It also acted to inhibit the increase of potassium efflux produced by ACh, suggesting that at this site, as opposed to the skeletal muscle endplate, it tends to block rather than potentiate ACh action. This effect was reversed by high calcium. Gage (1965) found that ethanol acted reversibly to reduce twitch and tetanic tension in toad sartorius muscle. Butanol, hexanol, toluene, and lidocaine are potent and reversible inhibitors of the twitch of frog sartorius muscle (Bianchi and Bolton, 1967). It would seem that general and local anaesthetics can act within the muscle fibers to interfere with either the activation of muscle contraction or the contractile elements themselves.

In vascular smooth muscle, ethanol has direct vasoconstrictor effect on placental vessels (Ciuchta and Gautieri, 1963), and applied intraarterially, it constricts human blood vessels; the well known vasodilator action of orally ingested ethanol is evidently entirely reflex in origin (Fewings *et al.*, 1966). It is conceivable that the constrictor action might be at least in part secondary to release of neurotransmitter; however, it seems more likely to be related to mobilization of calcium into the myoplasm, perhaps on the same basis as that found with high concentrations of ethanol in skeletal muscle. Other general anaesthetics including cyclopropane and halothane also contract vascular smooth muscle (Price and Price, 1962). Halothane also produces vasodilatation and inhibits the vasoconstrictor action of noradrenaline (Black and McArdle, 1962), while cyclopropane enhances the contractile effects of catecholamines, angiotensin, histamine, and serotonin on the rabbit aorta (Price and Price, 1962). On bronchial smooth muscle, halothane causes an inhibition that is blocked by the β-adrenergic blocking agent MJ 1999 (Klide and Aviado, 1967), suggesting that the effect is indirect and mediated by catecholamine release. Similarly, intraarterial thiopental has a vasoconstrictor action on rabbit ear arteries which abolished after reserpinization and potentiated by cocaine (Burn and Hobbs, 1959), again suggesting that the thiopental acts mainly presynaptically to increase noradrenaline release.

Rang (1964) found that a variety of volatile anaesthetics, including ether, chloroform, trichloroethylene, halothane, and various halogenated compounds, caused excitation of intestinal smooth muscle. This excitation was of two kinds: (1) rapid transient contractions abolished by cocaine or lachesine, and presumably related to nerve stimulation, although unaffected by hexamethonium, and (2) slow sustained contractions, unaffected by cocaine and lachesine; this effect predominated among fluorinated ring compounds. The first type of action may be related to the increased nerve excitability caused by low concentrations of general anaesthetic agents, e.g. methanol, ethanol, acetone, and chloroform (Berney and Posternak, 1956), or perhaps, in part, to an effect to increase spontaneous ACh release, as occurs at the skeletal neuromuscular junction (see Table I).

The action of local anaesthetics at the skeletal neuromuscular junction is curare-like in that the EPP is reduced in amplitude. Furukawa (1957) found that in addition to the reduction of EPP amplitude by procaine there was a peculiar change in time course, with a plateau following a fast spikelike initial potential. Maeno (1966) found evidence that this effect was related to a selective prolongation of the time course of the increased sodium conductance associated with transmitter action, a conclusion that has since been substantiated by voltage clamp studies by Maeno *et al.* (1971). Lidocaine, on the other hand, has no such selective action (Maeno *et al.*, 1971), in agreement with the conclusion of Steinbach (1968a,b) that the action of lidocaine and some derivatives was to prolong both potassium conductance and sodium conductance in parallel, the complicated time course of the EPP being explainable on the basis of the double time constant of the muscle membrane revealed by Falk and Fatt (1964).

It has already been remarked that local anaesthetics as well as alcohols act to reduce twitch tension in skeletal muscle. They also act to inhibit contractures elicited by raised potassium, shifting the curve relating tension to potassium concentration to the right without any effect on membrane potential in low concentrations. At higher concentrations (e.g., 1 mM tetracaine), they can very substantially reduce the maximum potassium contracture. This effect is apparently mediated by intracellular action. In frog rectus abdominis muscle, the contracture produced by isotonic potassium sulfate is persistent and not seriously affected by chelation of extracellular calcium by EGTA, indicating that the contracture probably represents shuttling back and forth of intracellular calcium between sarcoplasmic reticulum and myoplasm. Yet tetracaine (1 mM) produces a fairly rapid and complete relaxation, which is not reversed by high external calcium (Feinstein and Paimre, 1969). In addition to their effect to inhibit twitches and potassium contractures, local anaesthetics can also, in sufficient concentration, themselves produce skeletal muscle contractures. Bianchi and Bolton (1967) first showed that lidocaine, procaine, and tetracaine all have this action, which from studies with varied pH, is to be attributed to the uncharged form of the compound. In contrast, local anaesthetic action per se in nerve is related to the intracellular concentration of the charged form (Narahashi *et al.*, 1970). The mechanism of the contracture seems to be release of calcium from the sarcoplasmic reticulum. Townsend (1967) observed that tetracaine, dibucaine, and lidocaine depressed the ability of isolated fragments of sarcoplasmic reticulum to accumulate calcium in the presence of ATP and the associated magnesium–calcium activated ATPase activity, and could actually release calcium previously accumulated with these fragments. This activity was a function of the uncharged form of the tertiary amine local anaesthetics. Contractures are also produced by compounds with local anaesthetic activity such as propranolol, diphenhydramine, and phenacaine, even in completely depolarized muscles (Feinstein and Paimre, 1969), as by ethanol (Yamaguchi, 1961). Propranolol has been shown to depress calcium accumulation by isolated sarcoplasmic reticulum (Scales and McIntosh, 1968). The two distinct actions of muscle contraction shared by general anaesthetics and the uncharged form of local anaesthetics can be reconciled without great difficulty. Presumably, the effects are mediated at the level of the sarcoplasmic reticulum. The inhibitory effect on potassium contractures and twitches can be attributed to a depression of the mechanism by which depolarization causes release of calcium into the myoplasm, perhaps simply by reduction of the amount of calcium bound, and thus raising the threshold for regenerative calcium mobilization, since calcium itself acts to release calcium. The second effect, to cause contraction, will follow from excessive displacement of calcium from the sarcoplasmic reticulum. This scheme is supported by the observation that the uncharged form of local anaesthetics tends to potentiate rather than inhibit contractures produced by caffeine, an agent (see below) which acts by displacing calcium from the sarcoplasmic reticulum (Bianchi and Bolton, 1967).

In smooth muscle, the direct action of local anaesthetics is to inhibit contraction, independently of whether contraction is produced by high potassium or drugs (Feinstein and Paimre, 1969). The contraction caused by adding calcium to calcium-free depolarized preparations is also antagonized, and this action is shared by ethanol and pentobarbital (Feinstein, 1966; Antonio *et al.*, 1970). There also exists a contracture-promoting action, as with volatile anaesthetics, but inhibition predominates at high concentrations of, for example, lidocaine (Feinstein and Paimre, 1969). The inhibition is most easily explained in terms of suppression of calcium permeation through the surface membrane (e.g., Feinstein and Paimre, 1969; Antonio *et al.*, 1970). Calcium-induced contractures in depolarized rat uterus, for example, are antagonized by local anaesthetics in an apparently competitive manner (Feinstein, 1966).

The excitatory action of procaine (Sanders, 1969) is unaffected by a large variety of receptor blocking agents but is blocked after depolarization by ouabain, while the noradrenaline contracture persists. This indicates a dependence upon the membrane potential and suggests that the excitatory effect reflects an increase in action potential activity by procaine, as observed in ureteral smooth muscle by Washizu (1968).

D. Miscellaneous Drugs

1. SKF 525-A

β-Diethylaminoethyldiphenylpropyl acetate hydrochloride (SKF 525-A) is a drug that has been described as having little or no pharmacological action of its own (Goldstein *et al.*, 1968), acting to potentiate neuromuscular blockade by competitive blocking agents by mobilization of molecules of the curarizing substances from nonspecific receptors (Bovet *et al.*, 1956). However, it is now evident that it has many pharmacological actions in common with local anaesthetics. It antagonizes depolarization at the endplate by succinylcholine and decamethonium (Suarez-Kurtz *et al.*, 1969); it depresses the muscle twitch of the directly stimulated sartorius muscle, with concomitant reduction of the amplitude of the muscle action potential; it inhibits the potassium contracture of sartorius and rectus abdominis muscles without affecting K depolarization; and it potentiates the caffeine induced contraction of sartorius muscle. All these effects are attributable to the uncharged form of the molecule (Suarez-Kurtz and Bianchi, 1970). It also has a local anaesthetic action which is apparently exerted by the charged form (Suarez-Kurtz and Bianchi, 1970).

Kalsner *et al.* (1970) have recently investigated the action of SKF 525-A on the response of aortic strips to potassium and excitatory drugs. It was possible to inhibit completely the response to potassium without interfering with the response to noradrenaline, suggesting that SKF 525-A selectively blocks some step by which extracellular or loosely bound calcium in contrast to tightly bound calcium, moves into the myoplasm. It is of considerable interest that SKF 525-A reduced responses to 5-HT almost as rapidly as to potassium, and also reduced the response to histamine, at a rate greater than that to noradrenaline but less than that to potassium. Responses to angiotensin were hardly affected at all by the drug, suggesting that angiotensin utilizes exclusively the firmly bound calcium fraction.

2. Haloalkylamines, Chlorpromazine, Cinnarizine, and Desipramine

The contractile response of smooth muscle to depolarization by potassium is reduced by phenoxybenzamine, dibenamine, chlorpromazine (Bevan *et al.*, 1963; Shibata and Carrier, 1967; Shibata *et al.*, 1968), desipramine (Hrdina and Garattini, 1967), and cinnarizine (Godfraind *et al.*, 1968), presumably by interference with calcium movement, since the effect is antagonized by raised calcium (Shibata *et al.*, 1968). In aortic strips, the block of potassium contracture by chlorpromazine, phenoxybenzamine, and dibenamine is not accompanied by any equivalent effect on contractures produced by angiotensin (Bevan *et al.*, 1963; Shibata and Carrier, 1967), suggesting that the mode of action is similar to that of SKF 525-A. These agents also act on the smooth muscle of taenia to inhibit the response to potassium and ACh, and at this site they do block the action of angiotensin, perhaps indicating that the calcium mobilized by angiotensin in gut is not as firmly bound as that in aortic strips. The inhibition of contraction was shown to be associated with reduction of ^{45}Ca uptake (Shibata *et al.*, 1968). They did not, however, inhibit contraction induced by barium. In mesenteric arteries, cinnarizine, in contrast to chlorpromazine, does not inhibit the contractile response to noradrenaline found in calcium-free, high potassium, depolarizing solution (Godfraind and Kaba, 1969) in which condition the response is presumably mediated by tightly bound calcium. It does, however, block responses to noradrenaline in normally polarized muscle (Godfraind and Kaba, 1969), suggesting that normally noradrenaline acts indirectly via depolarization and hence loosely bound calcium to elicit muscle contraction.

It should be noted that the concentration of phenoxybenzamine and chlorpromazine required for the nonspecific inhibition of contraction is considerably higher than that required for α-adrenergic blockade (Bevan *et al.*, 1963), and the effect is not shared by other α blockers, dihydroergotamine, and phentolamine (Shibata *et al.*, 1968).

3. Phosphodiesterase Inhibitors

The evidence for a role of cyclic AMP in mediating the relaxant effects on smooth muscle of β-adrenergic agents has already been discussed (see p. 82). One of the major points in its favor has been the potentiation of β-mediated effects by theophylline, well known as inhibitor of phosphodiesterase, the enzyme that inactivated cyclic AMP. Other agents that are inhibitors of the phosphodiesterase are papaverine (Triner *et al.*, 1970) and the benzothiazides, well known as antihypertensive agents, including diazoxide (Moore, 1968; Senft *et al.*, 1968). In guinea pig taenia coli, papaverine has been found to be very similar in its action to isoprenaline, causing suppression of action potential activity with little change in the membrane potential or in membrane conductance. The only differences are that papaverine is rather more effective in antagonizing a contracture produced by raised potassium than is isoprenaline, and it is not influenced by propranolol, which blocks the action of isoprenaline (Tashiro and Tomita, 1970). These results support the hypothesis that papaverine works simply by phosphodiesterase inhibition, there being a resting production of cyclic AMP even in the absence of β-adrenergic stimulation. If this resting cyclic AMP production were increased by raised potassium, it would account for papaverine having greater effectiveness than isoprenaline in the depolarized preparation.

In isolated rabbit ileum, theophylline and diazoxide have both been shown to potentiate the relaxant effect of isoprenaline, but has no effect on the inhibitory action of phenylephrine, which is a pure α-adrenoceptor agonist (Wilkenfeld and Levy, 1969). This observation supports the hypothesis that diazoxide, like theophylline, acts by inhibition of phosphodiesterase, since cyclic AMP is considered to be second messenger for the inhibitory response mediated by β-adrenoceptors but not that mediated by α-adrenoceptors.

Whether phosphodiesterase inhibition accounts fully for the inhibitory action of diazoxide on vascular smooth muscle has not yet been determined. Rhodes and Sutter (1972) have found, in addition to inhibition of spontaneous electrical activity and that induced by noradrenaline, 5-HT, and procaine, an action to antagonize calcium-induced contractures in depolarized anterior mesenteric vain, although noradrenaline induced contraction in ouabain-depolarized veins were not inhibited. However, whether dibutyrylcyclic AMP acts in the same way was not investigated. The finding of McNeill *et al.* (1969) that diazoxide is capable of inhibiting noradrenaline contractions in dog mesenteric artery, where isoprenaline does not act, indicates merely the absence of β-receptors in this tissue rather than the absence of the systems required for cyclic AMP production and effect on contraction.

4. DIPYRIDAMOLE AND NITRITE

Dipyridamole is a general smooth muscle relaxant, used for treatment of angina pectoris, that acts to potentiate markedly the vasodilator action of adenosine and adenosine triphosphate (Afonso and O'Brien, 1967). It is known to have effects on nucleotide metabolism, blocking the enzymic breakdown of adenosine, probably by blocking uptake into erythrocytes (Bunag *et al.*, 1964). These observations could explain the general relaxant effects of the drug, in view of the recent evidence that an adenine nucleotide may function as an inhibitory transmitter in intestine and vascular smooth muscle (see above, p. 93).

The mechanism by which nitrites and organic nitrates relax smooth muscle is unknown and appears not to have been the subject of much investigation, except to rule out selective action. The relaxant effect on a tissue is independent of how the tissue responds to various agonists such as ACh, histamine, noradrenaline, etc., and does not involve blockade of receptors to such compounds. Keatinge (1966) reported that nitrites relaxed and repolarized sheep carotid arterial strips which had been partially contracted and depolarized by either noradrenaline or histamine.

5. Caffeine

Caffeine appears to have three distinct actions on frog sartorius muscle contraction. In low concentrations (0.05–1 mM) caffeine acts primarily on sites related to coupling muscle contraction to the action potential (e.g., Sandow and Brust, 1966), the result being potentiation of twitch amplitude. Intermediate concentrations (1–5 mM) cause a contracture that does not require membrane depolarization and is not accompanied by membrane depolarization (Axelsson and Thesleff, 1958). At concentrations higher than 5 mM, caffeine causes an irreversible rigor. Frank (1962) demonstrated the requirement for calcium of reversible caffeine contractures. The irreversible contracture can occur in muscles in which surface calcium has been removed by chelation with EDTA, and probably reflects blockade of calcium uptake by the sarcoplasmic reticulum (Weber and Herz, 1968). The reversible caffeine contracture can be blocked by local anaesthetics in the charged but not the uncharged form, which potentiates caffeine action (Bianchi and Bolton, 1967). The action of caffeine is exerted intracellularly, where it acts to enhance influx and efflux of calcium (Bianchi, 1961, 1962), the increase of intracellular calcium being the trigger for contractile activity. Exactly how caffeine exerts this effect is still unclear. Lüttgau and Oetliker (1968) have put forward evidence that the T system is the site of drug action, and that the drug affects both activation and relaxation processes, acting to lower the threshold for potassium contracture and to impede relaxation. It may be relevant that caffeine acts on crayfish muscle fibers to provoke calcium action potential activity, an effect attributable to increased calcium permeability of the surface membrane (Chiarandini *et al.*, 1970). Howell (1969) has found that caffeine contractures persist in glycerol-treated muscle, which does not respond to electrical stimulation.

In smooth muscle, caffeine produces only a transient contraction followed by relaxation (e.g., A. V. Somlyo and Somlyo, 1968). The relaxation may owe something to phosphodiesterase inhibition, but it is of interest that similar results are observed with local and general anaesthetics which in high concentrations produce contractures in skeletal muscle but relax smooth muscle (see above).

6. RYANODINE

This alkaloid has the remarkable effect of causing a slowly developing contracture in skeletal muscle. The rate of development of contracture is increased by stimulation and can be affected by adrenergic agents; i.e., the rate of contracture development is a function of to what extent the muscle contractile activity is stimulated (e.g., N. L. Katz *et al.*, 1970). The pharmacology of ryanodine has recently been reviewed by Jenden and Fairhurst (1969), who concluded that "ryanodine interferes with intracellular Ca translocation mechanisms that normally lower the sarcoplasmic concentration and thereby effect relaxation."

In guinea pig or rat uterus, ryanodine was found to have no effect whatsoever (Haslett and Jenden, 1961). In intestinal smooth muscle, on the other hand, ryanodine caused a progressive increase in tone (Hillyard and Procita, 1958). This increase in tone was slowed by low calcium. Relaxing agents, including adrenaline and papaverine, restored tone to normal.

Heart muscle is again different; here ryanodine causes a progressive decline of contractile force (see Jenden and Fairhurst, 1969). Evidently, the mechanisms by which relaxation is effected vary in ryanodine sensitivity from one type of muscle to another.

REFERENCES

Abbs, E. T. (1966). *Brit. J. Pharmacol. Chemother.* **26**, 162–171.

Abbs, E. T., and Robertson, M. I. (1970). *Brit. J. Pharmacol.* **38**, 776–791.

Adrian, E. D., Feldberg, W., and Kilby, B. A. (1947). *Brit. J. Pharmacol. Chemother.* **2**, 56–58.

Adrian, R. H., Chandler, W. K., and Hodgkin, A. L. (1969). *J. Physiol. (London)* **204**, 207–230.

Afonso, S., and O'Brien, G. S. (1967). *Circ. Res.* **20**, 403–408.

Ahlqvist, R. P. (1948). *Amer. J. Physiol.* **153**, 586–600.

Ahlqvist, R. P. (1962). *Arch. Int. Pharmacodyn. Ther.* **139**, 38–41.

Ahlqvist, R. P. (1966). *J. Pharm. Sci.* **55**, 359–367.

Aiken, J. W., and Reit, E. (1969). *J. Pharmacol. Exp. Ther.* **169**, 211–223.

Akert, K., and Sandri, C. (1968). *Brain Res.* **7**, 286–295.

Akert, K., Sandri, C., and Pfenninger, K. (1968). *Proc. 4th Europ. Regional Conf. Electronic Microscopy, Rome.* Pp. 521–522.

Albuquerque, E. X., and Thesleff, S. (1968). *Acta Physiol. Scand.* **72**, 248–252.

Albuquerque, E. X., Sokoll, M. D., Sonesson, B., and Thesleff, S. (1968). *Eur. J. Pharmacol.* **4**, 40–46.

Ambache, N. (1946). *J. Physiol. (London)* **104**, 266–287.

Ambache, N. (1951). *Brit. J. Pharmacol. Chemother.* **6**, 51–67.

Ambache, N. (1955). *Pharmacol. Rev.* **7**, 467–494.

Ambache, N., and Freeman, M. A. (1968). *J. Physiol. (London)* **198**, 92P–94P.

Ambache, N., and Lessin, A. W. (1955). *J. Physiol. (London)* **127**, 449–478.

Ambache, N., and Zar, M. A. (1970). *J. Physiol. (London)* **210**, 761–783.

Ambache, N., Brummer, H. C., Rose, J. G., and Whiting, J. (1966). *J. Physiol. (London)* **185**, 77P–78P.

Andén, N.-E. (1964). *Acta Pharmacol. Toxicol.* **21**, 260–271.

Andén, N.-E., Magnusson, T., and Waldeck, B. (1964). *Life Sci.* **3**, 149–158.

Andersen, P., and Curtis, D. R. (1964). *Acta Physiol. Scand.* **61**, 85–99.

Antonio, A., Rocha e Silva, M., and Yashuda, Y. (1970). *Brit. J. Pharmacol.* **40**, 501–507.

Ariëns, E. J. (1964). "Molecular Pharmacology," Academic Press, New York.

Ariëns, E. J. (1967). *Ann. N.Y. Acad. Sci.* **139**, 606–631.

Armstrong, C. M., and Binstock, L. (1965). *J. Gen. Physiol.* **48**, 859–872.

Armstrong, D., Jepson, J. B., Keele, C. A., and Stewart, J. W. (1957). *J. Physiol. (London)* **135**, 350–370.

Austin, L., Chubb, I. W., and Livett, B. G. (1967). *J. Neurochem.* **14**, 473–475.

Axelsson, J. (1970). *In* "Smooth Muscle" (E. Bulbring *et al.*, eds.), pp. 289–315. Arnold, London.

Axelsson, J., and Högberg, S. G. R. (1967). *Acta Pharmacol. Toxicol.* **25**, Suppl. 4, 53.

Axelsson, J., and Holmberg, B. (1969). *Acta Physiol. Scand.* **75**, 149–156.

Axelsson, J., and Thesleff, S. (1958). *Acta Physiol. Scand.* **44**, 55–66.

Axelsson, J., and Thesleff, S. (1959). *J. Physiol. (London)* **149**, 178–193.

Axelsson, J., Gjone, E., and Naess, K. (1957). *Acta Pharmacol. Toxicol.* **13**, 319–336.

Ayitey-Smith, E., and Varma, D. R. (1970). *Brit. J. Pharmacol.* **40**, 186–193.

Bacq, Z. M., and Monnier, A. M. (1935). *Arch. Int. Physiol.* **40**, 467–484.

Bainbridge, J. G., and Brown, D. M. (1960). *Brit. J. Pharmacol. Chemother.* **15**, 147–151.

Baker, P. F., Blaustein, M. P., Hodgkin, A. L., and Steinhardt, R. A. (1969). *J. Physiol. (London)* **200**, 431–59.

Barger, G., and Dale, H. H. (1910). *J. Physiol. (London)* **41**, 19–59.

Barker, D. (1948). *Quart. J. Microsc. Sci.* **89**, 143–186.

Barrett, A. M., Crowther, A. F., Dunlop, D., Shanks, R. G., and Smith, L. H. (1968). *Naunyn-Schmiedebergs Arch. Exp. Pathol. Pharmakol.* **259**, 152–153.

Barstad, J. A. B. (1962). *Experientia* **18**, 579–580.

Bartels, J., Vogt, W., and Wille, G. (1968). *Naunyn-Schmiedebergs Arch. Exp. Pathol. Pharmakol.* **259**, 153–154.

Beani, L., Bianchi, C., and Ledda, F. (1966). *Brit. J. Pharmacol. Chemother.* **27**, 299–312.

Beck, L. (1964). *Tex. Rep. Biol. Med.* **22**, 375–409.

Beck, L. (1965). *Fed. Proc., Fed. Amer. Soc. Exp. Biol.* **24**, 1298–1310.

Beck, L., and Brody, M. J. (1961). *Angiology* **12**, 202–222.

Bell, C. (1968). *Brit. J. Pharmacol. Chemother.* **32**, 96–104.

Bennett, A., Buchnell, A., and Dean, A. C. B. (1966). *J. Physiol. (London)* **182**, 57–65.

Bennett, A., Friedmann, C. A., and Vane, J. R. (1967). *Nature (London)* **216**, 873–876.

Bennett, A., Murray, J. G., and Wylie, J. H. (1968). *Brit. J. Pharmacol. Chemother.* **32**, 339–349.

Bennett, M. R. (1966a). *Nature (London)* **211**, 1149–1152.

Bennett, M. R. (1966b). *J. Physiol. (London)* **185**, 132–147.

Bennett, M. R. (1967). *J. Gen. Physiol.* **50**, 2459–2475.

Bennett, M. R., and Rogers, D. C. (1967). *J. Cell Biol.* **33**, 573–596.

Bennett, M. R., Burnstock, G., and Holman, M. E. (1966a). *J. Physiol. (London)* **182**, 527–540.

Bennett, M. R., Burnstock, G., and Holman, M. E. (1966b). *J. Physiol. (London)* **182**, 541–548.

Benoit, P. R., and Mambrini, J. (1970). *J. Physiol. (London)* **210**, 681–695.

Bentley, G. A. (1966). *Brit. J. Pharmacol. Chemother.* **27**, 64–80.

Beránek, R., and Vyskočil, F. (1967). *J. Physiol. (London)* **188**, 53–66.

Berde, B., Boissonnas, R. A., Huguenin, R. L., and Stürmer, E. (1964). *Experientia* **20**, 42–43.

Berger, E., and Marshall, M. (1961). *Amer. J. Physiol.* **201**, 931–934.

Bergström, S., Carlson, L. A., and Weeks, J. R. (1968). *Pharmacol. Rev.* **20**, 1–48.

Bernard, C. (1856). *C. R. Acad. Sci.* **43**, 825–829.

Bernard, P. J., Rowe, H. M., and Gardier, R. W. (1968). *Arch. Int. Pharmacodyn. Ther.* **171**, 217–220.

Berney, J., and Posternak, J. (1956). *Helv. Physiol. Pharmacol.* **14**, C5.

Berti, F., and Usardi, M. M. (1964). *G. Arterioscler.* **2**, 261–265.

Bertler, Å., Hillarp, N.-Å., and Rosengren, E. (1961). *Acta Physiol. Scand.* **52**, 44–48.

Bevan, J. A., Osher, J. V., and Su, C. (1963). *J. Pharmacol. Exp. Ther.* **139**, 216–221.

Bianchi, C. P. (1961). *J. Gen. Physiol.* **44**, 845–858.

Bianchi, C. P. (1962). *J. Pharmacol. Exp. Ther.* **157**, 388–405.

Bianchi, C. P., and Bolton, T. C. (1967). *J. Pharmacol. Exp. Ther.* **157**, 388–405.

Birks, R. I. (1963). *Can. J. Biochem. Physiol.* **41**, 2573–2597.

Birks, R. I., and Cohen, M. W. (1968a). *Proc. Roy. Soc., Ser. B* **170**, 381–399.

Birks, R. I., and Cohen, M. W. (1968b). *Proc. Roy. Soc., Ser. B* **170**, 401–421.

Birks, R. I., and MacIntosh, F. C. (1961). *Can. J. Biochem. Physiol.* **39**, 787–827.

Birks, R. I., Burstyn, P. G. R., and Firth, D. R. (1968). *J. Gen. Physiol.* **52**, 887–908.

Bisset, G. W., and Clark, B. J. (1968). *Nature (London)* **218**, 197–199.

Bisson, G. M., and Muscholl, E. (1962). *Naunyn-Schmiedebergs Arch. Exp. Pathol. Pharmakol.* **244**, 185–194.

Blaber, L. C. (1970). *J. Pharmacol. Exp. Ther.* **175**, 664–672.

Blaber, L. C., and Bowman, W. C. (1963). *Brit. J. Pharmacol. Chemother.* **20**, 326–344.

Black, G. W., and McArdle, L. (1962). *Brit. J. Anaesth.* **34**, 2–10.

Blakely, A. G. H., Brown, G. L., and Ferry, C. B. (1963). *J. Physiol. (London)* **167**, 505–514.

Blaschko, H. (1939). *J. Physiol. (London)* **96**, 500.

Blaustein, M. P. (1968). *J. Gen. Physiol.* **51**, 293–307.

Blaustein, M. P., and Goldman, D. E. (1968). *J. Gen. Physiol.* **51**, 279–291.

Blinks, J. R. (1966). *J. Pharmacol. Exp. Ther.* **151**, 221–235.

Blinks, J. R. (1967). *Ann. N.Y. Acad. Sci.* **139**, 673–685.

Blioch, L., Glagoleva, I. M., Liberman, E. A., and Nenashev, V. A. (1968). *J. Physiol.* (*London*) 199, 11–37.

Bogdanski, D. F., and Brodie, B. B. (1966). *Life Sci.* 5, 1563–1569.

Boissonnas, R. A., Guttmann, S., and Jaguenod, P. A. (1960). *Helv. Chim. Acta* 43, 1349–1358.

Bondareff, W., and Gordon, B. (1966). *J. Pharmacol. Exp. Ther.* 153, 42–47.

Born, G. V. R. (1970). *In* "Smooth Muscle" (E. Bülbring *et al.*, eds.), pp. 418–450. Arnold, London.

Born, G. V. R., and Bülbring, E. (1956). *J. Physiol.* (*London*) 131, 690–703.

Boullin, D. J. (1967). *J. Physiol.* (*London*) 189, 85–89.

Boura, A. L. A., and Green, A. F. (1959). *Brit. J. Pharmacol. Chemother.* 14, 536–548.

Boura, A. L. A., and Green, A. F. (1965). *Annu. Rev. Pharmacol.* 5, 183–212.

Boura, A. L. A., Copp, F. C., Duncombe, W. G., Green, A. F., and McCoubrey, A. (1960). *Brit. J. Pharmacol. Chemother.* 15, 265–270.

Bovet, D. (1951). *Ann. N.Y., Acad. Sci.* 54, 407–437.

Bovet, D., Bovet-Nitti, F., Bettschart, A., and Sconamiglio, W. (1956). *Helv. Physiol. Pharmacol. Acta* 14, 430–440.

Bowman, W. C., and Hall, M. T. (1970). *Brit. J. Pharmacol.* 38, 399–415.

Bowman, W. C., and Nott, M. W. (1969). *Pharmacol. Rev.* 21, 27–72.

Bowman, W. C., and Nott, M. W. (1970). *Brit. J. Pharmacol.* 38, 37–49.

Bowman, W. C., and Rand, M. J. (1961). *Lancet* 1, 480–481.

Bowman, W. C., and Raper, C. (1962). *Nature* (*London*) 193, 41–43.

Bowman, W. C., and Raper, C. (1964). *Brit. J. Pharmacol. Chemother.* 23, 184–200.

Bowman, W. C., and Raper, C. (1965). *Brit. J. Pharmacol. Chemother.* 24, 98–109.

Bowman, W. C., and Raper, C. (1967). *Ann. N.Y. Acad. Sci.* 139, 741–753.

Bowman, W. C., and Zaimis, E. (1958). *J. Physiol.* (*London*) 144, 92–107.

Bowman, W. C., Hemsworth, B. A., and Rand, M. J. (1962). *Brit. J. Pharmacol. Chemother.* 19, 198–218.

Boyd, H., Burnstock, G., Campbell, G., Jowett, A., O'Shea, J., and Wood, M. (1963). *Brit. J. Pharmacol. Chemother.* 20, 418–435.

Boyd, I. A., and Forrester, T. (1968). *J. Physiol.* (*London*) 199, 115–135.

Bozler, E. (1948). *Experientia* 4, 213–218.

Brading, A. F., and Jones, A. W. (1969). *J. Physiol.* (*London*) 200, 387–401.

Brading, A. F., Bulbring, E., and Tomita, T. (1969a). *J. Physiol.* (*London*) 200, 621–635.

Brading, A. F., Bulbring, E., and Tomita, T. (1969b). *J. Physiol.* (*London*) 200, 637–654.

Briggs, A. H. (1962). *Amer. J. Physiol.* 203, 849–852.

Brodie, B. B., and Kuntzman, R. (1960). *Ann. N.Y. Acad. Sci.* 88, 939–943.

Brodie, B. B., Costa, E., Groppetti, A., and Matsumoto, C. (1968). *Brit. J. Pharmacol.* 34, 648–658.

Brody, M. J. (1968). *Fed. Proc., Fed. Amer. Soc. Exp. Biol.* 27, 756.

Brody, M. J., Du Charme, D. W., and Beck, L. (1967). *J. Pharmacol. Exp. Ther.* 155, 84–90.

Brooks, V. B. (1956). *J. Physiol.* (*London*) 134, 264–77.

Brown, A. M. (1967). *J. Physiol.* (*London*) 191, 271–288.

Brown, G. L., and Gillespie, J. S. (1957). *J. Physiol.* (*London*) 138, 81–102.

Brown, M. C., and Matthews, P. B. C. (1960). *J. Physiol.* (*London*) 150, 332–346.

Brownlee, G., and Johnson, E. S. (1965). *Brit. J. Pharmacol. Chemother.* 24, 689–700.

Bueding, E., Butcher, R. W., Hawkins, J., Timms, A. R., and Sutherland, E. W. (1966). *Biochim. Biophys. Acta* **115**, 173–178.
Bülbring, E. (1957). *J. Physiol. (London)* **135**, 412–425.
Bülbring, E., and Burn, J. H. (1938). *J. Physiol. (London)* **91**, 459–473.
Bülbring, E., and Burnstock, G. (1960). *Brit. J. Pharmacol. Chemother.* **15**, 611–624.
Bülbring, E., and Gershon, M. D. (1967). *J. Physiol. (London)* **192**, 823–846.
Bülbring, E., and Kuriyama, H. (1963a). *J. Physiol. (London)* **166**, 29–58.
Bülbring, E., and Kuriyama, H. (1963b). *J. Physiol. (London)*. **166**, 59–74.
Bülbring, E., and Kuriyama, H. (1963c). *J. Physiol. (London)* **169**, 198–212.
Bülbring, E., and Tomita, T. (1966). *J. Physiol. (London)* **185**, 24P–25P.
Bülbring, E., and Tomita, T. (1967). *J. Physiol. (London)* **189**, 299–315.
Bülbring, E., and Tomita, T. (1968a). *J. Physiol. (London)* **194**, 74P–76P.
Bülbring, E., and Tomita, T. (1968b). *J. Physiol. (London)* **196**, 137P–139P.
Bülbring, E., and Tomita, T. (1969a). *Proc. Roy. Soc., Ser. B* **172**, 89–102.
Bülbring, E., and Tomita, T. (1969b). *Proc. Roy. Soc. Ser. B* **172**, 103–119.
Bülbring, E., and Tomita, T. (1969c). *Proc. Roy. Soc., Ser. B* **172**, 121–136.
Bülbring, E., and Tomita, T. (1970). *J. Physiol. (London)* **210**, 217–232.
Bülbring, E., Goodford, P. J., and Setekleiv, J. (1966). *Brit. J. Pharmacol. Chemother.* **28**, 296–307.
Bülbring, E., Brading, A. F., Jones, A. W., and Tomita, T., eds. (1970). "Smooth Muscle." Arnold, London.
Bull, G., and Hemsworth, B. A. (1963). *Nature (London)* **199**, 487–488.
Bunag, R. D., Douglas, C. R., Imai, S., and Berne, R. M. (1964). *Circ. Res.* **15**, 83–88.
Burgen, A. S. V. (1970). *Annu. Rev. Pharmacol.* **10**, 7–18.
Burgen, A. S. V., and Iversen, L. L. (1965). *Brit. J. Pharmacol. Chemother.* **25**, 34–49.
Burgen, A. S. V., and Spero, L. (1968). *Brit. J. Pharmacol.* **34**, 99–115.
Burgen, A. S. V., and Spero, L. (1970). *Brit. J. Pharmacol.* **40**, 492–500.
Burks, T. F., and Long, J. P. (1966a). *J. Pharm. Sci.* **55**, 1383–1386.
Burks, T. F., and Long, J. P. (1966b). *Amer. J. Physiol.* **211**, 619–625.
Burks, T. F., and Long, J. P. (1967a). *Brit. J. Pharmacol. Chemother.* **30**, 229–239.
Burks, T. F., and Long, J. P. (1967b). *J. Pharmacol. Exp. Ther.* **156**, 267–276.
Burn, J. H. (1932). *J. Pharmacol. Exp. Ther.* **46**, 75–95.
Burn, J. H. (1963). *Bull. Johns Hopkins Hosp.* **112**, 167–182.
Burn, J. H., and Gibbons, W. R. (1965). *J. Physiol. (London)* **181**, 214–223.
Burn, J. H., and Hobbs, R. (1958). *Lancet* **1**, 1112–1115.
Burn, J. H., and Rand, M. J. (1958). *J. Physiol.* **144**, 314–336.
Burn, J. H., and Rand, M. J. (1959). *Nature (London)* **184**, 163–165.
Burn, J. H., and Rand, M. J. (1960). *Brit. J. Pharmacol. Chemother.* **15**, 47–55.
Burn, J. H., and Tainter, M. L. (1931). *J. Physiol. (London)* **71**, 169–193.
Burns, J. J., Salvador, R. A., and Lemberger, L. (1967). *Ann. N.Y. Acad. Sci.* **139**, 833–840.
Burnstock, G. (1958). *J. Physiol. (London)* **143**, 165–182.
Burnstock, G. (1960). *Nature (London)* **186**, 727–728.
Burnstock, G. (1970). *In* "Smooth Muscle" (E. Bülbring *et al.*, eds.), pp. 1–69. Arnold, London.
Burnstock, G., and Holman, M. E. (1962a). *J. Physiol. (London)* **160**, 446–460.
Burnstock, G., and Holman, M. E. (1962b). *J. Physiol. (London)* **160**, 461–469.

Burnstock, G., and Holman, M. E. (1964). *Brit. J. Pharmacol. Chemother.* **23**, 600–612.

Burnstock, G., and Holman, M. E. (1966). *Pharmacol. Rev.* **18**, 481–493.

Burnstock, G., and Prosser, C. L. (1960). *Proc. Soc. Exp. Biol. Med.* **103**, 269–270.

Burnstock, G., Holman, M. E., and Prosser, C. L. (1963). *Physiol. Rev.* **43**, 482–527.

Burnstock, G., Holman, M. E., and Kuriyama, H. (1964a). *J. Physiol. (London)* **172**, 31–49.

Burnstock, G., Campbell, G., Bennett, M. R., and Holman, M. E. (1964b). *Int. J. Neuropharmacol.* **31**, 163–166.

Burnstock, G., Campbell, G., Satchell, D., and Smythe, A. (1970). *Brit. J. Pharmacol.* **40**, 668–688.

Butcher, R. W. (1970). *Advan. Biochem. Psychopharmacol.* **3**, 173–183.

Butcher, R. W., and Sutherland, E. W. (1962). *J. Biol. Chem.* **237**, 1244–1250.

Butcher, R. W., Ho, R. J., Meng, H. C., and Sutherland, E. W. (1965). *J Biol Chem.* **240**, 4515–4523.

Callingham, B. A., and Burgen, A. S. V. (1966). *Mol. Pharmacol.* **2**, 37–42.

Campbell, A. G. M., Dawes, G. S., Fishman, A. P., Hyman, A. I., and Perks, A. M. (1968). *J. Physiol. (London)* **195**, 83–96.

Campbell, G. (1970). *In* "Smooth Muscle" (E. Bülbring *et al.*, eds.), pp. 451–495. Arnold, London.

Carlsson, A. (1964). *Progr. Brain Res.* **8**, 9–27.

Carlsson, A. (1966). *Pharmacol. Rev.* **18**, 541–549.

Carlsson, A., and Lindqvist, M. (1962). *Acta Physiol. Scand.* **54**, 87–91.

Carlsson, A., and Waldeck, B. (1965). *Acta Pharmacol. Toxicol.* **22**, 293–300.

Carlsson, A., and Waldeck, B. (1966). *Acta Physiol. Scand.* **67**, 471–480.

Carlsson, A., Rosengren, E., Bertler, Å., and Nilsson, J. (1957). *In* "Pyschotropic Drugs" (S. Garattini and V. Ghetti, eds.), pp. 363–372. Elsevier, Amsterdam.

Carlsson, A., Falck, B., and Hillarp, N.-Å. (1962). *Acta Physiol. Scand.* **56**, Suppl. 196, 1–28.

Carlsson, A., Corrodi, H., and Waldeck, B. (1963a). *Helv. Chim. Acta* **46**, 2271–2285.

Carlsson, A., Hillarp, N.-Å., and Waldeck, B. (1963b). *Acta Physiol. Scand.* **59**, Suppl. 215, 1–38.

Carlsson, A., Fuxe, K., Hamberger, B., and Lindqvist, M. (1966). *Acta Physiol. Scand.* **67**, 481–497.

Carlyle, R. F. (1963). *Brit. J. Pharmacol. Chemother.* **21**, 137–149.

Carpenter, F. G. (1963). *J. Physiol. (London)* **204**, 727–731.

Carpenter, F. G. (1967). *J. Physiol. (London)* **188**, 1–12.

Carpenter, F. G., and Rand, S. A. (1965). *J. Physiol. (London)* **180**, 371–382.

Cass, R., and Spriggs, T. L. B. (1961). *Brit. J. Pharmacol. Chemother.* **17**, 442–450.

Cavallito, C. (1967). *Ann. N.Y. Acad. Sci.* **144**, 900–911.

Chiarandini, D. J., Rueben, J. P., Brandt, P. W., and Grundfest, H. (1970). *J. Gen. Physiol.* **55**, 640–664.

Chidsey, C. A., Harrison, D. C., and Braunwald, E. (1962). *Proc. Soc. Exp. Biol. Med.* **109**, 488–490.

Choksey, H. K., and Jindal, M. N. (1965). *Arch. Int. Pharmacodyn. Ther.* **157**, 339–346.

Ciani, S., and Edwards, C. (1963). *J. Pharmacol. Exp. Ther.* **142**, 21–23.

Ciuchta, H. P., and Gautieri, R. F. (1963). *J. Pharm. Sci.* **52**, 974–978.

Clark, A. W., Mauro, A., Longenecker, H. E., Jr., and Hurlbut, W. P. (1970). *Nature (London)* **225**, 703–705.

Clarke, D. E. (1970). *Brit. J. Pharmacol.* **38**, 1–11.
Clegg, P. C. (1966). *Nature (London)* **209**, 1137–1139.
Clement, D., Vanhoutte, P., and Leusen, I. (1969). *Arch. Int. Physiol.* **77**, 73–87.
Cliff, W. J. (1967). *Lab. Invest.* **17**, 599–615.
Coceani, F., and Wolfe, L. S. (1966). *Can. J. Physiol. Pharmacol.* **44**, 933–950.
Coceani, F., Pace-Asciak, C., Volta, F., and Wolfe, L. S. (1967). *Amer. J. Physiol.* **213**, 1056–1064.
Collier, B., and Katz, H. S. (1970). *Brit. J. Pharmacol.* **39**, 428–438.
Collier, B., and MacIntosh, F. C. (1969). *Can. J. Physiol. Pharmacol.* **47**, 127–135.
Cooke, J. D., and Quastel, D. M. J. (1969). *Abstr. Int. Biophys. Congr., 3rd,* 1900 p. 99.
Cooke, J. D., and Quastel, D. M. J. (1973). In preparation.
Corrado, A. P., and Ramos, A. D. (1960). *Rev. Brasil. Biol.* **20**, 43–50.
Corrado, A. P., Ramos, A. O., and De Escobar, C. T. (1959). *Arch. Int. Pharmacodyn. Ther.* **121**, 380–394.
Costa, E., Chang, C. C., and Brodie, B. B. (1964). *Pharmacologist* **6**, 174.
Costantin, L. L. (1970). *J. Gen. Physiol.* **55**, 703–715.
Couteaux, R. (1958). *Exp. Cell Res., Suppl.* **5**, 294–322.
Creed, K. E., and Wilson, J. A. F. (1969). *Aust. J. Exp. Biol. Med. Sci.* **47**, 135–144.
Crout, J. R., Alpers, H. S., Tatum, E. L., and Shore, P. A. (1964). *Science* **145**, 828–829.
Curtis, D. R., and Ryall, R. W. (1966). *Exp. Brain Res.* **2**, 81–96.
Cuthbert, A. W. (1967). *Bibl. Ana.* **8**, 11–15.
Cuthbert, A. W., and Sutter, M. C. (1964). *Nature (London)* **202**, 15.
Cuthbert, A. W., and Sutter, M. C. (1965). *Brit. J. Pharmacol. Chemother.* **25**, 592–601.
Dale, H. H. (1937). *J. Mt. Sinai Hosp., New York* **4**, 401–415.
Dale, H. H., and Laidlaw, P. P. (1910). *J. Physiol. (London)* **41**, 318–344.
Dale, H. H., and Laidlaw, P. P. (1911). *J. Physiol. (London)* **43**, 182–195.
Daniels, E. G., Hinman, J. W., Leach, B. E., and Muirhead, E. E. (1967). *Nature (London)* **215**, 1298–1299.
Davidson, W. J., and Innes, I. R. (1970). *Brit. J. Pharmacol.* **39**, 175–181.
Davies, B. N., and Withrington, P. G. (1968). *Brit. J. Pharmacol. Chemother.* **32**, 136–144.
Davies, B. N., Horton, E. W., and Withrington, P. G. (1966). *J. Physiol. (London)* **188**, 38P–39P.
Davies, B. N., Horton, E. W., and Withrington, P. G. (1968). *Brit. J. Pharmacol. Chemother.* **32**, 127–135.
Davis, R., and Koelle, G. B. (1967). *J. Cell Biol.* **34**, 157–171.
Day, M. D. (1962). *Brit. J. Pharmacol. Chemother.* **18**, 421–439.
Day, M. D., and Rand, M. J. (1962). *J. Pharm. Pharmacol.* **14**, 541–549.
Day, M. D., and Rand, M. J. (1963). *J. Pharm. Pharmacol.* **15**, 221–224.
Day, M. D., and Rand, M. J. (1964). *Brit. J. Pharmacol. Chemother.* **22**, 72–86.
DeIraldi, A. P., and De Robertis, E. (1963). *Int. J. Neuropharmacol.* **2**, 231–239.
de la Lande, I. S., and Waterson, J. G. (1968). *Brit. J. Pharmacol.* **34**, 8–18.
de la Lande, I. S., Cannall, V. A., and Waterson, J. G. (1966). *Brit. J. Pharmacol. Chemother.* **28**, 255–272.
de la Lande, I. S., Frewin, D., and Waterson, J. G. (1967). *Brit. J. Pharmacol. Chemother.* **31**, 82–93.

del Castillo, J., and Katz, B. (1955). *Nature* (*London*) **175**, 1035.

del Castillo, J., and Katz, B. (1956). *Progr. Biophys. Biophys. Chem.* **6**, 121–170.

del Castillo, J., and Katz, B. (1957). *Proc. Roy. Soc., Ser. B* **146**, 339–356.

Dengler, H. J. (1964). *Proc. Int. Pharmacol. Meet., 2nd, 1963* Vol. 3, pp. 261–275.

Dengler, H. J., and Reichel, G. (1957). *Naunyn-Schmiedebergs Arch. Exp. Pathol. Pharmakol.* **232**, 324–326.

de Schaepdryver, A. F., Bogaert, M., Delannois, A. L., and Bernard, P. (1963). *Arch. Int. Pharmacodyn. Ther.* **142**, 243–259.

Detwiler, P. B., and Standaert, F. G. (1969). *Fed. Proc., Fed. Amer. Soc. Exp. Biol.* **28**, 669.

Dhattiwala, A. S., Jindal, M. N., and Kelkar, V. V. (1970). *Brit. J. Pharmacol.* **39**, 738–747.

Dixon, W. E., and Brodie, T. G. (1903). *J. Physiol.* (*London*) **29**, 97–173.

Dockry, M., Kernan, R. P., and Tangney, A. (1966). *J. Physiol.* (*London*) **186**, 187–200.

Dodge, F. A., Jr., and Rahamimoff, R. (1967). *J. Physiol.* (*London*) **193**, 419–433.

Dorward, P., and Holman, M. E. (1967). *Aust. J. Exp. Biol. Med. Sci.* **45**, 48P.

Douglas, W. W. (1951). *J. Physiol.* (*London*) **112**, 20P.

Douglas, W. W. (1968). *Brit. J. Pharmacol.* **34**, 451–474.

Douglas, W. W., Kanno, T., and Sampson, S. R. (1967). *J. Physiol.* (*London*) **188**, 107–120.

Dudel, J., and Kuffler, S. W. (1961). *J. Physiol.* (*London*) **155**, 514–529.

Durbin, R. P., and Jenkinson, D. H. (1961). *J. Physiol.* (*London*) **157**, 74–89.

Eakins, K. E., and Katz, R. L. (1966). *Brit. J. Pharmacol. Chemother.* **26**, 205–211.

Eakins, K. E., and Katz, R. L. (1967). *J. Pharmacol. Exp. Ther.* **157**, 524–531.

Eakins, K. E., Karim, S. M. N., and Miller, J. D. (1970). *Brit. J. Pharmacol.* **39**, 556–563.

Ebashi, S., and Endo, M. (1968). *Progr. Biophys. Mol. Biol.* **18**, 125–183.

Eccles, R. M., and Libet, B. (1961). *J. Physiol.* (*London*) **157**, 484–503.

Edwards, C., and Ikeda, K. (1962). *J. Pharmacol. Exp. Ther.* **138**, 322–328.

Ehinger, B. (1966). *Acta Physiol. Scand.* **67**, Suppl. 268, 1–35.

Ehinger, B., and Sporrong, B. (1968). *Experientia* **24**, 265–266.

Ehinger, B., Falck, B., and Sporrong, B. (1966). *Bibl. Anat.* **8**, 35–45.

Ehinger, B., Falck, B., and Sporrong, B. (1967). *Bibl. Anat.* **8**, 35–45.

Ehinger, B., Falck, B., and Sporrong, B. (1970). *Z. Zellforsch. Mikrosk. Anat.* **107**, 508–521.

Ehrenpreis, S., Fleisch, J. H., and Mittag, T. W. (1969). *Pharmacol. Rev.* **21**, 131–181.

Ehrlich, P. (1913). *Lancet* **2**, 445–451.

Eisenberg, R. S., and Gage, P. W. (1969). *J. Gen. Physiol.* **53**, 279–297.

Eisenfeld, A. J., Axelrod, J., and Krakoff, L. (1966). *J. Pharmacol. Exp. Ther.* **156**, 107–113.

Eisenfeld, A. J., Landsberg, L., and Axelrod, J. (1967). *J. Pharmacol. Exp. Ther.* **158**, 378–385.

Eliasson, R., and Posse, N. (1965). *Int. J. Fert.* **10**, 373–377.

Elmqvist, D. (1965). *Acta Physiol. Scand.* **64**, 340–344.

Elmqvist, D., and Feldman, D. S. (1965a). *J. Physiol.* (*London*) **181**, 487–497.

Elmqvist, D., and Feldman, D. S. (1965b). *J. Physiol.* (*London*) **181**, 498–505.

Elmqvist, D., and Feldman, D. S. (1966). *Acta Physiol. Scand.* **67**, 34–42.

Elmqvist, D., and Josefsson, J. O. (1962). *Acta Physiol. Scand.* **54**, 105–110.

Elmqvist, D., and Lambert, E. H. (1968). *Fed. Proc., Fed. Amer. Soc. Exp. Biol.* **27**, 236.

Elmqvist, D., and Quastel, D. M. J. (1965a). *J. Physiol.* (*London*) **177**, 463–482.

Elmqvist, D., and Quastel, D. M. J. (1965b). *J. Physiol.* (*London*) **178**, 505–529.

Elmqvist, D., Hofmann, W. W., Kugelberg, J., and Quastel, D. M. J. (1964). *J. Physiol.* (*London*) **174**, 417–434.

Endo, M., Tanaka, M., and Ogawa, Y. (1970). *Nature* (*London*) **228**, 34–36.

Erdmann, W. D., and Heye, D. (1958). *Naunyn-Schmiedebergs Arch. Exp. Pathol. Pharmakol.* **232**, 507–521.

Erdös, G. (1966). *Advan. Pharmacol.* **4**, 1–90.

Erspamer, V., subed. (1966). "Handbuch der experimentellen Pharmakologic," Vol. 19. Springer Publ., New York.

Evans, D. H. L., and Schild, H. O. (1953). *J. Physiol.* (*London*) **122**, 63P.

Evans, D. H. L., and Schild, H. O. (1957). *Nature* (*London*) **180**, 341–342.

Evans, D. H. L., Schild, H. O., and Thesleff, S. (1958). *J. Physiol.* (*London*) **143**, 474–485.

Falk, G., and Fatt, P. (1964). *Proc. Roy. Soc., Ser B* **160**, 69–123.

Farmer, J. B., and Campbell, I. K. (1967). *Brit. J. Pharmacol. Chemother.* **29**, 319–328.

Farnebo, L. O., and Hamberger, B. (1970). *J. Pharmacol. Exp. Ther.* **172**, 332–341.

Fastier, F. N., McDowall, M. A., and Waal, H. (1959). *Brit. J. Pharmacol. Chemother.* **14**, 527–535.

Fatt, P., and Ginsborg, B. L. (1958). *J. Physiol.* (*London*) **142**, 516–543.

Fatt, P., and Katz, B. (1951). *J. Physiol.* (*London*) **115**, 320–370.

Fatt, P., and Katz, B. (1952). *J. Physiol.* (*London*) **117**, 109–128.

Fatt, P., and Katz, B. (1953). *J. Physiol.* (*London*) **120**, 171–204.

Feinstein, M. B. (1966). *J. Pharmacol. Exp. Ther.* **152**, 516–524.

Feinstein, M. B., and Paimre, M. (1967). *Nature* (*London*) **214**, 151–153.

Feinstein, M. B., and Paimre, M. (1969). *Fed. Proc., Fed. Amer. Soc. Exp. Biol.* **28**, 1643–1648.

Feldberg, W., and Lewis, G. P. (1965). *J. Physiol.* (*London*) **178**, 239–251.

Feldman, D. S., Lieberman, J. S., and Levere, R. D. (1969). *Fed. Proc., Fed. Amer. Soc. Exp. Biol.* **28**, 670.

Feltz, A., and Mallart, A. (1970). *Brain Res.* **22**, 264–267.

Feng, T. P. (1936). *Chin. J. Physiol.* **10**, 513–528.

Feng, T. P. (1938). *Chin. J. Physiol.* **13**, 119–140.

Ferreira, S. H., and Van, J. R. (1967). *Nature* (*London*) **216**, 868–873.

Ferry, C. B. (1966). *Physiol. Rev.* **46**, 420–456.

Fewings, J. D., Hanna, M. J. D., Walsh, J. A., and Whelan, R. F. (1966). *Brit. J. Pharmacol. Chemother.* **27**, 93–106.

Fischer, J. E., Horst, W. D., and Kopin, I. J. (1965). *Brit. J. Pharmacol. Chemother.* **24**, 477–484.

Fischer, J. E., Weise, V. K., and Kopin, I. J. (1966). *J. Pharmacol. Exp. Ther.* **153**, 523–529.

Fleckenstein, A., and Bass, H. (1953). *Naunyn-Schmiedebergs Arch. Exp. Pathol. Pharmakol.* **220**, 143–156.

Fleckenstein, A., and Burn, J. H. (1953). *Brit. J. Pharmacol. Chemother.* **8**, 69–78.

Fleckenstein, A., and Stockle, D. (1955). *Naunyn-Schmiedebergs Arch. Exp. Pathol. Pharmakol.* **224**, 401–415.

Folkow, B. (1964a). *Circ. Res.* **15**, Suppl. 1, 19–29.
Folkow, B. (1964b). *Circ. Res.* **15**, Suppl. 1, 279–285.
Follenfant, M. J., and Robson, R. D. (1970). *Brit. J. Pharmacol.* **38**, 792–801.
Fonnum, F. (1967). *Biochem. J.* **103**, 626–670.
Fonnum, F. (1968). *Biochem. J.* **109**, 389–398.
Foster, R. W. (1969). *Brit. J. Pharmacol.* **35**, 418–427.
Foulks, J. G., and Perry, F. A. (1966). *J. Physiol.* (*London*) **185**, 355–381.
Frank, G. B. (1962). *J. Physiol.* (*London*) **163**, 254–268.
Fujii, R., and Novales, R. (1969). *Amer. Zool.* **9**, 453–463.
Funaki, S. (1967). *Bibl. Anat.* **8**, 5–10.
Funaki, S., and Bohr, D. F. (1964). *Nature* (*London*) **203**, 192–194.
Furchgott, R. F. (1954). *J. Pharmacol. Exp. Ther.* **111**, 265–284.
Furchgott, R. F. (1955). *Pharmacol. Rev.* **7**, 183–265.
Furness, J. B., Campbell, G. R., Gillard, S. M., Malmfors, T., Cobb, J. L. S., and Burnstock, G. (1970). *J. Pharmacol. Exp. Ther.* **174**, 111–122.
Furshpan, E. J. (1956). *J. Physiol.* (*London*) **134**, 689–697.
Furukawa, T. (1957). *Jap. J. Physiol.* **7**, 199–212.
Fuxe, K., and Sedvall, G. (1964). *Acta Physiol. Scand.* **64**, 75–86.
Gaddum, J. H. (1937). *Proc. Roy. Soc., Ser. B* **121**, 598–601.
Gaddum, J. H. (1953). *J. Physiol.* (*London*) **119**, 363–368.
Gaddum, J. H., and Hameed, K. A. (1954). *Brit. J. Pharmacol. Chemother.* **9**, 240–248.
Gaddum, J. H., and Paasonen, M. K. (1955). *Brit. J. Pharmacol. Chemother.* **10**, 474–483.
Gaddum, J. H., and Picarelli, Z. P. (1957). *Brit. J. Pharmacol. Chemother.* **12**, 323–328.
Gage, P. W. (1965). *J. Pharmacol. Exp. Ther.* **150**, 236–243.
Gage, P. W., and Armstrong, C. M. (1968). *Nature* (*London*) **218**, 363–365.
Gage, P. W., and Hubbard, J. I. (1966). *J. Physiol.* (*London*) **184**, 353–375.
Gage, P. W., and Quastel, D. M. J. (1965). *Nature* (*London*) **206**, 625–626.
Gage, P. W., and Quastel, D. M. J. (1966). *J. Physiol.* (*London*) **185**, 95–123.
Galindo, A. (1970). *Fed. Proc., Fed. Amer. Soc. Exp. Biol.* **29**, 280.
Garattini, S., and Valzelli, L. (1965). "Serotonin." American Elsevier, New York.
Gerschenfeld, H. M., and Stefani, E. (1966). *J. Physiol.* (*London*) **185**, 684–700.
Gershon, M. D. (1967a). *Brit. J. Pharmacol. Chemother.* **29**, 259–279.
Gershon, M. D. (1967b). *J. Physiol.* (*London*) **189**, 317–327.
Gershon, M. D. (1970). *In* "Smooth Muscle" (E. Bülbring *et al.*, eds.), pp. 496–524. Arnold, London.
Gershon, M. D., and Ross, L. L. (1966a). *J. Physiol.* (*London*) **186**, 451–476.
Gershon, M. D., and Ross, L. L. (1966b). *J. Physiol.* (*London*) **186**, 477–492.
Gershon, M. D., Drakontides, A. B., and Ross, L. L. (1965). *Science* **149**, 197.
Gertner, S. B., Paasonen, M. K., and Giarman, N. J. (1959). *J. Pharmacol. Exp. Ther.* **127**, 268–275.
Gesler, R. M., Lasher, A. V., Hoppe, J. O., and Steck, E. A. (1959). *J. Pharmacol. Exp. Ther.* **125**, 323–329.
Gessa, G. L., Costa, E., Kuntzman, R., and Brodie, B. B. (1962). *Life Sci.* **1**, 353–360.
Gewirtz, G. P., and Kopin, I. J. (1970). *Nature* (*London*) **227**, 406–407.
Gill, E. W., and Rang, H. R. (1966). *Mol. Pharmacol.* **2**, 284–297.

Gillespie, J. S. (1962a). *J. Physiol.* (*London*) 162, 54–75.
Gillespie, J. S. (1962b). *J. Physiol.* (*London*) 162, 76–92.
Gillespie, J. S. (1964). *In* "Pharmacology of Smooth Muscle" (E. Bülbring, ed.), pp. 81–85. Pergamon, Oxford.
Gillespie, J. S., and Hamilton, D. N. H. (1966). *Nature* (*London*) 212, 524–525.
Gillespie, J. S., Hamilton, D. N. H., and Hosie, J. A. (1970). *J. Physiol.* (*London*) 206, 563–590.
Gillis, C. N., and Paton, D. M. (1967). *Brit. J. Pharmacol. Chemother.* 9, 309–318.
Gilman, A. G., and Rall, T. W. (1970). *In* "Action of Hormones: Genes to Population" (P. P. Foa, ed.), p. 87–128. Thomas, Springfield, Illinois.
Gissen, A. J., Karis, J. H., and Nastuk, W. L. (1966). *J. Amer. Med. Ass.* 197, 770–774.
Glick, G., Wechsler, A. S., and Epstein, S. E. (1968). *J. Clin. Invest.* 47, 511–520.
Godfraind, T., and Kaba, A. (1969). *Brit. J. Pharmacol.* 36, 549–560.
Godfraind, T., Kaba, A., and Polster, P. (1968). *Arch. Int. Pharmacodyn. Ther.* 172, 235–239.
Goffart, M., and Ritchie, J. M. (1952). *J. Physiol.* (*London*) 116, 357–371.
Gokhale, S. D., Gulati, O., Kelkar, L. V., and Kelkar, V. V. (1966). *Brit. J. Pharmacol. Chemother.* 27, 332–346.
Goldberg, A. L., and Singer, J. J. (1969). *Proc. Nat. Acad. Sci. U.S.* 64, 134–141.
Goldstein, A., Aronov, L., and Kalman, S. (1968). "The Principles of Drug Action," p. 250. Harper, New York.
Goodford, P. J. (1970). *In* "Smooth Muscle" (E. Bülbring *et al.*, eds.), pp. 100–121. Arnold, London.
Gordon, P., Haddy, F. J., and Lipton, M. A. (1958). *Science* 128, 531–532.
Gordon, P., Haddy, F. J., and Lipton, M. A. (1959). *Fed. Proc., Fed. Amer. Soc. Exp. Biol.* 18, 397.
Graham, J. D. P., and Katib, H. Al. (1966). *Brit. J. Pharmacol. Chemother.* 28, 1–14.
Gray, J. A. B., and Diamond, J. (1957). *Brit. Med. Bull.* 13, 185–8.
Green, R. D., and Miller, J. W. (1966). *J. Pharmacol. Exp. Ther.* 152, 42–50.
Gross, F. (1968). *Acta Endocrinol.* (*Copenhagen*), *Suppl.* 124, 41–64.
Gulati, O. D., Parikh, H. M., Ringe, S. Y., and Sherlekar, M. L. (1970). *Brit. J. Pharmacol.* 40, 689–701.
Gyermek, L. (1964a). *Arch. Int. Pharmacodyn. Ther.* 150, 570–581.
Gyermek, L. (1964b). *J. Med. Chem.* 7, 280–284.
Gyermek, L., and Bindler, E. (1962a). *J. Pharmacol. Exp. Ther.* 135, 344–348.
Gyermek, L., and Bindler, E. (1962b). *J. Pharmacol. Exp. Ther.* 138, 159–164.
Hackett, J. T., and Quastel, D. M. J. (1971). Unpublished experiment.
Hackett, J. T., Leung, H., and Quastel, D. M. J. (1971). Unpublished observations.
Haddy, F. J., and Scott, J. B. (1968). *Physiol. Rev.* 48, 688–707.
Haefely, W., Hürlimann, A., and Thoenen, H. (1965a). *Biochem. Pharmacol.* 14, 1393.
Haefely, W., Thoenen, H., and Hürlimann, A. (1965b). *Life Sci.* 4, 913–918.
Haeusler, G., Thoenen, H., Haefely, W., and Hürlimann, A. (1968). *Naunyn-Schmiedebergs Arch. Pharmakol. Exp. Pathol.* 261, 389–411.
Häggendal, J. (1963). *Acta Physiol. Scand.* 59, 261–268.
Häggendal, J. and Malmfors, T. (1969). *Acta Physiol. Scand.* 75, 33–38.
Hagiwara, S., and Naka, K. (1964). *J. Gen. Physiol.* 48, 141–162.

Hagiwara, S., and Nakajima, S. (1966). *J. Gen. Physiol.* **49**, 793–806.

Hagiwara, S., and Saito, N. (1959). *J. Physiol.* (*London*) **148**, 161–179.

Hagiwara, S., and Takahashi, K. (1967). *J. Gen. Physiol.* **50**, 583–601.

Haigh, A. L., Lloyd, S., and Pickford, M. (1965). *J. Physiol.* (*London*) **178**, 563–576.

Hamberger, B. (1967). *Acta Physiol. Scand.* **71**, Suppl. 295, 1–56.

Hamberger, B., Norberg, K. A., and Olson, L. (1967). *Acta Physiol. Scand.* **69**, 1–12.

Harris, E. J., and Ochs, S. (1966). *J. Physiol.* (*London*) **187**, 5–21.

Harry, J. (1963). *Brit. J. Pharmacol. Chemother.* **20**, 399–417.

Harvey, S. C., and Nickerson, M. (1954). *J. Pharmacol. Exp. Ther.* **112**, 274–290.

Haslett, W. L., and Jenden, D. J. (1961). *J. Cell. Physiol.* **57**, 123–133.

Hasselbach, W. (1964). *Progr. Biophys. Mol. Biol.* **14**, 167–227.

Hebb, C. O. (1963). *In* "Handbuch der experimentellen Pharmakologie" (G. B. Koelle, ed.), Vol. 15, p. 55–88. Springer-Verlag, Berlin and New York.

Hebb, C. O., Krnjević, K., and Silver, A. (1964). *J. Physiol.* (*London*) **171**, 504–513.

Hedqvist, P., and Stjärne, L. (1969). *Acta Physiol. Scand.* **76**, 270–283.

Henderson, V. E., and Roepke, M. H. (1934). *J. Pharmacol. Exp. Ther.* **51**, 97–111.

Henning, M., and Johnsson, G. (1967). *Acta Pharmacol. Toxicol.* **25**, 373–384.

Henning, M., and van Zwieten, P. A. (1968). *J. Pharm. Pharmacol.* **20**, 409–417.

Hertting, G., and Schiefthaler, T. (1963). *Naunyn-Schmiedebergs Arch. Exp. Pathol. Pharmakol.* **246**, 13–16.

Hertting, G., and Suko, J. (1966). *Brit. J. Pharmacol. Chemother.* **26**, 686–696.

Hertting, G., Axelrod, J., and Whitby, L. G. (1961). *J. Pharmacol. Exp. Ther.* **134**, 146–153.

Hertting, G., Potter, L. T., and Axelrod, J. (1962). *J. Pharmacol. Exp. Ther.* **136**, 289–292.

Hess, A., and Pilar, G. (1963). *J. Physiol.* (*London*) **169**, 780–789.

Hidaka, T., and Kuriyama, H. (1969a). *J. Gen. Physiol.* **53**, 471–486.

Hidaka, T., and Kuriyama, H. (1969b). *J. Physiol.* (*London*) **201**, 61–71.

Hille, B. (1966). *Nature* (*London*) **210**, 1220.

Hille, B. (1967). *J. Gen. Physiol.* **50**, 1278–1302.

Hillyard, I. W., and Procita, L. (1958). *J. Pharmacol. Exp. Ther.* **123**, 140–144.

Hinke, J. A. M., Wilson, M. L., and Burnham, S. C. (1964). *Amer. J. Physiol.* **206**, 211–217.

Hodgkin, A. L., and Horowicz, P. (1960a). *J. Physiol.* (*London*) **153**, 370–385.

Hodgkin, A. L., and Horowicz, P. (1960b). *J. Physiol.* (*London*) **153**, 386–403.

Hodgkin, A. L., and Horowicz, P. (1960c). *J. Physiol.* (*London*) **153**, 404–412.

Hodgkin, A. L., and Huxley, A. F. (1952). *J. Physiol.* (*London*) **116**, 449–472.

Hofmann, W. W. (1969). *Amer. J. Physiol.* **216**, 621–629.

Hofmann, W. W., Feigen, G. A., and Genther, G. H. (1962). *Nature* (*London*) **193**, 175–176.

Hokfelt, T. (1967). *Acta Physiol. Scand.* **69**, 125–126.

Hokfelt, T. (1968). *Z. Zellforsch. Mikros. Anat.* **91**, 1–74.

Hollands, B. C. S., and Vanov, S. (1965). *Brit. J. Pharmacol. Chemother.* **25**, 307–316.

Holman, M. E. (1967). *Circ. Res.* **20, 21**, Suppl. 3, 71–81.

Holman, M. E. (1969). *Ergeb. Physiol. Biol. Chem. Exp. Pharmakol.* **61**, 137–177.

Holman, M. E. (1970). In "Smooth Muscle" (E. Bülbring, et al., eds.), pp. 244–288. Arnold, London.

Holman, M. E., Kasby, C. B., and Suthers, M. B. (1967). Aust. J. Exp. Biol. Med. Sci. 45, 50P.

Holman, M. E., Kasby, C. B., Suthers, M. B., and Wilson, J. A. (1968). J. Physiol. (London) 196, 111–132.

Holmes, S. W., and Horton, E. W. (1968). J. Physiol. (London) 195, 731–741.

Holton, P. (1959). J. Physiol. (London) 145, 494–504.

Holtz, P., Osswald, W., and Stock, K. (1960). Naunyn-Schmiedebergs Arch. Exp. Pathol. Pharmakol. 239, 14–28.

Horaćkova, M., Nonner, W., and Stämpfli, R. (1968). Proc. Int. Union Physiol. Sci. 7, 594.

Horst, W. D., Kopin, I. J., and Ramey, E. R. (1968). Amer. J. Physiol. 215, 817–822.

Horton, E. W. (1965). Experientia 21, 113–118.

Horton, E. W., Main, I. H. M., and Thompson, C. J. (1965). J. Physiol. (London) 180, 514–528.

Hotta, Y., and Tsukui, R. (1968). Nature (London) 217, 867–869.

Howell, J. N. (1969). J. Physiol. (London) 201, 515–533.

Hrdina, P., and Garattini, S. (1967). J. Pharm. Pharmacol. 19, 667–673.

Hubbard, J. I. (1961). J. Physiol. (London) 158, 507–517.

Hubbard, J. I. (1970). Progr. Biophys. Mol. Biol. 21, 35–124.

Hubbard, J. I., and Løyning, Y. (1966). J. Physiol. (London) 185, 205–223.

Hubbard, J. I., and Willis, W. D. (1962a). J. Physiol. (London) 163, 115–137.

Hubbard, J. I., and Willis, W. D. (1962b). Nature (London) 193, 1294–1295.

Hubbard, J. I., and Willis, W. D. (1968). J. Physiol. (London) 194, 381–407.

Hubbard, J. I., Schmidt, R. F., and Yokota, Y. (1965). J. Physiol. (London) 181, 810–829.

Hubbard, J. I., Jones, S. F., and Landau, E. M. (1968a). J. Physiol. (London) 194, 381–407.

Hubbard, J. I., Jones, S. F., and Landau, E. M. (1968b). J. Physiol. (London) 196, 75–86.

Hubbard, J. I., Llinás, R., and Quastel, D. M. J. (1969a). "Electrophysiological Analysis of Synaptic Transmission." Arnold, London.

Hubbard, J. I., Wilson, D. F., and Miyamoto, M. (1969b). Nature (London) 223, 531–533.

Hudgins, P. M., and Weiss, G. B. (1968). J. Pharmacol. Exp. Ther. 159, 91–97.

Hughes, F. B. (1955). J. Physiol. (London) 130, 123–130.

Hughes, J., and Vane, J. R. (1967). Brit. J. Pharmacol. Chemother. 30, 46–66.

Hunt, C. C. (1952). Fed. Proc., Fed. Amer. Soc. Exp. Biol. 11, 75.

Hurwitz, L., Battle, F., and Weiss, G. B. (1962). J. Gen. Physiol. 46, 315–332.

Hutter, O. F. (1961). In "Nervous Inhibition" (E. Florey, ed.), pp. 114–123. Pergamon, Oxford.

Hutter, O. F., and Loewenstein, W. R. (1955). J. Physiol. (London) 130, 559–571.

Huxley, A. F., and Taylor, R. E. (1958). J. Physiol. (London) 144, 426–441.

Huxley, H. E. (1964). Nature (London) 202, 1067–1071.

Ichikawa, S., and Bozler, E. (1955). Amer. J. Physiol. 182, 92–96.

Imai, S., and Takeda, K. (1967a). J. Pharmacol. Exp. Ther. 156, 557–564.

Imai, S., and Takeda, K. (1967b). Nature (London) 213, 1044–1045.

Ingenito, A. J., Barrett, J. P., and Procita, L. (1970). *J. Pharmacol. Exp. Ther.* **175**, 593–599.

Innes, I. R., and Kohli, J. D. (1969). *Brit. J. Pharmacol.* **35**, 383–393.

Innes, I. R., and Kosterlitz, H. W. (1954). *J. Physiol. (London)* **124**, 25–43.

Innes, I. R., and Krayer, O. (1958). *J. Pharmacol. Exp. Ther.* **124**, 245–251.

Inoue, F., and Frank, G. B. (1962). *J. Pharmacol. Exp. Ther.* **136**, 190–196.

Inoue, F., and Frank, G. B. (1967). *Brit. J. Pharmacol. Chemother.* **30**, 186–193.

Iversen, L. L. (1963). *Brit. J. Pharmacol. Chemother.* **21**, 523–537.

Iversen, L. L. (1965a). *Advan. Drug. Res.* **2**, 1–46.

Iversen, L. L. (1965b). *Brit. J. Pharmacol. Chemother.* **25**, 18–33.

Iversen, L. L. (1967). "The Uptake and Storage of Noradrenaline in Sympathetic Nerves." Cambridge Univ. Press, London and New York.

Iversen, L. L., and Kravitz, E. A. (1966). *Mol. Pharmacol.* **2**, 360–362.

Iversen, L. L., Glowinski, J., and Axelrod, J. (1965). *J. Pharmacol. Exp. Ther.* **150**, 173–183.

Jacobowitz, D. (1965). *J. Pharmacol. Exp. Ther.* **149**, 358–364.

Jelliffe, R. W. (1962). *J. Pharmacol. Exp. Ther.* **135**, 349–353.

Jenden, D. J., and Fairhurst, A. S. (1969). *Pharmacol. Rev.* **21**, 1–26.

Jenkinson, D. H. (1957). *J. Physiol. (London)* **138**, 438–444.

Jenkinson, D. H. (1960). *J. Physiol. (London)* **152**, 309–324.

Jenkinson, D. H., and Morton, I. K. M. (1965). *Nature (London)* **205**, 505–506.

Jenkinson, D. H., and Morton, I. K. M. (1967a). *Ann. N.Y. Acad. Sci.* **139**, 762–771.

Jenkinson, D. H., and Morton, I. K. M. (1967b). *J. Physiol. (London)* **188**, 373–386.

Jenkinson, D. H., and Morton, I. K. M. (1967c). *J. Physiol. (London)* **188**, 387–402.

Jenkinson, D. H., Stamenovic, B. A., and Shitaker, B. D. L. (1968). *J. Physiol. (London)* **195**, 743–754.

Jéquier, E. (1965). *Helv. Physiol. Pharmacol. Acta* **23**, 163–179.

Jindal, M. N., and Deshpande, V. R. (1960). *Brit. J. Pharmacol.* **15**, 506–509.

Johansson, B., and Bohr, D. F. (1966). *Amer. J. Physiol.* **210**, 801–806.

Johansson, B., Jonsson, O., Axelsson, J., and Wahlstrom, B. (1967). *Circ. Res.* **21**, 619–633.

Johnson, G. E., (1964). *Acta Physiol. Scand.* **60**, 181–188.

Jonason, J., Rosengren, E., and Waldeck, B. (1963). *Acta Physiol. Scand.* **60**, 136–140.

Jones, S. F., and Kwanbunbumphen, S. (1968). *Life Sci.* **7**, 1251–1255.

Jonsson, G., Hamberger, B., Malmfors, T., and Sachs, C. (1969). *Eur. J. Pharmacol.* **8**, 58–72.

Kahlson, G., and Rosengren, E. (1965). *Annu. Rev. Pharmacol.* **5**, 305–320.

Kalsner, S., and Nickerson, M. (1969a). *Brit. J. Pharmacol.* **35**, 428–439.

Kalsner, S., and Nickerson, M. (1969b). *Brit. J. Pharmacol.* **35**, 440–455.

Kalsner, S., Nickerson, M., and Boyd, G. N. (1970). *J. Pharmacol. Exp. Ther.* **174**, 500–508.

Kao, C. Y. (1966). *Pharmacol. Rev.* **18**, 997–1049.

Kao, C. Y. (1967). *In* "Cellular Biology of the Uterus" (R. M. Wynn, ed.), pp. 386–448. North Holland Publ., Amsterdam.

Karis, J. H., Gissen, A. J., and Nastuk, W. L. (1966). *Anesthesiology* **27**, 42–51.

Karlin, A. (1969). *J. Gen. Physiol.* **54**, Part 2, 245–264.

Kato, M., and Fujimori, B. (1965). *J. Pharmacol. Exp. Ther.* **149**, 124–131.

Katz, B. (1962). *Proc. Roy. Soc. Ser. B* **155**, 455–477.

Katz, B. (1966). "Nerve, Muscle and Synapse." McGraw-Hill, New York.

Katz, B. (1969). "The Release of Neural Transmitter Substances." Thomas, Springfield, Illinois.
Katz, B., and Miledi, R. (1963). *J. Physiol. (London)* **168**, 389–422.
Katz, B., and Miledi, R. (1965a). *Proc. Roy. Soc., Ser. B* **161**, 483–495.
Katz, B., and Miledi, R. (1965b). *Proc. Roy. Soc., Ser. B* **161**, 496–503.
Katz, B., and Miledi, R. (1967a). *Proc. Roy. Soc., Ser. B* **167**, 8–22.
Katz, B., and Miledi, R. (1967b). *J. Physiol. (London)* **189**, 535–544.
Katz, B., and Miledi, R. (1967c). *J. Physiol. (London)* **192**, 407–436.
Katz, B., and Miledi, R. (1967d). *Nature (London)* **215**, 651.
Katz, B., and Miledi, R. (1968a). *J. Physiol. (London)* **195**, 481–492.
Katz, B., and Miledi, R. (1968b). *J. Physiol. (London)* **199**, 729–743.
Katz, B., and Miledi, R. (1969a). *J. Physiol. (London)* **203**, 459–487.
Katz, B., and Miledi, R. (1969b). *J. Physiol. (London)* **203**, 689–706.
Katz, B., and Miledi, R. (1970a). *J. Physiol. (London)* **207**, 789–801.
Katz, B., and Miledi, R. (1970b). *Nature (London)* **226**, 962–963.
Katz, B., and Thesleff, S. (1957a). *J. Physiol. (London)* **137**, 267–278.
Katz, B., and Thesleff, S. (1957b). *J. Physiol. (London)* **138**, 63–80.
Katz, N. L., Ingenito, A., and Procita, L. (1970). *J. Pharmacol. Exp. Ther.* **171**, 242–248.
Katz, R. L. (1964). *Anesthesiology* **25**, 653–661.
Keatinge, W. R. (1964). *J. Physiol. (London)* **174**, 184–205.
Keatinge, W. R. (1966). *Circ. Res.* **18**, 641–649.
Keatinge, W. R. (1968a). *J. Physiol. (London)* **194**, 169–182.
Keatinge, W. R. (1968b). *J. Physiol (London)* **174**, 183–200.
Khairallah, P. A., and Page, I. H. (1963). *Ann. N.Y. Acad. Sci.* **104**, 212–221.
Khairallah, P. A., Page, I. H., Bumpus, F. M., and Türker, R. K. (1966). *Circ. Res.* **19**, 247–254.
Kirpekar, S. M., and Cervoni, P. (1963). *J. Pharmacol. Exp. Ther.* **141**, 59–70.
Kirpekar, S. M., and Misu, Y. (1967). *J. Physiol. (London)* **188**, 219–234.
Kirpekar, S. M., and Wakada, A. R. (1968). *J. Physiol. (London)* **194**, 595–609.
Kirpekar, S. M., and Wakada, A. R. (1970). *Brit. J. Pharmacol.* **39**, 533–541.
Kirpekar, S. M., Prat, J. C., and Yamamoto, H. (1970). *J. Pharmacol. Exp. Ther.* **172**, 342–350.
Kirschner, L. B., and Stone, W. E. (1951). *J. Gen. Physiol.* **34**, 821–834.
Kirshner, N. (1962). *J. Biol. Chem.* **237**, 2311–2317.
Klide, A. M., and Aviado, D. M. (1967). *J. Pharmacol. Exp. Ther.* **158**, 28–35.
Knoll, J., and Vizi, E. S. (1970). *Brit. J. Pharmacol.* **40**, 554P.
Kobayashi, M. (1965). *Amer. J. Physiol.* **208**, 715–719.
Kobayashi, M., and Irisawa, H. (1964). *Amer. J. Physiol.* **206**, 205–210.
Koelle, G. B. (1962). *J. Pharm. Pharmacol.* **14**, 65–90.
Koelle, G. B. (1970). *In* "The Pharmacological Basis of Therapeutics" (L. S. Goodman and A. Gilman, eds.), 4th ed., pp. 442–465. Macmillan, New York.
Kopin, I. J., Gordon, E. K., and Horst, W. D. (1965). *Biochem. Pharmacol.* **14**, 753–760.
Kopin, I. J., Hefling, G., and Gordon, E. K. (1962). *J. Pharm. Pharmacol.* **138**, 38–40.
Kordaš, M. (1969). *J. Physiol. (London)* **204**, 443–502.
Korey, S., de Braganza, B., and Nachmansohn, D. (1951). *J. Biol. Chem.* **189**, 705–715.
Kosterlitz, M. W., and Robinson, J. A. (1955). *J. Physiol. (London)* **129**, 18P.
Kosterlitz, H. W., and Watt, A. J. (1965). *J. Physiol. (London)* **177**, 11P.

Kottegoda, S. R. (1969). *J. Physiol.* (*London*) **200**, 687–712.
Kottegoda, S. R. (1970). *In* "Smooth Muscle" (E. Bülbring *et al.*, eds.), pp. 525–541. Arnold, London.
Kraatz, H. G., and Trautwein, W. (1957). *Naunyn-Schmiedebergs Arch. Exp. Pathol. Pharmakol.* **231**, 419–439.
Krauss, K. R., Kopin, I. J., and Weise, V. K. (1970). *J. Pharmacol. Exp. Ther.* **172**, 282–288.
Krejčí, I., Kupková, B., and Vávra, I. (1967). *Brit. J. Pharmacol. Chemother.* **30**, 397–505.
Krnjević, K. (1969). *Fed. Proc., Fed. Amer. Soc. Exp. Biol.* **28**, 113.
Krnjević, K., and Miledi, R. (1958). *J. Physiol.* (*London*) **141**, 291–304.
Krnjević, K., and Phillis, J. W. (1963a). *Brit. J. Pharmacol. Chemother.* **20**, 471–490.
Krnjević, K., and Phillis, J. W. (1963b). *J. Physiol.* (*London*) **166**, 328–350.
Kuba, K. (1969). *Jap. J. Physiol.* **19**, 762–774.
Kuba, K. (1970). *J. Physiol.* (*London*) **211**, 551–570.
Kuffler, S. W., and Vaughan Williams, E. M. (1953a). *J. Physiol.* (*London*) **121**, 289–317.
Kuffler, S. W., and Vaughan Williams, E. M. (1953b). *J. Physiol.* (*London*) **121**, 318–340.
Kuffler, S. W., Hunt, C. C., and Quilliam, J. P. (1951). *J. Neurophysiol.* **14**, 29–54.
Kuno, M. (1964). *J. Physiol.* (*London*) **175**, 81–99.
Kupferman, A., and Gillis, C. N. (1970). *J. Pharmacol. Exp. Ther.* **171**, 214–222.
Kuriyama, H. (1963). *J. Physiol.* (*London*) **169**, 213–228.
Kuriyama, H. (1970). *In* "Smooth Muscle" (E. Bülbring *et al.*, eds.), pp. 366–395. Arnold, London.
Kuriyama, H., and Tomita, T. (1970). *J. Gen. Physiol.* **55**, 147–162.
Kuriyama, H., Osa, T., and Toida, N. (1966). *Brit. J. Pharmacol. Chemother.* **27**, 366–376.
Kuriyama, H., Osa, T., and Toida, N. (1967a). *J. Physiol.* (*London*) **191**, 225–238.
Kuriyama, H., Osa, T., and Toida, N. (1967b). *J. Physiol.* (*London*) **191**, 239–255.
Kuriyama, H., Osa, T., and Toida, N. (1967c). *J. Physiol.* (*London*) **191**, 257–270.
Kuschinsky, G., Lindmar, R., Lüllman, H., and Muscholl, E. (1960). *Naunyn-Schmiedebergs Arch. Exp. Pathol. Pharmakol.* **240**, 242–252.
Kwant, W. O., and Seeman, P. (1969). *Biochim. Biophys. Acta* **193**, 338–349.
Lambert, D. H., and Parsons, R. L. (1970). *J. Gen. Physiol.* **56**, 309–321.
Landau, E. M. (1969). *J. Physiol.* (*London*) **203**, 281–299.
Laskowski, M., and Thies, R. (1970). *Fed. Proc., Fed. Amer. Soc. Exp. Biol.* **29**, 715.
Laurence, D. R., and Webster, R. A. (1961). *Lancet* **1**, 481–482.
Law, H. D., and du Vigneaud, V. (1960). *J. Amer. Chem. Soc.* **82**, 4579–4581.
Lee, C. Y. (1970). *In* "Smooth Muscle" (E. Bülbring *et al.*, eds.), pp. 558–588.
Leitz, F. H. (1970). *J. Pharmacol. Exp. Ther.* **173**, 152–157.
Levitt, M., Spector, S., Sjoerdsma, A., and Udenfriend, S. (1965). *J. Pharmacol. Exp. Ther.* **148**, 1–8.
Levy, B. (1966). *J. Pharmacol. Exp. Ther.* **151**, 413–422.
Levy, B., and Wilkenfeld, B. E. (1969). *Eur. J. Pharmacol.* **5**, 227–234.
Lewis, G. P., and Reit, E. (1966). *Brit. J. Pharmacol. Chemother.* **26**, 444–460.
Lightman, S. L., and Iversen, L. L. (1969). *Brit. J. Pharmacol.* **37**, 638–649.
Liley, A. W., and North, K. A. K. (1953). *J. Neurophysiol.* **16**, 509–527.

Lindgren, P., and Uvnäs, B. (1954). *Amer. J. Physiol.* **176**, 68–76.
Lindmar, R., and Muscholl, E. (1961). *Naunyn-Schmiedebergs Arch. Exp. Pathol. Pharmakol.* **242**, 214–227.
Lindmar, R., and Muscholl, E. (1964). *Naunyn-Schmiedebergs Arch. Exp. Pathol. Pharmakol.* **247**, 469–492.
Lindmar, R., Löffelholz, K., and Muscholl, E. (1967). *Experientia* **23**, 933–934.
Lindmar, R., Löffelholz, K., and Muscholl, E. (1968). *Brit. J. Pharmacol. Chemother.* **32**, 280–294.
Lloyd, D. P. C. (1942). *J. Neurophysiol.* **5**, 153–164.
Lloyd, S., and Pickford, M. (1961). *Brit. J. Pharmacol. Chemother.* **16**, 129–136.
Lloyd, S., and Pickford, M. (1967). *J. Physiol. (London)* **192**, 43–52.
Lockett, M. F. (1950). *Brit. J. Pharmacol. Chemother.* **5**, 485–496.
Loizou, L. A. (1971). *Brit. J. Pharmacol.* **41**, 41–48.
Longenecker, H. E. Jr., Hurlbut, W. P., Mauro, A., and Clark, A. W. (1970). *Nature (London)* **225**, 701–703.
Lucchesi, B. R., Whitsitt, L. S., and Stickney, J. L. (1967). *Ann. N.Y., Acad. Sci.* **139**, 940–951.
Lundberg, A. (1958). *Physiol. Rev.* **38**, 21–40.
Lunde, P. K. M., Waaler, B. A., and Walløe, L. (1969). *Acta Physiol. Scand.* **72**, 331–337.
Lüttgau, H. C. (1963). *J. Physiol. (London)* **168**, 679–697.
Lüttgau, H. C., and Oetliker, M. (1968). *J. Physiol. (London)* **194**, 51–74.
MacIntosh, F. C., Birks, R. I., and Sastry, P. B. (1956). *Nature (London)* **178**, 1181.
McCubben, J. W., Kaneko, Y., and Page, I. H. (1962). *Circ. Res.* **11**, 74–83.
McDougal, M. D., and West, G. B. (1954). *Brit. J. Pharmacol. Chemother.* **9**, 131–137.
McIntyre, A. R., King, R. E., and Dunn, A. L. (1945). *J. Neurophysiol.* **8**, 292–307.
McKinstry, D. N., Koenig, E., Koelle, W. A., and Koelle, G. B. (1963). *Can. J. Biochem. Physiol.* **41**, 2599–2609.
McNeil, J. H., Barnes, R. V., Davis, R. S., and Hook, J. B. (1969). *Can. J. Physiol. Pharmacol.* **47**, 663–670.
Maeno, T. (1966). *J. Physiol. (London)* **183**, 592–606.
Maeno, T., Edwards, C., and Hashimura, S. (1971). *J. Neurophysiol.* **34**, 32–46.
Magazanik, L. G., and Vyskočil, F. (1970). *J. Physiol. (London)* **210**, 507–518.
Maître, L. (1965). *Naunyn-Schmiedebergs Arch. Exp. Pathol. Pharmakol.* **251**, 160–161.
Malik, K. V., and Ling, G. M. (1969). *Circ. Res.* **25**, 1–9.
Mallart, A., and Feltz, A. (1968). *C. R. Acad. Sci.* **268**, 2724–2726.
Malmfors, T. (1965a). *Acta Physiol. Scand.* **64**, Supppl. 248, 1–93.
Malmfors, T. (1965b). *Acta Physiol. Scand.* **65**, 259–267.
Malmfors, T. (1967). *Circ. Res.* **20, 21**, Suppl. 3, 25–42.
Malmfors, T. (1969). *Pharmacology* **2**, 138–150.
Malmfors, T., and Abrams, W. B. (1970). *J. Pharmacol. Exp. Ther.* **174**, 99–110.
Malmfors, T., and Sachs, C. (1968). *Eur. J. Pharmacol.* **3**, 89–92.
Manthey, A. A. (1970). *J. Gen. Physiol.* **56**, 407–420.
Marchbanks, R. M. (1968). *Biochem. J.* **106**, 87.
Marks, B. H., Samorajski, T., and Webster, E. J. (1962). *J. Pharmacol. Exp. Ther.* **138**, 376–381.
Marois, R. L., and Edwards, C. (1969). *Fed. Proc., Fed. Amer. Soc. Exp. Biol.* **28**, 669.

Marsden, C. D., and Meadows, J. C. (1970). *J. Physiol.* (*London*) **207**, 429–448.
Marshall, J. M. (1963). *Amer. J. Physiol.* **204**, 732–738.
Marshall, J. M. (1964). *In* "Pharmacology of Smooth Muscle" (E. Bülbring, ed.), pp. 143–153. Pergamon, Oxford.
Marshall, J. M. (1967). *Fed. Proc.* **26**, 1104–1110.
Marshall, J. M., and Csapo, A. I. (1961). *Endocrinology* **68**, 1026–1035.
Martin, A. R. (1955). *J. Physiol.* (*London*) **130**, 114–132.
Martin, A. R. (1966). *Physiol. Rev.* **46**, 51–66.
Masland, R. L., and Wigton, R. S. (1940). *J. Neurophysiol.* **3**, 269–275.
Matthews, E. K. (1966). *Brit. J. Pharmacol. Chemother.* **26**, 552–566.
Matthews, E. K., and Quilliam, J. P. (1964). *Brit. J. Pharmacol. Chemother.* **22**, 415–440.
Matthews, E. K., and Sutter, M. C. (1967). *Can. J. Physiol. Pharmacol.* **45**, 509–520.
Maxwell, R. A., Povalski, H., and Plummer, A. J. (1959). *J. Pharmacol. Exp. Ther.* **125**, 175–183.
Maxwell, R. A., Wastila, W. B., and Eckhardt, S. B. (1966). *J. Pharmacol. Exp. Ther.* **151**, 253–261.
Mellander, S., and Johansson, B. (1968). *Pharmacol. Rev.* **20**, 117–196.
Melmon, K. L., and Cline, M. J. (1967). *Amer. J. Med.* **43**, 153–160.
Melmon, K. L., Cline, M. J., Hughes, T., and Nies, A. S. (1968). *J. Clin. Invest.* **47**, 1295–1302.
Miledi, R. (1966). *Nature* (*London*) **212**, 1233–1234.
Miledi, R. (1967). *J. Physiol.* (*London*) **192**, 379–406.
Miledi, R., and Slater, C. R. (1966). *J. Physiol.* (*London*) **184**, 473–498.
Mines, G. R. (1911). *J. Physiol.* (*London*) **42**, 251–266.
Mitchell, J. R., and Oates, J. A. (1970). *J. Pharmacol. Exp. Ther.* **172**, 100–107.
Moore, P. F. (1968). *Ann. N.Y. Acad. Sci.* **150**, 256–260.
Mottram, D. R. (1970). *Brit. J. Pharmacol.* **40**, 157–158.
Mottram, D. R., and Graham, J. D. P. (1970). *J. Pharm. Pharmacol.* **22**, 316–317.
Muscholl, E. (1961). *Brit. J. Pharmacol. Chemother.* **16**, 352–359.
Muscholl, E. (1966a). *Annu. Rev. Pharmacol.* **6**, 107–128.
Muscholl, E. (1966b). *Pharmacol. Rev.* **18**, 551–559.
Muscholl, E., and Maître, L. (1963). *Experientia* **19**, 658–659.
Nachmansohn, D. (1964). *In* "New Perspectives in Biology," pp. 192–204. Elsevier, Amsterdam.
Nachmansohn, D., and Machado, A. L. (1943). *J. Neurophysiol.* **6**, 397–403.
Nagasawa, J., Douglas, W. W., and Schulz, R. A. (1970). *Nature* (*London*) **227**, 407–409.
Nagatsu, T., Levitt, M., and Udenfriend, S. (1964). *J. Biol. Chem.* **239**, 2910–2917.
Nakajima, A., and Horn, L. (1967). *Amer. J. Physiol.* **213**, 25–30.
Narahashi, T., Anderson, N. C., and Moore, J. W. (1967). *J. Gen. Physiol.* **50**, 1413–1428.
Narahashi, T., Frazier, D. T., and Yamada, M. (1970). *J. Pharmacol. Exp. Ther.* **171**, 32–44.
Nastuk, W. L. (1967). *Fed. Proc.* **26**, 1639–1646.
Nastuk, W. L., and Hodgkin, A. L. (1950). *J. Cell. Physiol.* **35**, 39–74.
Natori, R. (1965). *Mol. Biol. Muscular Contraction* pp. 190–196.
Nauss, K. M., and Davies, R. E. (1966). *Biochem. Z.* **345**, 173–187.
Nickerson, M. (1949). *Pharmacol. Rev.* **1**, 27–101.
Nickerson, M. (1956). *Nature* (*London*) **178**, 697–698.
Nickerson, M. (1957). *Pharmacol. Rev.* **9**, 246–259.

Nickerson, M., and Gump, W. S. (1949). *J. Pharmacol. Exp. Ther.* **97**, 25–47.
Nickerson, M., and Hollenberg, N. K. (1967). *Physiol. Pharmacol.* **4**, Part D, 243–305.
Nikodijevic, B., Creveling, C. R., and Udenfriend, S. (1963). *J. Pharmacol. Exp. Ther.* **140**, 224–228.
Nishi, S. (1970). *Fed. Proc., Fed. Amer. Soc. Exp. Biol.* **29**, 1957–1965.
Nonomura, Y., Hotta, Y., and Ohashi, H. (1966). *Science* **152**, 97–99.
Norberg, K.-A. (1964). *Int. J. Neuropharmacol.* **3**, 379–382.
Norberg, K.-A. (1967). *Brain Res.* **5**, 125–170.
Nybäck, H., Borzecki, Z., and Sedvall, G. (1968). *Eur. J. Pharmacol.* **4**, 395–403.
Oates, J. A., Gillespie, L., Udenfriend, S., and Sjoerdsma, A. (1960). *Science* **31**, 1890–1891.
Ohashi, H., and Ohga, A. (1967). *Nature (London)* **216**, 291–292.
Okamoto, M., and Baker, T. (1969). *Fed. Proc., Fed. Amer. Soc. Exp. Biol.* **28**, 669.
Okamoto, M., Song, S. K., Riker, W. F., and Longenecker, H. (1970). *Fed. Proc., Fed. Amer. Soc. Exp. Biol.* **29**, 280.
Oomura, Y., and Tomita, T. (1961). *Tohoku J. Exp. Med.* **73**, 398–415.
Orbeli, L. A. (1923). *Bull. Int. Sci. (Leshaft)* **6**, 194–197.
Otsuka, M., and Endo, M. (1960). *J. Pharmacol. Exp. Ther.* **128**, 273–282.
Otsuka, M., and Nonomura, Y. (1963). *J. Pharmacol. Exp. Ther.* **140**, 41–45.
Overton, E. (1901). "Studien über die Narcose Zugleich ein Beitrag zur Allgemeinen Pharmokologie." Fischer, Jena.
Owman, C. (1965). *Int. J. Neuropharmacol.* **3**, 105–112.
Ozeki, M., Freeman, A. R., and Grundfest, H. (1966). *J. Gen. Physiol.* **49**, 1319–1334.
Page, I. H. (1954). *Hypertension; Humoral Neurogenic Factors, Ciba Found. Symp., 1953* pp. 3–30.
Page, I. H., and Bumpus, F. M. (1961). *Physiol. Rev.* **41**, 331–390.
Palaič, D., and Khairallah, P. A. (1967). *J. Pharm. Pharmacol.* **19**, 396–397.
Palaič, D., and Khairallah, P. A. (1968). *Life Sci.* **7**, 169–172.
Panisset, J. C., Biron, P., and Beaulines, A. (1966). *Experientia* **22**, 394–395.
Papasova, M. P., Nagai, T., and Prosser, C. L. (1968). *Amer. J. Physiol.* **214**, 695–702.
Parrot, J. L., and Thouvenot, J. (1966). *In* "Handbuch der experimentellen Pharmakologie" (Rocha e Silva, M., ed.), Vol. 18, Part 1, pp. 202–224. Springer-Verlag, Berlin and New York.
Paterson, G. (1963). *Biochem. Pharmacol.* **12**, 85.
Paton, W. D. M. (1957). *Brit. J. Pharmacol. Chemother.* **12**, 119–127.
Paton, W. D. M., and Vane, J. R. (1963). *J. Physiol. (London)* **165**, 10–46.
Paton, W. D. M., and Vizi, E. S. (1969). *Brit. J. Pharmacol.* **35**, 10–28.
Paton, W. D. M., and Zar, A. M. (1965). *J. Physiol. (London)* **179**, 85–86P.
Paton, W. D. M., and Zar, A. M. (1968). *J. Physiol. (London)* **194**, 13–34.
Peart, W. S. (1965). *Pharmacol. Rev.* **17**, 143–182.
Peterson, H. (1936). *Z. Biol.* **97**, 393–398.
Phillis, J. W. (1970). "The Pharmacology of Synapses." Pergamon, Oxford.
Phillis, J. W., and Tebēcis, A. K. (1968). *Nature (London)* **217**, 1076–1077.
Pickles, V. R. (1967). *Biol. Rev.* **42**, 614–652.
Pierce, J. V. (1968). *Fed. Proc., Fed. Amer. Soc. Exp. Biol.* **27**, 52–57.
Pilar, G. (1969a). *Fed. Proc., Fed. Amer. Soc. Exp. Biol.* **28**, 670.

Pilar, G. (1969b). "Acetylcholine Release at Synapses." *NATO Advan. Study Inst.*, *1969* 45–60.

Potter, L. T. (1966). *Pharmacol. Rev.* **18**, 439–451.

Potter, L. T. (1970). *J. Physiol. (London)* **206**, 145–166.

Potter, L. T., and Axelrod, T. (1963a). *J. Pharmacol. Exp. Ther.* **140**, 199–206.

Potter, L. T., and Axelrod, T. (1963b). *J. Pharmacol. Exp. Ther.* **142**, 299–305.

Potter, L. T., Glover, V. A. S., and Saelens, J. K. (1968). *J. Biol. Chem.* **243**, 3864–3870.

Price, M. L., and Price, H. L. (1962). *Anesthesiology* **23**, 16–20.

Quastel, D. M. J. (1962). Ph.D. Dissertation, McGill University, Montreal.

Quastel, D. M. J., and Hackett, J. T. (1971). *Fed. Proc., Fed. Amer. Soc. Exp. Biol.* **30**, 2022.

Quastel, D. M. J., Hackett, J. T., and Cooke, J. D. (1971). *Science* **172**, 1034–1036.

Quinn, G. P., Shore, P. A., and Brodie, B. B. (1959). *J. Pharmacol. Exp. Ther.* **127**, 103–109.

Rahamimoff, R. (1968). *J. Physiol. (London)* **195**, 471–481.

Ramwell, P. W., and Shaw, J. E. (1967). *In* "Prostaglandins" (S. Bergström and B. Samuelsson, eds.), pp. 283–292. Wiley (Interscience), New York.

Ramwell, P. W., and Shaw, J. E. (1968). *In* "Prostaglandins. Symposium of the Worcester Foundation for Experimental Biology." Wiley (Interscience), New York.

Ramwell, P. W., and Shaw, J. E. (1970). *Recent Progr. Horm. Res.* **26**, 139–173.

Ramwell, P. W., Shaw, J. E., and Kucharski, J. (1965). *Science* **149**, 1390–1391.

Rand, M. J., and Varma, B. (1970). *Brit. J. Pharmacol.* **38**, 758–770.

Rand, M. J., and Wilson, J. (1967). *Eur. J. Pharmacol.* **1**, 200–209.

Rang, H. P. (1964). *Brit. J. Pharmacol.* **22**, 356–365.

Redfern, P., Lundh, H., and Thesleff, S. (1970). *Eur. J. Pharmacol.* **11**, 263–265.

Reid, W. D., Stefano, F. J. E., Kurzepa, S., and Brodie, B. B. (1969). *Science* **164**, 437–439.

Rhodes, H. J., and Sutter, M. C. (1971). *Can. J. Physiol. Pharmacol.* **49**, 276–287.

Rhodin, J. A. G. (1967). *J. Ultrastruct. Res.* **18**, 181–223.

Riker, W. F., Jr., and Okamoto, M. (1970). *Annu. Rev. Pharmacol.* **9**, 173–208.

Riker, W. F., Jr., and Wescoe, W. C. (1946). *J. Pharmacol. Exp. Ther.* **88**, 58–66.

Ritchie, A. K., and Goldberg, A. M. (1970). *Science* **169**, 489–490.

Ritchie, J. M., and Armett, C. J. (1963). *J. Pharmacol. Exp. Ther.* **139**, 201–207.

Robert, A., Nezamis, J. E., and Phillips, J. P. (1967). *Amer. J. Dig. Dis.* **12**, 1073–1076.

Roberts, D. V., and Thesleff, S. (1968). *Acta Anaesthesiol. Scand.* **9**, 165–172.

Rocha e Silva, M. (1966). *In* "Handbuch der experimentellen Pharmakologie" (Rocha e Silva, ed.), Vol. 18, Part 1, pp. 225–237. Springer-Verlag, Berlin and New York.

Rocha e Silva, M., Valle, J. R., and Picarelli, Z. P. (1953). *Brit. J. Pharmacol. Chem. Ther.* **8**, 378–388.

Rogers, D., and Burnstock, G. (1966). *J. Comp. Neurol.* **126**, 625–652.

Rosell, S., Kopin, I. J., and Axelrod, J. (1964). *Nature (London)* **201**, 301.

Rosenthal, J. (1969). *J. Physiol. (London)* **203**, 121–134.

Rubin, R. P. (1970). *Pharmacol. Rev.* **22**, 389–428.

Rudinger, J. (1968). *Proc. Roy. Soc., Ser. B* **170**, 17–26.

Rudolph, A. M., Kurland, M. D., Auld, D. A. N., and Paul, M. H. (1959). *Amer. J. Physiol.* **197**, 617–623.

Rutledge, C. O., and Weiner, N. (1967). *J. Pharmacol. Exp. Ther.* **157**, 290–302.

Salerno, P. R., and Coon, J. M. (1949). *J. Pharmacol. Exp. Ther.* **95**, 240–255.

Samuelsson, B. (1965). *Angew. Chem., Int. Ed. Engl.* **4**, 410–416.

Sandberg, F., Ingelmann-Sundberg, A., and Rydén, G. (1963). *Acta Obstet. Gynecol. Scand.* **42**, 269–278.

Sanders, H. D. (1969). *Can. J. Physiol. Pharmacol.* **47**, 218–221.

Sandow, A. (1965). *Pharmacol. Rev.* **17**, 265–320.

Sandow, A., and Brust, M. (1966). *Biochem. Z.* **345**, 232–247.

Scales, B., and McIntosh, D. A. D. (1968). *J. Pharmacol. Exp. Ther.* **160**, 261.

Schacter, M. (1964). *Annu. Rev. Pharmacol.* **4**, 281–292.

Schacter, M. (1968). *Fed. Proc., Fed. Amer. Soc. Exp. Biol.* **27**, 49–51.

Schanker, L. S., and Morrison, A. S. (1965). *Int. J. Neuropharmacol.* **4**, 27–39.

Schapiro, S. (1958). *Acta Physiol. Scand.* **42**, 311–375.

Schatzman, H. J. (1961). *Pflüegers Arch. Gesamte Physiol. Menschen Tiere* **274**, 295–310.

Schatzman, H. J. (1964). *Ergeb. Physiol., Biol. Chem. Exp. Pharmakol.* **55**, 28–130.

Schild, H. O. (1947). *Brit. J. Pharmacol. Chemother.* **2**, 189–206.

Schild, H. O. (1964). *In* "Pharmacology of Smooth Muscle" (E. Bülbring, ed.), pp. 95–104. Pergamon, Oxford.

Schild, H. O. (1969). *Brit. J. Pharmacol.* **36**, 329–349.

Schmidt, J. L., and Fleming, W. W. (1963). *J. Pharmacol. Exp. Ther.* **139**, 230–237.

Schmitt, H., and Schmitt, H. (1960). *Arch. Int. Pharmacodyn. Ther.* **125**, 30–47.

Schueler, F. W. (1955). *J. Pharmacol. Exp. Ther.* **115**, 127–143.

Schumann, H. J. (1958). *Naunyn-Schmiedeberg Arch. Exp. Pathol. Pharmacol.* **233**, 296–300.

Scroop, G. C., and Whelan, R. F. (1968). *Aust. J. Exp. Biol. Med. Sci.* **46**, 563–572.

Sedvall, G. (1964). *Acta Physiol. Scand.* **62**, 101–109.

Sedvall, G., and Thorson, J. (1963). *Biochem. Pharmacol.* **12**, Suppl., 65–66.

Seeman, P. (1966). *Int. Rev. Neurobiol.* **9**, 145–221.

Seeman, P., Chau, M., Goldberg, M., Sauks, T., and Saxs, L. (1971). *Biochim. Biophys. Acta* **225**, 185–193.

Semba, T., Fujii, K., and Kimura, N. (1964). *J. Physiol. Soc. Jap.* **14**, 319–327.

Senft, G., Munske, K., Schultz, G., and Hoffmann, M. (1968). *Naunyn-Schmiedebergs Arch. Exp. Pathol. Pharmakol.* **259**, 344–359.

Shanes, A. M., Freygang, W. H., Grundfest, H., and Amatniek, E. (1959). *J. Gen. Physiol.* **42**, 793–802.

Shelley, H. (1955). *Brit. J. Pharmacol. Chemother.* **10**, 26–35.

Shibata, S., and Briggs, A. H. (1966). *J. Pharmacol. Exp. Ther.* **153**, 466–470.

Shibata, S., and Carrier, O. (1967). *Can. J. Physiol. Pharmacol.* **45**, 587–596.

Shibata, S., Carrier, O., and Frankenheim, J. (1968). *J. Pharmacol. Exp. Ther.* **160**, 106–111.

Somani, P., and Lum, B. K. B. (1965). *J. Pharmacol. Exp. Ther.* **147**, 194.

Somlyo, A. P., and Somlyo, A. V. (1968). *Pharmacol. Rev.* **20**, 197–272.

Somlyo, A. P., and Somlyo, A. V. (1970). *Pharmacol. Rev.* **22**, 251–353.

Somlyo, A. V., and Somlyo, A. P. (1968). *J. Pharmacol. Exp. Ther.* **159**, 129–145.

Somlyo, A. V., Sandberg, R. L., and Somlyo, A. P. (1965). *J. Pharmacol. Exp. Ther.* **149**, 106–112.

Somlyo, A. V., Haeusler, G., and Somlyo, A. P. (1970). *Science* **169**, 490–491.

Sourkes, T. L. (1954). *Arch. Biochem. Biophys.* **51**, 444–456.

Spector, S. (1966). *Pharmacol. Rev.* **18**, 599–610.

Spector, S., Sjoerdsma, A., and Udenfriend, S. (1965). *J. Pharmacol. Exp. Ther.* 147, 86–95.

Speden, R. N. (1970). *In* "Smooth Muscle" (E. Bülbring *et al.*, eds.), pp. 558–588. Arnold, London.

Speden, R. N. (1969). *Aust. Exp. Biol. Med. Sci.* 47, 553–564.

Stafford, A. (1963). *Brit. J. Pharmacol.* 21, 361–367.

Standaert, F. G. (1964). *J. Pharmacol. Exp. Ther.* 143, 181–186.

Standaert, F. G., and Roberts, J. (1967). *Ann. N.Y. Acad. Sci.* 139, 815–820.

Staszewska-Barczak, J., and Konopka-Rogatko, K. (1967). *Bull. Acad. Pol. Sci., Ser. Sci. Biol.* 15, 503–510.

Steedman, W. M. (1966). *J. Physiol. (London)* 186, 382–400.

Steinbach, A. B. (1968a). *J. Gen. Physiol.* 52, 144–161.

Steinbach, A. B. (1968b). *J. Gen. Physiol.* 52, 162–180.

Steinberg, M. I., and Smith, C. B. (1970). *J. Pharmacol. Exp. Ther.* 173, 176–192.

Stephenson, R. P. (1956). *Brit. J. Pharmacol. Chemother.* 11, 379–393.

Stitzel, R., and Lundborg, P. (1967). *Brit. J. Pharmacol. Chemother.* 29, 99–104.

Stjärne, L. (1961). *Acta Physiol. Scand.* 51, 224–229.

Stjärne, L. (1964). *Acta Physiol. Scand.* 62, Suppl. 228, 1–97.

Stjärne, L. (1966). *Acta Physiol. Scand.* 67, 441–454.

Stovner, J. (1958a). *Acta Pharmacol. Toxicol.* 14, 317–332.

Stovner, J. (1958b). *Acta Pharmacol. Toxicol.* 15, 55–69.

Su, C., and Bevan, J. A. (1967). *Bibl. Anat.* 8, 30–34.

Su, C., Bevan, J. A., and Ursillo, R. C. (1964). *Circ. Res.* 15, 20–27.

Suarez-Kurtz, G., and Bianchi, C. P. (1970). *J. Pharmacol. Exp. Ther.* 172, 33–43.

Suarez-Kurtz, G., Paulo, L. P., and Fonteles, M. C. (1969). *Arch. Int. Pharmacodyn. Ther.* 177, 185–195.

Sutherland, E. W., and Rall, T. W. (1960). *Pharmacol. Rev.* 12, 265–299.

Sutherland, E. W., and Robison, G. A. (1966). *Pharmacol. Rev.* 18, 145–161.

Sutherland, E. W., Øye, I., and Butcher, R. W. (1965). *Recent Progr. Horm. Res.* 21, 623–642.

Sutherland, E. W., Robison, G. A., and Butcher, R. W. (1968). *Circulation* 3, 279–306.

Suzuki, T., and Vogt, W. (1965). *Naunyn-Schmiedebergs Arch. Exp. Pathol. Pharmakol.* 252, 68–78.

Symposium On Histamine. (1965). Various authors. *Fed. Proc., Fed. Amer. Soc. Exp. Biol.* 24, 1293–1352.

Symposium On Histamine. (1967). Various authors. *Fed. Proc., Fed. Amer. Soc. Exp. Biol.* 26, 211–240.

Symposium. Various authors. Vasoactive Peptides (1968). *Fed. Proc., Fed. Amer. Soc. Exp. Biol.* 27, 49–99.

Tafuri, W. L., and Raick, A. (1964). *Z. Naturforsch B* 19, 1126–1128.

Tainter, M. L., and Chang, D. K. (1927). *J. Pharmacol. Exp. Ther.* 30, 193–207.

Takeuchi, A., and Takeuchi, N. (1959). *J. Neurophysiol.* 22, 395–411.

Takeuchi, A., and Takeuchi, N. (1960). *J. Physiol. (London)* 154, 52–67.

Takeuchi, A., and Takeuchi, N. (1961). *J. Physiol. (London)* 155, 46–58.

Takeuchi, N. (1963). *J. Physiol. (London)* 167, 141–155.

Tashiro, N., and Tomita, T. (1970). *Brit. J. Pharmacol.* 39, 608–613.

Taxi, J., and Droz, B. (1966). *C.R. Acad. Sci., Ser.* 263, 1237–1240.

Thaemert, J. C. (1963). *J. Cell Biol.* 16, 361–377.

Thesleff, S. (1955). *Acta Physiol. Scand.* 34, 218–231.

Thesleff, S. (1956). Acta Physiol. Scand. 37, 335–349.

Thesleff, S. (1958). Acta Anaesthesiol. Scand. 2, 69–79.

Thies, R. E., and Brooks, V. B. (1961). Fed. Proc. Fed. Amer. Soc. Exp. Biol. 20, 569–578.

Thoenen, H., Tranzer, J. P., Hürlimann, A., and Haefely, W. (1966). Helv. Physiol. Pharmacol. Acta 24, 229–246.

Townsend, L. (1967). Ph.D. Dissertation (quoted in Feinstein and Paimre, 1969).

Tranzer, J. P., and Thoenen, H. (1968). Experientia 24, 155–156.

Trautwein, W., and Dudel, J. (1958). Pflüegers Arch. Gesante Physiol. Manschen Tiere 266, 324–334.

Trendelenburg, U. (1956a). Brit. J. Pharmacol. Chemother. 11, 74–80.

Trendelenburg, U. (1956b). J. Physiol. (London) 135, 66–72.

Trendelenburg, U. (1963). Pharmacol. Rev. 15, 225–276.

Trendelenburg, U. (1966a). Pharmacol. Rev. 18, 629–640.

Trendelenburg, U. (1966b). J. Pharmacol. Exp. Ther. 151, 95–102.

Trendelenburg, U. (1967). Ergeb. Physiol., Biol. Chem. Exp. Pharmakol. 59, 1–85.

Trendelenburg, U., and Pfeffer, R. I. (1964). Naunyn-Schmiedebergs Arch. Exp. Pathol. Pharmakol. 248, 39–53.

Trendelenburg, U., Muskus, A., Fleming, W. W., and Gomez, B. (1962a). J. Pharmacol. Exp. Ther. 138, 170–180.

Trendelenburg, U., Muskus, A. Fleming, W. W., and Gomez, B. (1962b). J. Pharmacol. Exp. Ther. 138, 181–193.

Trendelenburg, U., Maxwell, R. A., and Pluchino, S. (1970). J. Pharmacol. Exp. Ther. 172, 91–99.

Triner, L., Vulliemoz, Y., Schwartz, I., and Nahas, G. G. (1970). Biochem. Biophys. Res. Commun. 40, 64–69.

Türker, R. K., and Karahüseyinoglu, E. (1968). Experientia 24, 921–922.

Tuttle, R. S. (1966). Fed. Proc., Fed. Amer. Soc. Exp. Biol. 25, 1593–1595.

Tuttle, R. S. (1967). Amer. J. Physiol. 213, 620–624.

Twarog, B. M. (1954). J. Cell Physiol. 44, 141–164.

Twarog, B. M. (1960). J. Physiol. (London) 152, 236–242.

Twarog, B. M. (1967). J. Gen. Physiol. 50, 157–169.

Twarog, B. M. (1968). Advan. Pharmacol. 6D, 613.

Uvnäs, B. (1966). Fed. Proc., Fed. Amer. Soc. Exp. Biol. 25, 1618–1622.

Vane, J. R. (1957). Brit. J. Pharmacol. Chemother. 12, 344–349.

Vane, J. R. (1960). Adrenergic Mech., Ciba Found. Symp., 1960 p. 356.

Van Orden, L. S. III, Bensch, K. G., and Giarman, N. J. (1967). J. Pharmacol. Exp. Ther. 155, 428–439.

Van Orden, L. S. III, Bloom, F. E., Barnett, R. J., and Giarman, N. J. (1966). J. Pharmacol. Exp. Ther. 154, 185–199.

Verity, M., and Bevan, J. A. (1967). Bibl. Anat. 8, 60–65.

Vizi, E. S. (1968). Naunyn-Schmiedebergs Arch. Exp. Pathol. Pharmakol. 259, 199–200.

von Euler, U. S., and Eliasson, R. (1967). "Prostaglandins." Academic Press, New York.

von Euler, U. S., and Lishajko, F. (1963a). Acta Physiol. Scand. 57, 468–480.

von Euler, U. S., and Lishajko, F. (1963b). Acta Physiol. Scand. 59, 454–461.

von Euler, U. S., and Lishajko, F. (1963c). Int. J. Neuropharmacol. 2, 127–134.

von Euler, U. S., and Lishajko, F. (1966). Acta Physiol. Scand. 68, 257–262.

von Euler, U. S., and Lishajko, F. (1968a). Acta Physiol. Scand. 73, 78–92.

von Euler, U. S., and Lishajko, F. (1968b). *Acta Physiol. Scand.* **74**, 501–506.
von Euler, U. S., Stjärne, L., and Lishajko, F. (1963). *Life Sci.* **11**, 878–885.
Ward, C. O., and Gautieri, R. F. (1968). *J. Pharm. Sci.* **57**, 287–292.
Warnick, J. E., and Albuquerque, E. X. (1970). *Fed. Proc., Fed. Amer. Soc. Exp. Biol.* **29**, 715.
Washizu, Y. (1966). *Comp. Biochem. Physiol.* **19**, 713–728.
Washizu, Y. (1968). *Comp. Biochem. Physiol.* **27**, 121–126.
Waud, D. R. (1968). *Pharmacol. Rev.* **20**, 49–88.
Waugh, W. H. (1962a). *Circ. Res.* **11**, 264–276.
Waugh, W. H. (1962b). *Circ. Res.* **11**, 927–940.
Weber, A. (1966). *Curr. Top. Bioenerg.* **1**, 203–254.
Weber, A., and Herz, R. (1968). *J. Gen. Physiol.* **52**, 750–759.
Webster, M. E., and Pierce, J. V. (1963). *Ann. N.Y. Acad. Sci.* **104**, 91–97.
Weight, F. F., and Votava, J. (1970). *Science* **170**, 755–758.
Welsh, J. H. (1968). *Advan. Pharmacol.* **6A**, 171–188.
Werman, R. (1969). *Comp. Biochem. Physiol.* **30**, 997–1017.
Whitby, L. G., Hertting, G., and Axelrod, J. (1960). *Nature (London)* **187**, 604–605.
Whittaker, V. P. (1965). *Progr. Biophys. Mol. Biol.* **15**, 39–96.
Whittaker, V. P. (1968). *Biochem. J.* **109**, 20–21.
Wilkenfeld, B. E., and Levy, B. (1969). *Fed. Proc., Fed. Amer. Soc. Exp. Biol.* **28**, 741.
Wilson, A. B. (1964). *J. Pharm. Pharmacol.* **16**, 834–835.
Wilson, J. (1970). *Brit. J. Pharmacol.* **40**, 159–160.
Wilson, R., and Long, C. (1960). *Lancet* **2**, 262.
Wolfe, D. E., Potter, L. T., Richardson, K. C., and Axelrod, J. (1962). *Science* **138**, 440–444.
Wolfe, D. E., and Potter, L. T. (1963). *Anat. Rec.* **145**, 301.
Wooley, P., and Waring, H. (1958). *Aust. J. Biol. Exp. Med. Sci.* **36**, 447–456.
Woolley, D. W., and Campbell, N. K. (1960). *Biochim. Biophys. Acta* **40**, 543–544.
Wurtman, R. J., Kopin, I. S., and Axelrod, J. (1964). *Endocrinology* **73**, 63–74.
Yamaguchi, T. (1961). *J. Fac. Sci., Hokkaido Univ., Ser. 6* **14**, 522–535.
Zar, M. A. (1966). Ph.D. Thesis, Oxford (quoted by Gershon, 1970).

2

EFFECTS OF NUTRITIONAL DEFICIENCY ON MUSCLE

KARL MASON

I. Introduction

Approximately one-half the total body weight of the animal body is represented by muscle. In no other tissue is there required such a complexity of metabolic interactions and structural organization for maintenance and functional activity. Yet, despite these remarkably intricate intracellular mechanisms and the influences exerted by the related peripheral nerves, vascular bed, and supporting tissues, muscle cells as a whole are remarkably resistant to undernutrition and to specific nutritional deficiencies which may exert rather profound effects upon other tissues and organs of the body. Furthermore, the morphological changes that do occur after nutritional deficiencies possess no features that clearly distinguish them from those induced by ischemia, trauma, denervation, heat, chemical toxins, steroid hormones, or hereditary factors. It is true that from animal to animal, and even in the same animal, there may be observed rather wide differences in the histopathological picture, depending in large part upon a combination of factors: (1) the particular muscle or group of muscles examined, (2) the developmental maturity of the fibers when affected, (3) the relative acuteness or chronicity of the deficiency state, and (4) the relative balance between degenerative and regenerative process in the muscle fibers. There is general agreement among investigators that muscle reacts to various types of injury in so similar a manner that no clear-cut diagnostic criterion capable of pinpointing the causative agent exists. For excellent descriptions of the reaction of muscle to injury, the reader is referred to Adams *et al.* (1962). It may be added that the somewhat limited observations to date on the ultrastructure of muscle in experimental myopathies have not revealed distinctive or diagnostic features, as has been true of various neuromuscular diseases of man.

Chronic undernutrition results in a general decrease in size of muscle fibers, not unlike that of disuse atrophy, and restoration to normal size and function follows return to full nutriture. Deficiencies of essential nutrients that lead to general debility and decreased intake of food may likewise result in some degree of simple atrophy of muscle. Typical histopathological lesions are observed in only a very limited number of deficiency states. Among these, only in the case of deficiency of vitamin E, with or without an associated deficiency of selenium, is there loss of structural integrity of skeletal, cardiac, and smooth muscle to variable degrees in a wide array of animal species. A discussion of the structural and biochemical changes observed in these deficiency states, represented by a rather extensive literature of the past 40 years, will naturally occupy a considerable segment of this chapter. The remainder will deal with more limited information concerning the effects of deficiency of water-soluble vitamins, choline, tryptophan, potassium, and magnesium, the only other nutritional factors shown to play a role in the metabolism of skeletal or cardiac muscle, or both, to the extent that in their absence structural alterations become apparent.

II. Skeletal Muscle

A. Vitamin E Deficiency

Vitamin E, of which α-tocopherol is the prototype, was first established as an antisterility vitamin through studies on the laboratory rat. Shortly thereafter, H. M. Evans and Burr (1928) observed a paralysis in the suckling young of female rats fed a diet low in vitamin E, but ten years elapsed before degenerative lesions in the skeletal muscles were recognized (Olcott, 1938). In the meantime, adult rats maintained on low-E diets for many months were shown to develop a gradual paralysis affecting the hind extremities and associated with extensive degeneration of skeletal muscles (Ringsted, 1935; Einarson and Ringsted, 1938; H. M. Evans *et al.*, 1938). Paralysis and muscle lesions in guinea pigs and rabbits reared on certain diets were observed during this same period by several investigators; but confusion existed regarding dietary factors involved until C. C. Mackenzie and McCollum (1940) and Shimotori *et al.* (1940) presented clear-cut evidence that deficiency of vitamin E was solely responsible. The reader is referred to excellent reviews of the literature on this subject by Pappenheimer (1943), C. G. Mackenzie (1953), West (1963), and Telford (1971).

The fact that all laboratory and farm animals that have been depleted of vitamin E early in life, whether birds or mammals, have manifested nutritional muscular dystrophy, suggests that the term "antidystrophy vitamin" might have been a more appropriate term than "antisterility vitamin." Actually, the antisterility role of α-tocopherol has been demonstrated only for both sexes of the rat, hamster, mouse, and guinea pig. Efforts to study reproductive functions in other laboratory animals depleted of vitamin E have been thwarted by onset of nutritional muscular dystrophy before attainment of sexual maturity. An approach to this problem seemed possible, following observations that α-tocohydroquinone, an oxidation–reduction product of α-tocopherol, was quite effective in preventing muscular dystrophy in vitamin E-deficient rabbits (J. B. Mackenzie *et al.*, 1950), rats (J. B. Mackenzie and Mackenzie, 1959) and Syrian hamsters (West and Mason, 1955) but not in preventing fetal resorption in the vitamin E-deficient rat (J. B. Mackenzie and Mackenzie, 1953). It thus appeared that prevention of muscular dystrophy by α-tocohydroquinone might permit study of the effects of vitamin E deficiency on reproduction in the rabbit and guinea pig. However, when put to the test, neither reproductive impairment nor muscular dystrophy was observed (Mason and Mauer, 1959). This was in accord with other observations, contrary to the earlier report of J. B. Mackenzie and Mackenzie (1953), that α-tocohydroquinone does have antisterility activity (to a somewhat lesser degree than α-tocopherol) in the male Syrian hamster (Mason and Mauer, 1957) and in the female rat (J. B. Mackenzie and Mackenzie, 1960).

As yet, we have no clear understanding of the role of vitamin E in reproductive processes or in maintaining the integrity of muscle—skeletal, cardiac, or smooth. As indicated in later sections of this chapter, interrelationships between vitamin E dietary unsaturated fatty acids, selenium, and sulfur-containing amino acids have, over the years, created a decidedly confusing and complex picture of dietary factors influencing muscle of various types and also other tissues. The fact that various synthetic antioxidants can substitute for vitamin E in preventing certain deficiency manifestations, especially those dependent upon or accentuated by dietary unsaturated fatty acids, strongly supports its long accepted role as an intracellular antioxidant. There has also accumulated evidence that another function relates to maintenance of the phospholipids of intracellular membranes, particularly those of lysosomes, thus controlling release of hydrolytic enzymes, which in excess exert damaging effects on cell metabolism and structure. An extensive literature on selenium interrelationships (see p. 178) indicates that selenium forms a specific complex with serum proteins, possibly a lipoprotein complex, which functions as a carrier of vitamin E, facilitating its absorption, blood transport, and transfer across cell membranes (Desai and Scott, 1965). This selenium–vitamin E complex is also thought to play some unknown role related to the metabolism of sulfur-containing amino acids, particularly cystine (M. L. Scott, 1966; Desai, 1968; Thompson and Scott, 1969). Whether vitamin E has functions in cell metabolism as a component of certain enzyme systems and apart from its antioxidant functions has long been a controversial question. Exhaustive and critical reviews of the literature pertaining to antioxidant and other possible functions of vitamin E have been presented by Alfin-Slater and Morris (1963) and by Green and Bunyan (1969).

By way of introduction to gross and histopathological alterations of skeletal muscle induced by vitamin E deficiency, a few general remarks may be pertinent. First the terms muscular dystrophy, nutritional myodegeneration, nutritional muscular dystrophy, and nutritional myopathy have been applied interchangeably to states of muscular weakness, with or without obvious paralysis, associated with histological demonstrable lesions of the skeletal musculature in animals deficient in vitamin E.

In the case of ruminants the term white muscle disease is also commonly used. These various designations imply that the disorder represents a true myopathy, affecting muscle primarily, rather than its nervy supply.

Second, the lesions of skeletal muscle observed in the many laboratory and farm animals present a somewhat confusing picture of normal fibers intermingled with degenerating and regenerating fibers combined with extensive cellular invasion and fibrotic reaction. Furthermore, since these fibers are seen in various profiles, depending upon the plane of section, there is provided a rather inadequate concept of events occurring in the muscle fiber as a whole.

Third, skeletal muscle reacts to vitamin E deficiency in much the same manner that it does to other noninflammatory types of injury. As mentioned previously, this may be modified by a variety of factors, such as intensity of reaction, chronological age of the muscle, functional differences, balance between degenerative and regenerative processes, and species differences in metabolic needs.

Lesions of skeletal muscles constitute a universal finding in all laboratory animals that have been subjected to vitamin E depletion. These include the rat, rabbit, guinea pig, Syrian hamster, mouse, mink, Florida cotton rat, dog, cat, Rhesus monkey, chick, turkey poult, duckling, and quail. Of these, the first five mentioned and the chick have received the most extensive study. Because of the tenacious manner in which vitamin E is stored and held by body tissues, it has been customary in experimental studies to employ newly weaned animals and even to place mother and suckling young on low-E diets in order to reduce initial storage to a minimum. In avian forms, it is the newly hatched chick that is used. This means that the states of muscular dystrophy observed in most laboratory animals, and in farm animals also, have been produced at periods comparable to infancy and adolescence. In only a limited number of species have chronic states of dystrophy in the adult been studied.

1. SYMPTOMATOLOGY

H. M. Evans and Burr (1928) called attention to an unusual type of paralysis occurring in the offspring of rats given just sufficient vitamin E to permit the successful completion of gestation and lactation. The young usually appear vigorous and healthy prior to the eighteenth day of life, but during the ensuing week there appears in all or in certain members of the litter a flaccid paralysis of the hind legs and often a flexor contraction of the forelimbs (Fig. 1). Some affected animals

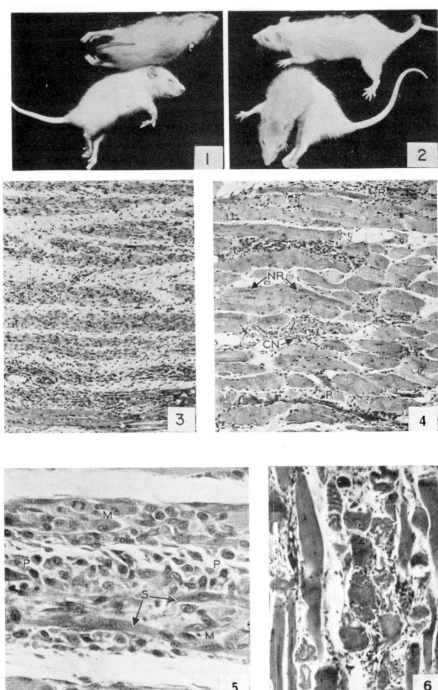

become lethargic and succumb within a few days, perhaps because of extensive involvement of respiratory muscles; others, in which the paresis is mild or moderate, show spontaneous recovery. Vitamin E given as late as the seventeenth day of lactation prevents paralysis. The affected animals, and also many grossly unaffected members of the litter, exhibit microscopically widespread lesions of the skeletal muscles that vary considerably in their intensity and in the extent to which individual muscles or portions of muscles are involved. When continued on low-E diets, the muscles of the young animals that recover spontaneously show rapid repair and little or no evidence of degenerative change until months later when the manifestations of chronic vitamin E deficiency appear.

When lactating mice and their offspring are fed low-E diets (with 18% lard and 2% cod liver oil) which produce late-lactation paralysis in the rat, no symptoms appear even though muscle lesions are found histologically (Pappenheimer, 1942; Tobin, 1950). However, if the lard is replaced by increased amounts of cod liver oil, gross paralysis appears and the muscle lesions are much more severe and widespread (Tobin, 1950).

The onset of dystrophy in the rabbit, which often progresses rather rapidly, has been divided into three stages by C. G. Mackenzie and McCollum (1940). There is an initial stage characterized by a twofold increase in the urinary output of creatine with no change in creatinine excretion, usually followed by retarded growth and a decline in food

---

Fig. 1. Late-lactation paralysis in 22-day-old rats, showing flaccid paralysis of hind limbs and flexor contraction of fore limbs.

Fig. 2. Paralysis of adult rats after chronic vitamin E deficiency. Rats are litter mates 360 days old. In one rat (above), paralysis was arrested at the spraddle stage by vitamin E therapy (243 to 360 days) with only partial recovery from paresis. Other rat (lower), which received no therapy, shows adductor contracture and atrophy of muscles of trunk and lower extremity. From Mason and Emmel (1945).

Fig. 3. Skeletal muscle from young rat such as shown in Fig. 1. Note extensive segmental necrosis of fibers separated by zones of edema and leukocytic infiltration. A few normal fibers are present (bottom) (× 80). From Mason (1952).

Fig. 4. Enlarged view of two adjacent segments from Fig. 3 transformed into densely cellular zones composed of macrophages (P), many muscle nuclei (M), and syncytial fusion of the latter to form basophilic, multinucleate, spindle-shaped strands (S) (× 360). From Mason (1952).

Fig. 5. Skeletal muscle from Syrian hamster 200 days deficient in vitamin E, showing segmental coagulation necrosis (CN), nuclear rowing (NR), and regenerative reactions (R) (× 80). From Mason (1952).

Fig. 6. Skeletal muscle of hamster showing extensive contraction-clot formation and varying degrees of necrosis of affected segments. Other fibers show internal nuclear rowing (× 100). From West and Mason (1958).

intake. In the second stage, there is stiffness of the forelegs, some head retraction, slowness in righting when placed on the side, and accentuation of the earlier manifestations. In the final phase, the animals have great difficulty in regaining an erect position when prone or supine, show pronounced loss of body tonus when picked up, and may become completely prostrate several days before death. Spontaneous recovery such as observed in weanling rats never occurs, but there is a remarkably rapid amelioration of all symptoms and manifestations of dystrophy if vitamin E therapy is instituted before these phenomena are too far advanced. In the guinea pig, as described in the classic studies of Goettsch and Papenheimer (1931), the course of events is much the same as in the rabbit.

In the Syrian hamster, depletion of vitamin E from early life produces relatively little indication of muscle weakness or functional impairment until after 10 or 12 months of deficiency, even though muscle lesions are evident histologically during the second month of deficiency and become progressively more severe as deficiency progresses. There are certain resemblances to the rat subjected to chronic vitamin E deficiency, except that for a comparable stage of depletion the microscopic lesions are much more extensive (West and Mason, 1955).

The paralysis occurring in rats reared for prolonged periods (6–8 months) on low-E diets has been studied by many investigators and is described in detail in the monograph of Einarson and Ringsted (1938). Briefly, there occurs an adductor weakness in the hind legs, leading to a spraddling or waddling type of gait (Fig. 2) with slight incoordination. As weakness and atrophy gradually extend to other muscles of the lower limbs, pelvic girdle, and trunk, the gait becomes more incoordinated and ataxic. Finally, with loss of ability to walk, the animals drag the hind quarters along while supporting themselves weakly by the forelegs in moving about the cage, or lie most of the time on the side with hind legs contracted and spastic (Fig. 2). Progressive hypasthenia and hypalgesia of the tail and lower extremities also occur. Einarson and Ringsted (1938) and Einarson (1952) consider the clinical picture to be one of a neuropathy superimposed upon an earlier myopathy, but others have regarded it as a true myopathy with some secondary sensory loss (see p. 173). In early phases, the paresis is reversible with vitamin E therapy; however, if spraddle of the hind limbs has persisted for some time, therapy does not result in full recovery (Fig. 2), probably because of previously established abductor contractures. A similar paralysis has been observed by Menschik *et al.* (1949) in mice fed vitamin E-deficient diets for 9 to 24 months. Furthermore, lesions of skeletal muscle have been observed, in the absence of very obvious disabilities in stance or locomotion, in various other laboratory animals such as the Florida cotton rat (Telford, 1971), Rhesus monkey (Mason and Telford, 1947; Dinning and Day, 1957), and dog (Brinkhous and Warner, 1941; Kaspar and Lombard, 1963; Hayes *et al.*, 1969, 1970). The same is true of chronic deficiency in the adult cat (Mason and Dju, 1963); however, if deficiency is induced early in life, there occurs sudden onset of inability to stand or to assume normal posture when placed prone or on the side, much like the disability described in the vitamin E-deficient rabbit (Pappenheimer and Goettsch, 1934); and, as in the latter, the response to vitamin E therapy with either α-tocopherol or α-tocohydroquinone is very dramatic.

2. HISTOPATHOLOGY

In the myopathy of suckling rats the skeletal muscles appear pale, moist, devoid of luster, and sometimes streaked and gritty. Microscopically, as described by Pappenheimer (1939, 1948) and others, there is widespread contraction clot formation and coagulation necrosis of muscle fiber segments, interstitial edema, and extensive invasion of macrophages and mononuclear cells suggesting mild inflammatory reaction (Fig. 3). Involved segments show loss of striations, transformation of myofibrils and sarcoplasm into a homogeneous coagulum, and irregular distribution of nuclei; some nuclei become pyknotic and degenerate, others survive. As necrotic material is removed by phagocytosis and lysis, the fiber segment becomes replaced by a densely packed mixture of muscle nuclei and invading cells, mostly macrophages (Fig. 4). Electron microscopically, earliest changes involve the myofilaments, especially the Z lines, and also components of the sarcoplasmic reticulum, mitochondria and sarcolemma, which are much more resistant to injury (Rumery and Hampton, 1959). This stands in contrast to the much earlier involvement of mitochondria in skeletal muscle of the adult rat and rabbit (p. 172) and in smooth muscle of the rat (p. 198).

Associated with these degenerative changes are much more interesting reparative cell reactions which warrant description at this point, since they occur also to varying degrees in association with muscle lesions of essentially all animal species referred to later in this chapter. In suckling rats the discontinuous type of regeneration predominates, and is characterized by the following cellular reactions. Certain muscle nuclei with their cytoplasmic investment, usually located along the periphery of the cell mass which replaces the necrotic coagulum, acquire cell membranes, and become presumptive myoblasts comparable to those arising during myogenesis. After a period of mitotic activity they become post-mitotic myoblasts, related possibly to the appearance of contractile proteins in the cytoplasm. These myoblasts then fuse to form elongated, bandlike, basophilic myotubes (Fig. 4) in which myofibrils soon appear and increase in number between centrally aligned nuclei and a limiting membrane, or sarcolemma. Typical of the continuous or terminal budding type of regeneration is the formation of multinucleated, basophilic outgrowths of sarcoplasm from intact segments of the muscle fibers. With continued growth and elongation into the degenerating fiber segment, myotubes such as those described above form the basis for regeneration of damaged fibers.

In both types of regeneration subsequent events are alike. With increased production of contractile elements and of myoglobin, basophilia diminishes and nuclei acquire a subsarcolemmic position and orderly distribution throughout the fibers. In this manner, there is at least partial restitution of muscle lost in degenerative processes. Brief reference should be made to the possible role that may be played in these regenerative reactions by mononuclear satellite cells, first described by Mauro (1961). These cells, located between the basement and plasma membranes of skeletal muscle fibers, are said to represent 10–20% of the sarcolemmic nuclei observed with light microscopy (Muir *et al.*, 1965). It is of interest that satellite cells are not present in cardiac muscle, which lacks the capacity to regenerate new fibers after injury. On the basis of more recent information concerning the nature and role of satellite cells in normal development and regeneration of skeletal muscle, much of which has been recorded in an excellent monograph (Mauro *et al.*, 1970), it presumed that these cells serve as an important source of myoblasts in the regenerative reactions described above.

Compared to the delayed and very gradual appearance of muscle lesions in rats depleted of vitamin E after weaning, the widespread degenerative changes seen in the deficient suckling rat and the spontaneous recovery observed if paresis is slight or moderate in degree constitute rather dramatic reactions of skeletal muscle to a specific deficiency state. These reactions can be related undoubtedly to the fact that normally there occurs a very rapid postnatal increase in the number of muscle fibers up to about 3 weeks of age, after which further increase of muscle mass is due to increased size of muscle fibers (Chiakulas and Pauly, 1965). Hence, these phenomena occur during a period in which proliferative activity and biochemical maturity are predominant, and an adequacy of vitamin E is particularly critical for their successful completion. Presumably, the biochemical maturity is most important in this context.

Vitamin E deficiency in infantile mice results in a rather mild degree of myodegeneration unless accentuated by increased dietary intake of unsaturated fatty acids (Pappenheimer, 1942; Tobin, 1950). Guinea pigs and rabbits reared from early age on vitamin E-deficient diets show a less explosive type of reaction in skeletal muscles than does the suckling rat, but the cellular reactions are much the same. There is, however, a greater tendency toward replacement of injured fibers by adipose and fibrous tissue, indicating perhaps that the considerable re-generation of new fibers observed may fail to fully replace those lost through degeneration. Vitamin E therapy effects rather rapid repair, structurally and biochemically. By proper regulation of vitamin E intake, it is possible to induce alternate states of dystrophy and recovery for long periods of time, or to maintain a state of chronic dystrophy compati-ble with good growth and vigor (Mackenzie, 1942). In states of chronic dystrophy, the lesions are characterized by a predominance of healthy muscle fibers between which are interspersed fibers showing degenera-tive and regenerative changes of a less acute type, resembling those of chronic deficiency in the rat.

Muscle lesions observed in the Syrian hamster reared on low-E diets, in terms of rate of onset and intensity of reaction, are somewhat inter-mediate between the acute types seen in the suckling rat, guinea pig, and rabbit and those observed after chronic vitamin E deficiency in the adult rat. They are particularly suitable for study of certain cellular reactions of dystrophic muscle. Furthermore, fixed spreads of the hamster cheek pouch permit study of individual fibers *in toto* for relatively long distances. This provides much more information regarding the extent to which segments of fibers are involved in histopathological changes and the variability in the nature of these changes than is possible through the study of histologic sections of muscle alone. The following descrip-tion is based upon the observations of West and Mason (1955, 1958) and a review by West (1963).

Exclusive of internal nuclear rowing, discussed below, the predomi-nant features of myodegeneration are coagulation necrosis and contrac-tion clot formation (Figs. 5 and 6), often referred to as hyaline or Zenker's necrosis. The usual sequence of events involve (1) irregular disposition of muscle nuclei; (2) breakdown of myofibrils and formation of an amorphous sarcoplasmic coagulum (Fig. 7); (3) phagocytosis of this coagulum by invading macrophages and variable numbers of leukocytes; and (4) the resultant dense aggregates of muscle nuclei and phagocytic cells (Figs. 8–10) sometimes referred to as *Muskelzel-lenschlauche* of Waldeyer. In these zones the continuous or terminal budding type of regeneration is commonly seen at the junction of unin-jured and necrotic fiber segments (Figs. 8–10), and the discontinuous

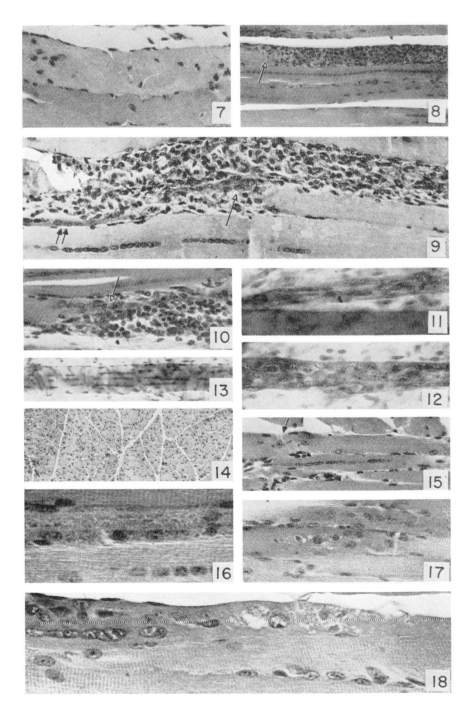

type seen in subsarcolemmic and other portions of the cellular conglomerate (Fig. 9). In the long, slender, cheek pouch fibers regeneration

---

Fig. 7. Skeletal muscle of vitamin E-deficient Syrian hamster. Typical coagulation necrosis as seen prior to invasion of mononuclear cells. Note irregular disposition of many muscle nuclei ($\times$ 200).

Fig. 8. Skeletal muscle of vitamin E-deficient Syrian hamster. Area of dense cellularity (*Muskelzellenschlauche* of Waldeyer) continuous with normal segment of same fiber (at arrow). Nuclear rowing in other fibers ($\times$ 100). From West and Mason (1958).

Fig. 9. Skeletal muscle of vitamin E-deficient Syrian hamster. Area similar to that of Fig. 8 in two adjacent fibers, showing continuous or terminal budding type of regeneration in the form of plasmodial outgrowth from undamaged fiber segment (single arrow), and discontinuous type of regeneration (double arrows) with fusion of myoblasts into elongated myotube. Internal nuclear rowing in other fiber ($\times$ 200).

Fig. 10. Skeletal muscle of vitamin E-deficient Syrian hamster. Similar to Figs. 8 and 9 showing plasmodial syncytium (arrow) at junction of *Muskelzellenschlauche* with normal segment of fiber ($\times$ 200). From West and Mason (1958).

Fig. 11. Skeletal muscle of vitamin E-deficient Syrian hamster. Elongated, spindle-shaped myoblasts in zone of early regeneration. Cell in mitosis cannot be identified. This figure and the following two figures are from unsectioned fibers observed in spreads of the cheek pouch ($\times$ 200). From West and Mason (1958).

Fig. 12. Skeletal muscle of vitamin E-deficient Syrian hamster. Later phase of regeneration, in which elongated myoblasts form a syncytium within the endomysial framework of the injured segment ($\times$ 200). From West and Mason (1958).

Fig. 13. Skeletal muscle of vitamin E-deficient Syrian hamster. Unsectioned fiber with numerous nuclear rows of variable length but with preservation of normal striations; other nuclei are in subsarcolemmic position ($\times$ 100). From West and Mason (1958).

Fig. 14. Skeletal muscle of vitamin E-deficient Syrian hamster. Section from muscle of hamster given 10 days of vitamin E therapy after development of lesions similar to those seen in Fig. 5. Necrosis has been arrested and eliminated, leaving only a few young, regenerating, basophilic fibers and many in which internal nuclear rowing (sublethal but not irreversible injury) had occurred prior to therapy and still persists ($\times$ 55). From West and Mason (1955).

Fig. 15. Skeletal muscle of vitamin E-deficient Syrian hamster. Segment showing focal breakdown of myofibrils and resulting granularity. Several enlarged and irregularly disposed nuclei lie near the junction (arrow) with the more normal segment of the fiber ($\times$ 200). From West and Mason (1958).

Fig. 16. Skeletal muscle of vitamin E-deficient Syrian hamster. Portion of a segment similar to that in Fig. 10 showing characteristic linear rowing of coarse granules ($\times$ 430). From West and Mason (1958).

Fig. 17. Skeletal muscle of vitamin E-deficient Syrian hamster. Segment showing enlargement and dissolution of irregularly disposed nuclei, with resultant vacuolation, yet with preservation of striations peripheral to the affected zone and in more normal portions of the fiber ($\times$ 200). From West and Mason (1958).

Fig. 18. Skeletal muscle of vitamin E-deficient Syrian hamster. A similar segment showing a more advanced phase of the same process (upper right) and an earlier phase (upper left) in which large vesicular nuclei are irregularly distributed or arranged in short rows ($\times$ 430). From West and Mason (1958).

usually takes the form of multinucleated syncytial strands (Figs. 11 and 12) which sometimes appear to be transformed into fibers with multiple rows of nuclei (Fig. 13) before attaining normal morphology.

The more usual type of internal nuclear rowing, characterized by single strings of nuclei located more or less centrally in fibers which otherwise usually appear normal, is a much more prominent feature of the hamster than of other experimental animals deprived of vitamin E. Such fibers are distinguishable from myotubes and later phases of regenerating fibers, in which similar nuclear rowing occurs, largely on the basis of dimensions of the fibers involved. Lacking evidence that in many regenerating fibers the nuclei fail to attain a subsarcolemmic location, it may be assumed that internal nuclear rowing, as seen in Figs. 8 and 9, represents a reaction of muscle fibers to relatively mild injury. It may even represent a process of dedifferentiation, possibly protective in nature, from which morphological recovery is possible when metabolic conditions become more favorable. In any case it may be said that in vitamin E deficient hamsters in which muscles of the trunk and limbs show rather extensive myodegeneration, the vast majority of fibers in the masseter muscle show only nuclear rowing, and that when the few areas of fiber necrosis are removed following vitamin E therapy (Fig. 14) internal rowing persists and is only partially elimi-nated after several months of therapy. There are also questions raised as to whether muscles with nuclei located internally, whether others are in a subsarcolemmic position or not, are functionally normal; also, whether muscles of branchial arch origin differ from skeletal muscle arising from paraxial mesoderm in their response to vitamin E deficiency. In this connection it is of interest to note that in their early studies on nutritional myodegeneration in guinea pigs and rabbits Goettsch and Pappenheimer (1931) mentioned that the masseter and tongue escaped injury despite rather devastating injury in skeletal muscle elsewhere. Since muscles of branchial arch origin have rarely been examined in routine studies of experimental myopathies, no adequate answer can be given.

Returning to the histopathology of myodegeneration in the hamster, there may be observed two other types of lesions occurring in fiber segments which, for some periods of time, may have shown only the early onset of internal rowing. One pertains to focal breakdown of myo-fibrils and conversion of fiber segments into a somewhat granular mass (Figs. 15 and 16). The other relates to an irregular distribution of en-larged, vesicular nuclei, or of vacuolar evidence of their dissolution (Figs. 17 and 18), often in fibers which are normally striated. Muscle lesions of this type, which are also rather characteristic of chronic vitamin E deficiency in the adult rat (Figs. 19 and 20), presumably represent a mild reaction of muscle cells to the deficiency state.

Muscle lesions of chronic vitamin E deficiency in adult rats, compared to those in the hamster, are much less extensive after comparable periods of deficiency; internal nuclear rowing is relatively infrequent and the rows are relatively short. Coagulation necrosis and associated regenerative reactions involve segments of somewhat scattered fibers. Less frequently seen are segments in which nuclei show vesicular enlargement, irregular distribution in small clusters or short chains, and progressive dissolution with formation of irregular vacuolar spaces in association with granular breakdown of myofibrils. This vacuolar type of degeneration is basically the same as that described for the hamster. As deficiency progresses to more advanced ages (10 to 18 months), the incidence of coagulation necrosis appears to diminish, while the vacuolar type of degeneration involves increasingly larger numbers of segments until it becomes the predominant type of reaction present (Figs. 19 and 20). At all stages of myodegeneration regenerating fibers are evident but are much fewer in relative number than in the dystrophic hamster. In both species, the large proportion of fibers showing degenerative changes often seems excessive in terms of the gross evidence of functional impairment observed in the animal.

Lesions in skeletal muscle following deprivation of vitamin E have also been observed in the dog (Brinkhous and Warner, 1941; Anderson *et al.*, 1939; Kaspar and Lombard, 1963; Hayes *et al.*, 1969), cat (Mason and Dju, 1963), monkey (Mason and Telford, 1947), mink (Mason and Hartsough, 1951; Stowe and Whitehair, 1963; 1964), and Florida cotton rat (Telford, 1971). The character of the lesions differs in no significant manner from those described previously in other laboratory species. Recent observations of particular interest pertain to the Rottnest quokka (*Setonix brachyurus*), a small nocturnal wallaby of Western Australia about the size of a fox terrier. When subjected to captivity and fed a high protein diet designed for sheep, after a month or two there develops a characteristic myopathy involving muscles of the lower extremity and pelvis in particular, and more cephalad muscles to a lesser degree. There is rapid and complete recovery after 2–3 weeks of vitamin E therapy (Kakulus, 1961), but selenium is ineffective in prevention or cure of this myopathy (Kakulus, 1963a). The muscle lesions are quite comparable to those observed in laboratory animals deficient in vitamin E (Kakulus and Adams, 1966), and the cellular reactions observed following tocopherol therapy closely resemble those seen in the hamster. On the other hand, the finding that decreasing the size of the enclosure markedly accentuates onset and severity of the myopathy (Kakulus,

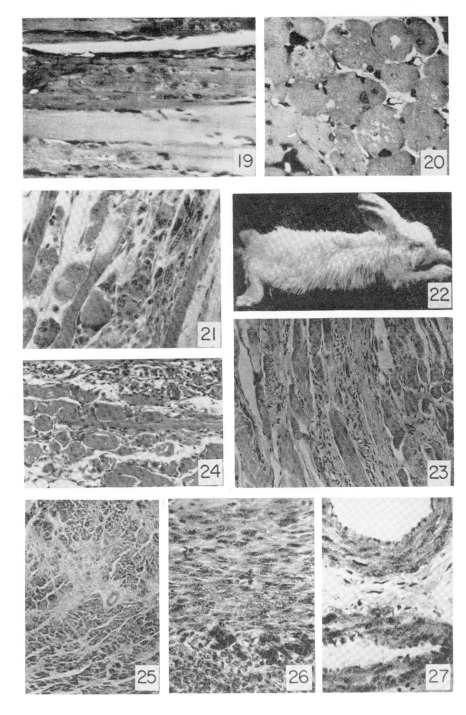

1963b) stands in contrast to other reports that immobilization of one extremity significantly reduces the severity of muscle lesions as compared to those seen in the contralateral limb in vitamin E-deficient mice (Pappenheimer and Goettsch, 1941) and in selenium-deficient lambs (S. Young and Keeler, 1962).

At this point, a few general comments are in order. First, the lesions described in different animals are associated with little or no inflammatory reaction, alterations in the vascular supply, or significant replacement by fibrous or adipose tissue. Second, muscle lesions are accentuated in rate of onset and severity by dietary unsaturated fatty acids, especially in the guinea pig and rabbit (and also farm animals). Third, in association with the muscle lesions of most experimental animals (monkey, hamster, cotton rat, rat, and mouse) there is observed in the degenerating fibers, and in many related macrophages, the occurrence of a fluorescent, acid-fast, insoluble, inert pigment. This is usually referred to as lipofuscin or ceroid pigment. An excellent description of its appearance and distribution in skeletal muscle has been given by Telford (1971). It increases in amount with increases in dietary unsaturated fatty acids.

Fig. 19. Longitudinal section of intercostal muscle of rat, after chronic vitamin E deficiency of 520 days, showing extensive vacuolar degeneration in fibers above and below a normal-appearing fiber (× 335).

Fig. 20. Cross section of same muscle, showing irregular shape and distribution of vacuolar spaces and of nuclei in fibers possessing a reduced complement of myofibrils (× 335).

Fig. 21. Intercostal muscle of vitamin C deficient guinea-pig showing early type of degenerative change, involving coagulation necrosis (left) and reactions suggestive of regeneration (upper right). From Dalldorf (1929).

Fig. 22. Rabbit deprived of choline for 200 days, showing pronounced plasticity of the hind legs. From Hove and Copeland (1954).

Fig. 23. Muscle from choline deficient rabbit, showing numerous degenerating fibers and some increase in fibrous connective tissue (× 120). From Hove and Copeland (1954).

Fig. 24. Skeletal muscle of guinea-pig fed diet deficient in the anti-stiffness factor, showing much necrosis and macrophagic invasion of muscle fibers (× 110). From Harris and Wulzen (1950).

Fig. 25. Portion of ventricular myocardium, showing areas of fibrotic replacement of cardiac muscle following prolonged vitamin E deficiency in the rat (× 35). From Mason and Emmel (1945).

Fig. 26. Portion of uterine myometrium of rat after prolonged vitamin E deficiency, showing extensive accumulation of brown, acid-fast, pigment granules in smooth muscle cells and in clusters of macrophages between circular and longitudinal muscle layers (× 225). From Mason and Emmel (1945).

Fig. 27. Similar accumulation of pigment granules within smooth muscle cells in the wall of small artery (above) and vein (below) of vitamin E deficient monkey (× 250). From Mason and Telford (1947).

It has also been observed in the central nervous system, cardiac muscle and smooth muscle of certain species subjected to vitamin E deficiency, and will be referred to again later (p. 196). It is considered to represent a lipoprotein or a polymerized product of unsaturated fatty acids having at least eighteen carbon atoms and two double bonds (Filer *et al.,* 1946). In the absence of α-tocopherol functioning as an antioxidant, such fatty acids cannot be adequately stabilized in the intracellular environment. Hence, this pigment product is considered a secondary result of the disturbed metabolic state of affected muscles, having no etiological relationship to the lesions observed.

3. ULTRASTRUCTURE

During the past 15 years, electron microscopy has made important contributions to a better understanding of neuronmuscular diseases in man and to delineation of several previously unrecognized entities (Walton, 1969). Its application to experimentally induced myopathies has been limited. Reference has been made previously to ultramicroscopical changes in young rats manifesting late lactation paralysis (p. 163). For the vitamin E-deficient adult rat, Howes *et al.* (1964) have presented a detailed description of the ultrastructural changes. Among the earliest alterations are: (1) appearance of potential myelin figures and of wavy membrane profiles, which include and surround portions of sarcoplasm, usually located near nuclei and Golgi zones; (2) irregularities in and loss of cristae of mitochondria; and (3) appearance of many small mitochondria, probably constituting a response to the degenerative process. In association with or following these changes, there appear many inclusion bodies, some containing a dense flocculent material and others revealing membrane condensations, myelin-like figures, and less dense particulate material. It is presumed that these reflect degrees of intracellular autolysis, and a protective mechanism of the cell concerned with the removal of damaged cytoplasmic components. Concurrently, there occurs disruption and streaking of Z line materials, and appearance in the sarcoplasm of fine particulate material resulting from disintegration of myofilaments, representing the source of much of the material noted in certain inclusion bodies. Also, there are focal alterations in terms of thickening and distortions of the nuclear membrane not unlike those occurring in other membrane systems. These observations are in general agreement with the studies of Van Vleet *et al.* (1967, 1968), who examined skeletal muscles of vitamin E-deficient rabbits. Early alterations are characterized by swelling of mitochondria, fragmentation of their cristae, and formation of free profiles and accumulation of dense granules within the mitochondria. Other evidence is also presented in support of the concept that vitamin E is necessary for maintenance of the biochemical and structural integrity of mitochondria. Later changes, considered to be secondary in nature, involve intramitochondrial calcification, appearance of liposomes and lipid products in the degenerating sarcoplasm, massive destruction of contractile elements, and localization of nuclei with investments of sarcoplasm adjacent to the altered sarcolemma. As mentioned earlier (p. 163), it is these muscle nuclei that may play an important role in regenerative processes.

In describing the predominantly discontinuous type of regeneration occurring in these muscles, Van Vleet *et al.* (1967) state that many of these sarcolemmic nuclei dedifferentiate to myoblasts which are characterized by abundant polysomes, scattered mitochondria, large nuclei with prominent nucleoli, and scattered free filamentous fragments. Such myoblasts proliferate rapidly by amitotic division and fuse to form a dense syncytium throughout the sarcolemmic tube in which fibrillogenesis soon appears. Free thin filaments become intermeshed with thick filaments to form rudimentary sarcomeres. Myofibrils appear to elongate by addition of free filaments to ends of formed sarcomeres. Maturation continues as band patterns develop, and fibrils become aligned in the regenerating fiber. These observations confirm and amplify those based on light microscopy previously described (p. 163).

4. MYOPATHY VERSUS NEUROPATHY

The muscle lesions of vitamin E deficiency in the various species studied have been universally regarded as purely myogenic in nature, except for conflicting opinions concerning the lesions associated with chronic vitamin E deficiency in adult rats. Certain investigators (Einarson and Ringsted, 1938; Einarson, 1952, 1953; Monnier, 1941) consider that the initial lesions are myogenic, but that with progressive deficiency there is superimposed a neurogenic type of lesion characterized by demyelinization of axons and gliosis in posterior spinal fasciculi and dorsal nerve roots, increasing accumulation of acid-fast pigment in anterior horn cells with simultaneous diminution in Nissl substance and, eventually, irreparable atrophy and sclerosis of sufficient numbers of these cells to superimpose a late spinal atrophy upon the earlier myopathic lesions. The demyelinization and gliosis in nerve roots and posterior columns, which have been confirmed by others (Luttrell and Mason, 1949; Malamud *et al.*, 1949), probably explain increasing degrees of hyperkinesia and hyperalgesia which occur during advanced stages of the late spinal atrophy upon the earlier myopathic lesions. Although subsequent investigators have differed with regard to involvement of anterior horn cells, there is general agreement that sensory neurons in the posterior columns of the spinal cord and medulla, especially the fasciculi gracilis and cuneatus, exhibit swelling and demyelinization, usually associated with considerable gliosis. Such changes, sometimes referred to as systemic axonal dystrophy, have been described in the rat (Einarson and Telford, 1960; Luttrell and Mason, 1949; Malamud *et al.*, 1949; Pentschew and Schwarz, 1962). Electron microscopic studies (Lampert and Pentschew, 1964; Lampert *et al.*, 1964) reveal extensive proliferation of mitochondria and accumulations of electron-dense bodies, supposedly granules of lipofuscin or ceroid pigment. The fact that this neuronal dystrophy occurs in dogs (Fig. 18) in which skeletal muscle lesions are rather minimal (Hayes *et al.*, 1969, 1970) supports the general belief that the neural lesions bear no particular relationship to muscle lesions, and that the latter represent a true myopathy. It is of interest that neural lesions have not been observed in the guinea pig, rabbit, calf, or lamb, where lesions of skeletal muscle have received much attention.

Considerable attention has also been given to peripheral nerves and motor end-plates in various species studied. Motor end-plates are reported to be unaffected (Rogers *et al.*, 1931; Pappenheimer, 1939) or to be reduced in severely dystrophic muscle, followed by return to normal upon recovery (Telford, 1941). An atrophy of muscle spindles has been reported (Einarson and Ringsted, 1938) that may also represent a secondary effect. Electromyographically, the dystrophic muscles at rest in the rabbit (Fudema *et al.*, 1960) and chick (Kodono and Hirose, 1969) show abnormalities of insertion and fibrillation potentials, and also sharp, or V waves. Also of interest is the report (Durack *et al.*, 1969) that, in the Rottnest quokka (see p. 169), skeletal muscles show increases in resting and insertional activity combined with frequent and prolonged myotonic discharges. Vitamin E therapy effects reversal of this myotonic state in about 1 week.

In the preceding section, an effort has been made to present in some detail the varied types of cellular reactions observed in skeletal muscle as they occur in different species of laboratory mammals at different ages and under different intensities of vitamin E depletion. The descriptions given reflect personal impressions of the author based upon histological study of the lesions in different species over many years and upon descriptions of the lesions by many other investigators whose separate contributions comprise a very extensive literature. There are obviously many imperfections in any attempt to detail specific cellular alterations and to relate them to any sequence or sequences of events presumed to occur within muscle fibers, especially when the evidence is based upon the static morphology presented by fixed and stained sections. It is hoped that the descriptions and interpretations made will, despite their shortcomings, serve to emphasize the many gaps that exist in our knowledge of the histopathology of muscle and to encourage others to contribute toward a more adequate understanding of the sequential events and intracellular mechanisms involved.

5. Biochemical Changes

Much of our information on chemical change in muscle after vitamin E deficiency is based upon the dystrophic rabbit and is represented by the pioneer studies of Goettsch and Brown (1932), Victor (1934), Morgulis and Spencer (1936), Fenn and Goettsch (1937), Morgulis *et al.* (1938), C. G. Mackenzie and McCollum (1940, 1941) Friedman and Mattill (1941), Houchin and Mattill (1942), and Kaunitz and Pappenheimer (1943). The subject has also been reviewed by C. G. Mackenzie (1953). It may be appropriate, however, at this point to present a brief resume of earlier and more recent studies concerning biochemical changes in dystrophic muscle.

Increased oxygen uptake of skeletal muscle after vitamin E deficiency, first demonstrated in rabbits (Victor, 1934), occurs also in the suckling rat, adult rat, hamster, and chick (Friedman and Mattill, 1941; Houchin and Mattill, 1942; Kaunitz and Pappenheimer, 1943), but, for unknown reasons, is not demonstrable in ducklings (Victor, 1934), which show much more pronounced muscle lesions than do chicks. The respiratory quotient and rate of glycolysis of muscle remains normal. Increased oxygen uptake may precede the appearance of histological lesions in young rats (Kaunitz and Pappenheimer, 1943) and rabbits (Hummel and Melville, 1951) and is restored to normal in rabbits within as little as 4 hours after intravenous and 10 hours after oral administration of α-tocopherol (as shown by comparison of muscle obtained by biopsy and necropsy from the same animal) even though lowered creatine and increased chloride content of the muscle is not improved (Houchin and Mattill, 1942). These observations suggest that the increased oxygen uptake of dystrophic muscle, which may reflect functions of tocopherol as an intracellular antioxidant, represents a primary response to vitamin E deficiency, whereas other biochemical changes are secondary to this or to some other metabolic disturbance. Efforts to study the effects of tocopherol on muscle slices *in vitro* have been hampered by difficulties inherent in getting a fat-soluble substance into an intracellular environment.

The major structural protein components of skeletal muscle (insoluble protein, actinomyosin, and soluble protein) are known to increase at different rates up to about the fifteenth day of life in the suckling rat, after which concentration of the insoluble fraction (scleroproteins) gradually decreases and that of actomyosin increases (Herrmann and Nicholas, 1948). It is therefore of interest that in rats showing late lactation paralysis of vitamin E deficiency between the eighteenth and twenty-fifth days of life, the concentration of these two fractions is distinctly reversed, and yet is restored to normal in the course of spontaneous recovery, which frequently occurs (Rumery *et al.,* 1955). Furthermore, a similar imbalance between these two protein fractions in dystrophic Syrian hamsters is restored to normal following as little as 5 days of vitamin E therapy (Mauer and Mason, 1958), during which most necrotic fibers would have been removed by phagocytosis and lysis and reasonably normal morphology restored. These protein components have not been studied in the more chronic dystrophy of adult rats. However, in the somewhat intermediate myopathy in vitamin E-deficient rabbits, the myofibrils are said to show either a loss of myosin or an alteration of the submicroscopic pattern in which myosin is organized prior to the loss of cross striations and other microscopic changes (Aloisi *et al.,* 1952). There is also evidence (Corsi, 1957) that in such muscles the myofibrils are reduced in number, and the extractible myosin significantly reduced in amount.

Attention has been given to other muscle proteins and to amino acids, especially in the dystrophic rabbit, in efforts to determine whether the loss of muscle mass in dystrophic muscle reflects impaired synthesis or accelerated breakdown of proteins or an imbalance of both processes. An increased concentration of ribonucleic and deoxyribonucleic acids, associated with increased urinary excretion of allantoin, in the monkey, rabbit, and rat are interpreted as reflecting an increased rate of turnover of nucleic acids in dystrophic muscle (J. M. Young and Dinning, 1951; Dinning and Day, 1957). This phenomenon is demonstrated by following the incorporation of formate ^{14}C into the nucleic acids of muscle in vitamin E-deficient rats (Dinning, 1955). There is a decrease in glycine prior to the onset of gross dystrophy and, after dystrophy is well advanced, an increase in concentration of free amino acids exclusive of the basic amino acids (Tallan, 1955; L. C. Smith and Nelson, 1957). Of particular interest are the observations that cathepsin and dipeptidase activities are increased (Weinstock *et al.,* 1955, 1956). These findings at least provide some important leads to the primary question that still remains unanswered.

A loss of muscle creatine and phosphocreatine associated with increased output of creatine in the urine is common to dystrophy of vitamin E deficiency and to many other conditions which result in general wasting or breakdown of muscle. There is evidence (Dinning and Fitch, 1958) that in vitamin E-deficient rabbits there is increased synthesis and increased rate of turnover of creatine in the muscle, as well as reduced ability of the muscle to retain creatine. In the rabbit, white muscle is said to show greater reduction in creatine, as well as more severe pathological lesions, than red muscle (Goettsch and Brown, 1932).

The degree of creatinuria, or the ratio of creatine to preformed creatinine excreted, provides a useful index of the severity of muscular dystrophy. Since creatinine excretion usually remains relatively constant, the creatine/creatinine ratio will increase as dystrophy progresses. This ratio may also be increased somewhat through a diminution in creatinine excretion, especially when the dystrophic state is chronic and prolonged (C. G. Mackenzie and McCollum, 1941; Hove and Hardin, 1952). This latter phenomenon is thought to be related to the rather considerable loss of total mass of skeletal muscle tissue under such conditions. Increased creatine and decreased creatinine excretion are also characteristic features of E-deficiency dystrophy in the monkey (Dinning and Day, 1957) and of choline-deficiency dystrophy in rabbits (Hove et al., 1957).

Vitamin E therapy causes a prompt reduction in creatinuria with restoration of normal excretion pattern in 4–5 days. Too little attention has been given to the creatine status of the muscle itself following therapy. In recent studies on deficient monkeys, Dinning and Day (1957) report that under conditions of therapy effective in bringing elevated concentrations of nucleic acids (RNA and DNA) to normal levels, muscle creatine is only partially restored to normal after several months of treatment.

It has been pointed out that dystrophic muscle is composed of muscle fibers showing varying degrees of sublethal injury, nuclear dissolution, necrosis and breakdown of cytoplasmic constituents, and also regenerative processes. It is therefore impossible to distinguish between those chemical abnormalities attributable to altered diffusion of electrolytes and other substances through the cell membrane, those due directly to intracellular disturbances of metabolism, and those attributable to actual disintegration of cellular constituents. The chemical changes observed are much like those of cortisone-induced dystrophy (Milman and Milhorat, 1953) but differ in certain respects from those of simple atrophy or denervation atrophy of muscle (Hines, 1952). They help in amplifying our picture of the dystrophic process even though they may, for the most part, represent secondary effects and provide no clue to the basic metabolic dysfunction in dystrophy.

Briefly stated, chemical analysis of muscle after vitamin E deficiency reveals a variable decrease in potassium, magnesium, creatine and creatine phosphate, acid-soluble phosphorus, total nitrogen, myosin and actomyosin, glutamine, and glycogen; there is, on the other hand, an increase in sodium, chloride, RNA, DNA, collagen, cholesterol, and fat. There is a considerable literature on altered levels of certain enzymes and enzymic systems in both muscle and serum, and on possible functions of tocopherol in enzymic processes. Present information points toward a tendency for respiratory enzymes to increase and glycolytic enzymes to remain unaffected in dystrophic muscle, but offers little toward explaining the cause of dystrophy. Studies on cathepsin and dipeptidase activities of muscle have been referred to earlier.

B. Vitamin E and Selenium Interrelationships

A few words regarding selenium may be of interest. According to Sharman (1960), it is sixty-sixth in order of abundance in the crust of the earth and is closely related to sulfur crystochemically and geochemically. Furthermore, it is the only element known to be absorbed by plants in amounts sufficient to cause death from poisoning in animals which consume them. Horses, cattle, and sheep grazing on crop plants and grasses where selenium in the soil is high and is bound to protein in the plant may develop chronic alkali disease, characterized by lethargy, anemia, stiff joints, lameness, rough coat, deformities of the hoofs, myocardial atrophy, and lesions of the liver and kidneys. If forage is limited primarily to milk vetch, in which selenium is in a more soluble form, a more acute disorder commonly referred to as blind staggers occurs. In neither disorder have there been described any specific lesions of the skeletal, cardiac, or smooth musculature. However, in acute selenium toxicosis produced in ewes by dietary administration of inorganic selenium, widespread gross and microscopic lesions of the ventricular myocardium occur (Maag and Glenn, 1967); yet, skeletal muscle is only slightly affected. By way of contrast, and as an example of dietary factors other than specific deficiencies exerting deleterious effects on muscle, reference may be made to extensive hyaline and granular degeneration of skeletal muscle described by Dewan *et al.* (1965) as occurring in skeletal and cardiac muscle of goats and calves fed fruits of the coyotillo plant (*Karwinskia humboldtiana*). Sheep and rabbits are unaffected. This disorder is commonly referred to as limberleg. A somewhat similar disorder occurs in calves and cattle fed *Cassia occidentalis* (stinkweed), but rabbits and sheep appear to be immune (Mercer *et al.,* 1967). These disorders are considered to be due to mycotoxins present in the plants involved. But this discussion has taken us away from our main topic; namely, deficiency of selenium in soils, in forage, in animals, and its interrelationships with vitamin E.

Interrelationships between vitamin E and selenium had their beginning with the observations that in rats a necrotic degeneration of the liver occurs on diets deficient in cystine (Weichselbaum, 1935) and on diets deficient in vitamin E and sulfur-containing amino acids (Schwarz, 1948, 1949). This soon led to involvement of a third factor, designated factor 3, absent in *Torula* yeast but present in brewers' yeast as grown in the United States. Hence, *Torula* yeast is commonly used as the protein component of diets designed to produce necrotic liver degeneration in the rat and exudative diathesis in chicks and poults. Both disorders are now considered primary manifestations of selenium deficiency, even though in part prevented by either vitamin E or sulfur-containing amino acids. A major advance was the discovery that selenium, bound in organic form, was the magical component of factor 3 (Schwarz and Foltz, 1957; Patterson *et al.,* 1957). Although it is not yet known whether selenium must be present in some specific organic form in the organism to exert its biological effects, sodium selenite and organoselenic compounds such as selenomethionene and selenocystine, as dietary constituents, can replace vitamin E in preventing many but not all of the manifestations of vitamin E deficiency.

These developments came at a period when much attention was being given to two other aspects of vitamin E functions: (1) the interrelationships between vitamin E and dietary unsaturated fatty acids and (2) the ability of many synthetic antioxidants having no structural resemblance to vitamin E to replace or substitute for it in many of its accepted functions. All three areas of investigation mentioned have played important roles in focusing upon the as yet unresolved question as to whether vitamin E has specific functions in the organism other than that of an intracellular antioxidant. Over the past 25 years, research in these three areas, and particularly in that pertaining to interrelationships between vitamin E and selenium, has not only contributed greatly to our understanding of myopathies in farm animals under field and experimental conditions, but has also significantly advanced our knowledge of the possible role of vitamin E and selenium in metabolic functions of mammalian and avian species. The reader is referred to excellent reviews of the early literature pertaining to interrelationships between vitamin E and selenium (Moore, 1962; M. L. Scott, 1962) and to understanding of their role in naturally occurring and experimentally induced myopathies of farm animals (Blaxter, 1962; Rosenfeld and Beath, 1964; Oksanen, 1967).

At this point, reference should be made to several recent reports pointing to a specific role of selenium as an essential nutrient, which may pave the way toward a better understanding of the heretofore complicated interrelationships between selenium and tocopherol. First, it has been reported (McCoy and Weswig, 1969) that second generation rats fed *Torula* yeast diets very low in selenium (0.02 ppm) and containing adequate vitamin E manifest a syndrome of poor growth, partial alopecia, sterility, and paleness of the iris and retina. Selenium therapy readily reverses these manifestations. No mention is made of skeletal muscle lesions in such animals. Second, in chicks reared on highly purified diets, with crystalline amino acids and sucrose as the major components and with levels of selenium ranging from 0.02 to 0.005 ppm, there occurs stunted growth, poor feathering, and degeneration of acinar cells of the pancreas (Thompson and Scott, 1969, 1970). Again, no reference is made to skeletal muscle lesions. This recalls earlier findings of DeWitt and Schwarz (1958) that mice on low selenium diets show degeneration of the pancreas. Moreover, in the absence of pancreatic lipases, selenium-deficient chicks are unable to absorb unhydrolyzed fats or vitamin E from the gut (Thompson and Scott, 1970). The evidence is suggestive of at least an indirect role of selenium in the intestinal absorption and transfer of vitamin E. Whether this, in turn, bears relationship to why tocopherol requirements for prevention of exudative diathesis in chicks are inversely related to the selenium content of the diet (Thompson and Scott, 1969) remains to be determined.

1. FARM ANIMALS

a. OCCURRENCE AND ETIOLOGY OF MYOPATHIES. For the past 50 years or more, an enzootic myopathy in suckling lambs and calves has been well recognized and has caused staggering economic losses to sheep and cattle raisers in the United States, Australia, and Europe, especially in the Scandinavian countries. The terms "stiff-lamb disease" and "white muscle disease" have been used to describe the disorder in lambs and calves, respectively; the latter term is now commonly used in reference to the myopathy in ruminants generally. Both skeletal and cardiac muscle are involved to varying degrees, with calcification of the musculature a rather frequent characteristic.

As early as 1934, stiff-lamb disease was produced experimentally in the offspring of ewes maintained on a diet composed mainly of cull beans and alfalfa hay, and a description was given of widespread, non-inflamatory, nonneurogenic lesions of the skeletal muscles (Willman *et al.*, 1934). These investigators at the Cornell University Agricultural Experiment Station subsequently demonstrated that vitamin E had a prophylactic effect when fed to the ewes and a curative effect when fed to affected lambs (Willman *et al.*, 1946; Whiting *et al.*, 1949). Similar observations were made on lambs in Michigan fed semipurified diets deficient in and supplemented by vitamin E (Culik *et al.*, 1951; Baci-galupo *et al.*, 1952). The first suggestion that deficiency of vitamin E might explain white muscle disease in calves came from Vawter and Records (1947), who carried out a careful study of the symptoms and the gross and microscopic pathology observed in suckling calves showing spontaneous occurrence of the disorder on cattle ranches in Nevada. Its first experimental production, reported by Blaxter *et al.* (1952) in Scotland, soon followed. Further confirmation came in the reports of MacDonald *et al.* (1952) and of Safford *et al.* (1954).

On the other hand, the hopeful and eager application of these findings led to many conflicting reports concerning the efficacy of vitamin E therapy. This proved relatively ineffective in certain geographic areas such as Oregon (Muth, 1955), Australia (Drake *et al.*, 1960), Finland (Oksanen, 1965), and Scotland (Sharman, 1954). This was due, in part at least, to inadequate dosages given. With ensuing recognition of se-lenium as an integral component of factor 3 and growing awareness of selenium deficiencies in soils and plants where these disorders were indigenous, studies carried out under controlled field conditions demon-strated the effectiveness of selenium in prevention and cure. Significant findings were: (1) that addition of as little as 0.1 ppm sodium selenite to a dystrophy-producing forage (of cull beans and alfalfa) of dams was much more effective than vitamin E in preventing lesions of skeletal and cardiac muscle in the suckling offspring, possibly reflecting differ-ences in placental transfer; (2) that both selenium and vitamin E in adequate dosage can prevent such lesions, and even reverse them if not too far advanced, in suckling lambs and calves; and (3) that sulfates such as present in gypsum fertilizers can interfere with the biological availability of selenium, either through decreased uptake of selenium by the forage plant or through interactions between sulfur and selenium after ingestion (Schubert *et al.*, 1961; S. Young *et al.*, 1961).

There has accumulated considerable evidence that dietary inade-quacies of selenium or vitamin E or of both may be responsible for disorders in skeletal muscles, and in some instances of cardiac muscle, observed in foals and pigs. The subject of myopathies in farm animals, causative factors, and therapeutic effects of selenium and vitamin E has been presented in several excellent reviews. That of Blaxter (1962) discusses the long recognized effects of cod liver oil and other fats high in unsaturated fatty acids in precipitating vitamin E-responsive myopathies on rations that are otherwise adequate, and the inability of selenium to provide protection. Hartley and Grant (1961), Muth (1963), and Oksanen (1967) present comprehensive reviews of various dietary factors which may relate to muscle disorders in calves, sheep, horses, and pigs, and responsiveness of these to selenium and vitamin E therapy. Cordy (1963) gives a good account of the gross and microscopic lesions in cattle and sheep. There is also a recent report (Trapp *et al.*, 1970) describing lesions of skeletal and cardiac muscle associated with naturally occurring outbreaks of dietary hepatic necrosis in thirty-seven different swine herds in Michigan, and recording an effective reduction of these conditions following dietary supplements of vitamin E or injections of selenium and vitamin E.

b. Symptomatology. White muscle disease, predominantly a myopathy of early life, has its onset in lambs 1–5 weeks and in calves 3–10 weeks after birth. Beef cattle herds, not dairy herds, are usually involved. Incidence is highest in early spring just prior to the availability of green pasturage, reflecting maintenance of dams during gestation on feeds, usually legumes, grown in certain geographic areas where soils and plants are deficient in selenium. The chief symptoms are muscular weakness, wobbly or staggering gait, stiffness of the hind legs in particular, arching of the back, reluctance to move about, and difficulty in suckling, swallowing, and arising from a prone position without assistance. There is eventual prostration and rather sudden death, usually attributed to cardiac failure. Under field conditions, it is considered that the forage and green pasturage generally provide adequate vitamin E and that the disorder represents a primary selenium deficiency. However, identical manifestations occur in lambs and calves reared on semisynthetic diets deficient in vitamin E, with or without an adequacy of selenium. Incidence and intensity of symptoms and corresponding muscle lesions are intensified by supplements of cod liver oil or other sources of unsaturated fatty acids. Under such conditions, selenium therapy affords no protection, but vitamin E and certain synthetic antioxidants such as DPPD (*n*-N′-diphenyl-*p*-phenylenediamine) and ethoxyquin (6 ethoxy-1,2-dihydro-2,2,4-trimethylquinone) are effective.

c. Histopathology. Grossly, there is first noted a vague pallor, translucency, and somewhat edematous appearance of the skeletal muscles. Later, whitish or yellowish streaks or patches are apparent. Most affected are suspensory muscles of the shoulders, extensors and adductors of the limbs, certain paravertebral muscles, and in some instances the tongue, intercostal muscles, and diaphragm. Bilateral symmetry is characteristic, and no associated lesions have been observed in the central nervous system. Calves are more prone than lambs to exhibit myocardial lesions.

Microscopically, lesions of skeletal muscle are much more widespread than suspected from gross observations and clinical signs. They are characterized by initial swelling, loss of striations, and coagulation or granular necrosis of individual or related groups of muscle fibers, in association with active regeneration, both continuous (plasmodial proliferation) and discontinuous (myotube formation) in type. When the myopathy is produced by semipurified diets deficient in vitamin E, whether or not accentuated by unsaturated fatty acids, the lesions are microscopically quite similar to those observed in laboratory animals. When selenium, rather than vitamin E, is the primary limiting factor, progressive calcification of the myofibrils in the dystrophic muscle is a quite characteristic feature. Calcium first appears at tiny interfibrillar beads regularly spaced in relation to fiber striations. These particles may continue to form until the fiber is filled to variable degrees with calcareous material in both striated and cardiac muscle. There exist differences of opinion as to whether calcification is an irreversible process or one which is reversible following appropriate therapy. Investigators are agreed that there are no indications of neuropathological changes.

d. BIOCHEMICAL CHANGES. As might be anticipated on the basis of the limited reaction of skeletal and cardiac muscle to nutritional deficiencies, biochemical findings in myopathies of farm animals usually conform to what is known regarding laboratory animals in general. Much attention has been given to alterations in plasma and tissue enzymes in the myopathy of lambs and sheep during the past decade. As in the case of laboratory animals, there always exists the question of whether increased plasma levels of any enzymes reflect a release from affected muscles, or from the liver or other body tissues, and also whether any observed change in enzyme content of plasma or of the muscle itself can be considered of etiological significance rather than a secondary result of myopathic changes due to unknown causes.

Considering first observations based on lambs reared under conditions where selenium rather than vitamin E is the limiting dietary factor, it can be said that: (1) with increasing severity of myopathy there occurs in the affected muscles a corresponding increase in levels of lyosomal enzymes such as β-glucuronidase, aryl sulfatase, and acid phosphatases (Whanger *et al.*, 1969, 1970), and 5'-nucleotidase (Arnold *et al.*, 1965), an enzyme similar to the lyososomal enzymes. These enzymes are probably associated with increased proteolytic and autolytic activity demonstrable in skeletal muscle from myopathic animals. On the other hand, the affected muscle shows significant decreases in cytoplasmic enzymes such as creatine phosphatase, lactic dehydrogenase, and glutamic–oxaloacetic transaminase (Whanger *et al.*, 1969). Moreover, there is a marked increase of such enzymes in the plasma of ewes and rams (Buchanan-Smith *et al.*, 1969) and of young pigs (Ewan and Wastell, 1970) reared on semipurified diets deficient in both selenium and vitamin E. Evidence available indicates that either vitamin E or selenium can prevent or reverse these changes and that muscle represents the primary source of elevated enzyme activities in the plasma.

2. Avian Species

a. Occurence and Etiology of Myopathies. The effects of deficiency of vitamin E and of selenium in the chick, turkey poult, and duckling are primarily those observed during the early post hatching period. Of particular interest are the striking differences in response of these three species. Skeletal muscle lesions occur in all three, but only in the duckling is there gross evidence of myopathy. Exudative diathesis has been produced in the chick and poult, encephalomalacia only in the chick, gizzard myopathy and cardiac muscle lesions only in the poult. Moreover, whereas cystine is protective against nutritional muscular dystrophy in the chick, it does not play a comparable role in the poult. There are rather complex interrelationships between vitamin E, selenium, sulfur amino acids, and unsaturated fatty acids in the genesis of these disorders.

Exudative diathesis, characterized by accumulation of serous fluid in subcutaneous tissues of the breast and elsewhere, and attributed to increased permeability of the capillary bed, can result from a deficiency of either vitamin E or selenium or both. Its late occurrence in the syndrome of uncomplicated selenium deficiency in the chick, and only after blood tocopherol levels become critically low, suggests that deficiency of vitamin E is primarily involved. However, under such conditions there occurs essentially no intestinal absorption of unhydrolyzed fats or vitamin E; hence, a direct deficiency of selenium and indirect deficiency of vitamin E exist.

Encephalomalacia, or ischemic necrosis of the cerebellum, has been observed only in the chick and occurs only when vitamin E-deficient diets contain certain levels of unsaturated fatty acids. Vitamin E in adequate dosage and certain synthetic antioxidants are protective; selenium is ineffective.

Nutritional muscular dystrophy can result from a deficiency of either vitamin E or selenium. Its incidence and severity are accentuated by deficiency of sulfur amino acids, especially cystine, and also by dietary unsaturated fatty acids. When diets low in vitamin E and selenium contain as much as 4% lard or its equivalent as linoleic acid, selenium supplements afford little or no protection, whereas either α-tocopherol or synthetic antioxidants are effective (Ewen and Jenkins, 1967; Jenkins and Ewen, 1967).

b. Histopathology. Lesions of skeletal muscle in chicks are characterized grossly, as first described by Dam *et al.* (1952), by white striations in the pectoral muscles and, to a lesser extent, in the thigh muscles. This greater involvement of white compared to red muscle is in accord with earlier findings in the vitamin E-deficient rabbit (Goettsch and Brown, 1932). There is no apparent disability in the affected chicks. There may occur spontaneous remission of lesions as chicks grown older (Nesheim and Scott, 1961), which is quite comparable to what occurs in the late-weaning paralysis of the rat (p. 163). Histologically, there is coagulation necrosis of muscle fibers, invasion of macrophages and leukocytes, and related cellular changes resembling those observed in the young rat, rabbit, and guinea pig. Electron microscopically, as described by Cheville (1966), there may be observed swelling and loss of cristae of mitochondria, dilation of sarcoplasmic reticulum, degeneration and streaming of Z lines, and focal myofibrillar destruction. There is no indication that the ultrastructural changes observed are attributable to vascular damage or to breakdown of lysosomes.

In the turkey poult reared on diets deficient in both vitamin E and selenium, there occurs, in addition to a rather unique gizzard myopathy described in a later section (p. 195), also typical exudative diathesis (Creech *et al.*, 1957) and myopathy of cardiac and skeletal muscle (Walter and Jensen, 1963; Ferguson *et al.*, 1964), the order of onset and prominence being somewhat in the sequence mentioned. According to M. L. Scott *et al.* (1967), who consider the skeletal and cardiac muscle lesions primarily selenium-responsive disorders, the skeletal muscle lesions are not demonstrable grossly; microscopically, they are characterized by acute coagulation necrosis (Zenker's degeneration) of the usual morphology, associated with moderate intramuscular edema and leukocytic infiltration. Neither nuclear proliferation nor fibrosis was observed.

Young ducklings reared for a few weeks on low-E diets exhibit a sudden onset of muscular weakness; there is awkward gait with feet turned inward, difficulty in raising the head and in righting after being placed supine, and eventual collapse (Pappenheimer and Goettsch, 1934; Pappenheimer *et al.*, 1939). A similar disorder occurs under field conditions. At necropsy, the skeletal muscles are pale, watery, and translucent. Microscopically, there is a widespread acute type of necrosis with active regeneration, not unlike that described for late-lactation paralysis in the rat. There is an increased water content and decreased creatine content of the dystrophic muscle. As in the chick, cardiac and smooth muscle are unaffected, which is quite in contrast to the situation in turkey poults.

C. Other Nutritional Deficiencies

1. VITAMIN C DEFICIENCY

Degeneration of skeletal muscle has long been recognized as a feature of human scurvy, though generally considered secondary to intramuscular hemorrhage, trauma, or cachetic atrophy. However, about 45 years ago, similar lesions were studied extensively in scorbutic guinea pigs and were considered an intrinsic part of the disease (Höjer, 1924; Meyer and McCormick, 1928; Dalldorf, 1929). Dalldorf calls attention to the high incidence of lesions in intercostal muscles (Fig. 21) and states that placing scorbutic animals for an hour daily in a slowly rotating barrel, so that they are obliged to exert themselves to remain on their feet, produces florid lesions in muscles of the extremities. More recently, Murray and Kodicek (1949) and Boyle and Irving (1951) have given good descriptions of the lesions in guinea pigs. The diets employed appear to have contained an adequacy of vitamin E.

The muscle lesions vary somewhat, depending upon the acuteness and chronicity of the scorbutic state, but have no features that clearly differentiate them from those of vitamin E deficiency in the guinea pig or rabbit. Some regeneration is usually evident, but the reparative response following vitamin C therapy has not been studied histologically. Murray and Kodicek (1949) call particular attention to the associated edematous and hyperplastic state of the connective tissue of affected muscle and the impaired vascularization. It is natural to wonder to what extent the lesions may be attributable to such changes in their supporting framework and to what degree they reflect loss of another important intracellular antioxidant, which because of its water-soluble properties, participates in metabolic functions of the muscle cell in a manner different from vitamin E.

2. OTHER VITAMIN DEFICIENCIES

An extensive literature on the effects of specific vitamin deficiencies other than those of vitamins E and C in various animal species reveals no evidence of directly related lesions of skeletal muscle. This question has been put to a critical test in the rat by C. G. Mackenzie (1953), who employed a single basal diet with purified vitamin supplements adjusted to maintain a state of chronic deficiency of specific vitamins up to 16–20 weeks, followed by a more acute deficiency for the remainder of a 20–30 week period. Skeletal muscle was analyzed for creatine and chloride content and examined histologically. Separate deficiency of thiamine, riboflavin, pantothenic acid, pyridoxine, vitamin A, or protein had no effect on the muscles, except for an increased chloride content after riboflavin deficiency. On the other hand, when vitamin E was simultaneously excluded from the diets, muscles of rats deficient in pyridoxine, in vitamin A, or in protein showed more pronounced muscle lesions, a greater chloride content, and a lower creatine content than those of rats deficient only in vitamin E. Undernutrition was excluded as a complicating factor. It has also been noted (Hove and Hardin, 1952) that low protein accentuates creatinuria in rats on low-E diets. The experiments referred to above also emphasize the importance in nutritional studies of assuring adequacy of all nutritional factors except the one under special study, and the difficulty in properly evaluating and interpreting results obtained by different investigators employing diets of rather different composition for the production of specific deficiency states.

A brief report on a myopathy in the rat attributed to vitamin B_6 deficiency (Margreth *et al.*, 1967) warrants mention. In weanling rats fed a B_6-deficient diet for 2–4 weeks, total phosphorylase activity of the gastrocnemius muscle decreased to about 30% of that in pair-fed controls, while glycolytic enzyme levels remained essentially normal. Electron microscopic studies showed scattered, predominantly longitudinally oriented, osmophilic deposits in intermyofibrillar spaces at the I band level in close relation to, or the products of, membranes of the sarcoplasmic reticulum. Morphology of mitochondria and distribution of glycogen were unaffected.

3. CHOLINE DEFICIENCY

Choline, a lipotropic agent, has an important function in the phospholipid turnover in liver and kidney and in transport of fatty acids from the liver to the fat depots. It can furnish methyl groups for the synthesis of methionine, cystine, or creatine, and can serve as a precursor of acetylcholine. Deficiency of choline causes fatty livers, kidney necrosis, and also lesions of skeletal muscle and cardiovascular system. Hove and Copeland (1954) first called attention to the skeletal muscle lesions in choline-deficient rabbits. This was not an accidental observation but the result of experiments designed to test the hypothesis that deficiency of choline may interfere with acetylcholine synthesis, which, in turn, may interfere with transmission of nerve impulses to the muscles or to their vascular supply. The data obtained provided some support for this interesting hypothesis. After choline deficiency of 70–100 days, with more than an adequacy of vitamin E provided, rabbits show weakness, flaccidity, and plasticity of the hind extremities such that when placed in unnatural positions they remain so for prolonged periods (Fig. 22). The muscles are paler than normal; histologically, many fibers show swelling, contraction clotting, loss of striations, and necrosis, associated with some increase in connective tissue (Fig. 23). The syndrome resembles that of chronic vitamin E deficiency in rabbits, yet high intake of vitamin E has neither a preventive nor curative effect. On the other hand, choline therapy of a few days duration dispels all signs of creatinuria and muscle weakness; as the authors state, "the legs could no longer be molded into bizarre positions, but instead had the feel of tightly coiled springs characteristic of normal rabbits." The observation that dietary choline lessened the vitamin E requirements of rabbits is also of interest.

Although Hove and Copeland (1954) indicate that muscular dystrophy of choline deficiency differs from that of vitamin E deficiency in that creatinuria appears more gradually and is associated with a decline in creatinine excretion, it may be noted that much the same picture has been observed in states of chronic vitamin E deficiency in rabbits (C. G. Mackenzie and McCollum, 1941) and in rats (Hove and Hardin, 1952). C. G. Mackenzie (1953) feels that the decline in creatinine excretion, which remains relatively constant in acute dystrophy, merely reflects a more pronounced loss in the total mass of skeletal muscle in chronic dystrophy.

In a subsequent report, Hove *et al.* (1957) refer to damaged heart muscle and valves in their choline-deficient rabbits, mentioning also that betaine hydrochloride at a 0.3% level in the diet prevented injury to skeletal muscle but not to cardiac muscle, whereas 0.12% choline chloride protected both. This may mean that skeletal muscle has a lower requirement for choline than cardiac muscle. It may be added that skeletal muscles of the betaine-supplemented rabbits were not entirely normal, since they had a glossy translucent and greenish white appearance in the fresh state but were histologically devoid of lesions. It is also of interest that methionine deficiency also caused gross paralysis and extensive lesions in skeletal muscles, preventable by either methionine or homocystine. No reference is made to cardiac lesions in these animals.

In rats fed a low-choline, low-protein diet for prolonged periods, there have been observed extensive lesions of skeletal muscles that are said to differ morphologically from those characteristic of vitamin E deficiency (Aloisi and Bonetti, 1952). Other investigations of choline deficiency in the rat, and in the mouse also, have made only casual reference to skeletal muscle injury but have reported rather extensive lesions of the cardiovascular system, as discussed later. The biochemical changes in skeletal muscle after choline deficiency in the rabbit are characterized by a pronounced increase in concentration of DNA and RNA, a decrease in total phosphorus, a decrease in total inorganic and acid-soluble phosphorus (but not acid-insoluble phosphorus), and a considerable decrease in ATP and creatine (Srivastava *et al.*, 1965). These changes are not significantly different from those observed in the vitamin E-deficient rabbit.

4. Antistiffness Factor

Guinea pigs fed a milk diet supplemented with minerals and known vitamins develop a peculiar wrist stiffness and skeletal muscle lesions of variable intensity and extent. There may also be considerable calcification, depending upon the inorganic constituents of the diet. Histologically, the lesions (Fig. 24) somewhat resemble those of vitamin E deficiency (Harris and Wulzen, 1950), but are not prevented by high vitamin E or accentuated by cod liver oil. Myocardial lesions also occur. Certain sterols such as ergostanyl acetate provide some protection. For further details, the reader is referred to the review by Krueger (1955). A similar calcinosis syndrome, with lesions in skeletal and cardiac muscle, has been described in cotton rats and attributed, at least in part, to dietary deficiency of manganese (Constant *et al.*, 1952; Constant and Phillips, 1954).

5. Potassium Deficiency

Deficiency of potassium has a rather specific effect on cardiac muscle. However, lesions of skeletal muscles in the absence of myocardial injury are reported in dogs fed a potassium-deficient diet, which, when fed to rats, produces only myocardial lesions (S. G. Smith *et al.*, 1950), again illustrating species differences in response to the same inadequacy of diet. The syndrome resembles familial periodic paralysis of man. A series of mild attacks are usually followed by a progressive paralysis and death due to respiratory failure. No lesions are found in the central nervous system. The disorder can be reversed by potassium therapy. The muscle lesions are widely scattered and characterized chiefly by contraction-clot necrosis with regenerative reactions; their somewhat minor character in proportion to the striking muscular weakness and disability of the animal suggests widespread functional impairment of muscle fibers in the absence of much structural change.

6. MAGNESIUM DEFICIENCY

Since magnesium ion is an essential component of a large number of metabolically important enzymes it is not surprising that dietary deficiency of magnesium can have a deleterious effect upon skeletal muscle. Various reports describing myocardial lesions in rats and calves depleted of magnesium early in life make reference to scattered foci of hyaline necrosis in skeletal muscle fibers. These lesions have been given more special consideration by Hegttveit (1969) who describes the occurrence of multifocal hyaline, vacuolar, granular, and floccular necrotic lesions, sometimes associated with fibroblastic proliferation in the second and third weeks of deficiency in the rat. Of particular interest is the absence of any regenerative reactions. Calcification is a distinctive feature of the lesions and sometimes occurs in otherwise intact fibers. It is postulated that calcium salts are deposited on a myofibrillar matrix or in linear rows of mitochondria. In cardiac muscle of deficient rats calcium deposition first occurs in mitochondria (Hegttveit *et al.,* 1964).

III. Cardiac Muscle

Cardiac muscle differs from skeletal muscle in certain structural, biochemical, and physiological features. Hence, it is of interest that it reacts to vitamin E deficiency in much the same manner as does skeletal muscle, yet responds quite differently from skeletal muscle to deficiencies of choline and potassium.

A. *Vitamin E Deficiency*

Myocardial lesions and electrocardiographic abnormalities have been observed in many animal species subjected to vitamin E deficiency.

Usually they occur in association with the skeletal muscle lesions and are accelerated or accentuated in much the same manner by dietary unsaturated fats. The range of response of cardiac muscle is even more limited than that of skeletal muscle; that is, the pattern of cellular response is qualitatively the same and lesions differ chiefly in the extent of myocardial involvement. The lesions occur as a myocardial necrosis (Fig. 25) in which there is observed a progressive vacuolation, loss of myofibrillar structure, nuclear pyknosis, and dissolution and gradual disintegration of cardiac muscle cells in isolated or confluent zones of the myocardium (chiefly ventricular and interventricular, but often auricular as well), associated with the presence of a few macrophages which do not actively invade the injured cells but assist in removal of the breakdown products. Edema and small hemorrhages are sometimes seen. Purkinje fibers are usually not involved. There is connective tissue replacement of the damaged muscle cells, for regenerative reactions are not observed in cardiac muscle.

Myocardial necrosis, preceding onset of liver necrosis and skeletal muscle lesions, occurs in mice fed diets low in vitamin E, selenium, and cystine (DeWitt and Schwarz, 1958). Myocardial lesions are also commonly associated with the myopathy in the Rottnest quokka maintained in captivity (see p. 169). Kakulus *et al.* (1968) describe a focal necrosis of myocardial fibers in subendocardial and subepicardial regions of the ventricles quite similar to those observed in farm animals deficient in selenium or vitamin E.

In the rabbit, the laboratory animal most extensively studied, there is increased oxygen consumption of the cardiac muscle (Gatz and Houchin, 1951), normal glycogen concentration, but marked reduction in creatine phosphate (Mulder *et al.*, 1954), and electrocardiographic abnormalities consisting of right axis deviation, notching of T_1 and T_2 waves and inversion of T_3 (Gatz and Houchin, 1951). The latter investigators, and also Bragdon and Levine (1949), give detailed descriptions of the myocardial lesions. Similar lesions observed in the rat, differing chiefly in the presence of considerable acid-fast pigment in muscle cells and macrophages, are not significantly modified by many months of vitamin E therapy (Mason and Emmel, 1945).

Gross and microscopic lesions of the myocardium are commonly observed in calves and lambs suffering from white muscle disease, whether produced under field conditions where selenium deficiency is primarily responsible, or by semipurified diets deficient in vitamin E. Calves are more prone to cardiopathy than lambs. Myocardial involvement alone may occur in calves, whereas skeletal lesions alone are more frequent in lambs (Cordy, 1963). However, depending upon experimental conditions, the reverse picture may occur in lambs (Nisbet *et al.*, 1959). Cardiac failure is generally considered to be the cause of death. This is supported by the progressive abnormalities in electrocardiographic changes and fall of blood pressure observed by Godwin and Fraser (1966) in selenium-deficient lambs, who also describe degeneration and calcification of Purkinje cells. Grossly, the lesions are characterized by conspicuous yellowish or gray foci variably distributed in subendocardial regions of the myocardium. Either ventricle and occasionally the atrial musculature may be involved. Microscopically, the degenerative changes are typical of coagulation necrosis, with individual or groups of fibers showing flocculation of the cytoplasm, pyknosis and disappearance of nuclei, and loss of fiber outlines. There is moderate leukocytic infiltration; calcareous infiltration is quite variable. Regenerative cell reactions, such as are common in skeletal muscle, have not been observed.

In avian species depleted of vitamin E and selenium, only the turkey poult shows cardiac muscle injury. As first described by M. L. Scott *et al.* (1967), onset of cardiac injury follows soon after the appearance of gizzard myopathy and prior to the onset of skeletal muscle myopathy. Unlike the predominance of ventricular involvement in calves and lambs, the myocardial lesions appear chiefly in the atria. They are characterized by an acute coagulation necrosis (Zenker's degeneration) associated with extensive intramural edema and hemorrhage, both subendocardial and subepicardial. No evidence of chronic type lesions was observed. The presence of hyaline bodies, noted also in the gizzard myopathy, is considered an expression of cellular anoxia induced by the vitamin E–selenium deficiency. Both factors are effective in preventing the myocardial damage.

B. Vitamin B_1 Deficiency

It has long been recognized that thiamine deficiency alters the function of cardiac muscle and results in myocardial lesions. Whether observed in the rat (Ashburn and Lowry, 1944), dog (Swank *et al.*, 1941), pig (Follis *et al.*, 1943; Cartwright *et al.*, 1945), fox (C. A. Evans *et al.*, 1942), or monkey (Rinehart and Greenberg, 1949), descriptions of the lesions are much the same; yet, the degree of involvement of ventricular and atrial musculature varies widely in these different reports. There is generally described a loss of striations, vacuolization, hyaline necrosis, pyknosis of nuclei, fragmentation of fibers, infiltration of mononuclear cells and macrophages, and eventual fibrotic replacement, with no evidence of regenerative reactions. Usually, no reference is made to the conduction fibers (Purkinje cells), which in the monkey, are said (Rinehart and Greenberg, 1949) to show hypertrophy of nuclei and hydropic degeneration of the cytoplasm. The only electron microscopic study to date (Bózner *et al.*, 1969) describes an increase in number, enlargement, and loss of granules of the mitochondria. This is associated with a marked decrease in pyruvate dehydrogenase activity. Beginning about 4 hr after injection of thiamine, granules reappear in some mitochondria, while others show lamellar degenerative changes. After about 12 hr, there is restoration of mitochondrial morphology and enzyme activity. It is thought that thiamine causes synthesis of new enzymes by enzymic induction.

C. Choline Deficiency

The cardiac musculature of the rat is peculiarly susceptible to choline deficiency, especially when dietary fats are high. Kesten *et al.* (1945) observed sudden death from heart failure during the first week of feeding choline deficient diets high in ethyl laurate, and describe a widespread interstitial myocarditis most marked in the subendocardial regions but usually not involving the conduction system. There was mild edema, considerable cellular infiltration, and variable amount of necrosis of heart muscle cells. The possibility of potassium deficiency, which results in a similar type of lesion, was excluded.

According to Wilgram *et al.* (1954) and Wilgram and Hartroft (1955), the myocardial lesions are accentuated or intensified by many other types of dietary fats, are initiated by the appearance of considerable lipid in the cardiac muscle cells before swelling and necrosis occur, are always associated with a fatty liver, are not seen in the absence of a fatty liver, and are usually associated with lesions in the larger blood vessels. The latter begin with lipid accumulations in the endothelium and intima followed by medial necrosis and variable degrees of calcification. Fat released following lysis of cardiac muscle cells is taken up by macrophages, which disappear subsequent to healing and replacement fibrosis. It is also of interest that male rats are much more susceptible than females, and that the natural resistance of the female rat is greatly reduced by treatment with androgens and growth hormone.

In mice, myocardial lesions are less readily produced by choline deficiency than in rats, are accentuated by marginal intake of proteins providing sulfur-containing amino acids, and apparently do not show the sex differences observed in rats (Williams and Aronsohn, 1956; Meader and Williams, 1957).

D. Other Nutritional Deficiencies

1. TRYPTOPHAN DEFICIENCY

Among the specific amino acid deficiencies studied, muscle lesions have been noted only after tryptophan deficiency in rats. These involve only the cardiac musculature, according to E. B. Scott (1955), although a separation and shredding of fibers in both cardiac and smooth muscle have been reported by others (Adamstone and Spector, 1950). Scott describes cytoplasmic vacuolation and karyolysis followed by cloudy swelling and hyalinization of cardiac muscle fibers, with the usual connective tissue replacement and formation of myocardial scars.

2. POTASSIUM DEFICIENCY

Of the various inorganic constituents of the diet, only potassium seems to be indispensible for maintaining the integrity of muscle; this relationship pertains almost exclusively to cardiac muscle and appears to be linked in some manner with certain of the B vitamins. Attention has been called to effects of thiamine deficiency on the myocardium. Thomas *et al.* (1940) present evidence that cardiac lesions occur in the rat only when there is a simultaneous deficiency of potassium and an inadequacy of B vitamins, particularly B_2. Of particular interest are the observations of Follis (1958) that in the rat a combined deficiency of potassium and thiamine, each of which results in cardiac lesions if absent from the diet, produces no cardiac lesions but results in lesions of skeletal muscle that are not produced by single deficiency of these factors. It has also been reported (S. G. Smith *et al.*, 1950) that a potassium-deficient diet producing rapid myocardial necrosis in the rat causes only lesions of skeletal muscles in the dog. Follis (1958), to whom the reader is referred for further details, considers that the myocardial lesions of potassium and thiamine deficiency are histologically rather different from those of vitamin E deficiency in the rat. Lesions of cardiac muscle also occur in rats if, after a period of protein depletion, a potassium-deficient diet is used during protein repletion (Cannon *et al.*, 1953). Furthermore, feeding a diet low in sodium prevents the lesions, even in the presence of prolonged potassium deficiency; conversely, the lesions are accentuated by high sodium levels in the repletion diet. It is reasoned that with such imbalances between these two ions, intracellular enzymes are injured to the degree that necrobiosis and coagulation necrosis of the muscle fibers occur.

3. MAGNESIUM DEFICIENCY

It is well recognized that scattered foci of necrosis and calcification throughout the myocardium are common in rats and calves depleted of magnesium early in life. The literature on this subject is reviewed by Hegttveit *et al.* (1964) in a report presenting an excellent description of myocardial lesions observed in deficient rats, based upon light and electron microscopy. These randomly scattered lesions are characterized by focal necrosis of cardiac muscle cells with replacement by macrophages and leukocytes, loss of striations, granularity, vacuolation, and calcification. Electron microscopy reveals swelling and vacuolation of mitochondria, fragmentation of myofibrils and calcification. The first evidence of calcification is the appearance of dense particulate material on cristae of mitochondria. Possible implications of these changes in disturbed oxidative phosphorylation in cardiac muscle are discussed.

IV. Smooth Muscle

Vitamin E Deficiency

Smooth muscle is remarkably resistant to structural alterations as a result of nutritional deficiencies. In fact, necrosis of the smooth muscle of the gizzard in turkey poults fed diets deficient in vitamin E and selenium represents an outstanding exception to this rule. This unusual myopathy was first described by Jungherr and Pappenheimer (1937) as a patchy hyaline necrosis associated with slight inflammatory reactions and replacement fibrosis. No other lesions were observed in the deficient poults. It had since been shown that poults reared on *Torula* yeast diets deficient in vitamin E and selenium may also show exudative diathesis (Creech *et al.,* 1957) as well as both gizzard myopathy and lesions of skeletal muscle (Walter and Jensen, 1963). Furthermore, M. L. Scott *et al.* (1967), employing similar diets, report that myopathy of the gizzard is the first symptom to appear, followed by cardiac muscle lesions and later by nutritional muscular dystrophy. They consider selenium deficiency to be the primary factor in the genesis of the gizzard lesion and suggest that the diet used by Jungherr and Pappenheimer contained more than enough selenium to prevent exudative diathesis but less than the amount required to prevent gizzard myopathy in poults. They report that the earliest microscopic change in muscle cells of the gizzard is the appearance of cytoplasmic vacuoles, with or without hyaline bodies, and pronounced edema separating the degenerating muscle bundles. The presence of hyaline bodies is considered an expression of cellular anoxia induced by the deficiency state. Later, there occurs extensive proliferation and haphazard distribution of nuclei interspersed between areas where fibrotic reactions predominate.

In certain other experimental animals, deficiency of vitamin E results in alterations in smooth muscle cells that are more metabolic than structural, and are related in large part to the dietary content of unsaturated fatty acids. As first described by Martin and Moore (1936) and later confirmed by other investigators (Barrie, 1938; Hessler, 1941; Demole, 1941; Mason and Emmel, 1945), rats reared on vitamin E-deficient diets exhibit a brownish discoloration of the uterus due to accumulation of brownish yellow, insoluble, iron-free pigment granules within the smooth muscle cells (Fig. 26). Similar alterations of a somewhat lesser degree of intensity were subsequently observed (Martin and Moore, 1939; Mason and Emmel, 1945; Ruppel, 1949) in smooth muscle cells of the vagina, oviducts, seminal vesicle, prostate, vas deferens, ureter, bronchi, splenic capsule and trabeculae, small intestine, and certain blood vessels. The simultaneous occurrence of this same pigment in association with lesions of skeletal and cardiac muscle, its extensive accumulation within macrophages of involved tissues, and its wide dissemination to various constituents of the reticuloendothelial system in the rat were also described (Mason and Emmel, 1945).

This pigment, considered to be either a lipoprotein or a product of polymerized unsaturated fatty acids, requires for its production in the vitamin E-deficiency state the dietary presence of fatty acids possessing at least eighteen carbon atoms and two or more unsaturated bonds (Filer *et al.*, 1946). It is frequently referred to as ceroid, a term originally applied by Lillie *et al.* (1942) to a pigment observed in cirrhotic livers of rats reared on low-protein diets and considered to be the product of unstable peroxides of unsaturated fatty acids. There is an extensive literature on this pigment, much of which has been reviewed by Hartroft and Porta (1965), who have also introduced the term "interceroid" for pigment substances formed as fat-soluble precursors of the completely insoluble ceroid. The pigment granules are usually identified in tissues by their acid-fast or PAS-positive staining reactions or by their yellowish brown fluorescence. Under some circumstances, the pigment may be PAS-positive but show little or no acid-fastness.

There are interesting species and organ differences with respect to the site of accumulation of this pigment, presumably reflecting rather striking differences in the content and utilization of lipids in smooth muscle cells generally. In the monkey, pigment is abundant in the smooth muscle of blood vessels (Fig. 27) and occurs also in that of the intestine, gall bladder, urinary bladder and bronchi (Mason and Telford, 1947). In the Syrian hamster, it is quite abundant in blood vessels, intestine, and urinary bladder, but absent from the uterus, as true also of the monkey. After vitamin E deficiency in the dog, pigment occurs primarily in the intestinal muscularis, but also to some degree in that of blood vessels, urinary bladder, bile duct, uterus, lung, and spleen (Cordes and Mosher, 1966; Hayes *et al.*, 1969, 1970). For reasons difficult to explain, this pigment has not been observed in any smooth muscle of the vitamin E-deficient guinea pig, rabbit, Florida cotton rat, or various avian species and farm animals deficient in vitamin E, selenium, or both.

This intracellular accumulation of pigment granules appears not to significantly impair physiological activity of smooth muscle cells, yet its occurrence may be influenced by the metabolic state of the cell. As an example, uterine pigment does not occur in vitamin E-deficient rats ovariectomized before puberty, but does occur if rats so treated are given estrogen treatment (Atkinson *et al.*, 1949). Once accumulated within the intracellular environment, prolonged vitamen E therapy at high dosage levels effects only a very limited reduction of pigmentation, reflecting perhaps its effect upon the preceroid or interceroid types of pigment present.

Much attention has been given to the nature of this pigment as it occurs in the myometrium of the vitamin E-deficient rat. Fine and coarse granular lipid material makes its appearance first in the region of the Golgi apparatus at each pole of the nucleus. As these granules accumulate, the myofibrils are displaced peripherally, often with resultant distension and distortion of the cell. On the basis of extensive applications of histochemical reactions, Gedigk and Fischer (1959) consider that there occur two major types of intracellular pigment: (1) lipopigments representing unsaturated lipids undergoing oxidation and polymerization, particularly in the perinuclear zones of the cell where protein synthesis and other metabolic processes are very active; and (2) lipopigments of the mesenchymal type (ceroid), appearing in muscle cells and macrophages after absorption of fat-containing substances to which proteins may be added secondarily. According to Gedigk and Wessel (1964), after prolonged vitamin E deficiency, muscle fibers degenerate, releasing pigment granules which break up into fragments and are phagocytized by macrophages. This is the only statement recorded relative to actual degeneration of smooth muscle cells after vitamin E deficiency. If this is the primary process whereby macrophages acquire pigment, the great abundance observed in such cells during much earlier stages of deficiency would imply much earlier and more extensive cell breakdown than has been observed. An alternate explanation is that macrophages of the myometrium acquire much the same lipids as do associated smooth muscle cells, and that the lipids undergo much the same alterations in both.

Electron microscopically, Gedigk and Wessel (1964) find no evidence of transformation of cytoplasmic organelles into pigment granules. On the other hand, Linder (1957) considers that the membrane system of mitochondria participates in the formation of pigment granules. He states also that as granules accumulate there is a reduction in the number of myofibrils, but finds no evidence of ultrastructural changes in the myofibrils themselves. These findings are in accord with those of Howes *et al.* (1964), who describe focal membrane thickening and distortion in mitochondria, flocculent aggregates in the endoplasmic reticulum, membrane-limited globular aggregates of finely particulate material, and bodies containing electron-dense material and varying numbers of membrane forms in whorls or in parallel array. The latter are probably representative of true ceroid.

It is not possible, as yet, to discriminate between ultrastructural changes in smooth muscle related to ceroid pigment formation and those reflecting unrelated effects of vitamin E deficiency. Yet, the problem is relatively simple compared to that in skeletal muscle where such a multiplicity of changes has been observed, and where comparisons between ultramicroscopic alterations in the rat (Howes *et al.*, 1964) in which ceroid is deposited and in the rabbit (Van Vleet *et al.*, 1967, 1968) in which it is not offer little in terms of the mode of its occurrence.

V. General Considerations

The subject matter of this chapter has, of necessity, been limited to a review of the nutritional factors that result in morphological and biochemical alterations in skeletal, cardiac, and smooth muscle in laboratory and farm animals under laboratory or field conditions. As an addendum to this discussion, brief reference should be made to morphological and biochemical alterations observed in genetically determined myopathies of different animal species. During the past two decades, such myopathies have been identified in the mouse, Syrian hamster, sheep, goat, cattle, pig, chick, duck, and turkey. In most instances the lesions observed do not differ significantly from those described after deficiency of vitamin E, selenium, or both. This again indicates the rather limited range of response of skeletal and cardiac muscle to nutritionally induced or genetically determined injury.

It is also natural to wonder whether nutritionally induced myopathies in animals have any counterpart in man. The rather limited evidence at hand warrants brief consideration. In human infants dying of the kwashiorkor type of protein–calorie malnutrition, the size of fibers in the sartorius muscle at 8–16 months of age is said to be reduced to that of the newborn infant, and the ratio of collagen to muscle tissue considerably increased (Montgomery, 1962). The latter is attributed to the greater withdrawal of protein from muscle than from connective tissue to meet other needs of the protein-deficiency state. Montgomery also describes and illustrates microscopic muscle lesions typical of coagulative or hyaline necrosis in these infants. It is presumed that these degenerative lesions reflect the extremely low intake of protein and calories over many months. Somewhat more marked degenerative changes have been described by van Bogaert *et al.* (1962) in biopsy and necropsy specimens from eleven infants, 6 months to $3\frac{1}{2}$ years of age suffering from kwashiorkor. These observations generally confirm earlier findings of Vincent and Radermecker (1958). Whether such muscle lesions can be ascribed, at least in part, to deficiency of vitamin E, selenium, or some other specific essential nutrient can be no more than speculative until more definitive data are available.

The report of Pappenheimer and Victor (1946) first called attention to the possibility that under situations where there is obvious malabsorption of fats and of vitamin E in man there can be observed accumulations of acid fast pigment in the esophageal and intestinal smooth musculature, and elsewhere, quite comparable to that found in the smooth muscle of experimental animals deprived of vitamin E. Since then, considerable evidence has been presented that on the basis of this criterion there exists some evidence that a condition of vitamin E deficiency may occur in man as a secondary result of nutritional and related disorders. In support of this postulate may be cited the possible association of vitamin E deficiency in the so-called brown bowel syndrome (Toffler *et al.*, 1963; Bauman *et al.*, 1968) and the high incidence of ceroid associated with liver disease in population groups in Thailand (Nye and Chittayasothorn, 1967). Particularly impressive is the careful and detailed follow-up study of a chronic alcoholic, diabetic subject demonstrably deficient in vitamin E, with evidence of chronic pancreatic disease and steatorrhea, manifesting at necropsy not only the brown bowel syndrome but also microscopic changes in skeletal muscle and the nervous system suggestive of those encountered in experimental vitamin E deficiency (Bauman *et al.*, 1968). The reader is referred to this same report for a more comprehensive review of the earlier literature on this subject.

Many hopes have been raised that vitamin E might have some beneficial effects upon myopathies of man. A carefully controlled clinical trial extending over a period of 3 years (Harris and Mason, 1956) has demonstrated that neither α-tocopherol nor α-tocohydroquinone, which is a potent antidystrophic agent in experimental animals deprived of vitamin E, exerted any beneficial effects upon young children suffering from the Duchenne type of progressive muscular dystrophy.

REFERENCES

Adams, R. D., Denny-Brown, D., and Pearson, C. M. (1962). "Diseases of Muscle. A Study in Pathology," 2nd ed. Harper (Hoeber), New York.

Adamstone, F. B., and Spector, H. (1950). *AMA Arch. Pathol.* 49, 173.

Alfin-Slater, R. B., and Morris, R. S. (1963). *Advan. Lipid Res.* 1, 183–210.

Aloisi, M., and Bonetti, E. (1952). *Arch. Sci. Biol.* (*Bologna*) 36, 205.

Aloisi, M., Ascenzi, A., and Bonetti, E. (1952). *J. Pathol. Bacteriol.* 64, 321.

Anderson, H. D., Elvehjem, C. A., and Gonce, J. E., Jr. (1939). *Proc. Soc. Exp. Biol. Med.* 42, 750.

Arnold, M. A., Weswig, P. H., Muth, O. H., and Oldfield, J. E. (1965). *Proc. Soc. Exp. Biol. Med.* 118, 75.

Ashburn, L. L., and Lowry, J. V. (1944). *AMA Arch. Pathol.* 37, 27.

Atkinson, W. B., Kaunitz, H., and Slanetz, C. A. (1949). *Ann. N.Y. Acad. Sci.* 52, 68.

Bacigalupo, F. A., Culik, R., Luecke, R. W., Thorp, F., and Johnston, R. L. (1952). *J. Anim. Sci.* 11, 609.

Barrie, M. M. O. (1938). *Biochem. J.* 32, 2134.

Bauman, M. B., DiMase, J. D., Oski, F., and Senior, J. R. (1968). *Gastroenterology* 54, 93.

Blaxter, K. L. (1962). *Proc. Nutr. Soc.* 21, 211.

Blaxter, K. L., Watts, P. S., and Wood, W. A. (1952). *Brit. J. Nutr.* 6, 125.

Boyle, P. E., and Irving, J. T. (1951). *Science* 114, 572.

Bózner, A., Knieriem, H. T., Meessen, H., and Reinauer, H. (1969). *Virchows Arch., B* 2, 125.

Bragdon, J. H., and Levine, H. D. (1949). *Amer. J. Pathol.* 25, 265.

Brinkhous, K. M., and Warner, E. D. (1941). *Amer. J. Pathol.* 17, 81.

Buchanan-Smith, J. G., Nelson, E. C., and Tillman, A. D. (1969). *J. Nutr.* 99, 387.

Cannon, P. R., Frazier, L. E., and Hughes, R. H. (1953). *Metab. Clin. Exp.* 2, 297.

Cartwright, G. E., Wintrobe, M. M., Buschke, W. H., Follis, R. H., Jr., Sukusta, A., and Humphreys, S. (1945). *J. Clin. Invest.* 24, 268.

Cheville, N. F. (1966). *Pathol. Vet.* 3, 208.

Chiakulas, J. J., and Pauly, J. E. (1965). *Anat. Rec.* **152**, 55.
Constant, M. A., and Phillips, P. H. (1954). *J. Nutr.* **52**, 165 and 327.
Constant, M. A., Phillips, P. H., and Angevine, D. M. (1952). *J. Nutr.* **47**, 327.
Cordes, D. O., and Mosher, A. H. (1966). *J. Pathol. Bacteriol.* **92**, 197.
Cordy, D. R. (1963). *In* "Muscular Dystrophy in Man and Animals" (G. H. Bourne and M. N. Golarz, eds.), pp. 499–514. Hafner, New York.
Corsi, A. (1957). *Sperimentale* **107**, 328.
Creech, B. G., Feldman, G. L., Ferguson, T. M., Reid, B. L., and Couch, J. R. (1957). *J. Nutr.* **62**, 83.
Culik, R., Bacigalupo, F. A., Thorp, F., Luecke, R. W., and Nelson, R. H. (1951). *J. Anim. Sci.* **10**, 1006.
Dalldorf, G. (1929). *J. Exp. Med.* **50**, 293.
Dam, H., Prange, I., and Søndergaard, R. (1952). *Acta Pathol. Microbiol. Scand.* **31**, 172.
Demole, V. (1941). *Schweiz. Med. Wochenschr.* **71**, 1251.
Desai, I. D. (1968). *Brit. J. Nutr.* **22**, 645.
Desai, I. D., and Scott, M. L. (1965). *Arch. Biochem. Biophys.* **110**, 309.
Dewan, M. L., Henson, J. B., Dollahite, J. W., and Bridges, C H.. (1965). *Amer. J. Pathol.* **46**, 215.
DeWitt, W. B., and Schwarz, K. (1958). *Experientia* **14**, 28.
Dinning, J. S. (1955). *J. Biol. Chem.* **212**, 735.
Dinning, J. S., and Day, P. L. (1957). *J. Nutr.* **63**, 393; *J. Exp. Med.* **105**, 395.
Dinning, J. S., and Fitch, C. D. (1958). *Proc. Soc. Exp. Biol. Med.* **97**, 109.
Drake, C., Grant, A. B., and Hartley, W. J. (1960). *N.Z. Het. J.* **8**, 4.
Durack, D. T., Gubbay, S. S., and Kakulas, B. A. (1969). *Austr. J. Exp. Biol. Med. Sci.* **47**, 581.
Einarson, L. (1952). *Acta Psychiat. Scand., Suppl.* **78**, 1.
Einarson, L. (1953). *J. Neurol., Neurosurg. Psychiat.* **16**, 98.
Einarson, L., and Ringsted, A. (1938). "Effect of Chronic Vitamin E Deficiency on the Nervous System and the Skeletal Musculature in Adult Rats." Levin & Munksgaard, Copenhagen.
Einarson, L., and Telford, I. R. (1960). *Biol. Skr. Dan. Videnskab. Selskab* **11**, No. 2, 1.
Evans, C. A., Carlson, W. E., and Green, R. G. (1942). *Amer. J. Pathol.* **18**, 79.
Evans, H. M., and Burr, G. O. (1928). *J. Biol. Chem.* **76**, 273.
Evans, H. M., Emerson, G. A., and Telford, I. R. (1938). *Proc. Soc. Exp. Biol. Med.* **38**, 625.
Ewan, R. C., and Wastell, M. E. (1970). *J. Anim. Sci.* **31**, 343.
Ewen, L. M., and Jenkins, K. J. (1967). *J. Nutr.* **93**, 470.
Fenn, W. O., and Goettsch, M. (1937). *J. Biol. Chem.* **120**, 41.
Ferguson, T. M., Omar, E. M., and Couch, J. R. (1964). *Tex. Rep. Biol. Med.* **22**, 902.
Filer, L. J., Jr., Rumery, R. E., and Mason, K. E. (1946). *Trans. First Conf. Biol. Antioxidants, 1946* Josiah Macy Jr. Foundation, New York, pp. 67–77.
Follis, R. H., Jr. (1958). "Deficiency Disease." Thomas, Springfield, Illinois.
Follis, R. H., Jr., Miller, M. H., Wintrobe, M. M., and Stein, H. J. (1943). *Amer. J. Pathol.* **19**, 341.
Friedman, I., and Mattill, H. A. (1941). *Amer. J. Physiol.* **131**, 595.
Fudema, J. J., Oester, Y. T., Fizzell, J. A., and Gatz, A. J. (1960). *Amer. J. Physiol.* **198**, 123.

Gatz, A. J., and Houchin, O. B. (1951). *Anat. Rec.* **110**, 249.

Gedigk, P., and Fischer, R. (1959). *Arch. Pathol. Anat. Physiol.* **332**, 431.

Gedigk, P., and Wessel, W. (1964). *Arch. Pathol. Anat. Physiol.* **337**, 367.

Godwin, K. O., and Fraser, F. J. (1966). *Quart. J. Exp. Physiol.* **51**, 94.

Goettsch, M., and Brown, E. F. (1932). *J. Biol. Chem.* **97**, 549.

Goettsch, M., and Pappenheimer, A. W. (1931). *J. Exp. Med.* **54**, 145.

Green, J., and Bunyan, J. (1969). *Nutr. Abstr. Rev.* **39**, 321.

Harris, P. L., and Mason, K. E. (1956). *Amer. J. Clin. Nutr.* **4**, 402.

Harris, P. N., and Wulzen, R. M. (1950). *Amer. J. Pathol.* **26**, 595.

Hartley, W. J., and Grant, A. B. (1961). *Fed. Proc., Fed. Amer. Soc. Exp. Biol.* **20**, 679.

Hartroft, W. S., and Porta, E. A. (1965). *Amer. J. Med. Sci.* **250**, 324.

Hayes, K. C., Nielsen, S. W., and Rousseau, J. E., Jr. (1969). *J Nutr.* **99**, 196.

Hayes, K. C., Rousseau, J. E., Jr., and Hegsted, D. M. (1970). *J. Amer. Vet. Med. Ass.* **157**, 64.

Heggtveit, H. A. (1969). *Ann. N.Y. Acad. Sci.* **162**, 758.

Heggtveit, H. A., Herman, Y., and Mishra, R. K. (1964). *Amer. J. Pathol.* **45**, 757.

Herrmann, H., and Nicholas, J. S. (1948). *J. Exp. Zool.* **107**, 165.

Hessler, W. (1941). *Z. Vitaminforsch.* **11**, 9.

Hines, H. M. (1952). *Proc. Med. Conf. Muscular Dystrophy Ass. Amer., 1st and 2nd, 1952* p. 89.

Höjer, J. A. (1924). *Acta Paediat.* **3**, Suppl. 2, 1.

Houchin, O. B., and Mattill, H. A. (1942). *J. Biol. Chem.* **146**, 301 and 309.

Hove, E. L., and Copeland, D. H. (1954). *J. Nutr.* **53**, 391.

Hove, E. L., and Hardin, J. O. (1952). *J. Nutr.* **48**, 193.

Hove, E. L., Copeland, D. H., Herndon, J. F., and Salmon, W. D. (1957). *J. Nutr.* **63**, 289.

Howes, E. L., Jr., Price, M. H., and Blumberg, J. M. (1964). *Amer. J. Pathol.* **45**, 599.

Hummel, J. P., and Melville, R. S. (1951). *J. Biol. Chem.* **191**, 391.

Jenkins, K. J., and Ewen, L. M. (1967). *Can. J. Biochem Physiol.* **45**, 1873.

Jungherr, E., and Pappenheimer, A. M. (1937). *Proc. Soc. Exp. Biol. Med.* **37**, 520.

Kakulas, B. A. (1961). *Nature (London)* **191**, 402.

Kakulas, B. A. (1963a). *Aust. J. Sci.* **25**, 313.

Kakulas, B. A. (1963b). *Nature (London)* **198**, 673.

Kakulas, B. A., and Adams, R. D. (1966). *Ann. N.Y. Acad. Sci.* **138**, 90.

Kakulas, B. A., Owen, C. G., Papadimitriou, J. M., and Durack, D. T. (1968). *Proc. Aust. Ass. Neurol.* **5**, 565.

Kaspar, L. V., and Lombard, L. S. (1963). *J. Amer. Vet. Med. Ass.* **143**, 284.

Kaunitz, H., and Pappenheimer, A. M. (1943). *Amer. J. Physiol.* **138**, 328.

Kesten, H. D., Salcedo, Jr., J., and Stetten, D. (1945). *J. Nutrition* **29**, 171.

Kodono, H., and Hirose, M. (1969). *Jap. J. Vet. Sci.* **31**, 187.

Krueger, H. (1955). *Amer. J. Phys. Med.* **34**, 185.

Lampert, P., and Pentschew, A. (1964). *Acta Neuropathol.* **4**, 158.

Lampert, P., Blumberg, J. M., and Pentschew, A. (1964). *J. Neuropathol. Exp. Neurol.* **23**, 60.

Lee, Y. C. P., Kuha, K. T., Visscher, M. B., and King, J. T. (1962). *Amer. J. Physiol.* **203**, 1103.

Lillie, R. D., Ashburn, L. L., Sebrell, W. H., Daft, F. S., and Lowry, J. V. (1942). *Pub. Health Rep.* **57**, 502.

Linder, E. (1957). *Beitr. Pathol. Anat. Allg. Pathol.* **117**, 1.

Luttrell, C. N., and Mason, K. E. (1949). *Ann. N.Y. Acad. Sci.* **52**, 113.

Maag, D. D., and Glenn, M. W. (1967). *In* "Symposium on Selenium in Biomedicine" (O. H. Muth, ed.), pp. 127–140. Avi Publ. Co., Westport, Connecticut.

McCoy, K. E. M., and Weswig, P. H. (1969). *J. Nutr.* **98**, 383.

MacDonald, A. M., Blaxter, K. L., Watts, P. S., and Wood, W. A. (1952). *Brit. J. Nutr.* **6**, 164.

Mackenzie, C. G. (1942). *Proc. Soc. Exp. Biol. Med.* **49**, 313.

Mackenzie, C. G. (1953). *In* "Symposium on Nutrition" (R. M. Heriott, ed.), pp. 136–178. Johns Hopkins Press, Baltimore, Maryland.

Mackenzie, C. G., and McCollum, E. V. (1940). *J. Nutr.* **19**, 345.

Mackenzie, C. G., and McCollum, E. V. (1941). *Proc. Soc. Exp. Biol. Med.* **48**, 642.

Mackenzie, J. B., and Mackenzie, C. G. (1953). *Proc. Soc. Exp. Biol. Med.* **84**, 388.

Mackenzie, J. B., and Mackenzie, C. G. (1959). *J. Nutr.* **67**, 223.

Mackenzie, J. B., and Mackenzie, C. G. (1960). *J. Nutr.* **72**, 322.

Mackenzie, J. B., Rosenkranz, H., Ulick, S., and Milhorat, A. T. (1950). *J. Biol. Chem.* **183**, 655.

Malamud, N., Nelson, M. M., and Evans, H. M. (1949). *Ann. N.Y. Acad. Sci.* **52**, 135.

Margreth, A., Shiaffino, S., and Salviati, G. (1967). *Excerpta Med. Found. Int. Congr. Ser.* **154**, p. 13.

Martin, A. J. P., and Moore, T. (1936). *Chem. Ind. (London)* **55**, 236.

Martin, A. J. P., and Moore, T. (1939). *J. Hyg* **39**, 643..

Mason, K. E. (1952). *Proceedings Med. Conf. Muscular Dystrophy Ass. Amer, 1st and 2nd, 1952* p. 25.

Mason, K. E., and Dju, M. Y. (1963). *Fed. Proc., Fed. Amer. Soc. Exp. Biol.* **22**, 319.

Mason, K. E., and Emmel, A. F. (1945). *Anat. Rec.* **92**, 33.

Mason, K. E., and Hartsough, G. R. (1951). *J. Amer. Vet. Med. Ass.* **99**, 72.

Mason, K. E., and Mauer, S. I. (1959). *Anat. Rec.* **133**, 307.

Mason, K. E., and Telford, I. R. (1947). *AMA Arch. Pathol.* **43**, 363.

Mauer, S. I., and Mason, K. E. (1958). *Anat. Rec.* **130**, 429.

Mauro, A. (1961). *J. Biophys. Biochem. Cytol.* **9**, 493.

Mauro, A., Shafiq, S. A., and Milhorat, A. T. (1970). *Excerpta, Med. Found. Int. Congr. Ser.* **154**, 1.

Meader, R. D., and Williams, W. L. (1957). *Amer. J. Anat.* **100**, 167.

Menschik, Z., Munk, M. K., Rogalski, T., Rymaszewski, O., and Szczesniak, T. J. (1949). *Ann. N.Y. Acad. Sci.* **52**, 94.

Mercer, H. D., Neal, F. C., Himes, J. A., and Edds, G. T. (1967). *J. Amer. Vet. Med. Ass.* **151**, 735.

Meyer, A. W., and McCormick, L. M. (1928). *Stanford Univ. Publ., Univ. Ser., Med. Sci.* **2**, No. 2, 129.

Milman, A. E., and Milhorat, A. T. (1953). *Proc. Soc. Exp. Biol. Med.* **84**, 654.

Monnier, M. (1941). *Z. Vitaminforsch.* **11**, 235.

Montgomery, R. D. (1962). *J. Clin. Pathol.* **15**, 511.

Moore, T. (1962). *Proc. Nutr. Soc.* **21**, 179.

Morgulis, S., and Spencer, H. C. (1936). *J. Nutr.* **173**, 191.

Morgulis, S., Wilder, V. M., Spencer, H. C., and Eppstein, S. H. (1938). *J. Biol. Chem.* **124**, 755.

Muir, A. R., Kanji, A. H. M., and Allbrook, D. B. (1965). *J. Anat. (London)* **99**, 435.

Mulder, A. G., Gatz, A. J., and Tigerman, B. (1954). *Amer. J. Physiol.* **179**, 246.

Murray, P. D. F., and Kodicek, E. (1949). *J. Anat.* **83**, 158.

Muth, O. H. (1955). *J. Amer. Vet. Med. Ass.* **126**, 355.

Muth, O. H. (1963). *J. Amer. Vet. Med. Ass.* **142**, 272.

Nesheim, M. C., and Scott, M. L. (1961). *Fed. Proc., Fed. Amer. Soc. Exp. Biol.* **20**, 674.

Nisbet, D. I., Butler, E. J., and Macintyre, I. J. (1959). *J. Comp. Pathol. Ther.* **69**, 339.

Nye, S. W., and Chittayasothorn, K. (1967). *Amer. J. Pathol.* **51**, 287.

Oksanen, H. E. (1965). *Acta Vet. Scand.* **6**, Suppl. 2.

Oksanen, H. E. (1967). In "Symposium on Selenium in Biomedicine" (O. H. Muth, ed.), pp. 215–229. Avi Publ. Co., Westport, Connecticut.

Olcott, H. S. (1938). *J. Nutr.* **15**, 221.

Pappenheimer, A. M. (1939). *Amer. J. Pathol.* **15**, 179.

Pappenheimer, A. M. (1942). *Amer. J. Pathol.* **18**, 169.

Pappenheimer, A. M. (1943). *Physiol. Rev.* **23**, 37.

Pappenheimer, A. M. (1948). "On Certain Aspects of Vitamin E Deficiency." Thomas, Springfield, Illinois.

Pappenheimer, A. M., and Goettsch, M. (1934). *J. Exp. Med.* **59**, 35.

Pappenheimer, A. M., and Goettsch, M. (1941). *Proc. Soc. Exp. Biol. Med.* **47**, 268.

Pappenheimer, A. M., and Victor, J. (1946). *Amer. J. Pathol.* **22**, 395.

Pappenheimer, A. M., Goettsch, M., and Jungherr, E. (1939). *Storrs (Conn). Agr. Exp. Sta. Bull.* **229**.

Patterson, E. L., Milstrey, R., and Stokstad, E. L. R. (1957). *Proc. Soc. Exp. Biol. Med.* **95**, 617.

Pentschew, A., and Schwarz, K. (1962). *Acta Neuropathol.* **1**, 313.

Rinehart, J. F., and Greenberg, L. D. (1949). *AMA Arch. Pathol.* **48**, 89.

Ringsted, A. (1935). *Biochem. J.* **29**, 788.

Rogers, W. M., Pappenheimer, A. M., and Goettsch, M. (1931). *J. Exp. Med.* **54**, 167.

Rosenfeld, I., and Beath, O. A. (1964). "Selenium." Academic Press, New York.

Rumery, R. E., and Hampton, J. C. (1959). *Anat. Rec.* **133**, 1.

Rumery, R. E., Mauer, S. I., and Mason, K. E. (1955). *J. Exp. Zool.* **129**, 495.

Ruppel, W. (1949). *Naunyn-Schmiedebergs Arch. Exp. Pathol. Pharamakol,* **206**, 584.

Safford, J. W., Swingle, K. F., and Marsh, H. (1954). *Amer. J. Vet. Res.* **15**, 373.

Schubert, J. R., Muth, O. H., Oldfield, J. E., and Remmert, L. F. (1961). *Fed. Proc., Fed. Amer. Soc. Exp. Biol.* **20**, 689.

Schwarz, K. (1948). *Hoppe-Seyler's Z. Physiol. Chem.* **281**, 101.

Schwarz, K. (1949). *Ann. N.Y. Acad. Sci.* **52**, 225.

Schwarz, K., and Foltz, C. M. (1957). *J. Amer. Chem. Soc.* **79**, 3292.

Scott, E. B. (1955). *Amer. J. Pathol.* **31**, 1111.

Scott, M. L. (1962). *Nutrition Abstr. Rev.* **32**, 82.

Scott. M. L. (1966). *Ann. N.Y. Acad. Sci.* **138**, 82.

Scott, M. L., Olson, G., Krook, L., and Brown, W. R. (1967). *J. Nutr.* **91**, 573.

Sharman, G. A. M. (1954). *Vet. Rec.* **66**, 275.

Sharman, G. A. M. (1960). *Proc. Nutr. Soc.* **19**, 169.

Shimotori, N., Emerson, G. A., and Evans, H. M. (1940). *J. Nutr.* **19**, 547.

Smith, L. C., and Nelson, S. R. (1957). *Proc. Soc. Exp. Biol. Med.* **94**, 644.

Smith, S. G., Black-Schaffer, B., and Lasater, T. E. (1950). *AMA Arch. Pathol.* **49**, 185.

Srivastava, U., Devi, A., and Sarkar, N. K. (1965). *J. Nutr.* **86**, 298.

Stowe, H. D., and Whitehair, C. K. (1963). *J. Nutr.* **81**, 287.

Stowe, H. D., and Whitehair, C. K. (1964). *Amer. J. Vet. Res.* **25**, 1542.

Swank, R. L., Porter, R. R., and Yeomans, A. (1941). *Amer. Heart J.* **22**, 154.

Tallan, H. H. (1955). *Proc. Soc. Exp. Biol. Med.* **89**, 553.

Telford, I. R. (1941). *Anat. Rec.* **81**, 171.

Telford, I. R. (1971). "Experimental Muscular Dystrophies in Animals." Thomas, Springfield, Illinois.

Thomas, R. M., Mylon, E., and Winternitz, M. C. (1940). *Yale J. Biol. Med.* **12**, 345.

Thompson, J. N., and Scott, M. L. (1969) *J Nutr* **97**, 335.

Thompson, J. N., and Scott, M. L. (1970). *J. Nutr.* **100**, 797.

Tobin, C. E. (1950). *AMA Arch. Pathol.* **50**, 385.

Toffler, A. H., Hulkill, P. B., and Spiro, H. M. (1963). *Ann. Intern. Med.* **58**, 872.

Trapp, A. L., Keahey, K. K., Whitenack, D. L., and Whitehair, C. K. (1970). *J. Amer. Vet. Med. Ass.* **157**, 289.

van Bogaert, L., Radermecker, M., Janssen, P., and Gatera, F. (1962). *Acta Neuro-pathol.* **1**, 363.

Van Vleet, J. F., Hall, B. V., and Simon, J. (1967). *Amer. J. Pathol.* **51**, 815.

Van Vleet, J. F., Hall, B. V., and Simon, J. (1968). *Amer. J. Pathol.* **52**, 1067.

Vawter, L. R., and Records, E. (1947). *J. Amer. Vet. Med. Ass.* **110**, 152.

Victor, J. (1934). *Amer. J. Physiol.* **108**, 229.

Vincent, M., and Radermecker, M. A. (1958). *Ann. Soc. Belge Med. Trop.* **38**, 487.

Walter, E. D., and Jensen, L. S. (1963). *J. Nutr.* **80**, 327.

Walton, J. N. (1969). "Disorders of Voluntary Muscle," 2nd ed. Little, Brown, Boston, Massachusetts.

Weichselbaum, T. E. (1935). *Quart. J. Exp. Physiol.* **25**, 363.

Weinstock, I. M., Goldrich, A. D., and Milhorat, A. T. (1955). *Proc. Soc. Exp. Biol. Med.* **88**, 257.

Weinstock, I. M., Goldrich, A. D., and Milhorat, A. T. (1956). *Proc. Soc. Exp. Biol. Med.* **91**, 302.

West, W. T. (1963). *In* "Muscular Dystrophy in Man and Animals" (G. H. Bourne and M. N. Golarz, eds.), pp. 366–405. Hafner, New York.

West, W. T., and Mason, K. E. (1955). *Amer. J. Phys. Med.* **34**, 223.

West, W. T., and Mason, K. E. (1958). *Amer. J. Anat.* **102**, 323.

Whanger, P. D., Weswig, P. H., Muth, O. H., and Oldfield, J. E. (1969). *J. Nutr.* **99**, 331.

Whanger, P. D., Weswig, P. H., Muth, O. H., and Oldfield, J. E. (1970). *Amer. J. Vet. Res.* **31**, 965.

Whiting, F., Willman, J. P., and Loosli, J. K. (1949). *J. Anim. Sci.* **8**, 234.

Wilgram, G. F., and Hartroft, W. S. (1955). *Brit. J. Exp. Pathol.* **36**, 298.

Wilgram, G. F., Hartroft, W. S., and Best, C. H. (1954). *Brit. Med. J.* **2**, 1.

Williams, W. L., and Aronsohn, R. B. (1956). *Yale J. Biol. Med.* **28**, 515.

Willman, J. P., Asdell, S. A., and Olafson, P. (1934). *N.Y., Agr. Exp. Sta., Ithaca, Bull.* **603**, 1.

Willman, J. P., Loosli, J. K., Asdell, S. A., Morrison, F. B., and Olafson, P. (1946). *Cornell Vet.* **36**, 200.

Young, J. M., and Dinning, J. S. (1951). *J. Biol. Chem.* **193**, 743.

Young, S., and Keeler, R. F. (1962). *Amer. J. Vet. Res.* **23**, 966.

Young, S., Hawkins, W. W., and Swingle, K. F. (1961). *Amer. J. Vet. Res.* **22**, 416 and 419.

VIRUS INFECTIONS

T. M. BELL AND E. J. FIELD

I. Introduction

Until recently, it was believed that muscle tissue in general was not very susceptible to virus infection. Levaditi (1926) pointed out that infections of mesodermal derivatives (*les mésodermoses*) are produced, by and large, by bacteria, fungi, or protozoa, i.e. by microorganisms visible in the light microscope, while those of the ectoderm (*les ectodermoses*) are due most often to filterable viruses. During the past decade, evidence has accumulated that suggests that involvement of muscle tissue during viral infections is not uncommon, and that cardiac muscle in particular is susceptible to a wide range of viruses. Indeed, due to changing patterns of virus infection, viral myocarditis appears to be an increasingly important problem (Grist and Bell, 1969).

The vast majority of viruses enter the host via the respiratory and alimentary tracts, either as droplet infections or by fecal/oral spread. When a person becomes infected, the initial stages of recovery are due to the action of nonspecific virus-induced inhibitors such as interferon, rather than to the production of specific antibodies (Glasgow, 1965; Hilleman, 1965). Most viral infections, whether clinical or subclinical, induce the formation of circulating antibodies which prevent reinfection with the same virus. These antibodies are also present in the blood of a child born to an immune mother. As general standards of living and hygiene have improved, the fecal/oral spread of viruses among infants has been reduced. Thus, instead of having an adult population immune to the bulk of the common viruses, there is now a nonimmune, susceptible, population. Similarly, children being born to nonimmune mothers are at the mercy of any virus at an age when their own body defenses are not fully developed. The implications of these changes will be discussed more fully in the succeeding sections.

II. Effects of Direct Intramuscular Inoculations

When a 10% suspension of normal sterile brain tissue is injected into a muscle it produces an interstitial inflammatory infiltration with lymphocytes and some polymorphs. Rustigian and Pappenheimer (1949), as a preliminary to their work on the local intramuscular effects of various viruses, using a standard 0.2 ml inoculum, studied the evolution of the "lesion" produced by normal brain tissue. They reported an early and often quite intense local reaction with edema and polymorph and mononuclear infiltration. Muscle fibers adjoining the injection site sometimes showed necrosis; but restitution followed, so that little sign of the reaction might persist after a few days. They found, too, considerable variation in the intensity of the reaction provoked by individual normal brain preparations. This obviously makes more difficult still the assessment of the significance of the relatively mild reactions described after intramuscular inoculation of certain viruses.

If a suspension of brain infected with herpes simplex or rabies virus is injected intramuscularly, general infection results in a wide range of laboratory animals. An inoculum of the order of 0.5–1.0 ml of suspension is commonly employed for larger animals, and 0.2 ml for mice. Experiments with similar volumes of india ink or Weed's Prussian blue solution show that a so-called intramuscular injection spreads widely in the tissue spaces between muscle fiber bundles and in the lymphatics of the fascial planes. It is indeed because of this immediate spread into the septal lymphatics that absorption from a therapeutic intramuscular injection can take place rapidly.

III. Adenoviruses

A. Historical Note

The adenoviruses were originally isolated from naturally degenerating cultures of hypertrophied tonsils and adenoids (Rowe *et al.,* 1953) and from cases of acute respiratory disease in military recruits (Hilleman and Werner, 1954). Since then a large number of antigenic types have been identified, some of which cause a range of respiratory diseases, mainly in children (T. M. Bell, 1965a). Members of the group have also been recovered from other animal species, although the individual types appear to infect only one species. All the types that infect mammals share a common soluble antigen detected in the complement fixation test. A rise in antibody to this group antigen denotes a concurrent infection, while high levels of antibody indicate that there has been a recent adenovirus infection. Two other important characteristics of the group are their tendency to remain latent for long periods following primary infection and the ability of some types to induce the formation of tumors in newborn hamsters and mice.

B. Natural Infections of Man

A small number of cases of acute myocarditis in infants less than 2 years of age have been associated with adenovirus type 7 (Sohier *et al.*, 1965). The first case, recorded by Drouhet (1957), was an infant who died of pneumonia. The initial symptoms were pharyngitis and severe diarrhea, followed by respiratory distress and irregular heart rhythm. Adenovirus type 7 was recovered post mortem from the lungs and feces. Five other infants infected with adenovirus type 7 also showed electrocardiographic changes indicative of myocardial damage (Chany *et al.*, 1958; Van Zaane and Van der Veen, 1962). In these five cases, the cardiopathies disappeared after some time and the patients made a full recovery.

Three cases of pleurodynia have been associated with adenovirus infection, two due to type 2 and one unidentified (T. M. Bell, 1965b). One patient, also suffering from pneumonia, had a rise in antibody to coxsackievirus B2, which is a common cause of pleurodynia. However, in the absence of any evidence of coxsackievirus infection, the pleurodynia in the other two cases may have been due to the adenovirus infection.

C. Experimental Infections

Intraperitoneal inoculation of 7-day-old white mice with 3×10^3 PFU (plaque forming units) of mouse adenovirus resulted in a fatal illness involving the heart, renal cortex, liver, and adrenals (Blailock *et al.*, 1967, 1968). Animals were killed at intervals up to 21 days post inoculation for virus titrations, electron microscopy, and histological examination. Virus was first detected in the heart, kidney, and blood on day 6, with the greatest quantity in the heart (2.3×10^4 PFU/gm). The highest titers in the heart (greater than 4.0×10^5 PFU/gm) occurred on days 9 and 11, in the blood (1.7×10^4 PFU/gm) on day 9, and in the kidneys (greater than 4.0×10^4 PFU/gm) on day 9. The titers in the heart and blood then fell, and virus was not detectable after day 13.

No gross changes were seen at autopsy until day 7. Thereafter scattered off-white areas were observed in the myocardium and on the pericardial and epicardial surfaces. By light microscopy, focal and confluent areas of necrosis and fragmentation of muscle fibers were found beginning on day 6. Most lesions contained intranuclear inclusion bodies in the myocytes. At first, the cellular infiltrate comprised small numbers of polymorphs, but in the later specimens most infiltrates consisted of macrophages, lymphocytes, and plasma cells. In addition to the myocarditis, lesions of the valves were also present. As the myocardial lesion progressed, slight fibrosis appeared, but the most striking feature of the healing lesions was a marked dystrophic calcification in the necrotic areas. This reaction was most prominent in the material from mice 17 days post inoculation. All the animals had either been previously killed or died by day 21. Intranuclear inclusions were noted in large numbers from day 6 to day 12, and a few were still present on day 15. The inclusions resembled those described in infected tissue culture preparations (Ginsberg and Dingle, 1965), being initially small and eosinophilic, but becoming larger and exhibiting varying degrees of basophilia.

By electron microscopy, developing adenovirions were seen in the nuclei of myocytes, fibrocytes, and endothelial cells lining the capillaries, the ventricular endothelium, the heart valves, and the ascending aorta. A variety of nuclear changes was observed. Some nuclei were enlarged and contained a rounded mass of filamentous material, the margins of which were occasionally associated with small numbers of adenovirions. A number of cells showed little or no reaction to the presence of large numbers of intranuclear virions. In the infected myocytes, there was frequently dilatation of the sarcotubular system and swelling of the mitochondria. Severe damage resulted from rupture of the nuclear membrane and spillage of the virions into the cytoplasm. Disruption of the cytoplasmic membrane was also seen and extracellular virus was only found in the immediate vicinity of cells undergoing lysis.

IV. Herpesviruses

A. Historical Note

The herpesvirus group was originally defined to include herpes simplex, varicella, pseudorabies, herpes B, and virus III of rabbits (Andrewes, 1954). Since then the group has been enlarged by the addition of a large number of strains isolated from other mammals and birds (Andrewes and Pereira, 1967). Among the newly recognized members are two that have been associated with sporadic cases of myocarditis. These are cytomegalovirus (Weller *et al.*, 1960) and the EB virus, which appears to be the cause of a large proportion of the cases of classic infectious mononucleosis (Henle *et al.*, 1968).

Herpes simplex virus does not infect muscle cells even following direct inoculation (Field, 1952; Plummer and Hackett, 1966). If the masseter muscle of a rabbit so inoculated is examined when encephalitis has become established, the muscle appears normal or perhaps a little swollen and rather gelatinous. Histologically, many muscle fibers lose their striations and take on a uniform eosin staining. Scattered columns of lymphocytes are present between the fibers, but polymorphs are absent. Accumulations of lymphocytes may be found around small nerve bundles, the fibers of which, however, remain unharmed. No herpetic inclusion bodies are present, and as a whole, the changes are nonspecific and resemble those found when a suspension of normal brain is inoculated in a similar way; i.e., there is no specific "take" in the muscle fibers (Field, 1952).

B. Natural Infections of Man

Myocardial involvement has been reported in cases of infectious mononucleosis (Gore and Saphir, 1947) and varicella (Hackel, 1953), as well as neonatal cytomegalic inclusion disease (Seifert, 1965). Six of nine persons who came to autopsy with the presumptive diagnosis of infectious mononucleosis showed evidence of a myocarditis (Gore and Saphir, 1947). Hackel (1953) reported the presence of focal interstitial inflammatory lesions in the hearts of seven patients who were suffering from varicella at the time of death. Examination of the hearts of healthy persons who died suddenly of traumatic causes revealed no similar lesions. Hackel suggested from these findings that myocardial lesions might occur without clinical manifestations in nonfatal cases of chicken pox.

Cytomegalic inclusion disease, while normally affecting infants in the neonatal period, is also occasionally seen in older children. Seifert (1965) describes five cases in children aged 3 months or less and one in a 10-year-old boy, all of whom presented with myocardial lesions in addition to interstitial pneumonia. A typical case occurred in a male infant who became ill during the third month of life with violent vomiting. This was followed by diarrhea, anemia, and dyspnea on feeding. The child died on the fourteenth day of illness. At autopsy, the heart appeared normal apart from being moderately dilated and of poor tone. Microscopically, there were widespread changes in both ventricles, comprising patchily distributed areas of fibrous replacement of focal degeneration, giant cell systems, and inflammatory interstitial infiltrates. In most areas of softening, three layers were seen. At the center there were the remains of sarcolemmal cylinders, young fibrous tissue, and isolated capillaries with swollen endothelium. In the surrounding zone were heart muscle fibers in various stages of myolysis; the fibers were swollen with blurred edges and showed a lumpy disintegration of the cytoplasm with loss of striations. Some of the nuclei of these muscle fibers were pyknotic and clumped together, others were clearly bigger than usual and appeared as in typical giant cells. The outermost zone of the focus contained interstitial infiltrates of lymphocytes, histiocytes, and solitary plasma cells. This zone was delimited by normal myocardium with fully striated fibers. In the remaining parts of the heart muscle some smaller vessels were surrounded by scattered lymphocytes, and there were isolated sections of capillary with swollen endothelium containing nuclear inclusion bodies. Giant cells and inclusion bodies were also found in sections of lung, spleen, adrenals, and kidneys. Typical cytomegalovirus particles were demonstrated in lung and liver cells by electron microscopic examination.

Cytomegaloviruses have been isolated from cultures of kidneys taken at autopsy from two very young children with congenital heart defects (Benyesh-Melnick *et al.*, 1964). It was considered that the cytomegalovirus infection was also congenital and might have been related to the heart defects.

C. Experimental Infections

A recently recognized canine herpesvirus has been found to produce a severe and usually fatal illness when inoculated into newborn (24–36-hr-old) puppies (Carmichael *et al.*, 1965). The pathological picture is one of disseminated focal necrosis and hemorrhage. Usually, the heart is moderately dilated and the valves are swollen. Frequently, there are petechiae on the valves and inner heart surfaces. Interstitial focal necrosis has been observed in the atria, where there are areas of interstitial edema, myolysis, and many Anitschkow's myocytes.

During the routine passage of virus III in rabbits by the intratesticular route, it was observed that most animals developed a severe myocarditis (Pearce, 1960). Prior to inoculation, all the animals were bled by cardiac puncture, and it was considered that this trauma was the reason for the virus settling in the heart. Experiments showed that virus III was capable of causing pericarditis, endocarditis, and myocarditis if the heart of the infected animal was subjected to stress such as intravenous inoculation of 50 ml of a 20% solution of gum acacia in normal saline. Only minimal cardiac lesions were observed in control animals infected with virus III but not receiving any other trauma.

V. Poxviruses

A. Historical Note

Variola, the most important poxvirus, has been the cause of epidemics of smallpox for at least 3000 years (Downie, 1965). Early attempts to control the disease involved the inoculation of susceptible individuals with vesicular fluid from mild cases. This variolation, as it was called, produced a modified form of smallpox with a greatly reduced mortality rate. In 1798, Jenner reported that vaccination with the virus of cowpox gave as good protection as variolation, and from this observation developed the first standard use of a live virus vaccine. At present, an artificially derived virus, vaccina, is used for vaccination. It has been claimed that vaccinia virus originated from smallpox, but many people consider that a derivation from cowpox is more probable (Andrewes and Pereira, 1967).

B. Natural Infections of Man

Death during smallpox is a summation of the toxic effects of the viral infection, and heart failure is a common feature in patients who die during the first week of illness (Downie, 1965). Gore and Saphir (1947) recorded typical myocarditis in one of nine cases of smallpox who came to autopsy. The most conclusive evidence of the importance of myocardial involvement in smallpox is found in the report of an outbreak involving unvaccinated nurses (Anderson et al., 1951).

During an outbreak of smallpox in Glasgow in 1950, five nurses who were unvaccinated died. Three had typical confluent smallpox, one had primary hemorrhagic smallpox, and the fifth had a normal vesicular rash until the third day when it became hemorrhagic. A striking feature of all five cases was myocarditis. The patients were all young, of good physique, gave no history of antecedant cardiac disease, and had little if any of the usual secondary fever from septic absorption (presumably the results of antibiotic therapy). Nevertheless, all five developed myocardial involvement, manifested by tachycardia, toneless apical sounds, and irregular heart rhythm. Restlessness was noticeable in all these cases, but the patients, being nurses, were disinclined to ask for help—for example, in reaching for a sputum cup. Exertions of this kind may well have accentuated their cardiac damage. They retained their lucidity of mind for a considerable time, but often did not follow instructions to rest, and sedatives did not always ensure the desired relaxation. All five patients died with acute cardiac failure.

More recent reports of acute cardiac disease 5–14 days following smallpox vaccination have suggested that cardiological complications of vaccination should be borne in mind (MacAdam and Whitaker, 1962). A typical case occurred in a 39-year-old man who died suddenly 11 days following primary vaccination after complaining of chest pain (Finlay-Jones, 1964). At autopsy, a severe local reaction was found together with diffuse myocarditis. There was a mixed infiltrate of mononuclear cells, lymphocytes, eosinophiles, and neutrophiles, mainly in the edematous intermuscular septums and extending among degenerating muscle fibers. There was also vascular congestion and edema of the lungs, but no evidence of encephalitis.

C. Experimental Infections

Inoculation of the dorsal fat pads of 5–7-day-old mice with 10^5 PFU of vaccinia virus resulted in a severe and often fatal illness. (Rabin *et al.*, 1965). Approximately 40% of the animals died between the fourth and seventh days post inoculation, and 30% had variable histological evidence of myocarditis. Gross lesions were apparent by the fifth day as gray-yellow streaks over the ventricular surface of the heart. Histologically, the myocardial fibers initially appeared fragmented, with the loss of striations and pyknosis or disappearance of the nuclei. There were also small numbers of inflammatory cells. More extensive foci of myocardial damage were characterized by dense infiltrates of mononuclear cells and polymorphs. Calcification of the myocardial fibers and proliferation of fibroblasts were observed in the extensive lesions. At the site of the subcutaneous inoculation, a lesion sometimes developed that involved the underlying skeletal muscle. Large amounts of virus were present in the heart, titers of 10^6 PFU/gm being found on post inoculation days 4 and 6. At the same times, the titers in the blood were 10^3 and $10^{2.2}$, respectively.

In the electron microscope, it was seen that particles were present in large numbers in areas of necrosis, but were scanty or absent in sections with little or no myocardial damage. Round, immature particles were abundant in the cytoplasm of myocytes, where they often appeared in dense clusters between myofibrils and in close proximity to mitochondria. Evidence of cellular damage was almost always seen in myocytes containing clusters of virus particles. Early pathological changes consisted of dilatation of the sarcoplasmic reticulum, while more extensive changes involved mitochondria, formation of numerous dense osmiophilic bodies, and necrosis of the cell. Developing and mature virus particles were occasionally observed in cells other than myocytes.

VI. Picornaviruses

A. *Historical Note*

The picornavirus group was proposed by the International Subcommittee on Virus Nomenclature (1963) to incorporate all viruses with biochemical and biophysical properties similar to those of the enteroviruses. Included within this extensive group were the polioviruses, coxsackieviruses A and B, echoviruses, rhinoviruses, encephalomyocarditis (EMC) viruses, foot-and-mouth disease viruses, and a collection of less well-defined viruses of lower animals. The name was constructed from "pico" meaning small and RNA, which is their nucleic acid.

Those members of the group that affect man are usually subdivided into the rhinoviruses (common cold viruses) and the enteroviruses, and it is among the latter subgroup that we find the majority of the viruses that naturally invade muscle tissue. The further subdivision of the enteroviruses into polio-, coxsackie-, and echoviruses is historical rather than natural, as there is considerable overlapping of properties. However, the principal distinguishing features of the subdivisions should be noted, with the proviso that they are not absolute and that final identification of an enterovirus can only be made by serological tests. Polioviruses alone paralyse monkeys, grow well in tissue cultures of human and simian origin, but have no effect on mice; group A coxsackieviruses produce a generalized myositis giving flaccid paralysis in mice, but do not grow in tissue cultures; group B coxsackieviruses produce lesions in the central nervous system giving spastic paralysis in mice and grow well in tissue cultures; while the echoviruses only grow in tissue cultures.

B. *Polioviruses*

When polioviruses were regarded as exclusively neurotropic, little attention was devoted to possible early changes in muscles that might be of importance for an understanding of the disease process. In more recent years, muscular changes have been sought and recorded.

Chor (1933), Horányi-Hechst (1935), Sanz-Ibanez (1945), and especially Carey and his co-workers (1944; Carey, 1943, 1944) claim to have demonstrated early changes in muscle motor end plates of humans and animals infected with poliovirus. In the monkey, Carey (1943) found some end plates to be retracted into ball-like masses as early as the first day of paralysis and to stain deeply with gold chloride. He found many to be small and dense, while others were large and granular. He estimated that some 20% of end plates were absent on the first day, but this must be accepted with reserve because of the well known capriciousness of the gold chloride method, in which many of the factors determining the success of impregnation are as yet undetermined. However, within 2–4 days Carey reported 50% of the end plates to have disappeared and that this was followed by centripetal changes in the motor nerve fibers. This sequence is different from that which occurs when a nerve fiber is sectioned, for then changes proceed distally. Carey also found the muscle fibers themselves to show lesions beginning in the vicinity of the sole plates and spreading into the adjacent sarcoplasm. Transient inclusion masses were found in the neighborhood of the end plates, but only in the early stages of the infection. It must be admitted, however, that the assessment of such masses in metallic-impregnated material is difficult. There were also lymphocytic infiltrations in the muscles. Other workers have not found such alterations but have drawn attention to the occurrence of sporadic sarcosporidiosis (Hurst, 1929). While there can be no doubt that the main attack of poliovirus is on the anterior horn cell (Bodian, 1948, 1949), it is possible that disturbance here may manifest itself very rapidly (within a matter of a few hours or even minutes) at the highly labile synapse of motor end plate before changes along the course of the intervening nerve fiber appear.

Hassin (1943) described histological changes in the intercostal muscles, diaphragm, and pectorales major and minor of a boy of 12 who succumbed to poliomyelitis. Varying degrees of homogenization, waxy degeneration, loss of striation, and disruption of muscle fibers were found, together with some infiltration of the myocardium, but the changes do not seem very convincing. Nevertheless, there is some evidence that muscle may be affected early and apparently directly in poliovirus infections. Caughey and Malcolm (1950), for instance, found that electromyographic studies failed to reveal motor potentials in muscles that were in spasm; curare failed to relieve spasm, which could, moreover, be seen in muscles devoid of voluntary power. They therefore suggested that it originated, to some extent at least, within the muscle itself. Guyton and Reeder (1950) performed anterior rhizotomy on dogs and observed the development of muscle tenderness to pressure and stretch such as occurs in poliomyelitis, suggesting that these features might arise from local pathological changes in the muscle, possibly following the accumulation of metabolic products. Successful attempts to isolate poliovirus from muscle have been relatively few in number; Jungeblut and Steevens (1950) succeeded in one of thirteen cases in isolating the virus from puncture biopsy specimens of muscle.

Myocardial involvement has also been reported in poliomyelitis. Saphir and Wile (1942) described perivascular foci of lymphocytes, occasional monocytes, and adventitial cells in the heart muscle of fatal cases of poliomyelitis in which myocarditis had not been diagnosed clinically. They point out that myocarditis might, therefore, be suspected when circulatory failure occurs unexpectedly in the course of poliomyelitis infection. Gore and Saphir (1947), in their survey, found myocarditis to be present at autopsy in thirteen of ninety-four fatal cases of poliomyelitis. Ludden and Edwards (1949), from a survey of the literature, found lesions in fourteen of thirty-five hearts of fatal cases of polio and suggested that myocarditis might be suspected in every patient who is seriously ill with acute poliomyelitis. In general, the lesions have no specific character and resemble those of Fiedler's myocarditis (Fiedler, 1900) discussed in Section X. Dogopol and Cragan (1948) point out that if many slides are examined, evidences of myocarditis are more commonly found in fatal cases of polio. For example, Spain et al. (1950) found twelve out of fourteen hearts to show changes when several blocks were examined from each specimen. Jurrow and Dogopol (1952) in a later series of cases found thirteen of twenty-eight hearts to be affected. An important finding was that made by Jungeblut and Edwards (1951), who demonstrated the presence of poliovirus in three of five hearts from fatal cases. This suggests that the presence of poliovirus in heart muscle is not rare, and it seems that it may or may not lead to demonstrable local damage.

A mild to moderate interstitial inflammation of the myocardium is therefore probably common in poliomyelitis and does not as a rule produce serious changes in the actual muscle fibers, although electrocardiographic (ECG) changes may be detectable [12% of cases in Bradford and Anderson's (1950) series]. Woodward et al. (1960) described a case of type 1 polio in a 19-month-old child who was hospitalized on the fourth day of illness with vomiting and tachycardia. Cardiomegaly, pulmonary congestion, and hepatomegaly were also present. After 3 weeks, the heart was smaller, and it had returned to normal after 6 weeks. Electrocardiography showed signs of nonspecific myocardial changes.

It should be noted that as the widespread vaccination against poliovirus during the past decade has virtually eliminated poliomyelitis from large areas of the world, much of the above is now purely of academic interest.

C. Group A Coxsackieviruses

1. GENERAL PROPERTIES

The group A coxsackieviruses were first isolated by Dalldorf and Sickles (1948), and to date, twenty-four distinct antigenic types with similar properties have been identified. Of these, type 23 is antigenically identical to the previously recognized echovirus type 9, the prototype strain of which does not infect newborn mice, while some strains of type 7 produce a paralytic disease of man and monkeys similarly to the polioviruses (Dalldorf and Melnick, 1965). Although the strains of coxsackievirus type A23 are now officially classified as echovirus type 9, they should be considered in this section, as no other echovirus has been definitely associated with infection of muscle.

This subgroup has received little attention in comparison with the other picornaviruses due to the relative difficulty involved in its primary isolation in newborn mice. However, it has been shown to cause outbreaks of herpangina, aseptic meningitis, undifferentiated febrile illness, and common cold, in addition to sporadic cases of more serious illnesses (Dalldorf and Melnick, 1965). It may also be found in the feces of healthy children without any associated disease. Since the virtual elimination of poliomyelitis, more attention is now being focussed on the sporadic cases of severe, and sometimes fatal, diseases caused by this group of viruses.

2. NATURAL INFECTIONS OF MAN

Although most strains of A coxsackievirus give rise to a rapidly fatal skeletal myositis when inoculated into newborn mice, infections of man appear only to affect heart muscle. Over the last decade, evidence has been accumulating that indicates that several strains can produce a severe and often fatal myocarditis in patients of all ages (Wright *et al.*, 1963; Cherry *et al.*, 1967; Monif *et al.*, 1967; E. J. Bell and Grist, 1968; Grist and Bell, 1968, 1969). The most comprehensive series has been collected by Grist and Bell (1969) and amounts to a total of seventeen cases due to types A1, A4, A9, A16, or A23 and ranging in age from less than 1 year to 74 years. The majority of the patients were, however, aged 1 year or less, and only eight survived, some taking as long as 5 months to recover. The degree of myocardial damage observed at autopsy varied from minimal to extensive.

Wright *et al.* (1963) described the case of a 7-week-old infant who died 4 days after the onset of the clinical manifestations of irritability, respiratory distress, tachycardia, cyanosis, and convulsions. The gross pathological findings were suggestive of encephalitis and myocarditis, but microscopical examination of the tissues revealed only enteritis, lymphohistiocytic arachnoiditis, and minimal interstitial myocarditis. Coxsackievirus type A16 was, however, recovered from the heart muscle in a greater concentration than from blood or bowel contents, but was absent in the brain.

At the other extreme is the case described by Monif *et al.* (1967) in which a 34-year-old man developed mild exertional dyspnea. Three days later, he had fever, malaise, nausea, and vomiting, and on the following day, he was treated for what was assumed to be gastroenteritis. On the sixth day of illness, he was hospitalized with complete heart block and marked respiratory distress. On examination there was peripheral cyanosis, a temperature of 104°F, respiratory rales one-quarter of the way up both posterior hemithoraxes. ECG showed an idioventricular rhythm at a rate of approximately 40, with multifocal ventricular beats and complete atrioventricular dissociation. Ventricular fibrillation developed that was resistant to conversion, and the patient died 24 hr after admission.

At autopsy, there was marked congestion of the lungs, liver, and spleen, with massive centrilobular necrosis of the liver. There was 60 ml of serous fluid in the pericardium and marked right-sided dilatation of the heart. The epicardial surface revealed focal discrete mottling of the underlying myocardium. Throughout the entire myocardium, and involving the papillary muscles, numerous discrete, slightly raised, ovoid grayish white plaques, often rimmed with hemorrhage, were observed. The largest lesions were soft and necrotic, while the adjacent myocardium was firm, but pale. Microscopically, there was extensive focal and diffuse inflammatory cell infiltrate, edema and congestion of the interstitium, and massive focal necrosis of cardiac muscle fibers. The infiltrate was composed of lymphocytes, plasma cells, neutrophiles, eosinophiles, histiocytic cells with ovoid nuclei, and occasional prominent clumps of chromatin. The numbers of neutrophiles, eosinophiles, and plasma cells varied from site to site, depending on the amount of myocardial necrosis present. Coxsackievirus type A23 (echovirus type 9) was isolated from the heart, but not from the spleen, liver, lung, muscles, or intestines.

3. EXPERIMENTAL INFECTION OF MICE

By definition, all group A coxsackieviruses produce a diffuse myositis when inoculated parenterally into baby mice, while some strains also affect the myocardium in older animals (Lerner, 1965). Animals infected within 48 hr of birth become paralyzed after an incubation period of 2–6 days, and death follows rapidly. If the inoculation is postponed until the animals are a few days older a proportion will survive and recover from the flaccid paralysis. The pathology of the disease in baby mice is characterized by lesions that affect the entire musculature, while sparing all other organs (Dalldorf and Melnick, 1965). Histopathological studies by Armstrong *et al.* (1950), Gifford and Dalldorf (1951), Melnick and Godman (1951), Godman *et al.* (1952a,b), and Aumonier (1953) have built up a comprehensive picture of the disease at the cellular level.

A. PATHOLOGY OF BABY MICE. The first stage is a degeneration of the muscle fibers, preeminently segmental; i.e., abruptly demarcated regions of obvious change may occur in the course of an otherwise normal muscle fiber. Accentuation of A disks is an early and transient or inconstant feature and is soon accompanied by proliferation of sarcolemmal nuclei. The Z disks appear unaltered at first, but later, they, too, disappear when myofibrillar thickening takes place. The longitudinally running myofibrils become accentuated and then obviously thickened, passing off on either side into still normal-appearing muscle. The sarcoplasm in which the myofibrils are embedded becomes strongly and uniformly eosinophilic, though at first it may present areas of mottled basophilia. Exactly the same sort of changes may be observed in the neighborhood of a septic focus in muscle and the changes do not seem to be specific. Affected segments may be folded or fractured in appearance, perhaps due to an increased rigidity as compared with adjacent unaffected and still contractile fibers. Karyolysis and karyorhexis take place and the affected segment becomes converted into a waxlike, homogenous mass. Regeneration changes appear very early and this is an outstanding feature of the pathology in these newborn mice. The course of the degenerative changes in the muscle fibers is also very rapid, and many hyalinized segments are to be found on the first day of clinical signs of paralysis.

Inflammation is the natural sequel to degeneration and follows its usual course with edema and cellular outpouring (histiocytes, polymorphs, and lymphocytes), reaching its height in about 24 hours, at which stage mononuclear cells predominate. Individual virus strains vary in the degree of inflammation they evoke. Mononuclear scavenger cells remove degenerated muscle protoplasm, leaving sarcolemmal tubes, which are the directing scaffolding into which regenerating muscular buds can later grow. The sarcolemmal tubes are seen to be filled with many nuclei derived from proliferated sarcolemmal cells, histiocytes, and beginning ingrowing muscle sprouts. The sarcolemmal tubes found in coxsackievirus infections are similar to the *Sarcolemmschlauche* described by Waldeyer (1865) in nonvirus destruction and also to the reconstitution of muscle discussed in Chapter 5. Phagocytes, from whatever source they are derived, laden with detritus, emerge from the tubes, and a reaction is found in the local lymph nodes. Preparations for regeneration begin very early, even before inflammation has had a chance to clear away the necrotic material.

Regeneration takes place to an extraordinary degree, despite the persistence of a high virus titer in the affected muscle (Melnick and Godman, 1951), and originates possibly in part from some of the proliferated and surviving sarcolemmal nuclei within the tubes, but mainly from the adjacent ends of surviving muscle segments. It seems that regeneration is therefore chiefly of the continuous type, rather than of the discontinuous, or embryonal, type discussed in Chapter 5.

At the ends of surviving muscle fibers bordering on the necrotic segments, there is a well marked proliferation of nuclei, which become elongated and tend to arrange themselves in line, acquiring as they do so a clear and deeply staining mantle of cytoplasm. In this way, a syncytial protoplasmic strand is formed which grows into the sarcolemmal tube, now mostly cleared of debris. Often, the muscular outgrowth takes place at the periphery of the healthy muscle fiber, but sometimes the healthy segment tapers off into a terminal syncytial outgrowth. The muscle plasmodial slips that grow in often bifurcate. Nuclear division seems to be by amitosis, while division of the sarcolemmal nuclei is mitotic. The sarcolemmal tubes clearly exert a guiding influence on the ingrowing muscle sprouts in much the same way as neurilemmal tubes do upon axis cylinder sprouts during nerve regeneration. There may be much cellularity at the height of the regeneration process, so that the microscopic appearance may come to resemble a myosarcoma. It seems that many of the muscle sprouts fail to find sarcolemmal tubes that can be occupied, and so come to nothing.

Restitution of damaged muscle fibers is for the most part complete about a week after the onset of signs. The muscle sprouts are at first thin and strongly basophilic, with centrally placed nuclei in line. In general, maturation changes follow the course occurring in normal ontogeny. Thus, longitudinal striae appear first; usually by the second day after the onset of signs they are definite, i.e., even before marked ingrowth into the sarcolemmal tubes has had time to take place. At the same time, the sarcoplasm increases in amount and becomes eosinophilic as myohemoglobin begins to accumulate within it. The longitudinal striations increase in number and show along their course accentuations that are the beginnings of A disks. The nuclei, which are still centrally placed, begin to space out from one another, and about the fifth day, Z disks and sarcolemmas begin to appear, so that by the end of the week, the muscle fibers look like well developed young structures. Only their slimness and the central location of their nuclei indicate their newness. Remnants of destroyed muscle tissue may still be scattered about. The process of striated muscle fiber degeneration and regeneration consequent upon coxsackievirus infection is essentially the same as that which follows fiber degeneration from other causes, e.g., direct intramuscular injection of toxic chemicals (Forbus, 1926), ischemia (Le Gros Clark, 1946). The time scale on which these changes occur is a rapid one, partly no doubt because they are taking place here in neonatal animals. The temporal sequence of changes in older animals is summarized in Chapter 5.

It has been pointed out above that a striking feature of the muscle changes is the discontinuous nature of the initial lesions, affected segments being found along the course of apparently normal fibers. Some histochemical observations of the changes brought about in muscle by coxsackievirus infection have been made, but they are largely unconnected, and a coordinated picture must wait upon a fuller development and interpretation of techniques. It has, for example, been recorded that birefringence is increased in the early stages of infection, but that no definite A disks can be seen; that basophilia is increased in the hylinized segments, but that this is apparently not associated with either DNA or RNA; that phosphate and phosphatase are variable; that lipids may appear as tiny granules; that with increasing length of infection there is a potassium and creatinine depression (Gifford and Dalldorf, 1949) and sodium elevation (Gädeke and Walenberger, 1952); and that glycogen content is diminished (Albrecht, 1954). On the other hand, it has been claimed that alkaline and acid phosphatase activity are unaltered (Kausche *et al.*, 1951) and that also unaltered are both inorganic and total phosphorus levels (Albrecht and Sauthoff, 1954). Ferric iron has been found in affected muscles (Godman *et al.*, 1952b). While these authors and also Pette (1952) have claimed to demonstrate calcium in the lesions, their results have not been substantiated (Sauthoff, 1956). Albrecht and Gädeke (1956) have claimed an increase in inorganic phosphorus in paralyzed muscle, while the acid-soluble phosphorus and phospholipids remain unaltered, and suggest that the change is associated with interference with the formation of the phosphates so important as a source of energy in muscular contraction. The change is not found in the first 24 hr of infection. The biochemical changes in muscle may be so severe as to lead to kidney lesions in infected suckling mice resembling those seen in "crush kidney" when striated muscle has been extensively damaged (Gädeke, 1952).

B. ELECTRON MICROSCOPY. A recent study by Bienz-Isler and her colleagues (1970; Bienz *et al.*, 1970) has detailed the events that lead to the eventual disintegration of muscle cells infected with coxsackievirus type A1. For convenience, the authors arbitrarily divide the sequence into five stage (Bienz-Isler *et al.*, 1970):

1. The first stage, which involves only the nucleus, is a general characteristic of coxsackievirus infections. The peripheral chromatin condenses into electron-dense masses with some small electron-lucent areas, which sometimes contain granular inclusions. The perinuclear space also begins to enlarge.

2. These nuclear changes become more pronounced during the second stage, and the nuclei assume a lobed shape. The light areas in the dark chromatin masses enlarge and develop granular or fibrillar inclusions. Simultaneously, cytoplasmic degeneration also becomes apparent. The A and I filaments no longer run strictly parallel, while the H and M bands disappear, although the I band and Z disk are not yet affected. The endoplasmic reticulum becomes vacuolated, most prominently in the peripheral regions of the fiber.

3. By the third stage, cellular destruction is obvious. The contractile material is fully disorganized, although filaments are still recognizable together with fragments of Z disks. Individual fibers differ in the number of calcium inclusions and the degree of mitochondrial degeneration present.

4. In the fourth stage, the nuclei present deep invaginations of the nuclear membrane, the outer layer of which shows marked redundancy. This leads to the formation of large vacuoles (or blebs) and channel systems, which penetrate deeply into the cytoplasm or even to the sarcolemma.

5. The nucleus disintegrates completely in the fifth stage, and its contents leak into the cytoplasm. The mitochondria are swollen and devoid of cristae, while phagocytes invade the damaged muscle fibers.

Only the intranuclear granular and fibrillar inclusions and the vacuolation of the endoplasmic reticulum are specific viral changes. The authors (Bienz *et al.*, 1970) suggest that viral RNA could be synthesized in the nucleus, with the granular inclusions representing viral precursor material. Immunofluorescent studies (Bienz *et al.*, 1969) indicate that viral coat protein is synthesized in the cytoplasm. Complete virions are transported from nuclear pores to the cell surface, via the channel system, and are continuously released from the cell surface, thereby spreading infection to neighboring cells.

C. Myocarditis in Adult Mice. Lerner and his colleagues has investigated the effects of intraperitoneal inoculation of several strains of coxsackievirus A9 into mice ranging in age from 1 day to 1 year (Lerner and Shaka, 1962; Lerner *et al.*, 1962; Lerner, 1965; Wilson *et al.*, 1965, 1969). In newborn mice, lesions are only found in skeletal muscle (Lerner and Shaka, 1962) and are identical with those described previously, whereas virus is only rarely found in the heart, which always appears normal. As the age of the mouse at inoculation increases, the severity of skeletal myositis decreases, and a greater number of animals survives. At the same time, virus can be more frequently recovered from the heart, and scattered foci of myocarditis may become demonstrable (Lerner *et al.*, 1962; Wilson *et al.*, 1969). In mice aged 1 year, skeletal muscle is unaffected, but many strains of coxsackievirus A9 multiply in heart muscle, sometimes giving rise to a subclinical focal myocarditis (Lerner, 1965; Wilson *et al.*, 1965). This reversal of susceptibility of cardiac and striated muscle from birth to middle age is predominantly due to cellular alteration with maturation of the mouse (Lerner, 1965).

The sequence of events observed microscopically was as follows (Lerner, 1965): (1) three days after inoculation the heart was normal; (2) scattered foci of myocardial involvement, not grossly visible, appeared by the sixth day; (3) the myocardial fibers were ruptured and fragmented; (4) there was slight hemorrhage and an infiltrate composed of polymorphs, lymphocytes, histiocytes, plasma cells, and occasional large mononuclear cells; (5) on the ninth day there were small necrotic foci of myocardial fibers that showed loss of sarcoplasm and infiltration by inflammatory cells, mainly histiocytes; (6) by day 16, only occasional small areas of patchy interstitial myocardial fibrosis were observed, and no histological traces of infection could be found after 106 days.

Although the coxsackievirus A9 infections in adult mice were subclinical, they produced an increase in heart weight, which was further increased by exercise. Similarly, the proportion of mice in which viral replication occurred (and its degree) was also increased by exercise (Lerner, 1965). Concurrent inoculation of hydrocortisone produced the same effects as exercise.

D. Group B Coxsackieviruses

1. GENERAL PROPERTIES

The group B coxsackieviruses are readily separated from the group A strains on the basis of the disease pattern produced in baby mice. Only six distinct antigenic types have been recognized and most strains will grow well in primate tissue cultures. Large outbreaks of such diseases as Bornholm disease, aseptic meningitis, and undifferentiated febrile illness, together with small outbreaks and sporadic cases of pericarditis, myocarditis, and severe systemic illness of infants have been associated with all six types (Dalldorf and Melnick, 1965).

Because of the extent of the morbidity produced by these diseases and the ease of isolation and identification of the group B coxsackieviruses, they have been extensively studied in many parts of the world. Although most of the illnesses are self limiting and complete recoveries are made, the cardiac disease produced by these viruses is becoming increasingly significant.

2. NATURAL INFECTIONS OF MAN

A. BORNHOLM DISEASE (EPIDEMIC MYALGIA, PLEURODYNIA, OR DEVIL'S GRIP). The disease takes its name from the epidemic on the island of Bornholm described by a Danish general practitioner whose family developed the condition while on holiday there in 1930, and his monograph on the subject appeared in 1934 (Sylvest, 1930a,b, 1934). About the same time, Pickles described cases in Yorkshire, England (Warin, 1956). Actually, the disease has been known for much longer, with descriptions from the last century from Norway (Daae, 1872; Homann, 1872) and Iceland (Finsen, 1874), while Windorfer (1963a,b) has pointed out an even earlier Danish report by Hannaeus (1735). Outbreaks have since been recorded throughout the world, and when adequately investigated, have all been associated with group B coxsackievirus infections (Dalldorf and Melnick, 1965).

Clinically, Bornholm disease is characterized by the sudden onset of severe stabbing pain, usually in the chest or abdomen, less commonly in the trunk or head and neck. A moderate pyrexia almost invariably accompanies the pain. Sometimes these symptoms may be preceded by malaise, anorexia, and other vague prodromal symptoms. The pain is most usually located in the thorax, especially in adults, and may be extremely severe. Abdominal pain occurs in approximately half of the cases, more commonly in children, and may be the only presenting symptom. Tenderness may be present in the affected region and appears to be superficial, indicating localization in the muscular wall rather than in the viscera. The disease is usually self limiting, and full recovery is made in 1–2 weeks.

The condition is not fatal, but the findings in muscle biopsy have been reported upon by Lépine *et al.* (1952). In two adult cases, they found the same sort of lesion in muscle fragments removed from the site of maximal affection as occur in newborn mice infected with coxsackievirus. Thus, in one case, there were localized interstitial mononuclear cellular infiltrations with few polymorphs, hyaline degeneration with loss of striation of muscle fibers, followed by necrosis and the clearance of sarcolemmal tubes by phagocytes. Coxsackievirus was demonstrated by passage to suckling mice. A second case showed disappointingly few lesions, but this may well have been due to the chances of biopsy. Welborn (1936), for example, found no pathological changes in a slip of latissimus dorsi muscle removed at biopsy from a case during an epidemic in Cincinnati.

B. MYOPERICARDITIS. Fatal cases of neonatal myocarditis caused by group B coxsackieviruses were first recognized in South Africa (Javett *et al.*, 1956; Gear and Measroch, 1958). Since then, many reports have described similar single cases and small outbreaks in which the infant was infected either *in utero* or in the immediate post natal period (Montgomery *et al.*, 1955; Kibrick and Benirschke, 1956, 1958; Delaney and Fukunaga, 1958; Hosier and Newton, 1958; Suckling and Vogelpoel, 1958; Simenhoff and Uys, 1958; Naudé *et al.*, 1958; Dömök and Molnár, 1960; Moossy and Geer, 1960; Woodward *et al.*, 1960; Fechner *et al.*, 1963; Artenstein *et al.*, 1964; 1965; Jennings, 1966). Reports were also made of sporadic cases of nonfatal myocarditis in older children (Lukacs and Romhanyi, 1960; Babb *et al.*, 1961; Connolly, 1961; Pollen, 1963; Sanyal *et al.*, 1965) and adults (Null and Castle, 1959; Woodward *et al.*, 1960; Glajchen, 1961; Swann, 1961; Artenstein *et al.*, 1964; 1965; Smith, 1966; Helin *et al.*, 1968; Sainani *et al.*, 1968). A much more commonly observed complication of group B coxsackievirus infection of the older patient is nonspecific benign pericarditis (Lewes and Lane, 1961; Dalldorf and Melnick, 1965). Smith (1966) pointed out that although myocarditis could occur as a single disease syndrome, cases presenting as pericarditis always showed evidence of concomitant myocardial involvement. He therefore proposed that the syndrome previously called "acute benign nonspecific pericarditis" should be known as "acute coxsackievirus myopericarditis."

Although myopericarditis may occur in the absence of any other symp toms, it is more usually part of a complex syndrome. This is especially true in the neonate, where the coxsackievirus may also affect the brain, spinal cord, lungs, liver, spleen, pancreas, kidneys, and adrenals (Hosier and Newton, 1958; Kibrick and Benirschke, 1958; Fechner *et al.*, 1963; Artenstein *et al.*, 1964). Virus may be recovered from all these organs as well as from blood, spinal fluid, and feces. Artenstein and his colleagues (1964) measured the virus content of several organs in tissue cultures and suggested that active viral replication takes place in the liver (titer of $10^{5.5}$ TCD_{50}/.gm) and probably also in heart muscle (titer of 5×10^4 TCD_{50}/gm). Dömök and Molnár (1960), who used newborn mice for their titrations, found titers of $10^{5.5}$ LD_{50}/gm in lung; $10^{6.5}$ LD_{50}/gm in central nervous system, liver, spleen, and pancreas; and as high as $10^{9.5}$ LD_{50}/gm in heart muscle. Antigenic types B2 through B5 are most frequently encountered in these cases with an occasional case due to type B1. Type B6 has not so far been recognized as a cause of either Bornholm disease or of myopericarditis.

The disease is most severe in very young infants, who usually present with fever, inflamed throat, respiratory distress, cyanosis, hepatomegaly, and systolic murmur. In a proportion of cases, there is also a rash, jaundice, enlarged heart, and sometimes terminal pulmonary hemorrhage (Delaney and Fukunaga, 1958; Hosier and Newton, 1958; Kibrick and Benirschke, 1958; Suckling and Vogelpoel, 1958; Artenstein *et al.*, 1964; Jennings, 1966). Many of the mothers of these infants had a typical coxsackievirus infection at the time of labor, and the onset of the disease in their children was within a few days of birth. In most outbreaks, the mortality rate nears 100%, although in one report, where central nervous system disease was more obvious than myocarditis, it was only 5% (Dömök and Molnár, 1960). The duration of illness varies from 2 to 11 days, and death is often sudden and unexpected.

Adult cases of myopericarditis can present a wide range of symptoms, including headache, dizziness, pharyngitis, nausea, chills, fever, chest and abdominal pain, backache, generalized myalgias, hepatomegaly, cardiomegaly, tachycardia, and pulmonary congestion (Woodward *et al.*, 1960; Connolly, 1961; Glajchen, 1961; Artenstein *et al.*, 1964). All patients have ECG changes, which, according to Sanyal and his colleagues (1965), are commonly seen as flattening of the T waves and/or S–T segments, indicating myocardial damage. The ECG remains abnormal for quite long periods, but most patients make a full recovery, and only rarely do ECG changes persist after 7 months (Connolly, 1961; Helin *et al.*, 1968). Death is a rare event and may be due to a summation of the viral myocarditis with coronary heart disease (Sanyal *et al.*, 1965; Smith, 1966).

As a result of the high neonatal mortality the pathology of coxsackie-virus myopericarditis has been well documented. At autopsy, the lungs show congestion and hemorrhage (Kibrick and Benirschke, 1958) or are collapsed (Artenstein *et al.*, 1964), while the heart may appear macroscopically normal (Artenstein *et al.*, 1964) or grossly enlarged (Moossy and Geer, 1960; Jennings, 1966). Occasionally, there may be congestion of the abdominal viscera (Fechner *et al.*, 1963) or even gastrointestinal hemorrhage (Kibrick and Benirschke, 1958). The most important microscopic lesion is a patchy necrosis of the myocardium, which generally involves the whole heart, but which varies in extent and intensity at different sites, often being most marked in the ventricular walls and interventricular septum (Kibrick and Benirschke, 1958; Simenhoff and Uys, 1958; Fechner *et al.*, 1963; Artenstein *et al.*, 1964).

Simenhoff and Uys (1958) have documented the histological changes caused by type B3 in great detail. They described four babies who became ill 5–8 days after birth and died at intervals up to 11 days. Two days after onset necrotic muscle fibers were clearly visible, but stained poorly, and in some only the outlines remained. At this stage, significant nuclear changes were noted, mainly consisting of pyknosis and karyorrhexis, but also including karyolysis with release and distribution of chromatin debris. After 6 days, the process of cellular autolysis had advanced. While some recognizable necrotic fibers persisted, many had disintegrated completely, leaving only scattered masses of chromatin. By the ninth day, calcification of dead fibers had occurred. Cellular dissolution was complete by the eleventh day, few recognizable fibers remained, and the necrotic areas were more clearly demarcated from the surrounding healthy muscle. At this stage, the inflammatory reaction was the only indication of what had occurred. Polymorphs were the principal inflammatory cells. They appeared in the earliest lesions, although necrosis existed in their absence. After 6 days, they were present in maximal numbers, many in the elongated form of migrating cells. Even in the earliest lesions (2 days), some histiocytes of Anitschow were seen. The polymorphs persisted, but diminished in numbers, while lymphocytes, plasma cells and histiocytes increased to roughly equal proportions. Eosinophiles appeared late and remained relatively scanty. Fibroblasts were evident on day 9, and by day 11, they dominated the picture. The speed of the repair processes was striking, even in an infant where such processes are normally rapid. Although many of the inflammatory foci extended to the pericardium and endocardium, there was no evidence of fibrin exudation or thrombosis at these sites. Similar changes have been described in the other studies where types B2 and B4 have been the causal agents as well as type B3. In most of these reports, the predominant inflammatory cells were mononuclear (Kibrick and Benirschke, 1958; Moossy and Geer, 1960; Woodward et al., 1960; Fechner et al., 1963; Artenstein et al., 1964), sometimes accompanied by lymphocytes (Hosier and Newton, 1958; Sanyal et al., 1965) and polymorphs (Delaney and Fukunaga, 1958; Jennings, 1966).

3. EXPERIMENTAL INFECTIONS

A. IN MONKEYS. Sun and his colleagues (1967b) inoculated a group of cynomolgous monkeys intravenously with 10^5 TCD_{50} of a strain of coxsackievirus B4 isolated from a fatal encephalomyocarditis in a 10-day-old infant. A chronic valvulitis showing fibrous thickening of the valves with or without verrucous formation was noted in the mitral and aortic valves in eight of eleven animals. Histologically, scattered areas of focal myocarditis were seen in all eleven monkeys. The lesions were in varying stages of development. Early changes consisted of segmented disintegration of muscle fibers associated with infiltration by numerous small round and polymorphonuclear cells. In the more chronic stages, lesions were fairly well circumscribed and exhibited a fusiform nodular pattern composed of giant multinucleated myocytes and mononuclear inflammatory cells. Scattered areas of perivascular inflammation were frequently seen in chronically infected animals. These lesions consisted of disrupted collagen fibers and a few mononuclear cells. Most of the myocytes in the vicinity of the lesions had large, bizarre hyperchromatic nuclei. Immunofluorescent staining usually demonstrated coxsackievirus B4 antigen within the perinuclear cytoplasm of the myofibers. The mononuclear cells also frequently contained fluorescent antigen within their cytoplasm.

Electron microscopic examination of the myocardium demonstrated ultrastructural changes consistent with focal myofiber damage. Mitochondria of the affected myocytes were reduced in number, swollen, and contained disrupted cristae with a decrease in electron density of their matrix. The extent of damage varied. In some myocytes the myofibrils lost their periodic bands, whereas in others, extensive dissolution of myofibrils was seen. The degenerative areas were occupied by loose ground substance, lysosomes, glycogen, randomly dispersed vescicles, and swollen mitochondria. The tubules of the endoplasmic reticulum of the damaged myofibers were swollen and dilated. In addition to the smooth endoplasmic reticulum, several membranes with ribosomal particles on their surfaces could be identified. These membranes were seen most frequently between the myofibrils.

Lipid droplets were noticeable in some regions of the myocardium, usually associated with mitochondria and apparently bound by a single membrane. Several extremely large vescicles bound by double membranes, ranging up to several microns in diameter, were occasionally seen among the degenerating myofibers. Osmiophilic particles were sometimes present inside the vesicles.

B. IN MICE. When baby mice are infected with group B coxsackieviruses, a spastic paralysis results (Dalldorf and Melnick, 1965). The principal lesions are degeneration of the brown fat of the interscapular fat pad and a patchy dissolution of the cerebral cortex, while gross muscle necrosis is uncommon. The lesions in striated muscle are focal and limited, but are similar in histological detail to those produced by the A strains. However, the most important experimental lesion is myocarditis produced only in some strains of mice (Grodums and Dempster, 1959). Types B1 (Gifford and Dalldorf, 1951; Melnick and Godman, 1951), B3 (Grodums and Dempster, 1959; Rabin *et al.*, 1964; Wilson *et al.*, 1969), and B4 (Pappenheimer *et al.*, 1950; Sun *et al.*, 1967a) show the greatest predilection for the myocardium, while B4 also produces valvulitis (Sun *et al.*, 1967a).

Newborn mice were less susceptible to myocardial involvement than 12-day-old animals, and susceptibility increased up to the age of 23 days; thereafter, it declined, but 6-month-old mice were still moderately susceptible (Grodums and Dempster, 1959). Mortality was low in older animals, reaching a maximum of around 3% although this could be increased to 6.5% by treatment with cortisone (Rabin *et al.*, 1964). However, even in untreated mice, myocarditis was present in all animals inoculated between 12 and 56 days of age (Grodums and Demster, 1959). The viremic phase of the illness was complete by the third day and maximal titers of virus in the heart were reached on day 4. Large quantities of infectious virus continued to be present in the heart until day 7, but no virus could be recovered on day 14 or thereafter (Rabin *et al.*, 1964). The heart weight to body weight ratios increased from day 9 through day 27 (Wilson *et al.*, 1969) and myocardial damage was so severe that it was apparent on gross inspection as yellow-white mottling of the ventricular surface of the heart (Rabin *et al.*, 1964). Histological evidence of myocardial damage was usually seen on the third day, but occasionally small foci of necrosis were apparent on day 2. The initial lesions consisted of a few scattered perivascular mononuclear cell infiltrates, usually associated with slightly enlarged segments of myocardial fibers exhibiting increased eosinophilia. Damaged myocardial fibers frequently appeared fragmented and swollen, and cross striations were not discernible. At later stages, infiltration of histiocytes, lymphocytes, plasma cells, and scattered polymorphs was a prominent feature. Marked fibrosis, dystrophic mineralization, continuing inflammation, and microscopic myocardial hypertrophy persisted for at least 6 months (Rabin et al., 1964; Wilson et al., 1969). No antibodies could be detected in mice sacrificed 185 days after infection, although these were detectable earlier (Wilson et al., 1969).

By immunofluorescence, coxsackievirus antigen was demonstrated in myocardial fibers adjacent to and within foci of myocardial damage. In large lesions, most fluorescence was seen around the periphery of the lesion. In electron micrographs, large groups of cytoplasmic particles compatible in appearance with coxsackievirus were seen within myocytes in a distribution similar to viral antigen. The earliest pathological changes observed in myocytes were swelling and disruption of the mitochondria and dilatation of the perinuclear spaces and endoplasmic reticulum (Rabin et al., 1964).

E. Other Picornaviruses

No description of muscle infections due to the picornaviruses would be complete without a short account of those produced by encephalomyocarditis (EMC) viruses. In man, this group causes only aseptic meningitis or a dengue-like illness without any symptoms of myocarditis (Warren, 1965). On the other hand, natural infections of such animals as swine (Murnane et al., 1960) and artificial inoculation of mice (Jungeblut and Steenberg, 1950; Craighead, 1966), mongooses, and monkeys (Jungeblut, 1950; Kilham et al., 1956) usually result in myocardial as well as central nervous system involvement. The relative extent of striated muscle and myocardial damage depends on the state of adaptation of the strain used and the route of inoculation (Jungeblut and Steenberg, 1950; Warren, 1965).

If a small amount (10^2–10^3 LD_{50}) of EMC virus is inoculated directly into the muscle of left hind limbs of mice, acute myositis results. There is a rapid fall in titer followed by a rise, and the titer reaches a higher level than that of the original inoculum 9 hours post infection. Virus only appears in the other hind limb when the level in the blood increases (Dickinson and Griffiths, 1966). By selective passage, it is possible to develop myocardiotropic and encephalotropic variants of EMC virus (Craighead, 1966).

The histology of the myocardial lesions was studied over a 45 day period in 12-week-old mice inoculated intraperitoneally with the myocardiotropic strain. The first signs seen 4 days after inoculation were focal myocytolysis localized to the subendocardial myocardium and the atria. An interstitial infiltrate of mononuclear cells and granulocytes was evident, but was not prominent. Extensive myocardial necrosis was evident by the tenth day, when changes were most prominent in subendocardial tissue, but were often also present in deeper levels of myocardium. Areas of myocardial destruction were usually confluent in animals that died between the fifth and tenth days. Variable numbers of histiocytes, lymphocytes, and unidentified mononuclear cells were present in the lesions, and rarely, focal interstitial collections of these cells were found scattered in the myocardium distant from sites of obvious myocytolysis. A variety of changes was found in the hearts of animals sacrificed late in the course of the experiment. Focal areas were evident in which only the myocardial stroma persisted. These foci contained scattered histiocytes, a few lymphocytes, and cells that resembled distorted myocardial fibers. Fibroblastic activity and capillary proliferation were evident in some sites. In a few animals, spontaneous death was attributed to filling of the ventricular cavities by thrombotic material. All portions of the myocardium were affected, but lesions were usually most extensive in the walls of the left ventricle. This was probably attributable to the relative mass of the ventricular muscle. Necrotic changes were not found in valve leaflets or the smooth muscle of the adjacent major vessels.

VII. Reoviruses

The reoviruses are a small group of ubiquitous viruses that infect a wide range of warm-blooded animals (Stanley, 1967). The prototype was originally classified as echovirus type 10, but differences between this type and the other enteroviruses led to the establishment of the reovirus group (Sabin, 1959). Only three antigenic types are known to infect mammals, and none has been associated with muscle disease in man. However, inoculation of newborn mice results in widespread necrosis of myocardium and skeletal muscle (Walters *et al.*, 1963, 1965; Stanley *et al.*, 1964; Hassan *et al.*, 1965; Bennette *et al.*, 1967). The degree and distribution of the muscle necrosis depends on the route of inoculation and the quantity and type of virus used. Large inocula give an acute disease in which death can sometimes be attributed to cardiac failure (Walters *et al.*, 1965; Bennette *et al.*, 1967). Lesions in skeletal muscle are more severe with type 3 as the infecting agent than with types 1 or 2. When small doses are given a chronic disease results in which muscle necrosis is much less severe and may even be absent (Stanley *et al.*, 1964; Walters *et al.*, 1965).

Oral (Walters *et al.*, 1963, 1965) or subcutaneous (Bennette *et al.*, 1967) infection usually results in more extensive muscle necrosis than intraperitoneal or intracerebral inoculation, although Hassan and his colleagues (1965) described one experiment in which intraperitoneal inoculation of type 1 produced very severe myocardial damage. The lesions are all similar and only vary in number and extent.

Extensive skeletal muscle involvement has been described following oral infection with large doses of type 1 and 3. In mice inoculated with type 1 (Walters *et al.*, 1965), no muscle lesions appeared until 11 days after infection when mild focal subacute interstitial myositis was found in the muscles of the thigh and cervical region. Similar lesions appeared in the tongue on the following day. The lingual glands also underwent necrosis and provoked an inflammatory reaction. The musculature of the scalp showed necrosis and lysis of the fasciculi with sarcolemmal cell proliferation. Foci of subacute panniculitis were observed in the surrounding fat. This became more diffuse 2 weeks after inoculation, but apart from periarterial collections of round cells in the involved areas and the presence of empty sarcolemmal tubes, no new lesions developed. Following inoculation of type 3 (Walters *et al.*, 1963), the muscle of the sacrospinalis group remained normal until the sixth to seventh days when many fasciculi, either singly or in isolated groups, showed a coarse eosinophilic granulation. In other areas, the muscle bundles became swollen, fragmented, and lost their striations. Their staining was pale pink first, but later became slightly basophilic. These changes were not accompanied by an inflammatory cellular exudate. Over the next week, the necrotic muscle became deeply basophilic and developed at first fine, then coarse granules which were black when stained with the van Kossa reagent. Segmented leukocytes, lymphocytes, macrophages, fibroblasts, and pyknotic sarcolemmal cells were not infrequently seen among the necrotic muscle at this stage. At 14 days, the granules had clumped into roughly circular, intensely basophilic bodies identical in appearance to those seen at this time in myocardium. Shrinkage of bundles was evident.

Myocardial damage was observed in the same animals following infection with type 3 (Walters *et al.*, 1963). The first signs were observed 8 days after inoculation, reaching maximum intensity by the tenth day. The initial lesions, consisting of small necrotic foci, were observed in papillary muscles. Later they were present beneath the endocardium and also deep within the myocardium and on the endocardium of both ventricles. The necrotic foci were composed of structureless and pale eosinophilic granular fragments of necrotic myocardial cells associated with edema, nuclear and cytoplasmic debris, a little fibrin, and a scant infiltration of lymphocytes and macrophages. Later, there was a fine basophilic stippling of the cytoplasm of many muscle cells and numerous eosinophilic hyaline bodies were found. By the tenth day, necrotic foci were still extending, but the granular necrosis was coarse and markedly basophilic. Fresh foci appeared in which the myocardial fibers swelled, lost their striations, and became fragmented. The nuclei underwent lysis, karyorrhexis, or pyknosis. Sometimes the fragmented cytoplasm was diminished to form small hyaline bodies, but more frequently they appeared as tiny circular basophilic granules. The cellular response was meager and lymphocytic and histiocytic in nature. Over superficial foci swelling of pericardial serosal cells was seen on the tenth day. The centers of many foci appeared less dense, and small capillaries together with a few fibroblasts were found. Clumping of necrotic debris into small, roughly circular, intensely basophilic bodies was seen on the twelfth day. These bodies were stained black by the van Kossa technique (for calcium) and were mauve–red with periodic acid–Schiff treatment. By day 13, the heart appeared normal.

All animals died 5–7 days following intraperitoneal inoculation of type 1 (Hassan *et al.*, 1965). Virus was detected in the blood and heart as early as 6 hr after infection. Titers increased to the third day and remained constant at 10^6 PFU/gm of heart and 10^4 PFU/ml of blood. Myocardial damage was often so severe, that by the fifth day it was apparent on gross inspection as gray-yellow mottling of the ventricular surface of the heart. Histological evidence of myocardial damage was usually observed on the fourth day, but rarely small foci of necrosis were visible as early as the third day. Damaged myocardial fibers initially appeared fragmented with loss of cross striations and sparse or no inflammatory cellular response. More extensive damage was observed in specimens taken between the fifth and seventh days. Focal myocardial necrosis with infiltration of mononuclear cells and scattered polymorphs constituted the main lesion. Small round or oblong inclusion bodies were infrequently found within myocardial fibers. Calcification of myocardial fibers and fibroblastic proliferation were noted in specimens examined after the fifth day. Liver, brain, and skeletal muscle also showed foci of necrosis, but lesions were not as extensive as those in myocardium.

By immunofluorescence, reovirus antigen was demonstrated in myocardial fibers adjacent to, and in foci of myocardial damage. Fluorescence was found in the cytoplasm, sometimes as a reticulumlike network surrounding the nucleus. The fluorescence was confluent and intense in damaged myocardial fibers. In the electron microscope, virus particles were seen in and around foci of myocardial damage. Particles were numerous in areas of necrosis, and scant or absent in sections showing little or no damage. Discrete nests of virus composed of particles embedded in a granular matrix were located in the cytoplasm of myocytes between myofibrils in proximity to mitochondria. Several nests in a single myocyte, varying in size upward from mitochondria were observed. Evidence of cellular damage was often present in myocytes containing virus particles. Necrosis of myocytes was frequently noted, but a less severe type of cellular injury was also present. This consisted of focal cytoplasmic degradation involving only a portion of the myocyte. The earliest pathological changes in myocytes were swelling and vesiculation of mitochondria and dilatation of sarcoplasmic reticulum.

VIII. Arboviruses

The arbovirus group is a heterogenous collection of ether-sensitive RNA viruses that have the common property of being transmitted by the bite of hematophagous arthropods (Casals and Reeves, 1965). By antigenic and biophysical characterization, the majority fall into three distinct groups, although it is usual to consider them as a single family. They cause a wide range of clinical illnesses, including yellow fever, encephalitis, hemorrhagic fevers, and the dengue syndrome. In none of these is muscle involvement a major factor, but degenerative changes occur in the myocardium in yellow fever (Bugher, 1951), while muscle and joint pains are a feature of Venezuelan equine encephalitis (Sanmartin-Barberi, et al., 1954), chikungunya (Robinson, 1955), Rift Valley fever (Weiss, 1957), dengue (Lumsden, 1958), and o'nyong-nyong (Shore, 1961). All arboviruses produce encephalitis when inoculated into baby mice; in fact, this is the routine method of isolation from clinical material. However, skeletal myositis and myocarditis have also been demonstrated with both newly isolated and highly mouse-adapted strains (Weinbren and Williams, 1958).

When Zika virus (a member of group B) is inoculated into mice between 1 and 5 days old, lesions are induced in skeletal muscle and the myocardium (Weinbren and Williams, 1958). In skeletal muscle, the fibers first become swollen over part of their length. The transition from normal to abnormal tissue is abrupt at this and all later stages. Next, there is a proliferation of nuclei in the affected areas and later many undergo karyorrhexis. The swollen sections of muscle fiber take on a hyaline appearance and exhibit greatly increased affinity for eosin.

The transverse striations become less distinct and eventually disappear. The hyaline sections of muscle fiber become colorless and wavy, eventually breaking into short sections with rounded ends. At about the time at which the fiber becomes wavy, the sarcolemma breaks down, and there is also a marked infiltration of phagocytes, which fill the space bounded by the sarcolemma of adjacent bundles. These changes are occasionally seen in isolated fibers, but more commonly in groups numbering between three and eight, thus leading to the production of roughly spherical foci of necrosis. This pathological process occurs in conjunction with an encephalitis and myocarditis, one of which terminates the life of the infant mouse before the myositis enters either a chronic or a regenerative phase.

The changes seen in the mycardium were in most respects similar to those described for skeletal muscle. The branching, syncytial nature of the cardiac muscle, combined with the curved nature of the heart wall rendered the study much more difficult than in skeletal muscle, which could be cut to show the majority of the fibers in longitudinal section. Multiple foci of necrosis could, however, be found and in these the characteristic changes were seen. In many cases a focus occurred in such a position in the section as to allow observation of longitudinally cut fibers, and in these, the wavy appearance of the hyaline fiber sections could be seen. In a few cases rupture of these segments had occurred. The process was rarely observed to have progressed beyond this point. In a few cases, there was between the muscle fibers extravasation of blood, which formed pockets over areas much greater in length than the diameter of the necrotic foci, although these were never wider than the space that might be occupied by three fibers.

Chikungunya virus (a member of group A) produces very similar myositis and myocarditis in addition to encephalitis in baby mice (Weinbren, 1958). A more recent and extensive study of a related member of group A, Semliki Forest virus, showed that 12 hr after intraperitoneal inoculation of 10^6 LD_{50}, there was two hundred times as much virus in the skeletal muscle as in the blood (Murphy *et al.*, 1970). The amount of virus in muscle only increased slightly above the 12 hr level, but the titer in brain rose to a maximum at 36 hr. Electron microscopy of muscle at 12 hr showed no abnormalities, with only a few cells infected. Twenty-four hours post inoculation, nearly all muscle cells examined were infected, although structure remained normal. By 36 hr muscle cell involvement was pronounced. Many cells were covered with virus particles, and larger accumulations of particles filled the cisternae of the endoplasmic reticulum. Mature virus was present in all extracellular spaces, perivascular spaces, and within the lumen of vessels. Viral nucleoids were not numerous in specimens taken up to 36 hr, but by 48 hr, there were large focal accumulations of nucleoids. Disruption of muscle cell architecture was first seen at 36 hr, and progressive degenerative changes were widespread by 48 hr. Within some cells, abnormal spaces between myofibril bundles in undifferentiated sarcoplasm resulted in a rarified appearance. Interfibrillar spaces increased, and the organization of myofilaments disintegrated in such cells. Accumulation of glycogen accompanied these changes, and when myofibril destruction was pronounced, massive accumulations of glycogen were present. At 48 hr the cytoplasm of some muscle cells was frankly necrotic, with organelle disruption, but nuclear changes were not prominent.

Ultrastructural studies of myocarditis produced in baby mice by another group A arbovirus (Aura virus) have recently been reported by Lascano and his colleagues (1970). Viral replication was demonstrated in cardiac myocytes with particles budding from the cell surface. Accumulations of particles surrounded by cellular membranes were also found within myocardial fibers. The earliest pathological changes were swollen mitochondria with disrupted cristae and vacuolation, dilatation of sarcotubules, and edema of cytoplasmic matrix, with separation of myofilaments and myofibrils. The myofibrils lost their periodic bands, and there was disarrangement of the myofilaments in more severely affected cells. The necrotic cells had dense cytoplasm containing remnants of damaged organelles embedded in a compact mass of disarranged myofilaments. Histiocytes were occasionally found surrounding the necrotic myocytes.

IX. Other Viruses

A. Influenza Viruses

Myocardial involvement is a relatively infrequent complication of influenza virus infection (Adams, 1959; Oseasohn et al., 1959; Woodward et al., 1960; Coltman, 1962). The majority of such cases occurred in young adults and were associated with type A2 strains during the prevalent outbreaks of "Asian" influenza.

Oseasohn and his colleagues (1959) described in detail their post mortem investigations carried out on thirty-three persons who died during the influenza outbreak in Cleveland, Ohio. Influenza type A2 was isolated from twenty-two of the thirty-three cases. Gross examination of hearts showed mainly nonspecific "toxic" changes, i.e., pale, flabby myocardium and bilateral dilatation. Subepicardial and subendocardial petechiae and hemorrhagic extravasation were seen in six cases. Preexisting cardiac abnormalities were present in fourteen patients. Microscopically, varying degrees of acute inflammatory disease were found in the myocardium of ten patients, eight of whom were under 40 years of age. The duration of illness was less than 4 days in two, 4–7 days in seven, and 8 days in one. The extent of involvement ranged from slight edema of the interstitial tissue accompanied by scattered foci of pleomorphic exudate, made up mainly of lymphocytes, monocytes, and occasional granular cells; to diffuse areas of heavy cellular infiltration with a more abundant exudate of the same type. Where the infiltrate was sparse and focal, the cells tended to be in a perivascular location, and the adjacent myocardium appeared normal. The heavier cellular collections were sufficient in some areas to separate the muscle bundles. The myocardial fibrils in the vicinity of the exudate were eosinophilic and had lost their cross striations. The infiltrate was found in all portions of the myocardium, including the subepicardial and subendocardial regions.

B. Lymphocytic Choriomeningitis

Theide (1962) described two cases of serologically proven lymphocytic choriomeningitis (LCM) in which limited cardiac involvement occurred. The first case was a 31-year-old male, who presented as a fairly typical case of LCM, and in whom attention was called to the heart by a labile pulse. ECG changes were present for over 1 month, but the only clinical signs were a short burst of tachycardia and palpitation. The second case, a 32-year-old male, presented with primary cardiac symptoms shortly after an illness compatible with acute mild lymphocytic choriomeningitis. In neither case did sequelae of myocardial inflammation follow.

C. Measles Virus

Myocardial damage is occasionally found in association with measles. Degen (1937) found myocarditis in four of ninety-one fatal cases of measles in the form essentially of an interstitial lymphocytic infiltration. L. J. Ross (1952) investigated seventy-one children aged 4–14 years hospitalized with measles. Electrocardiograms were taken for all the children, followed by physical examination of the heart. All the children recovered, and clinically there was little evidence of heart disease. The ECGs were abnormal in about 30% of the children, the frequency being higher in the younger children under eight years of age. The author suggested that the ECG abnormalities could have been the result of myocardial involvement.

D. Rabies Virus

Although rabies is essentially an acute and fatal encephalities, E. Ross and Armentrout (1962) described the pathology of the heart in a single fatal case. A 19-year-old woman was bitten on the right hand by a cat and developed rabies 52 days later. On the third day of hospitalization, she had a temperature of 104.9°F, peripheral cyanosis, tachycardia with a gallop rhythm, and vascular collapse. She died 14 hr later. At autopsy, the heart showed an acute exudative myocarditis characterized by diffuse collections, within interstitial zones, of an exudate composed chiefly of polymorphs, macrophages, plasma cells, and immature Aschoff myocytes. Capillary dilatation, edema, and localized pooling of fibrin were also prominent. There was focal evidence of acute degeneration of myocardial fibers. Rabies virus was isolated from the brain, in which Negri bodies were found.

E. An Unidentified Virus

Caulfield and his colleagues (1968) described the adventitious finding of viruslike particles in apparently normal muscle. A 72-year-old woman died of a ruptured aortic aneurism. During the preceeding 24 hr period she complained of muscle weakness and mild headache. Twelve hours after death, pieces of superior rectus muscle of the right eye were taken as control material for electron microscopic study. In thick sections, the muscle appeared normal. Thin sections showed the presence of viruslike particles 250–300 Å in diameter. Many clusters were in crystalline array. Virus particles were present in all muscle cells of the extraocular muscles examined, but were not seen in heart and pectoralis muscles. The size was compatible with picornavirus, but no material was available for virus isolation. There was no evidence at all of inflammation of muscle.

F. Polymyositis

Polymyositis is a relatively common disease of unknown etiology. The history and clinical features are often suggestive of an infectious etiology.

Muscle tissue biopsies obtained from a patient with a clinical and pathological diagnosis of chronic polymyositis were studied in the electron microscope (Chou, 1967). Aggregates of tubular filamentous structures resembling those of myxovirus nucleoprotein were demonstrated in both nuclei and cytoplasm. These structures were observed in three biopsies taken during an 18 month period.

The patient, a 66-year-old male, developed progressive weakness of all muscles, mild dysphagia, and muscle atrophy, most pronounced in the shoulder girdle and quadriceps. Histologically, the lesions seen in the muscle were characterized by necrosis of muscle fibers, phagocytosis, basophilic regenerating fibers, and diffuse mononuclear cell infiltrates with no relation to the blood vessels. In the electron microscope, most fibers appeared relatively normal. Fibers in the vicinity of the cellular infiltrate often showed degenerative alterations characterized by disruption, distortion, and disappearance of myofibrils. This was accompanied by numerous membranous bodies containing electron-dense granules.

Intracytoplasmic aggregates of filamentous structures were present in all three biopsies, and intranuclear clusters in the last two. At high magnification, the filaments in both nuclei and cytoplasm revealed distinct tubular structures that had hollow circular profiles on cross section and triple density in longitudinal sections. The diameter measured 200–230 Å (inner diameter 60–74 Å) in the cytoplasm and 100–200 Å (inner diameter 60–70 Å) in the nucleus. It was suggested that these are similar to myxoviruses.

X. Myocarditis

Fiedler (1900) first described isolated myocarditis in man, and the condition has been variously termed idiopathic, interstitial, isolated, primary, etc. Many cases have been recorded, and Saphir and Cohen (1957) recorded that the incidence appeared to be increasing. With improved methods of diagnosis, a greater proportion of these cases have been associated with viral infections, and Sainani and his colleagues (1968) pointed out that Fiedler's myocarditis could now be identified as being of an infectious (usually viral) etiology. We have seen in the previous sections that many viruses can occasionally infect the myocardium, and that the syndrome previously known as "nonspecific benign pericarditis" should probably be called "acute (coxsackie) virus myopericarditis" (Smith, 1966) or "postviral myopericarditis" (Adams, 1959) (See p. 227).

In their review, Rabin and Jenson (1967) point out that viral myocarditis in man and experimental animals usually occurs in association with generalized viral infection. Very young children and infants appear to be most susceptible to myocardial complications (Sanders, 1963), although from studies on a series of routine autopsies, Burch and his colleagues (1967, 1968) have shown that subacute coxsackievirus myocarditis is relatively common in older persons. The increasing incidence has been related to improvements in hygiene and general living standards (Grist and Bell, 1969). As fewer are infected with enteroviruses in childhood, more older persons experience primary infections, thereby increasing the incidence of adult myocarditis. Similarly, babies born to nonimmune mothers are highly susceptible to neonatal infections and more important, to transplacental infection. This neonatal and adult susceptibility to myocarditis is paralleled in the results found in experimental animals. Several factors can increase the severity of viral myocarditis once it is established. Anderson and his colleagues (1951) suggested that exercise can induce cardiac failure in smallpox patients, while Pearce (1960) showed that deprivation of oxygen to the heart in rabbits allowed several viruses to invade the myocardium.

An association of subclinical and chronic myocardial lesions with viral infections has been suggested by recent investigations. Burch and his colleagues (1967, 1968) tested the heart tissue from a series of routine autopsies (from stillbirth to middle age) for the presence of coxsackievirus antigen by immunofluorescence. A large proportion of the hearts contained coxsackievirus antigen, most of which also showed evidence of interstitial myocarditis. In conjunction with their results in experimental monkeys (Sun *et al.*, 1967b), the authors postulate that the coxsackie B viruses may play a part in the genesis of rheumatic heart disease or a similar disease state. A certain amount of experimental proof for this hypothesis has been provided by the work of Wilson and his colleagues (1969). Weanling mice infected with coxsackievirus B3 developed diffuse, virulent lesions with marked myocardial necrosis accompanying the inflammatory response. Marked fibrosis, dystrophic mineralization, continuing inflammation, and microscopic myocardial hypertrophy persisted for at least 6 months. All the mice remained apparently well during the investigation, and therefore subclinical coxsackievirus myocarditis resulted in permanent myocardial injury. No virus was demonstrable after day 6, and in mice sacrificed at day 185, no antibodies were found, although these were detectable earlier. A similar role has also been suggested for varicella virus (Hackel, 1953). Finally, Brown and Evans (1967) showed that significantly more mothers of infants with congenital heart disease had experienced infection with coxsackievirus than had matched control women. It therefore appears that acute and chronic viral myopericarditis is a complex syndrome that is likely to become more important in the future.

REFERENCES

Adams, C. W. (1959). *Amer. J. Cardiol.* 4, 56.

Albrecht, W. (1954). *Z. Naturforsch.* B 9, 583.

Albrecht, W., and Gädeke, R. (1956). *Z. Naturforsch.* B 11, 241.

Albrecht, W., and Sauthoff, R. (1954). *Z. Naturforsch.* B 9, 340.

Anderson, T., Foulis, M. S., Grist, N. R., and Landsman, J. B. (1951). *Lancet* 1, 1248.

Andrewes, C. H. (1954). *Nature (London)* 179, 620.

Andrewes, C. H., and Pereira, H. G. (1967). "Viruses of Vertebrates," 2nd ed. Baillière, London.

Armstrong, P. M., Wilson, F. H., McLean, W. J., Silverthorne, N., Clark, E. M., Rhodes, A. J., Knowles, D. S., Ritchie, R. C., and Donohue, W. L. (1950). *Can. J. Pub. Health* 41, 51.

Artenstein, M. S., Cadigan, F. C., and Buescher, E. L. (1964). *Ann. Intern. Med.* 60, 196.

Artenstein, M. S., Cadigan, F. C., and Buescher, E. L. (1965). *Ann. Intern. Med.* 63, 597.

Aumonier, F. J. (1953). *J. Roy. Microsc. Soc.* [3] 72, 218.

Babb, J. M., Stoneman, M. E. R., and Stern, H. (1961). *Arch. Dis. Childhood* 36, 551.

Bell, E. J., and Grist, N. R. (1968). *Scot. Med. J.* 13, 47.

Bell, T. M. (1965a). "An Introduction to General Virology." Heinemann, London.

Bell, T. M. (1965b). *Scot. Med. J.* 10, 276.

Bennette, J. G., Bush, P. V., and Steele, R. D. (1967). *Brit. J. Exp. Pathol.* 48, 251.

Benyesh-Melnick, M., Rosenberg, H. S., and Watson, B. (1964). *Proc. Soc. Exp. Biol. Med.* 117, 452.

Bienz, K., Bienz-Isler, G., Weiss, M., and Loeffler, H. (1969). *Brit. J. Exp. Pathol.* 50, 471.

Bienz, K., Bienz-Isler, G., Egger, D., Weiss, M., and Loeffler, H. (1970). *Arch. Gesamte Virusforsch.* 31, 257.

Bienz-Isler, G., Bienz, K., Weiss, M., and Loeffler, H. (1970). *Arch. Gesamte Virusforsch.* 31, 247.

Blailock, Z. R., Rabin, E. R., and Melnick, J. L. (1967). *Science* 157, 69.

Blailock, Z. R., Rabin, E. R., and Melnick, J. L. (1968). *Exp. Mol. Pathol.* 9, 84.

Bodian, D. (1948). *Bull. Johns Hopkins Hosp.* 83, 1.

Bodian, D. (1949). *Amer. J. Med.* 6, 563.

Bradford, H. A., and Anderson, L. L. (1950). *Ann. Intern. Med.* 32, 270.

Brown, G. C., and Evans, T. N. (1967). *J. Amer. Med. Ass.* 199, 183.

Bugher, J. C. (1951). *In* "Yellow Fever" (G. K. Strode, ed.), pp. 137–163. McGraw-Hill, New York.

Burch, G. E., Sun, S. C., Colcolough, H. L., Sohal, R. S., and DePasquale, N. P. (1967). *Amer. Heart J.* 74, 13.

Burch, G. E., Sun, S. C., Chu, K. C., Sohal, R. S., and Colcolough, H. L. (1968). *J. Amer. Med. Ass.* **203**, 1.

Carey, E. J. (1943). *Proc. Soc. Exp. Biol. Med.* **53**, 3.

Carey, E. J. (1944). *Amer. J. Pathol.* **20**, 961.

Carey, E. J., Massopust, L., Zeit, W., and Haushalter, E. (1944). *J. Neuropathol.* **3**, 121.

Carmichael, L. E., Squire, R. A., and Krook, L. (1965). *Amer. J. Vet. Res.* **26**, 803.

Casals, J., and Reeves, W. C. (1965). *In* "Viral and Rickettsial Infections of Man" (F. L. Horsfall and I. Tamm, eds.), 4th ed., pp. 580–582. Lippincott, Philadelphia, Pennsylvania.

Caughey, J. E., and Malcolm, D. S. (1950). *Arch. Dis. Childhood* **25**, 13.

Caulfield, J. B., Rebeiz, J., and Adams, R. D. (1968). *J. Pathol. Bacteriol.* **96**, 232.

Chany, C., Lépine, P., Lelong, M., LeTan-vinh, Satge, P., and Virat, J. (1958). *Amer. J. Hyg.* **67**, 367.

Cherry, J. D., Jahn, C. L., and Meyer, T. C. (1967). *Amer. Heart J.* **73**, 681.

Chor, H. (1933). *Arch. Neurol. Psychiat.* **29**, 344.

Chou, S-M. (1967). *Science* **158**, 1453.

Coltman, C. A. (1962). *J. Amer. Med. Ass.* **180**, 204.

Connolly, J. H. (1961). *Brit. Med. J.* **1**, 877.

Craighead, J. E. (1966). *Amer. J. Pathol.* **48**, 333.

Daae, A. (1872). *Nor. Mag. Laegevidensk.* **2**, 409.

Dalldorf, G., and Melnick, J. L. (1965). *In* "Viral and Rickettsial Infections of Man" (F. L. Horsfall and I. Tamm, eds.), 4th ed., pp. 474–512. Lippincott, Philadelphia, Pennsylvania.

Dalldorf, G., and Sickles, G. M. (1948). *Science* **108**, 61.

Degen, J. A. (1937). *Amer. J. Med. Sci.* **194**, 104.

Delaney, T. B., and Fukunaga, F. H. (1958). *N. Engl. J. Med.* **259**, 234.

Dickinson, L., and Griffiths, A. J. (1966). *Brit. J. Exp. Pathol.* **47**, 35.

Dogopol, V. B., and Cragan, M. D. (1948). *Amer. J. Pathol.* **24**, 713.

Dömök, I., and Molnár, E. (1960). *Ann. Paediat.* **194**, 102.

Downie, A. W. (1965). *In* "Viral and Rickettsial Infections of Man" (F. L. Horsfall and I. Tamm, eds.), 4th ed., pp. 932–964. Lippincott, Philadelphia, Pennsylvania.

Drouhet, V. (1957). *Ann. Inst. Pasteur, Paris* **93**, 138.

Fechner, R. E., Smith, M. G., and Middelkamp, J. N. (1963). *Amer. J. Pathol.* **42**, 493.

Fiedler, A. (1900). *Zentralbl. Inn. Med.* **21**, 212.

Field, E. J. (1952). *J. Pathol. Bacteriol.* **64**, 1.

Finlay-Jones, L. R. (1964). *N. Engl. J. Med.* **270**, 41.

Finsen, J. (1874). "Iagttagelser angaaende Sygdomsforholdene i Island," pp. 145–151. Reitzels, Copenhagen.

Forbus, W. D. (1926). *Arch. Pathol. Lab. Med.* **2**, 486.

Gädeke, R. (1952). *AMA Arch. Pathol.* **54**, 276.

Gädeke, R., and Walenberger, H. (1952). *Z. Naturforsch. B* **7**, 524.

Gear, J., and Measroch, V. (1958). *S. Afr. Med. J.* **32**, 1062.

Gifford, R., and Dalldorf, G. (1949). *Proc. Soc. Exp. Biol. Med.* **71**, 589.

Gifford, R., and Dalldorf, G. (1951). *Amer. J. Pathol.* **27**, 1047.

Ginsberg, H. S., and Dingle, J. H. (1965). *In* "Viral and Rickettsial Infections of Man" (F. L. Horsfall and I. Tamm, eds.), 4th ed., pp. 860–891. Lippincott, Philadelphia, Pennsylvania.

Glajchen, D. (1961). *Brit. Med. J.* **2**, 870.

Glasgow, L. A. (1965). *J. Pediat.* **67**, 104.

Godman, G. C., Bunting, G. H., and Melnick, J. L. (1952a). *Amer. J. Pathol.* **28**, 223.

Godman, G. C., Bunting, G. H., and Melnick, J. L. (1952b). *Amer. J. Pathol.* **28**, 583.

Gore, I., and Saphir, O. (1947). *Amer. Heart J.* **34**, 827.

Grist, N. R., and Bell, E. J. (1968). *Brit. Med. J.* **3**, 556.

Grist, N. R., and Bell, E. J. (1969). *Amer. Heart J.* **77**, 295.

Grodums, E. I., and Dempster, G. (1959). *Can. J. Microbiol.* **5**, 605.

Guyton, A. C., and Reeder, D. C. (1950). *Arch. Neurol. Psychiat.* **63**, 954.

Hackel, D. B. (1953). *Amer. J. Pathol.* **29**, 369.

Hannaeus, G. (1735). Dissertation, Copenhagen.

Hassan, S. A., Rabin, E. R., and Melnick, J. L. (1965). *Exp. Mol. Pathol.* **4**, 66.

Hassin, G. B. (1943). *J. Neuropathol. Exp. Neurol.* **2**, 293.

Helin, M., Savola, J., and Lapinleimu, K. (1968). *Brit. Med. J.* **3**, 97.

Henle, G., Henle, W., and Diehl, V. (1968). *Proc. Nat. Acad. Sci. U.S.* **59**, 94.

Hilleman, M. R. (1965). *Amer. J. Med.* **38**, 751.

Hilleman, M. R., and Werner, J. R. (1954). *Proc. Soc. Exp. Biol. Med.* **85**, 183.

Homann, C. (1872). *Nor. Mag. Laegevidensk.* **2**, 542.

Horányi-Hechst, B. (1935). *Deut. Z. Nervenheilk.* **137**, 1.

Hosier, D. M., and Newton, W. A. (1958). *AMA Amer. J. Dis. Child.* **96**, 251.

Hurst, E. W. (1929). *J. Pathol. Bacteriol.* **32**, 457.

International Subcommittee on Virus Nomenclature. (1963). *Virology* **19**, 114.

Javett, S. N., Heymann, S., Mundel, B., Pepler, W. J., Lurie, H. L., Gear, J., Measroch, V., and Kirsch, Z. G. (1956). *J. Pediat.* **48**, 1.

Jennings, R. C. (1966). *J. Clin. Pathol.* **19**, 325.

Jungeblut, C. W. (1950). *Bull. N.Y. Acad. Med.* [2] **20**, 571.

Jungeblut, C. W., and Edwards, J. E. (1951). *Amer. J. Clin. Pathol.* **21**, 601.

Jungeblut, C. W., and Steenberg, E. (1950). *AMA Arch. Pathol.* **49**, 574.

Jungeblut, C. W., and Steevens, M. A. (1950). *Amer. J. Clin. Pathol.* **20**, 701.

Jurrow, S. S., and Dogopol, V. B. (1952). *Amer. J. Pathol.* **28**, 566.

Kausche, G. A., Landschütz, C., and Sauthoff, R. (1951). *Z. Naturforsch. B* **6**, 445.

Kibrick, S., and Benirschke, K. (1956). *N. Engl. J. Med.* **255**, 883.

Kibrick, S., and Benirschke, K. (1958). *Pediatrics* **22**, 857.

Kilham, L., Mason, P., and Davies, J. N. P. (1956). *Amer. J. Trop. Med. Hyg.* **5**, 655.

Lascano, E. F., Berria, M. I., and Barrera Oro, J. G. (1970). *Arch. Gesamte Virusforsch.* **32**, 99.

Le Gros Clark, W. E. (1946). *J. Anat.* **80**, 24.

Lépine, P., Desse, G., and Sautter, V. (1952). *Bull. Acad. Nat. Med., Paris* [3] **136**, 66.

Lerner, A. M. (1965). *Progr. Med. Virol.* **7**, 97.

Lerner, A. M., and Shaka, J. A. (1962). *Proc. Soc. Exp. Biol. Med.* **111**, 804.

Lerner, A. M., Levin, H. S., and Finland, M. (1962). *J. Exp. Med.* **115**, 745.

Levaditi, C. (1926). "L'Herpès et le Zona." Masson, Paris.

Lewes, D., and Lane, W. F. (1961). *Lancet* **2**, 1385.

Ludden, T. E., and Edwards, J. E. (1949). *Amer. J. Pathol.* **25**, 357.

Lukacs, V. F., and Romhanyi, J. (1960). *Ann. Paediat.* **149**, 89.

Lumsden, W. H. R. (1958). *East Afr. Med. J.* **35**, 519.

MacAdam, D. B., and Whitaker, W. (1962). *Brit. Med. J.* **2**, 1099.

Melnick, J. L., and Godman, G. C. (1951). *J. Exp. Med.* **93**, 247.

Monif, G. R. G., Lee, C.-W., and Hsiung, G. D. (1967). *N. Engl. J. Med.* **277**, 1353.

Montgomery, J., Gear, J., Prinsloo, F. R., Kahn, M., and Kirsch, Z. G. (1955). *S. Afr. Med. J.* **29**, 608.

Moossy, J., and Geer, J. C. (1960). *AMA Arch. Pathol.* **70**, 615.

Murnane, T. G., Craighead, J. E., Mondragon, H., and Shelokov, A. (1960). *Science* **131**, 498.

Murphy, F. A., Harrison, A. K., and Collin, W. K. (1970). *Lab. Invest.* **22**, 318.

Naudé, W. duT., Selzer, G., and Kipps, A. (1958). *Med. Proc.* **4**, 397.

Null, F. C., and Castle, C. H. (1959). *N. Engl. J. Med.* **261**, 937.

Oseasohn, R., Adelson, L., and Kaji, M. (1959). *N. Engl. J. Med.* **260**, 509.

Pappenheimer, A. M., Daniels, J. B., Cheever, F. S., and Weller, T. H. (1950). *J. Exp. Med.* **92**, 169.

Pearce, J. M. (1960). *Circulation* **21**, 448.

Pette, H. (1952). *Monatsschr. Kinderheilk.* **100**, 155.

Plummer, G., and Hackett, S. (1966). *Brit. J. Exp. Pathol.* **47**, 82.

Pollen, R. H. (1963). *Amer. J. Cardiol.* **12**, 736.

Rabin, E. R., and Jenson, A. B. (1967). *Progr. Med. Virol.* **9**, 392.

Rabin, E. R., Hassan, S. A., Jenson, A. B., and Melnick, J. L. (1964). *Amer. J. Pathol.* **44**, 775.

Rabin, E. R., Phillips, C. A., Jenson, A. B., and Melnick, J. L. (1965). *Exp. Mol. Pathol.* **4**, 98.

Robinson, M. C. (1955). *Trans. Roy. Soc. Trop. Med. Hyg.* **49**, 28.

Ross, E., and Armentrout, S. A. (1962). *N. Engl. J. Med.* **266**, 1087.

Ross, L. J. (1952). *AMA Amer. J. Dis. Child.* **83**, 282.

Rowe, W. P., Huebner, R. J., Gilmore, L. K., Parrott, R. H., and Ward, T. G. (1953). *Proc. Soc. Exp. Biol. Med.* **84**, 570.

Rustigian, R., and Pappenheimer, A. M. (1949). *J. Exp. Med.* **89**, 69.

Sabin, A. B. (1959). *Science,* **130**, 1387.

Sainani, G. S., Krompotic, E., and Slodki, S. J. (1968). *Medicine (Baltimore)* **47**, 133.

Sanders, V. (1963). *Amer. Heart J.* **66**, 707.

Sanmartin-Barberi, Č., Groot, H., and Osborno-Mesa, H. (1954). *Amer. J. Trop. Med. Hyg.* **3**, 283.

Sanyal, S. K., Mahdavy, M., Gabrielson, M. O., Vidone, R. A., and Browne, M. J. (1965). *Pediatrics* **35**, 36.

Sanz-Ibanez, J. (1945). *Trab. Inst. Cajal Invest. Biol.* **37**, 259.

Saphir, O., and Cohen, N. A. (1957). *AMA Arch. Pathol.* **64**, 446.

Saphir, O., and Wile, S. A. (1942). *Amer. J. Med. Sci.* **203**, 781.

Sauthoff, R. (1956). *Z. Naturforsch. B* **11**, 241.

Seifert, G. (1965). *Ger. Med. Mon.* **10**, 437.

Shore, H. (1961). *Trans. Roy. Soc. Trop. Med. Hyg.* **55**, 361.

Simenhoff, M. L., and Uys, C. J. (1958). *Med. Proc.* **4**, 389.

Smith, W. G. (1966). *Brit. Heart J.* **28**, 204.

Sohier, R., Chardonnet, Y., and Prunieras, M. (1965). *Progr. Med. Virol.* **7**, 253.

Spain, D. M., Bradess, V. A., and Parsonet, V. (1950). *Amer. Heart J.* **40**, 336.

Stanley, N. F. (1967). *Brit. Med. Bull.* **23**, 150.

Stanley, N. F., Leak, P. J., Walters, M. N.-I., and Joske, R. A. (1964). *Brit. J. Exp. Pathol.* **45**, 142.

Suckling, P. V., and Vogelpoel, C. (1958). *Med. Proc.* **4**, 372.

Sun, S. C., Colcolough, H. L., Burch, G. E., DePasquale, N. P., and Sohal, R. S. (1967a). *Proc. Soc. Exp. Biol. Med.* **125**, 157.

Sun, S. C., Sohal, R. S., Burch, G. E., Chu, K. C., and Colcolough, H. L. (1967b). *Brit. J. Exp. Pathol.* **48**, 655.

Swann, N. H. (1961). *Ann. Intern. Med.* **54**, 1008.

Sylvest, E. (1930a). *Ugeskr. Laeger* **92**, 798.

Sylvest, E. (1930b). *Ugeskr. Laeger* **92**, 982.

Sylvest, E. (1934). "Epidemic Myalgia." Oxford Univ. Press, London and New York.

Theide, W. H. (1962). *Arch. Intern. Med.* **109**, 50.

Van Zaane, D. J., and Van der Veen, J. (1962). *Presse Med.* **70**, 1021.

Waldeyer, W. (1865). *Arch. Pathol. Anat. Klin. Med. Physiol.* **34**, 473.

Walters, M. N-I., Joske, R. A., Leak, P. J., and Stanley, N. F. (1963). *Brit. J. Exp. Pathol.* **44**, 427.

Walters, M. N.-I., Leak, P. J., Joske, R. A., Stanley, N. F., and Perret, D. H. (1965). *Brit. J. Exp. Pathol.* **46**, 200.

Warin, J. F. (1956). *Brit. J. Med. Pract.* **10**, 27.

Warren, J. (1965). *In* "Viral and Rickettsial Infections of Man" (F. L. Horsfall and I. Tamm, eds.), 4th ed., pp. 562–568. Lippincott, Philadelphia, Pennsylvania.

Weinbren, M. P. (1958). *Trans. Roy. Soc. Trop. Med. Hyg.* **52**, 259.

Weinbren, M. P. and Williams, M. C. (1958). *Trans. Roy. Soc. Trop. Med. Hyg.* **52**, 263.

Weiss, K. E. (1957). *Bull. Epizootic Dis. Afr.* **5**, 431.

Welborn, M. B. (1936). *Amer. J. Med. Sci.* **191**, 673.

Weller, T. H., Hanshaw, J. B., and Scott, D. E. (1960). *Virology* **12**, 130.

Wilson, F. M., Hashimi, A., and Lerner, A. M. (1965). *J. Bacteriol.* **90**, 546.

Wilson, F. M. Miranda, Q. R., Chason, J. L., and Lerner, A. M. (1969). *Amer. J. Pathol.* **55**, 253.

Windorfer, A. (1963a). *Deut. Med. Wochenschr.* **88**, 1077.

Windorfer, A. (1963b). *Med. Klinin (Münich)* **38**, 1.

Woodward, T. E., McCrumb, F. R., Carey, T. N., and Togo, Y. (1960). *Ann. Intern. Med.* **53**, 1130.

Wright, H. T., Landing, B. H., Lennette, E. H., and McAllister, R. M. (1963). *N. Engl. J. Med.* **268**, 1041.

PARASITIC INFECTIONS

P. C. C. GARNHAM

I. Introduction

Muscle is subject to infection with two types of animal parasites—protozoa and helminths. Both these infections are widespread in the animal kingdom, but they have nearly always been studied from the angle of the parasite living in muscle rather than that of the muscle infected with a parasite. The dramatic changes undergone by the parasite are undoubtedly responsible for this focusing of attention on one side of the picture, and it has led to a neglect of the changes occurring in the host; even gross physiological disturbances have only been studied in a few instances, let alone the minute histological effects. In hardly any examples is the exact site of development of the parasite known for certain, though it is probable that the muscle fiber is only exceptionally the actual host cell.

As a general rule, the parasite lives in muscle tissue without causing much local reaction. It often becomes encysted, and only on rupture of the cyst does any phagocytic response occur. The reaction is normally brief and localized, but if the infection persists for a long time with periodic rupture of cysts and invasion of new muscle fibers, a widespread inflammatory change eventually takes place in the tissue; thus, chronic myocarditis is the final result of a long-standing case of Chagas' disease.

All types of muscle may be the seat of parasitic infection, and it is not unusual for an organism to attack equally skeletal muscle, the heart, and the unstriated muscles of the alimentary tract. In a few cases, the parasite has a predilection for special tissues, such as the diaphragm, heart, or tongue. Many infections are asymptomatic, and in the case of man, they may often pass undetected, because muscle is not usually included in the histological examination of organs and tissues taken at autopsy. It is curious that the muscle of domestic animals, on the other hand, receives more attention, namely in the form of meat for human consumption. Here the flesh is subjected to a visual examination by meat inspectors as a routine, and parasitic infection of meat is one narrow aspect of the subject that has been recorded in great detail.

The parasites of muscle are by no means confined to human parasites or to parasites occurring in meat destined for human consumption. Their range extends to all classes of vertebrates and to many invertebrates; some of the most interesting examples are to be found, for instance, in insects. All of them exhibit the most important feature of parasite life, namely, the attempt of the parasite to live as harmoniously as possible with the host. It is only in exceptional circumstances, or at a late stage of evolution, before a state of natural tolerance has been achieved, that gross destruction of the tissue of the host takes place.

Muscle provides a unique environment for the parasite, in one respect apparently most unsuitable, in another very favorable. The repeated contraction and relaxation of muscle subjects the parasite to intense stress and strain. It is difficult to understand how the rapidly growing filarial larva, for instance, is able to pursue its development in the thoracic muscles of a mosquito in flight; one would expect that growth inevitably would be adversely affected. Mere muscular contraction must expose the parasite to much buffeting, while the periodic interruption of the flow of blood in the vessels supplying the tissue results in equal periods of anoxemia. On the other hand, muscle is the site of rapid metabolism. Here, carbohydrate metabolism is at a maximum, and amino acids and other metabolites are produced in quantity. The parasite has ample opportunity of finding the food it wants; those muscles that have a high metabolic rate are even more heavily parasitized than others; e.g., in trichinosis and cysticercosis, infection is intense in the diaphragm, intercostals, and heart. Little or nothing is known of these facts, and the subject is open to an interesting study by a parasitologist who is also a comparative physiologist.

It is possible to describe the effect of parasitic infections on muscle in three ways: (1) in relation to the type of muscle (striated, unstriated, and cardiac), (2) in relation to the different host animals, and (3) in relation to the parasite.

The parasitological approach (3) is the most convenient, and the following two sections are devoted to the protozoa and helminths, respectively. The more important, i.e., the more frequent infections, are described in some detail, while a few examples only are mentioned concerning the myxosporidia and the extensive group of nematode worms.

II. Protozoal Infections

Protozoal infections of muscle are either uncommon or are confined to single geographical areas or to special groups of animals. There is one exception, *Sarcocystis*, a parasite widespread in the animal kingdom and one with which we (as meat eaters) are in contact nearly every day of our lives. Meat infected with protozoa is rarely considered unfit for human consumption; but certain protozoal infections of domestic animals, such as *Theileria parva* causing East Coast fever in African cattle, give rise to gross deterioration in the quality of the flesh, and the carcass may have to be condemned. The parasite in such cases does not infect the muscle itself, and infections that have only an indirect effect on the muscle are excluded from consideration here.

Although most protozoa are easily identifiable in the musculature, differential diagnosis sometimes proves exacting, especially if the tissue had undergone post mortem changes. A particular difficulty is presented

by *Toxoplasma, Sarcocystis, Besnoitia,* and *Trypanosoma cruzi,* the cysts
of which can be confused. The size and shape offer criteria, as do the
individual parasites with their characteristic morphology. A useful guide
is provided in a paper by Salfelder *et al.* (1969) with respect to some
of these parasites found in cardiac muscle.

Each parasite is described from the point of view of its incidence,
its character and mode of entry into the muscle, and the changes it
produces in the tissues.

A. Sarcocystis

Of all parasitic infections of muscle, *Sarcocystis* must be the most
common; the parasite affects many mammals, some birds, lizards, and
fish. At least fifty species have been described, including one from man.
The organism is found in skeletal muscle, heart muscle, and in involun-
tary muscle, in which tissues the organism appears in a cystic form.
The cysts are often macroscopic and were easily seen by the early
workers who described the parasite in detail over a hundred years ago.
In spite of its frequency and the long period during which it has been
studied, our knowledge of the life cycle of *Sarcocystis* remains defective.
The classification of the organism is tentative, and even the mode of
entry into muscle is unknown. Recent observations by Rommel and Heydorn (1972) indicate that *Sarcocystis* belongs to the Coccidia (*Isospora*).

The parasite grows as a cylindrical or fusiform object, white in color,
and up to 2 inches in length. Its popular name is Miescher's tube, and
the spores inside the tube are known as Rainey's corpuscles. Another
name commonly used in veterinary medicine is balbiania, for the cysts
found in the esophagus of goats and sheep.

According to Scott (1943), the development of the parasite inside
a muscle cell takes place in the following way. An ameboid parasite
enters the cell, and the nucleus undergoes repeated division, giving rise
to a large number of sporoblasts. The sporoblasts themselves give rise
to ellipsoid and later banana-shaped spores. The parasite itself has mean-
while grown enormously in size and has become surrounded by a cyst
wall, which is derived primarily from the muscle and connective tissue
of the host. The mature spores measure between 10 and 15 mμ in length
and contain a nucleus, a vacuole, and a few granules. They exhibit
a peculiar type of movement. They are sometimes called sporozoites
or schizozoites. The sarcocyst may become calcified or it may rupture
into the surrounding tissue. It is thought that the spores are carried
by the blood stream to other parts of the body where they may invade

muscular tissue. On the other hand, spores have been found in nasal secretion and in the feces, and it is possible that the method of transmission from animal to animal takes place via the feces rather than through the consumption of infected meat—now confirmed by Rommel and Heydorn (1972) who demonstrated typical Isosporan oocysts.

The sarcocyst is usually incompletely divided into a number of compartments by septae passing inward from the cyst wall. *Sarcocystis* has given rise to difficulties of various sorts of diagnosis. Sometimes, when blood is taken from a cow for diagnosing East Coast fever, a sarcocyst may be accidentally penetrated by the needle, and the blood may become contaminated with the spores. In these circumstances, cattle malaria has been erroneously described with crescent-shaped gametocytes, resembling those of *Plasmodium falciparum* in man. Again, during the search for the exoerythrocytic forms of malaria, curious structures were found in the psoas muscles of infected monkeys (Fig. 1), which were at first thought to be the long sought tissue stage of malaria. A closer examination revealed that the parasites were in fact *Sarcocystis* (subsequently named *S. nesbitti*). Other rarer protozoal parasites of muscle may be confused with this organism, including *Toxoplasma* and *Besnoitia*.

The size and general morphology of sarcocysts vary from species

Fig. 1. Sarcocystis nesbitti cysts in skeletal muscle of *Macaca mulatta* (×60).

to species, but it is doubtful if all of the so-called species that have been described are really valid. Certainly, when a sarcocyst is transferred from one animal to another, a great change in morphology may take place.

The incidence of *Sarcocystis* may be extremely high. Sheep may be 100% infected. Lambs show a low rate, but sheep of over a year in age may be practically all infected (Awad, 1973). In cattle, swine, and horses, the incidence of the parasite likewise increases with age. The youngest age at which the infection has been detected is 6 weeks in calves and in lambs. The organism in sheep is known as S. *tenella,* and it is usually found in the form of nodules in the muscles of the esophagus (Fig. 2). In cattle, the species is called S. *fusiformis.* In cattle, it affects particularly the musculature of the esophagus, tongue, larynx, diaphragm, and skeletal muscles. The camel is commonly infected in India and in Egypt, the parasite being present chiefly in the esophagus and in the heart. The incidence in birds varies greatly. Chickens have been found to be infected with this parasite in up to 50% of tissues examined. The most affected muscles are those in the head, the neck, and the pelvic region. In the duck, the infection may be quite extensive; the cysts rupture, the spores are set free in the surrounding tissue, and the muscle fibers become greatly disorganized (Fig. 3). Sarcosporidiosis can be pathogenic to mice because of the intensity of the infection; nearly all the muscles become parasitized and the animals die. The infection, however, both in laboratory and in wild rodents is comparatively uncommon. It is occasionally encountered by accident in the

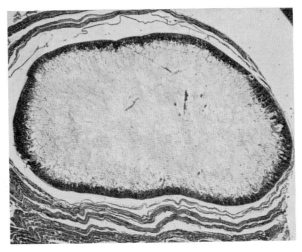

Fig. 2. Sarcocystis tenella in esophagus of sheep (×28).

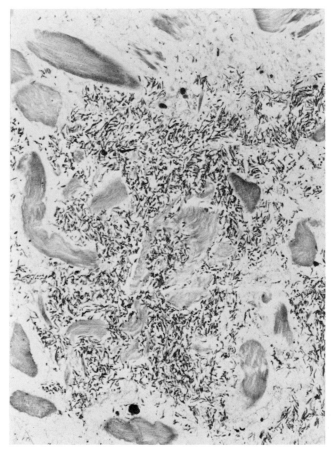

Fig. 3. *Sarcocystis* of wild duck. Ruptured cysts with numerous spores free in disorganized muscle tissue (×250). Courtesy of Professor W. Frank.

course of postmortem examination of white mice or, less often, of rabbits, Ball (1944) found a heavy infection of *Sarcocystis* in lizards in the mountains of southern California. The parasites were present in the muscles of the trunk region. Other records from reptiles are confined to the Mediterranean basin.

Sarcocystis muris was studied by Theobald Smith (1901), and successive passages were obtained in mice by feeding infected muscle from mouse to mouse, and in this way, the infection was maintained for 7 years.

Sarcocystis has rarely been reported from the brain of oxen and sheep, but the frequent presence of the allied sporozoan parasite *Frenkelia* (=M organism) in the brain (but not the muscle) of wild rodents, makes the former records suspect, as the two forms of cyst may easily be confused.

Sarcocystis has been reported about eighteen times in human beings. The sites of infection included the heart, larynx, tongue, and various skeletal muscles (Fig. 4). In most of these instances, the parasite was found accidentally, as in Mackinnon and Abbott's (1955) case from the Sudan, where the leg of the patient was removed on account of mycetoma. When sections of the diseased tissue were prepared, the neighboring muscle was found to contain large numbers of sarcocysts. It is possible that if human muscle were especially examined for the presence of this organism, many more infections might be revealed. On the other hand, if the mode of infection of *Sarcocystis* is via the feces, man is unlikely to contract this infection, unlike herbivorous animals, which graze on land heavily contaminated with their own excreta.

The pathogenicity of *Sarcocystis*, as a general rule, is low. Its presence may cause symptoms or even death if a vital organ, for instance the heart, is attacked, and apparently it may occasionally cause an allergic response. McGill and Goodbody (1957) described a fatal case in man associated with widespread periarteritis nodosa, which they suggest is an allergic reaction following rupture of the cysts. When the sarcocysts are present in very large numbers, the intensity of the infection affects the vital processes and may cause death, as in the case of *S. muris* in mice.

During the course of the development of the parasite in the muscle cell, there is usually no tissue response beyond compression of the neighboring fibers, but Destombes (1957) has described a severe inflammatory reaction in pigs in the muscle around the *Sarcocystis*. In this animal, the mature cyst does not grow to a size greater than a millimeter; an infiltration of adventitious cells, including numerous eosinophiles, is seen

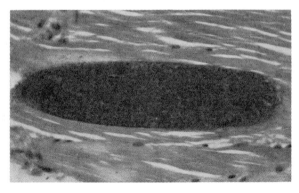

Fig. 4. Sarcocystis lindemanni in human skeletal muscle (MacKinnon and Abbott's case); longitudinal section (×350).

around the parasite, together with much nuclear debris, while beyond this zone there is an inflammatory reaction which extends far into the neighboring muscle fibers. The outer cells are chiefly polymorphonuclear leukocytes, basophiles, and again, many eosinophiles. Animals in which this condition has persisted for a long time show a fibrocalcareous myositis (myositis sarcosporidica).

Pugh (1950) described briefly the behavior of an infection in himself in which the peripheral muscles were heavily infected with *Sarcocystis* (Fig. 8). Fleeting swellings occurred over the infected parts, and eosinophilia was found in the blood. Tissue biopsies revealed a zone of reaction around the parasites. Twenty years later, he was still alive and well

It is strange that the presence of *Sarcocystis*, even in quite large num-
with no symptoms of the disease.
bers, can be tolerated by the host, although a highly toxic substance, sarcocystin, exists in the cysts. Paralysis of muscles in infected pigs has been attributed, however, to the escape of this substance from the cyst into the surrounding tissue, and it is probable that the pathological picture seen in an infected muscle is due to the action of the toxin (sarcocystin) liberated by the parasite.

The muscles of birds are sometimes excessively heavily parasitized with *Sarcocystis*, and yet no interference of flight seems to take place. In ducks, however, *S. rileyi* may cause a severe disease resulting in death of the animal (Erickson, 1940). Awad (1973) noted that the muscle of sheep and other domestic animals may be heavily infected, leading to a general loss of condition, with emaciation and gross deterioration in the quality of the flesh. In goats, the muscles of the larynx may become so heavily infiltrated with sarcocysts that respiration becomes difficult and eventually impossible. In an artificial infection of a lamb with *S. tenella*, Awad noted that the heart became involved and that the animal died suddenly from heart failure. *S. tenella*, like the other species, causes no cellular reaction in the early stages of infection of the muscle. In old infections, when the cysts have ruptured, a marked round cell infiltration occurs in the infected tissue, while caseation and calcification is the final fate of the parasite itself.

Meat is not normally condemned for human consumption unless the infection is highly generalized and the flesh is watery and heavily infiltrated.

In view of the prevalence of this parasite in the animal kingdom, it is desirable that its presence should be excluded when making any researches involving the behavior of muscular tissue. The latest researches on the life cycle of *Sarcocystis* have entirely changed our conception of the infection.

B. *Trypanosoma cruzi*

Trypanosoma cruzi is the most pathogenic protozoal parasite of human muscle, and the infection is found over extensive areas of the New World. It occurs in all South American countries from the Argentine up to Central America, Mexico, and, in its insect host and animal reservoirs, in the United States. *T. cruzi* causes the disease known as Chagas' disease, or South American trypanosomiasis. In some parts of Brazil the disease is so prevalent that 60% of the population may be infected. It is primarily a disease of wild animals, such as the armadillo in Brazil, the bat in Panama, oppossums in various parts of South America, and the cavy, or wild guinea pig, in Bolivia. In all these animals and in small laboratory animals artificially inoculated (particularly baby mice, puppies, and kittens), the muscles are infected in much the same way as in man. The involvement of any particular tissue, however, varies from place to place or from strain to strain in the same way as it does in human infections.

The trypanosome is introduced into the skin, eye, or mouth of man during the bite of an infected reduviid bug. The bug discharges metacyclic forms of the parasite in its excreta, and these infect the person. After a local development at the site of entry, the trypanosome enters the bloodstream and is carried to various organs and tissues of the body. It settles down particularly in the muscle cells of the heart, though frequently various skeletal muscles and other muscular structures are involved.

In a heart muscle cell (Fig. 5), the trypanosome loses its flagellum and becomes rounded into an amastigote, i.e., leishmanial form. The parasite then multiplies by binary fission and pushes the nucleus of the host cell to one side. The muscle fiber enlarges, and a pseudocyst is produced that contains many parasites. Such bodies were first seen by Chagas; he thought they were the result of schizogony and he therefore called the organism *Schizotrypanum cruzi*, almost certainly in a wrong interpretation of events. Flagella grow out of the leishmanial bodies, and the organism assumes an epimastigote form; this in turn becomes a trypanosome by the backward migration of the kinetoplast. The pseudocyst now ruptures and the trypanosomes are set free into the neighboring tissue where they invade new muscle cells or the blood. Meyer and Musacchio have grown amastigotes in tissue culture and observed them under the electron microscope (Fig. 6).

Up to this point, there has been no local reaction to the presence of the organism except for the enlargement of the invaded cells. Once

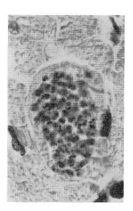

Fig. 5. Pseudocyst of *Trypanosoma cruzi* in heart muscle of experimentally infected mouse (×1000).

the pseudocyst has ruptured, however, the area rapidly becomes invaded by adventitious cells, consisting chiefly of lymphocytes and histiocytes. After repeated multiplication, a chronic inflammatory state of the heart muscle develops; this is followed by fatty and waxy degeneration, which is seen as mottling of the muscle near the endocardium. Finally, a thinning of the ventricle takes place, particularly at the apex of the heart where an aneurism may form.

This chronic condition of the heart, as a rule, does not develop fully for many years after infection; usually the individual is infected in childhood and the disease pursues an insidious course until between the ages of 40 and 50 years he may suddenly die of acute heart failure. In certain areas in Brazil, 8% of the population may show gross electrocardiographic abnormalities, while heart block and Adam–Stokes syndrome commonly occur. The electrocardiographic changes include ventricular extrasystoles and right bundle and arteriovenous blocks.

Fulminating myocarditis is a less common sequela of the disease, but has been seen in children in Bahia, Belém, Panama, and other places; it is practically always quickly fatal, or death follows in a few months from heart failure and generalized edema.

Other tissues beside the heart may become infected in a similar way. Pseudocysts are occasionally found in the skeletal muscles. However, it is in the alimentary tract that another type of lesion of considerable medical importance occurs. The muscle of the esophagus, stomach, duodenum, and colon becomes infected by the organism; pseudocysts develop, and on rupture give rise to interstitial inflammation. The adventitious cells spread to the local nervous plexus and extend far along the

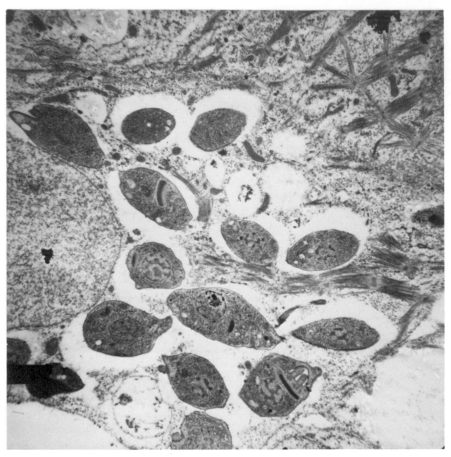

Fig. 6. *Trypanosoma cruzi* in amastigote (i.e., leishmanial) form in tissue culture of heart muscle cells (×11,400). Courtesy of Drs. Herta Meyer and Olivio de Musacchio.

course of these nerves, destroying them and thus grossly interfering with the normal peristaltic movements of the organ. The organ fails to empty, food collects above the lesion, and gross muscular hypertrophy takes place. The pathology of this condition has been described in great detail by Köberle (1956), Köberle and Nador (1956), and most recently, in summary by Köberle (1968). In their opinion, the heavy damage to Auerbach's plexus with necrosis of ganglion cells is the effect of a toxin set free upon rupture of the pseudocyst. The essential lesion is the great reduction in the number of ganglion cells, which was determined both in the heart and in the hollow muscular organs by actual counting in serial histological sections. Extensive denervation then fol-

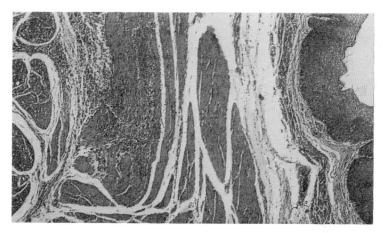

Fig. 7. Inflammatory changes in musculature of human esophagus in late stage of Chagas' disease (×30). Courtesy of Professor F. Köberle.

lows. This interpretation of the pathogenesis was confirmed by Tafuri (1970) in electron microscopic studies on laboratory-infected mice in Belo Horizonte. He showed extensive changes in both the Schwann and ganglion cells in the heart muscle.

The commonest part of the alimentary tract to be affected is the esophagus (Fig. 7), which becomes enormously dilated to give rise to the malady known as *mal d'engasgo* in which the patient vomits nearly all food and becomes emaciated. The colon is also commonly affected and becomes enormously hypertrophied; it is not unusual for a volvulus to ensue, often with fatal results. The stomach and duodenum

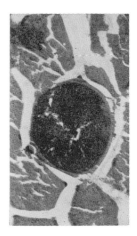

Fig. 8. Sarcocystis lindemanni in human skeletal muscle (Pugh's case); transverse section (×350).

are less often affected; sometimes the muscular wall of the ureter may be similarly infected and give rise to obstruction of the passage of urine. These conditions are known as megaesophagus, megacolon, megaureter, etc.

In the late form of Chagas' disease, it is usually impossible to demonstrate the organism in the affected tissue. The focal accumulations of adventitious cells, however, present a characteristic picture.

The pseudocysts of T. cruzi are up to 60 mμ in length; on superficial examination they could be mistaken for sarcocysts or pseudocysts of Toxoplasma, but their contents reveal the true nature of the parasite. The condition is unlikely to be met with in laboratory animals, though accidental cross infections have been known to take place in laboratories in Europe and North America where the trypanosome has been maintained. In endemic areas, of course, any suspicious changes involving muscular organs should be investigated to see if they could have a Chagasian etiology.

C. Leishmania donovani

Nests enclosing leishmanial bodies are frequently found in muscle in Trypanosoma cruzi infections. It might be thought, therefore, that the disease leishmaniasis itself would be accompanied by similar lesions, but invasion of the actual muscle fiber by this organism is unknown. In rare instances, the skeletal muscles may be infected with Leishmania donovani, but as Adler (1940) pointed out, the parasite lies in the connective tissue cells between the fibers of voluntary muscle; in such circumstances, the muscles show focal accumulations of lymphoid macrophage cells. C. Manson-Bahr (1957) has reported the presence of L. donovani in large numbers in the hearts from acute human cases occurring in an endemic area in East Africa. Undoubtedly, in this instance also, the parasite grows in lymphoid macrophage cells in the muscular tissue, proliferating in this site, but with no infection of the muscle fiber itself.

Leishmania enriettii causes severe cutaneous lesions of the nose of guinea pigs. The parasites, which are double the size of L. donovani, infiltrate into the deeper tissues to invade the muscles below.

D. Besnoitia besnoiti

A parasite of coccidial affinities is found commonly in cattle of the Transvaal and Rhodesia, primarily in the subcutaneous tissue, but often

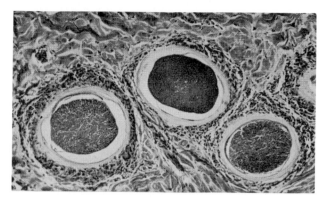

Fig. 9. *Besnoitia besnoiti* cysts in muscles of cow (×95). Courtesy of Dr. P. L. Leroux.

in such numbers that the cysts extend into musculature. For this reason and because of some similarity in morphology, the organism was originally placed in the genus *Sarcocystis.*

The cysts are large bodies, up to half a millimeter or even more in diameter, and are filled with narrow spores, resembling more the sporozoites of a malaria parasite than those of *Sarcocystis* (Fig. 9). They are easily distinguished from sarcocysts or other protozoa by their wide cyst wall. This is actually a double wall, a narrow and nucleated inner wall and a thick and hyaline outer wall up to 20 mμ in thickness.

Another species of *Besnoitia* (*B. saurianum*) is found in Central American lizards, and in British Honduras, the basilisk, is often extensively parasitized throughout its tissues and organs, including muscle. Signs of reaction are minimal.

E. Toxoplasma gondii

This parasite has a distribution as wide or wider than that of *Sarcocystis* and is a very common infection of man nearly everywhere in the world. Its life history and classification have only recently been determined. *Toxoplasma* was shown to be a typical coccidian, with oocysts like *Isospora* occurring in the feces of cats (Hutchinson *et al.,* 1970; Sheffield and Melton, 1970; Frenkel *et al.,* 1970). *Toxoplasma* is an important infection of laboratory animals; in fact, it was first described in domestic rabbits by Splendore (1909) in Brazil. It occurs commonly in dogs, cats, rodents, sheep, cattle, pigs, as well as various wild mammals, marsupials, and birds (including chickens). Numerous human cases have now been diagnosed in all continents, and if positive serological reactions truly indicate past or present infection, then the number of human infections must run into millions.

In acute toxoplasmosis, the organism exists in the proliferative form as a crescentic body 4–5 mμ long, with a single nucleus; in the chronic disease, the organisms are found in cysts of various sizes in different parts of the body. In both types, muscle may be the site of infection (Fig. 10), though much less commonly than in the brain. *Toxoplasma* is invariably intracellular at first, and it invades a variety of cells, including probably the muscle fiber. But the origin of the focus in muscle is perhaps more commonly the lymphoid macrophage cell lying between the fibers than a muscle cell itself.

Whatever the origin, the skeletal and cardiac muscles not infrequently show parasites. Frenkel (1953) has demonstrated the existence of a severe myositis of the diaphragm in acute infections in mice and also cysts in skeletal and heart muscle in old-standing infections in pigeons, guinea pigs, and other animals. In man, a severe myositis may occur, and Kass *et al.* (1952) describe such a case in which a biopsy of the gastrocnemius muscle showed degeneration of individual muscle fibers surrounded by lymphocytes, plasma, and mononuclear cells. In a second biopsy, the organism was found in the fibers. Later, there was gross destruction of the muscle fibers, the sarcous content of the fiber becoming swollen, hyaline, and usually eosinophilic, with much fragmentation. The sarcolemmal nuclei first increased in number and size, then became hyperchromatic, and finally were shrunken and necrotic.

In the less acute cases, cysts may be present only in small numbers and may be detected by accident, such as the single cyst found in the heart muscle of an American soldier drowned in the Panama Canal

Fig. 10. Cysts of *Toxoplasma gondii* in thigh muscle of mouse (6-month-old infection) (×350). See also Lainson (1958).

(Mantz *et al.*, 1949). Such infections, whether in man or animals, often give rise to difficulties in diagnosis, because the tissue for examination may well be inadequately prepared. If it is fresh, then diagnosis is easily made by inoculating a suspension into susceptible animals [e.g., the multimammate mouse Lainson (1958)], but a poorly stained section with cysts filled with small oval bodies might be mistaken for *Sarcocystis, Trypanosoma cruzi, Besnoitia,* or the "P" organism (Frenkel, 1956).

The type of myocarditis caused by *Toxoplasma* has been described by Bengtsson (1950). Small cysts lie within the unchanged muscle fiber; larger ones are not so clearly inside. But in both types there is a complete absence of local inflammation. Foci of histiocytes, mast cells, and lymphocytes, however, are found scattered through the heart muscle. Potts and Williams (1956) described a fatal case in a man where autopsy revealed a much enlarged heart with foci of lymphocytes, plasma cells, and mononuclears, occasionally perivascular, but more often generalized in the heart muscle.

There is an almost invariable custom in Paris and other French cities for mothers to feed their babies on raw meat juice. Desmonts *et al.* (1965) showed that raw meat is often heavily infected with cysts of *Toxoplasma gondii* and that the children quickly develop antibodies in the Sabin–Feldman test after being fed such material. In a parallel series of children who were kept off raw meat, no serological conversion took place. Lainson and Garnham (1960) demonstrated that the slightest cooking of meat was sufficient to sterilize the infection. Kean *et al.* (1969) reported an outbreak of clinical toxoplasmosis in a group of students in New York, who had eaten hamburgers, which in the United States are made of rare beef, often mixed with rare pork. The latter in particular is not uncommonly infected with *T. gondii.*

Carnivorous animals, such as dogs, cats, ferrets, and weasels, may easily become infected by eating prey with parasites in the muscle and organs (Jacobs, 1967), though the respective importance of this route and of the oocyst in transmission has yet to be determined.

Congenital transmission of *Toxoplasma* is well established, and in such cases the child is usually born dead or dies soon after birth. The muscles may show extensive involvement, particularly those of the heart. One of the earliest reports of human infection was made by Torres (1927) in Brazil; the newborn child showed permanent and generalized muscular contractures. At death 2 days later, sections of these muscles exhibited a condition of disseminated myositis, while the heart muscle was in a condition of acute myocarditis. All these lesions were associated with the presence of cysts. Nearly 30 years later, a whole series of such cases was described by Cardoso *et al.* (1956) from the same region.

F. Amebic Abscess

The parenteral distribution of *Entamoeba histolytica* may be widespread in practically every organ and tissue of the human body, though sites other than the liver and lungs are only rarely affected. The ameba is carried from the original site in the intestine by the bloodstream to a tissue or organ. There, perhaps because of some local damage, it starts to multiply; phagocytes accumulate and a small abscess forms. This abscess spreads by radial diffusion, the amebas being numerous at the advancing edge. Amebic infection of the skin, subcutaneous tissue, or bone may extend into the neighboring muscles, e.g., into the buttocks or thigh. A liver abscess due to this organism may rupture into the muscles of the diaphragm or into those of the anterior abdominal wall.

The ordinary amebic ulcers of the large intestine develop in the submucosa and open into the lumen of the bowel; sometimes, the organism progresses in the reverse direction and attacks the unstriated muscle, eventually reaching and penetrating the peritoneal coat of the intestine. Usually, there is little tissue reaction in the muscularis in spite of a heavy infection of amebas and extensive lysis of the muscle fibers.

G. Malaria Parasites and Other Hemosporidia

Myocardial degeneration of the heart may occur in chronic malaria when repeated attacks of the disease lead to anemia and general debility. Apart from this condition, a pernicious form of cardiac malara occasionally develops in *Plasmodium falciparum* infections, which usually has a rapidly fatal termination. In this species of malaria parasite, the growth of the parasite continues in the peripheral blood for only a short part of the cycle, and final maturation takes place in the capillaries of the internal organs. The presence of sticky infected corpuscles in vessels of narrow caliber leads to a slowing down of the circulation of the blood in certain organs, including the heart. The endothelial lining of such vessels suffers, the cells swell, fluid passes from the blood into the adjacent muscle, and the process of anoxemia and anoxia gradually increase. Stasis, thrombosis, and hemorrhages through the weakened endothelium follow, and large areas of cardiac muscle may be occupied by infarcts, while flame-shaped hemorrhages occur on the pericardial surface (Garnham, 1949).

Sections of the heart muscle from a fatal case of cardiac malaria show a characteristic picture. The smaller vessels are filled with erythrocytes containing mature pigmented schizonts of *P. falciparum,* together with numerous lymphoid macrophage cells. The muscle cells themselves never contain parasites, but according to Maegraith (1948), small hemorrhages are occasionally found in the muscle substance around the small veins.

The condition is primarily a case of the parasites blocking the blood vessels of the muscle. The fibers become involved only secondarily. Changes in the fibers usually occur as part of a fulminating process extending throughout the body. The secondary changes include the fatty degeneration of the muscle with a deposition of globules of irregular size clustered around the nucleus or distributed throughout the contractile substance. Similar changes are found in rhesus monkeys dying of *Plasmodium knowlesi* malaria.

Plasmodium coatneyi is another malaria parasite, which disappears into the internal organs to undergo schizogony, though, like *P. falciparum,* schizonts overflow into the peripheral blood toward the end of fulminating infections. The monkey parasite selects the capillary vessels of the ventricular myocardium as its principal site for sequestration (Garnham, 1965; Desowitz *et al.*, 1969; Miller *et al*, 1972); little necrosis of the heart muscle follows and, any accumulation of adventitious cells is minimal, probably because the course of the disease is too rapid (Fig. 11).

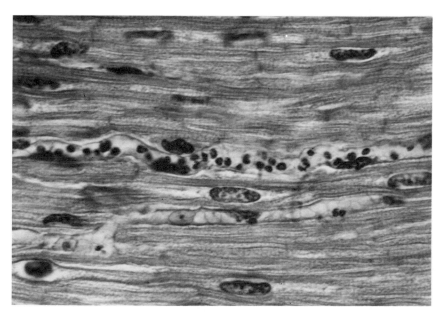

Fig. 11. Plasmodium coatneyi schizonts blocking a capillary in the heart muscle of a monkey (×820).

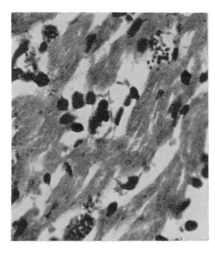

Fig. 12. Exoerythrocytic schizonts of *Plasmodium gallinaceum* in heart muscle of chick (×1200).

In avian malaria, exoerythrocytic schizogony takes place in a great variety of tissue cells, chiefly, however, in those of the lymphoid macrophage type, including the endothelium of the smaller blood vessels. In *Plasmodium gallinaceum* infections of chickens, invasion of such cells may be very extensive; the skeletal muscles also become involved during the third week of infection, while the heart muscle is frequently the site of exoerythrocytic schizogony of this parasite (Fig. 12). In neither of these conditions is any marked cellular reaction visible among the occluded vessels, but the nutrition of the surrounding muscle must inevitably suffer. A similar involvement of the heart muscle occurs in other species of avian malaria, and in lizard malaria also, where the actual development of the parasite takes place in the connective tissue of the subendothelial layer of the heart.

It has long been known that chickens suffering from chronic *Plasmodium juxtanucleare* malaria develop a pericarditis with effusion from which they die, and Al-Dabagh (1958) demonstrated the nature of the pathological changes in the muscle. Large accumulations of adventitious cells, particularly lymphocytes, plasma cells, and macrophages, are present between the muscle fibers of the heart, forming in places almost tumorlike masses. The muscle cells lose their striation and show other degenerative changes (Fig. 13).

Malaria parasites can also be found in the muscle of artificially infected mosquitoes during the sporogonic stage. Weathersby (1952) showed that when blood containing male and female gametocytes of

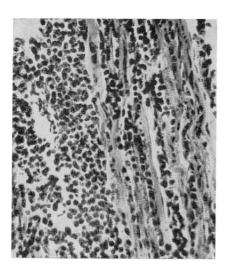

Fig. 13. Myocarditis in chick heart in *Plasmodium juxtanucleare* malaria (×350). Courtesy of Dr. M. A. Al-Dabagh.

P. gallinaceum is introduced into the hemocoele of a vector mosquito, fertilization follows and the oocysts grow to maturity, principally between the muscle fibers of the thoracic muscle of the insect. Even in natural infections, sporozoites are found in large numbers among the muscle fibers. In neither instance does the presence of the parasite excite any visible cellular reaction. Garnham *et al.* (1961) showed that *Hepatocystis kochi,* the common malaria parasite of African monkeys, underwent sporogonic development in the muscles and ganglia of *Culicoides* without causing any apparent damage to the insect.

Part of the exoerythrocytic cycle of *Leucocytozoon* takes place in the muscle cells of the heart. This hemosporidian has a cosmopolitan distribution in a wide variety of birds. In certain places, notably Canada (in ducks) and Scandinavia (in various game birds), the infection is of considerable veterinary importance. The birds become infected by the bites of *Simulium* flies; the sporozoites develop in different parts of the body, including the heart muscle (Fig. 14). The effect of the parasite on the host cell is quite extraordinary; with the growth of the organism, the nucleus and protoplasm of the host cell undergo an increase in size which can only be described as a case of gigantism. The host cell nucleus finally attains a diameter of 190 mμ. *Leucocytozoon* is able to evoke this intense stimulation of growth in a variety of cells, probably chiefly of lymphoid macrophage origin, but Huff (1942) suggests, on the evidence of the resemblance of the shape of the megalo-

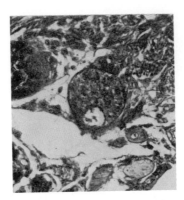

Fig. 14. Leucocytozoon megaloschizonts in heart muscle of bird (×350). Courtesy of Dr. Cowan.

schizonts and their nuclei, that the parasites may develop in the muscle cell itself. The megaloschizont grows to a size of nearly half a millimeter and causes extensive damage to the heart muscle. This exoerythrocytic stage in the heart muscle is ephemeral and can only be detected very early in the infection. As in most protozoal infections, the presence of the unruptured parasites causes no reaction on the part of the host, but when the schizont bursts, a local infiltration of phagocytes takes place (Fallis *et al.*, 1956). The pathogenicity of *Leucocytozoon* in birds results not so much from these lesions in the organs, but from the later heavy destruction of blood cells. On the other hand, Walker and Garnham (1972) described fatal cases of a *Leucocytozoon*-like infection in parakeets in England, in which death was undoubtedly caused by very heavy infiltration of the heart muscle by the megaloschizonts of the parasite before the blood had become invaded.

The musculature of other arthropods may also be infiltrated by the developmental stages of blood protozoa. The common hemogregarine *Hepatozoon balfouri* of jerboas undergoes sporogony in the mite *Haemolaelaps longipes*, and the zygotes, oocysts, and ruptured sporoblasts gradually usurp much of the body cavity; the muscle bundles are flattened and pushed aside, though there is little actual destruction. The effect of gregarines on invertebrate muscle is described by Vivier in Chapter 3 of Volume II.

H. Microsporidia and Myxosporidia

Insects and fish are not uncommonly affected by protozoa belonging to the subclass Cnidosporidia, which possess spores of a unique character. Well known examples are the microsporidians *Nosema bombycis*, causing pebrine or silkworm disease, and *Nosema apis*, causing bee disease, and the myxosporidians affecting goldfish, trout, and other fish of commercial importance. In silkworm disease, colonies of the parasite invade muscle fibers under the cuticle of the insect, as well as most other tissues. *Plistophora, Nosema,* and other microsporidians are common parasites of mosquitoes and simuliids. *Plistophora culicis* is found in *Aedes aegypti* and various sylvatic species of the genus in temperate and tropical latitudes; it is pathogenic for *Aedes aegypti* and at one time appeared to be a candidate for biological control of this and other important vectors of disease. The parasite develops in the fat body and various organs of the mosquito and may become disseminated throughout the tissues; the muscle is occupied by a mass of spores and developing forms (Fig. 15). *Nosema stegomyiae* is even more lethal and may completely wipe out colonies of *Anopheles stephensi* and *Anopheles gambiae*. Unfortunately, in nature a balance between host and parasites becomes quickly established, and the population remains stable.

The muscle is not a very common site of development of many of these parasites, but a few species have a special affinity for this tissue. Several genera of microsporidia attack the muscles of fish, frogs, and crustaceans and cause a hyaline degeneration of the muscle fibers

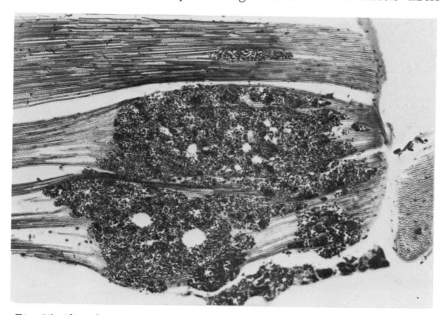

Fig. 15. *Plistophora culicis* in thoracic muscle of *Culex fatigans* (×318). From material kindly supplied by Dr. D. G. Reynolds.

(Grassé, 1953). *Glugea destruens* affects *Callionymus lyra; Glugea danilewskyi* affects the frog; and several species of *Thelohania* invade the muscles of fish and various invertebrates and cause their death. *Plistophora* attacks blennies and other fish, forming little protoplasmic masses between the fibrils, which develop into pansporoblasts. Another species of *Plistophora, P. myotrophica,* was so named because of its predilection for the muscular system of the host, the common toad *Bufo bufo*. Canning *et al.* (1964) showed that lysis of the myofibrils was followed by atrophy of the muscles and general emaciation of the toads, many of which died of infection. The spores of the microsporidian are eventually set free in spaces between the fibers, and phagocytes infiltrate the area, perhaps because the parasites do not become enclosed in a cyst wall (Fig. 16).

Microsporidian infection of mammals is rare in the musculature, though *Nosema cuniculi* has long been known in the brain and kidneys of rodents under the name of *Encephalitozoon*. Doby and Jeanne (1968), however, reported the discovery of a single "cyst" or colony of *Thelohania apodemi* in the muscle of a field mouse caught in France in a

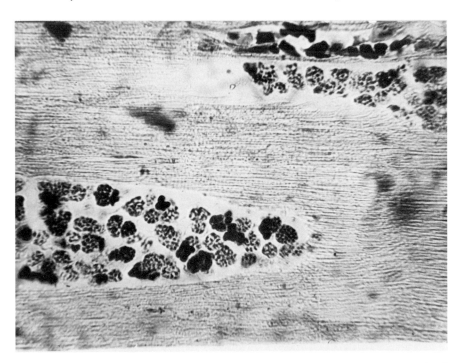

Fig. 16. Plistophora myotrophica in skeletal muscle of toad (×250). By kind permission of Canning *et al.* (1964).

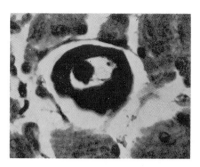

Fig. 17. Myxosporidian in heart muscle of sole (×1000). Courtesy of Mr. A. M. Qureshy.

locality where this parasite is often found in the brain of this rodent; the colony was 200 mμ long and contained typical sporoblasts and spores.

An important myxosporidian parasite of fish is *Unicapsula muscularis*, which invades the skeletal muscle fibers of the halibut. After repeated multiplication of the parasite, the fiber becomes almost entirely obliterated and is occupied by a mass of spores. The condition is known as wormy halibut and is of some economic importance. It is found off the Pacific Coast of North America. *Myxobolus pfeifferi* is very fatal to species of *Barbus* (*B. fluviatilis*). It develops in the muscle of these fish and causes tumors up to 7 cm in diameter. A fibrous corpuscle is formed by the host, and inside this corpuscle the parasites continue their development, with the formation of spores. Eventually the parasites die and all that remains is a fibrous nodule containing many spores. *Myxobolus arbiculatus* attacks the muscles of other fish. Kudo (1954) has demonstrated a most unusual effect on muscle due to *Myxobolus intestinalis*. This organism invades the intestinal wall of the black crappie and causes rupture of some of the circular muscle fibers which then swivel so as to lie at right angles to the main bundles and project into the submucosa. *Myxosoma catostomi* is a parasite of the muscle and connective tissue of *Catostomus cammersonii* (Fig. 17), while *Henneguya tegidiensis* becomes encapsulated in the muscles of the Welsh Gwyniad (Nicholas and Jones, 1959).

III. Helminthic Parasites

Unlike many of the protozoal infections of muscle, the helminthic infestations of this tissue are practically always part of a well known cycle of the parasite in which the exact stage and nature of the process are fully understood, even though the parasite may have undergone kaleidoscopic changes of host and of tissue or organ in a single host.

Certain helminths are obligatory parasites of muscle, and the successful progress of their life history depends upon the consumption of meat. When man forms a link in this chain, the matter assumes public health importance, and meat inspection is a routine procedure in sanitary practice. The presence of cysticerci of tapeworms is particularly looked for; different countries have various standards for the treatment or disposal of infected carcasses. Such procedures are well described in the monograph series of the World Health Organization (1957).

The prevention of trichinosis is also attempted in many countries by legal measures; thus, in England, it is a legal obligation for pig farmers to cook all food given to pigs. The parasite in pork can be killed by adequate cooking, by rapid freezing to —35°C., or by irradiation with X- and γ-rays (Gould et al., 1954).

A. Helminthic Parasites of the Muscle of Arthropods

Many species of filarial worms spend part of their developmental cycle in mosquitoes and other dipterous insects. The mosquito bites an infected individual and takes up the microfilaria either from the skin or the blood. The microfilariae in the course of 24 hours pass through the epithelial lining of the midgut of the insect into the hemocoele and thence to the muscles of the thorax. The larval forms of the parasite assume a typical sausage shape, measuring about 150 mμ in length at first, but growing in the next few days to a length of 1–5 mm or more (Fig. 18). Lavoipierre (1958) and a few other workers have demonstrated that in some species at least, the filarial larvae after arrival between the fibers of the thoracic muscles proceed to invade the fibers themselves, where they continue to grow. The final maturation of the parasite takes place outside the muscle cell. Whereas the thoracic muscles of arthropods are the muscles usually attacked by these parasites, a few species invade the leg muscles (*Dipetalonema viteae* in ticks, Chabaud, 1954) or the muscles of the abdomen or head. No evidence of marked tissue response is usually found, but in certain circumstances a reaction may occur. If the species of intermediate host is the incorrect one, or if the temperature is abnormally raised or lowered, the larvae undergo a chitinous encapsulation (Kartman, 1957). The chitin is apparently secreted by the tracheoles, and hardens to form a thick capsule around the worm. P. H. Manson-Bahr (1912) describes such parasites as mummies lying in their coffins in the thoracic muscles of *Stegomyia* mosquitoes.

Fig. 18. Larvae of *Mansonella ozzardi* in thoracic muscle of *Culicoides* sp. (×75). Courtesy of the late Professor J. J. C. Buckley.

Development of the worm in the muscles of insects is strikingly similar in all the different species or genera of filaria and in the different insect hosts. Lavoipierre (1958) gives a complete list of arthropods in whose muscles filarial larvae undergo development. This information is presented in Table I.

TABLE I
ARTHROPODS IN WHICH MUSCLES HARBOR VARIOUS FILARIAL LARVAE

Arthropod	Parasite	Primary host
Culex fatigans	*Conispiculum flavescens*	Lizard
Culicoides species	*Dipetalonema perstans*	Man
	Dipetalonema streptocerca	Man
Culicoides furens	*Mansonella azzardi*	Man
Culicoides nubeculosus	*Onchocerca cervicalis*	Horse
Culicoides pungens	*Onchocerca gibsoni*	Cattle
Ornithodorus tartakowskyi	*Dipetalonema viteae*	Merion
Forcipomyia velox	*Icosiella neglecta*	Frog
Simulium ornatum	*Onchocerca gutterosa*	Cattle
Simulium species	*Onchocerca volvulus*	Man
Anopheles sinensis and others	*Setaria digitata*	Sheep, goat, horse
Aedes and other mosquitoes	*Wuchereria bancrofti* and other species	Man
Chrysops species	*Loa loa*	Man

B. *Cysticercosis and Other Cestode Infections of Muscle*

One of the commonest parasitic infections of muscle is the larval stage of various species of tapeworms, which in their adult form usually live in the intestine of another host. The eggs pass out in the feces and are ingested by the intermediate host, a herbivorous or omnivorous vertebrate animal; on hatching, the embryos penetrate the intestinal wall and settle down in the muscles, which they reach via the bloodstream or lymphatics. In the muscle, the larva undergoes a metamorphosis into a cysticercus or other bladderlike structure, according to the genus of tapeworm. The life cycle is completed when the meat is consumed by the definitive host, in whom the cysticercus is digested out of the muscle. The uninjured head of the worm escapes into the small intestine to give rise to a mature tapeworm.

Various degrees of infestation of muscle occur, from a single cysticercus to a condition in which all the skeletal muscles of the body, together with the tongue and heart, are honeycombed with the parasite. Such infected meat, in the case of the pig, is known popularly as measly pork. It is probable that the development of this parasite in the muscle is extracellular, in the adipose connective tissue between the fibers; as it grows, it pushes aside the fibers, although the lateral pressure tends to make it assume an ovoid form. A tissue reaction always accompanies the formation of the cysticercus, whose outer wall is the product of the host. Particularly in abnormal hosts, such as cysticercosis in man, a further reaction is likely to occur. Leukocytes, eosinophiles, and cholesterol crystals may accumulate; finally, calcification of the whole lesion will ensue. The calcified bodies are easily seen in skiagrams of the muscles.

Taenia solium in the muscles of the pig has been known from the time of Aristophanes (450 b.c.). Today, the infection is widespread throughout the world, except in Mohammedan or Jewish lands where pork is not eaten. The larva is known as *Cysticercus cellulosae* and is a hyaline or opalescent body up to 15 mm long and 8 mm wide, containing the invaginated scolex (Fig. 19). These bodies are very long lived and may remain visible for 25 years. The following muscles, in order of frequency, are affected: heart, intercostals, neck and shoulder, diaphragm, and tongue. Man is normally the definitive host of the tapeworm, but he occasionally acts as the intermediate host also, when the larval form develops in the muscles and in other tissues. In the skeletal muscle, its presence is usually well tolerated, unless the infestation is heavy, when muscular fatigue, cramps, and general lassitude result. In

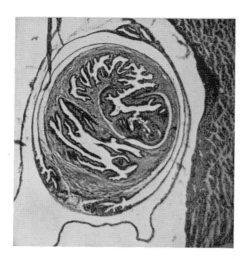

Fig. 19. *Cysticercus celluosae* in muscle of pig (×17).

the heart, the parasite provokes a myocarditis and the aortic and mitral valves become affected.

Taenia saginata is even more widespread in its distribution, the intermediate host being cattle. In Europe, this species of tapeworm is seven times as common as *T. solium* (Brumpt, 1949). *Cysticercus bovis* has the same appearance as *C. cellulosae*, but is slightly smaller in size (Fig. 20); it occurs in antelopes, giraffes, llamas, and other bovids,

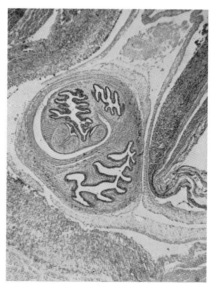

Fig. 20. *Cysticercus bovis* in muscle of ox (×20).

while man is the definitive host of this cestode also. In cattle, the following muscles are particularly affected: heart, masticatory muscles, tongue,
and diaphragm (in the ratio of $8:4:1:1$).

The larval stages of other cestodes are only rarely located in muscle.
The coenurus of *Multiceps serialis* is occasionally found in the muscles
of the back of man. Small foci of this infection exist, e.g., on the slopes
of Mount Kenya. The coenurus is a cystlike structure characterized by
the possession of numerous scolices.

The hydatid cyst of *Echinococcus granulosus* is occasionally found
in the heart and skeletal muscles of man and other mammals. Infection
takes place when the animal swallows the eggs of this tapeworm, whose
adult habitat is the small intestine of dogs, jackals, and wolves. The
adult worm is tiny, but the larva (the hydatid cyst) is very much
larger and is characterized by having a laminated layer and a germinal
layer from which issue brood capsules, each containing many scolices.
Echinococcus granulosus is particularly common in Iceland, Canada,
Argentina, South Africa, and South Australia. The large size of the cyst
interferes with the cardiac rhythm, and it may even rupture out of the
muscle into one of the chambers of the heart; in the skeletal muscle,
its pathogenicity is confined to pressure changes.

A different types of cestode invasion of muscle is seen in the immature
stages of *Diphyllobothrium* in the muscles of fish. The mature worm
lives in the intestine of man, dogs, cats, foxes, bears, and other animals;
the eggs pass out in the feces; a coracidium hatches in water and is
swallowed by a *Cyclops*, where it becomes a procercoid; and the *Cyclops*
is eaten by a freshwater fish, in whose muscles the parasite continues
its development as a plerocercoid or sparganum. Man becomes infected
by eating uncooked or improperly cooked fish. The following fish are
commonly found infected: pike, Miller's thumb, perch, salmon, trout,
grayling, and barbel. The infection is heaviest in the older fish. The
plerocercoid is easily visible to the naked eye, measuring up to 2 cm
in length, and is usually encapsulated. A fibrous wall is formed as a
result of a tissue response by the host. In other respects, the muscle
is unaffected and these plerocercoids seem to be harmless to the fish.
In man, this stage of the cestode (the sparganum) is rarely found, but
according to Faust (1949), the spargana proliferate in thousands in
the muscle fascia and elsewhere, and the infected area becomes
edematous and painful. Weinstein *et al.* (1954) described the pathology
of three human cases from Korea. A toxin is thought to diffuse through
the cuticle of the sparganum and affect directly the muscle in which
focal necrosis occurs. An inflammatory response follows with an accumulation of eosinophils, macrophages, and fibroblasts. A curious feature

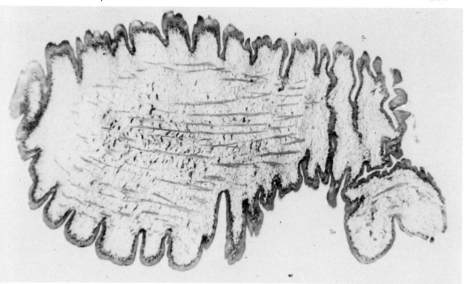

Fig. 21. Sparganum from human infection. Note muscle fibers in body of parasite (×50). Courtesy of Professor George Nelson.

is the presence sometimes of Charcot–Leyden crystals in the granuloma. Figure 21 represents a section of a sparganum from a human infection; as in many helminths, muscle fibers are present in the body of the parasite, which itself may lie in the muscle of the host.

C. Trichinosis

Trichinella spiralis is an extremely common parasite of the muscles of man, cat, rat, dog, pig, bear, mongoose, and other animals, and is of cosmopolitan distribution. Man contracts the infection by eating uncooked meat. The consumption of pork sausages is a particularly common method of infection. In some places—e.g., in the Liverpool outbreak described by Semple et al. (1954)—the incidence of the infection is much higher in women than men. This is because women in these localities have the habit of eating raw sausage meat; in men, occupation is an important factor, such as the handling of infected pork by slaughterhouse workers or butchers.

The adult nematode lives in the mucosa of the small intestine. Embryos are evacuated and migrate by the lymphatics or portal vein to the musculature of the body, particularly the diaphragm, intercostals, and the muscles of the throat, tongue, and eye. In the skeletal muscles, the parasites tend to congregate toward the tendinous extremities.

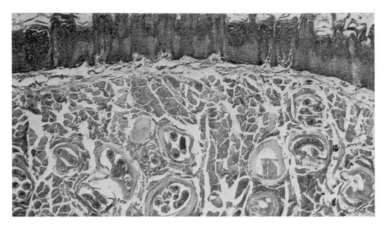

Fig. 22. *Trichinella spiralis* larvae in tongue of experimentally infected rat (×85).

Encystment of the larva follows (Fig. 22), though the exact location of this process is still debatable. It was originally thought that the larva penetrated the muscle cell, where an ellipsoidal cyst capsule was quickly formed, but the French school (Brumpt, 1949) has demonstrated that the larva inhabits the interfascicular connective tissue from which the cyst wall is derived.

The cysts are just visible to the naked eye as lemon-shaped bodies about half a millimeter in length. The long axis lies in the direction of the fibers, and the larva is tightly coiled up inside the cyst. The cysts remain viable for many years and have been found in man over 30 years after infection. Calcification of the older cysts frequently takes place.

The effect of the parasite depends upon the intensity of the infestation. If this is high, grave symptoms follow, including high fever and delirium, accompanied by severe muscle pains, difficulty in swallowing, eating, and breathing, and ending in cachexia and death between the second and seventh week. In light infections, the symptoms gradually decline and the patient passes into a state of premunition.

According to Faust (1949), the histological changes in the muscle are quite characteristic. The muscle fibers around the cysts lose their striae and the nucleus divides repeatedly; adjacent fibers swell and the connective tissue increases in amount. This hyperplasia continues at the expense of the muscle fibers, which gradually degenerate. The removal by biopsy of a small portion of deltoid or biceps muscle from near the tendinous attachment and the examination in a trichina press under low magnification will usually reveal the presence of the organism.

Digestion of a portion of muscle is recommended for the diagnosis of a suspected clinical infection. Small fragments are incubated for some hours with 1% acid pepsin; the mixture is then centrifuged and the deposit is examined with an inverted microscope in a hollow ground slide; the heavier larvae are then easily visible in the bottom of the cell.

Myocarditis due to *Trichinella spiralis* is common; an interstitial infiltrate composed of eosinophils and plasma cells accumulates beneath the pericardium, while granulomata, composed of histiocytes and giant cells surrounded by various adventitious cells, are found deeper in the myocardium (Ash and Spitz, 1945).

An outbreak of human trichinosis in Liverpool in 1953, accompanied by two fatalities, provided an opportunity to study the pathology of the disease in some detail. Kershaw *et al.* (1956) describe the distribution of the larvae in the different muscles of the body on a quantitative basis (number of larvae per gram of muscle) and showed that in a heavy infection, the muscles of the tongue, diaphragm, forearm, and calf were the most severely affected, whereas in a light infection, the infection was most intense in the limb muscles and absent in the tongue. The practice, therefore, of confining trichinoscopy to the diaphragm is unwise, and the somatic musculature should be included also. Paradoxically, the degree of pathogenicity of the organism is proportional to the resistance offered by the host. In the usual infections, the larvae encyst harmlessly in the skeletal muscles; in a fatal case, the larvae go to the heart muscle where a widespread focal destruction of fibers follows the intense inflammatory reaction, with disintegration of the parasite (Semple *et al.*, 1954). The chief damage to the heart is produced before encystment.

D. Trematode and Other Parasites of Muscles of Fish, Crustacea, and Mammals

Fish are affected by a large variety of helminths of all types. The plerocercoid stage of *Diphyllobothrium* has been described above. Immature stages of various flukes are found in the musculature of fish. The metacercariae of *Opisthorchis felineus* occur in the tench (Fig. 23). In the common human fluke of the Far East *Clonorchis sinensis*, the cercariae attack the fish (particularly the red fish *Carassius auratus*) and get through the scales and into the flesh where they encyst as metacercariae. A double capsule is formed in the muscle, the outer one being a tissue response by the host. Raw fish are eaten, the cyst walls

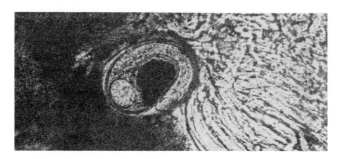

Fig. 23. Metacercariae of *Opisthorchis felineus* in muscle of fish; crush preparation (×75).

are dissolved in the duodenum, and the parasite continues its development in man, dogs, cats, badgers, mink, and other animals. A similar type of development occurs in the small fluke *Heterophyes heterophyes*. The metacercariae of this helminth encyst between the muscle fibers of various fish, including the mullet. The adult fluke lives normally in the intestine of man and other mammals; occasionally, it migrates from this site to the heart muscle (Fig. 24), where eggs are deposited and a myocarditis develops, with edema, hemorrhage, and myocardial degeneration ending in death.

Another fluke, *Paragonimus westermani*, spends part of its developmental stage in the muscle and other tissues of crabs and crayfish. The

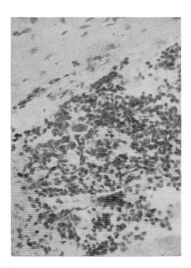

Fig. 24. *Heterophyes heterophyes* egg in human heart muscle with myocarditis (×300).

metacercariae become encysted in the muscles of the thoracic legs and in the cephalothorax itself. The cysts are pearly white and lie encapsulated in host tissue envelopes and derive nourishment from the host (Faust and Russell, 1958). The cyst is fixed to the muscle by a special fibrous attachment present at one end of the wall. Man becomes infected by eating half-pickled crustaceans, in which the worm is still alive. Normally, the fluke makes its way to the lungs of the human host, but sometimes it loses its way and settles down in one of the skeletal muscles instead.

The schistosomes commonly affect the smooth muscle in the walls

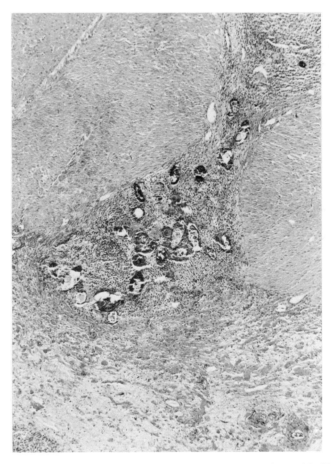

Fig. 25. *Schistosoma haematobium* eggs deposited in muscle of bladder wall of experimentally infected baboon (*Theropithecus gelada*) (×65). Courtesy of Drs. R. Kuntz, E. Myers and A. Cheever.

of the intestine and bladder. The eggs of *Schistosoma mansoni, S. haematobium,* and S. *japonicum* become deposited in large numbers in this tissue (Fig. 25) where they evoke an intense inflammatory reaction so that much of the muscle may be replaced by granulomata (Sadun et al., 1970). The eggs are seen in little groups, like a nidus of *Balantidium coli* trophozoites, but surrounded by lymphocytes, fibroblasts, polymorphonuclear leukocytes, eosinophils, and giant cells. The wall of the viscera becomes much thickened in chronic cases of the disease, though this reaction is less pronounced in the indigenous inhabitants of hyperendemic areas than in the newcomers who contract the disease (Gelfand, 1967).

Similar lesions may be produced in the heart muscle and pseudo abscesses develop around the eggs deposited in the myocardium. A severe myocarditis follows (Zahawi and Shukri, 1956). Ectopic infiltration of schistosome eggs is occasionally seen in the diaphragm and skeletal muscle.

Larval nematodes may be found in large numbers in the muscles of a variety of saltwater fish. Thus, the flesh of cod off the coast of Iceland is often found to be heavily infested with a small threadworm (Syme, 1949). The larva of different species of *Anisakis* may be found in the flesh of both freshwater and marine fishes, while the larvae of *Porrocaecum* and *Anisakis* are common in thin-walled cysts in the musculature of Icelandic cods (Grainger, 1959). No gross damage appears to be produced by these worms, though in the case of cestodes, Linton (1908) noted that the American butterfish when heavily parasitized with the cysts weighed much less than uninfected specimens; however, he found no evidence of any inflammatory or pathological condition of the muscle.

E. Nematode Parasites of Muscle

The nematodes form one of the largest classes in the animal kingdom, and it is therefore not surprising to find them well represented as parasites of muscle. *Trichinella* in man, filaria in mosquitoes, and larval nematodes in fish have been considered separately, but many more examples exist, and two more are mentioned here to illustrate larval and adult stages, respectively.

The roundworm *Ascaris devosi* lives in its adult form in the intestine of Viverridae. Its development in an intermediate host, the white mouse, has been studied by Sprent (1952), who showed that unlike the situation in *Ascaris lumbricoides,* this worm migrated in large numbers via the portal system into the general musculature, particularly in the anterior part of the body. A week after infection, the parasite may become encapsuled in the diaphragm or heart; after 3 weeks, white nodules imprisoning the third stage larvae are very conspicious in the somatic musculature. The nodules consist of a layer of fibroblasts, histiocytes, and inflammatory cells around the parasite. The ferret or other carnivorous mammal eats such an infected mouse and the larvae are set free in the intestine, where they grow to maturity.

Onchocerca is found in various species of mammals, including man, and in most animals, the adult form of the worm is enclosed in nodules lying in the subcutaneous tissue. In eland and cattle of tropical and South Africa, a species of *Onchocerca* is known (Leroux, 1947) that inhabits nodules projecting deeply into the muscles of the trunk of these animals (Fig. 26). In sections of such structures, the adult filarial worms are seen lying in a cyst surrounded by dense fibrous tissue compressing the muscle fibers in the vicinity. The tissues may be riddled with these nodules, and the meat is of inferior quality, but no specific lesions in the muscle are to be found. Another species of *Onchocerca* occurs in the tendinous insertion of the triceps muscle of antelope in African swamps.

Visceral larva migrans sometimes involves muscle and may be due to various species of nematodes. Second stage larvae of *Toxocara canis* were recovered by Dent *et al.* (1956) from the heart and skeletal muscle

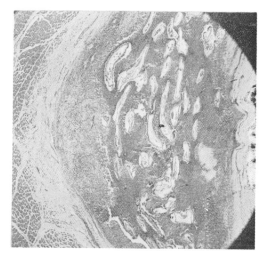

Fig. 26. Onchocerca nodule in muscle of cow (×22). Courtesy of Dr. P. L. Leroux.

of a child who had died from acute hepatitis; the infection was less intense in the muscle than in other organs (5 larvae per gram in muscle as compared with 60 larvae in the liver). The migrations of the *Toxocara* leave tracks in the muscle, which are accompanied by cellular infiltrates of lymphocytes and various leukocytes or plasma cells. Another form of larva migrans is due to infections with *Gnathostoma,* which produces abscesses in the somatic musculature during its migrations through the human body (Faust and Russell, 1965).

REFERENCES

Adler, S. (1940). *Trans. Roy. Soc. Trop. Med. Hyg.* **33,** 419.

Al-Dabagh, M. A. (1958). *Trans. Roy. Soc. Trop. Med. Hyg.* **53,** 8.

Ash, J. E., and Spitz, S. (1945). "Pathology of Tropical Diseases." Saunders, Philadelphia, Pennsylvania.

Awad, F. I. (1973). *Z. Parasiterk.* **42,** 43.

Ball, G. H. (1944). *Trans. Amer. Microsc. Soc.* **63,** 144.

Bengtsson, E. (1940). *Cardiologia* **17,** 289.

Brumpt, E. (1949). "Précis de Parasitologie," 6th ed. Masson, Paris.

Canning, E. U., Elkan, E., and Trigg, P. I. (1964). *J. Protozool.* **11,** 157.

Cardoso, R. A. de A., Guimarães, F. N., and Garcia, A. P. (1956). *Mem. Inst. Oswaldo Cruz* **54,** 571.

Chabaud, A. C. (1954). *Ann. Parasitol. Hum. Comp.* **29,** 42.

Dent, J. H., Nichols, R. L., Beaver, P. C., Carrera, G. M., and Staggers, R. J. (1956). *Amer. J. Pathol.* **32,** 777.

Desmonts, G., Couvreur, J., Alison, F., Baudelot, J., Gerbeaux, J., and Lelong, M. (1965). *Rev. Fr. Etud. Clin. Biol.* **10,** 952.

Desowitz, R. S., Miller, L. H., Buchanan, R. D., and Permapanich, B. (1969). *Trans. Roy. Soc. Trop. Med. Hyg.* **63,** 198.

Destombes, P. (1957). *Bull. Soc. Pathol. Exot.* **50,** 221.

Doby, J. M., and Jeanne, A. (1968). *Ann. Parasitol. Hum. Comp.* **43,** 619.

Erickson, A. B. (1940). *Auk* **57,** 514.

Fallis, A. M., Anderson, R. C., and Bennett, G. F. (1956). *Can. J. Zool.* **34,** 389.

Faust, E. C. (1949). "Human Helminthology," 3rd ed. Lea & Febiger, Philadelphia, Pennsylvania.

Faust, E. C., and Russell, P. F. (1965). "Craig and Faust's Clinical Parasitology." Lea & Febiger, Philadelphia, Pennsylvania.

Frenkel, J. K. (1953). *Amer. J. Trop. Med. Hyg.* **2,** 390.

Frenkel, J. K. (1956). *Ann. N.Y. Acad. Sci.* **64,** 215.

Frenkel, J. K., Dubey, J. P., and Milles, N. L. (1970). *Science* **167,** 893.

Garnham, P. C. C. (1949). *Ann. Trop. Med. Parsitol.* **43,** 47.

Garnham, P. C. C. (1965). *Omagiu Acad. Professor Dr. M. Ciuca Ed. Acad. Republ. Pop. Romane.* p. 199.

Garnham, P. C. C., Heisch, R. B., and Minter, D. M. (1961). *Trans. Roy. Soc. Trop. Med. Hyg.* **55,** 497.

Gelfand, M. (1967). "A Clinical Study of Intestinal Bilharziasis." Arnold, London.

Gould, S. E., Gomberg, H. J., and Bethel, F. H. (1954). *J. Amer. Med. Ass.* **154,** 653.
Grainger, J. N. R. (1959). *Parasitology* **49,** 121.
Grassé, P.-P., ed. (1953). "Traité de Zoologie," Vol. 1, p. 2. Masson, Paris.
Huff, C. G. (1942). *J. Infec. Dis.* **71,** 8.
Hitchinson, W. M., Dunachie, J., Siim, J. C., and Work, K. (1970). *Brit. Med. J.* **1,** 142.
Jacobs, L. (1967). *In* "Advances in Parasitology" (B. Dawes, ed.), Vol. 5. Academic Press, New York.
Kartman, L. (1957). *Rev. Brasil. Malariol. Doencas Trop., Publ. Avulsas* **5.**
Kass, E. H., Andrus, S. B., Adam, R. D., Turner, F. C., and Feldman, H. A. (1952). *AMA Arch. Intern. Med.* **89,** 759.
Kean, B. H., Kimball, A. C., and Christenson, W. (1969). *J. Amer. Med. Ass.* **208,** 1002.
Kershaw, W. E., St. Hill, C. A., Semple, A. B., and Davies, J. B. M. (1956). *Ann. Trop. Med. Parasitol.* **50,** 355.
Köberle, F. (1956). *Zentralbl. Allg. Pathol. Pathol. Anat.* **95,** 321.
Köberle, F. (1968). *Advan. Parasitol.* **6,** 79.
Köberle, F., and Nador, E. (1956). *Z. Tropenmed. Parasitol.* **7,** 259.
Kudo, R. R. (1954). "Protozoology," 4th ed. Thomas, Springfield, Illinois.
Lainson, R. (1958). *Trans. Roy. Soc. Trop. Med. Hyg.* **52,** 396.
Lainson, R., and Garnham, P. C. C. (1960). *Lancet* **2,** 71.
Lavoipierre, M. M. J. (1958). *Ann. Trop. Med. Parasitol.* **52,** 326.
Leroux, P. L. (1947). *Trans. Roy. Soc. Trop. Med. Hyg.* **41,** 8.
Linton, E. (1908). *Bull. U.S. Fish. Bur.* **28,** 1197.
McGill, R. J., and Goodbody, R. A. (1957). *Brit. Med. J.* **2,** 33.
Mackinnon, J. E., and Abbott, P. (1955). *Ann. Trop. Med. Parasitol.* **49,** 308.
Maegraith, B. G. (1948). "Pathological Processes in Malaria and Blackwater Fever." Oxford Univ. Press, London and New York.
Manson-Bahr, C. (1957). *Trans. Roy. Soc. Trop. Med. Hyg.* **51,** 371.
Manson-Bahr, P. H. (1912a). *Res. Mem. London School Hyg. & Trop. Med.* No. 1.
Mantz, F. A., Dailey, H. R., and Grocott, R. G. (1949). *Amer. J. Trop. Med. Hyg.* **29,** 895.
Miller, L. H., Chien, S., and Usami, S. (1972). *Amer. J. Trop. Med. Hyg.* **21,** 133.
Nicholas, W. L., and Jones, J. W. (1959). *Parasitology* **49,** 1.
Potts, R. E., and Williams, A. A. (1956). *Lancet* **1,** 483.
Pugh, A. M. (1950). *Trans. Roy. Soc. Trop. Med. Hyg.* **44,** 1.
Rommel, M. and Heydorn, A. (1972). *Berliner Munchener Tierarztliche Wochschr.* **85,** 143.
Sadun, E., von Lichtenberg, F., Cheever, A. W., Erickson, D. G., and Hickman, R. L. (1970). *Amer. J. Trop. Med. Hyg.* **19,** 427.
Salfelder, K., Werner, H., Leon, A., and de Liscano, T. R. (1969). *Z. Tropenmed. Parasitol.* **20,** 511.
Scott, J. W. (1943). *Wyo., Agr. Exp. Sta., Bull.* **259.**
Semple, A. B., Davies, J. B. M., Kershaw, W. E., and St. Hill, C. A. (1954). *Brit. Med. J.* **1,** 1002.
Sheffield, H. G., and Melton, M. L. (1970). *Science* **167,** 892.
Smith, T. (1901). *J. Exp. Med.* **6,** 1.
Splendore, A. (1909). *Bull. Soc. Pathol. Exot.* **2,** 462.
Sprent, J. F. A. (1952). *Parasitology* **42,** 244.

Syme, J. D. (1949). "Fish and Fish Inspection." Lewis, London.

Tafuri, W. L. (1970). *Amer. J. Trop. Med. Hyg.* **19**, 405.

Torres, C. M. (1927). *C. R. Soc. Biol.* **97**, 1778.

Walker, D., and Garnham, P. C. C. (1972) *Vet. Rec.* **91**, 70.

Weathersby, A. B. (1952). *J. Infec. Dis.* **91**, 198.

Weinstein, P. P., Krawczyk, H. J., and Peers, J. H. (1954). *Amer. J. Trop. Med. Hyg.* **3**, 112.

World Health Organization. (1957). *World Health Organ., Monogr.* **33.**

Zahawi, S., and Shukri, N. (1956). *Trans. Roy. Soc. Trop. Med. Hyg.* **50**, 166.

5

HISTOCHEMISTRY OF SKELETAL MUSCLE AND CHANGES IN SOME MUSCLE DISEASES

E. B. BECKETT and G. H. BOURNE
Revised by M. Nelly Golarz de Bourne and G. H. Bourne

I. Introduction

The last fifty years or so have seen the accumulation of a vast mass of information concerning the biochemical properties of the various components of normal striated muscle fibers. The muscle mitochondria (or sarcosomes) have been shown to possess the same types of enzymic activity as the mitochondria of other tissues; and on the other hand, the structural and enzymic properties of the components of the contractile elements of muscle fibers have been fully investigated with a view to elucidating the processes involved in muscular contraction.

The biochemistry of human skeletal muscle in pathological conditions has attracted far less attention, possibly because of the difficulty of obtaining suitably fresh material with which to work, and possibly also because of the difficulty involved in the interpretation of results when structural changes—e.g., a large increase in the collagenous tissue—have occurred in the muscle sample as a result of pathological processes.

An attempt was made to overcome this latter difficulty by Dreyfus et al. (1954) by relating results to noncollagen nitrogen, and they observed that in progressive muscular dystrophy there was a decrease in phosphorylase and aldolase activity in striated muscular tissues.

Compared with the literature available on the biochemistry of striated muscle, there is relatively little to be found dealing with histochemical aspects of this tissue. Indeed, for some time, the only detailed study of muscle carried out on anything other than motor end plates was that of Dempsey and associates in 1946, in which a range of histochemical properties of the tissue was investigated. There have, however, been a number of papers published in muscle enzyme histochemistry in the last ten years, and these have made a spectacular contribution to the differential diagnosis of myopathies.

II. Oxidative Enzymes

A. Normal Muscle

With adequate oxygen, muscle metabolism follows the aerobic pathway; pyruvate does not produce lactic acid, but instead undergoes oxidative decarboxylation with the production of a two-carbon substance that becomes linked to coenzyme A to form acetyl coA, which is also known by the name of active acetate. This combines with oxaloacetate, leading to the production of citric acid and so initiating the Krebs cycle. This cycle produces three molecules of carbon dioxide and five pairs of hydrogen atoms for each molecule of pyruvate that combines with oxaloacetate. The hydrogen atoms are transferred to hydrogen acceptors and eventually combine with oxygen to form water. The energy balance of this cycle is very favorable, so that thirty-six molecules of ATP are synthesized for each turn of the metabolic wheel. Enzymes that catalyze the Krebs cycle are either decarboxlases or dehydrogenases, and current histochemical techniques using the tetrazoliums as hydrogen acceptors have made it a realtively simple procedure to demonstrate the latter enzymes. The succinic dehydrogenase reaction was the first of these reactions to be applied to skeletal muscle, in which it was demonstrated by Malaty and Bourne (1953). The authors suggested that the reaction was contained in the sarcosomes. This has been amply confirmed by many authors. It is now known that all the enzymes of the Krebs cycle are present in sarcosomes (Azzone and Carafoli, 1960, using pigeon breast muscle and rat limb muscle). An extensive study of the succinic dehydrogenase (SDH) reaction in muscle was made by Ogata in 1958. He found different degrees of reaction in different muscles and discovered that these reactions were related to the proportions of small and large fibers present.

Muscles vary in the function they perform and in their metabolic reactions. Some muscles are continuously active, such as the postural (antigravity) muscles, or they are tonic (e.g., the diaphragm); others work from time to time in an explosive fashion. The characteristics of muscle can be altered by exercising it in a particular way or by crossing the nerves supplying the different muscles (Buller *et al.*, 1959).

Ogata divided muscles into three types, based largely on the SDH reaction. In any section of a muscle, the individual fibers can be seen to vary considerably in the intensity of the reaction which they give for SDH. The division of muscle fibers into two principal types was originally shown by Padykula and Herman (1955a), using an ATPase reaction. The larger fibers in the muscle, which also give a light or no reaction with fat stains, stain most intensely for ATPase and are called type 2 fibers. If the SDH technique is applied to muscle, these large type 2 fibers give only a light reaction. The small fibers, however, which give a good positive reaction for fat, a light ATPase reaction, and which are rich in myoglobin, give a very strong reaction. These are the type 1 fibers; they are the so-called slow twitch fibers, the type 2 fibers being known as the fast twitch fibers. A number of intermediate-reacting fibers for SDH are present in most muscles. The type 1 fibers under the electron microscope show more sarcosomes than the type 2 fibers, which is appropriate since SDH is largely present in these organelles.

W. K. Engel (1970) points out that because of this result, it is surprising that α-glycerophosphate dehydrogenase, which is concentrated in mitochondria, is less active in the type 1 fibers than in the type 2 fibers. Engel suggested that in the case of muscle, there is simply a difference in the amount of activity for this enzyme in the two types of sarcosomes.

Ogata divided muscles into (1) the gastrocnemius type, (2) the soleus type, and (3) the diaphragmatic type. In the gastrocnemius type of muscles, there were about even numbers of type 1 and type 2 fibers. He also identified in these muscles a third fiber, intermediate in size and histochemical reaction. In the cat, muscles belonging to this type are the cervical muscles, back muscles, and limb muscles. Ogata noted that in the species of animal studied by him, muscle located near the surface of the body is rich in white muscle fibers, while muscle situated more deeply in the body is rich in medium and type 1 fibers. Even in individual muscle, he claimed that the type 2 fibers were present in largest amounts near the surface of the muscle, with the type 1 fibers more frequent in the interior. The studies of G. Bourne and M. Golarz de Bourne (unpublished data) have demonstrated that the reverse arrangement occurs in the tail muscles of the shark.

The differentiation between red and white muscle was made as long ago as 1678 by Stefano and Lorenzi in the rabbit. Ranvier, in 1880, showed that most vertebrate muscles contained red (type 1) and white (type 2) fibers, and that the muscles that had a greater production of the former appeared red to the naked eye. Grutzner (1884) confirmed this in the muscles of man. Muscles belonging to the soleus type appear red to the naked eye. The greater proportion of the fibers are type 1 fibers and give an intense SDH reaction. There are some fibers, however, that are larger in size and have an enzyme activity comparable with medium fibers of the gastrocnemius. The diaphragmatic muscle contains fibers that have a size distribution similar to that of the gastrocnemius, but a number of the larger fibers show a stronger SDH reaction.

Dubowitz and Pearce (1960a,b,c), found in rat, pigeon and human muscles that type 1 fibers were rich in activity for fourteen oxidative enzymes but showed little phosphorylase activity. The reverse was shown by the type 2 fibers. The di- and triphosphopyridine (DPN and TPN) diaphorases are also found most commonly in the type 2 fibers and so is cytochrome oxidase.

W. K. Engel listed the histochemical differences (Table I) between

TABLE I
MUSCLE FIBER REACTIVITY

Reaction	Type 1	Type 2
DPNH dehydrogenase	High	Low
TPNH dehydrogenase	High	Low
Succinate dehydrogenase	High	Low
Cytochrome oxidase	High	Low
Dihydroorotic acid dehydrogenase	High	Low
Benzidine peroxidase (probably myoglobin)	High	Low
Menadione-mediated α-glycerophosphate dehydrogenase	Low	High
DPN-linked lactate dehydrogenase	Low	High
DPN-linked glycerophosphate	Low	High
Phosphorylase	Low	High
Glycogen	Low	High
UDPG-glycogen transferase (reversed without PMS and azide)	High	Low
Argyrophil reaction	Medium	Medium
ATPase, myofibrillar	Low	High
ATPase, edetic acid–low pH activated	High	Low
ATPase "West"	Medium	Medium
Antimyosin fluorescent antibody	Medium	Medium
Tyrosine	Medium	Medium
Esterase	High	Low
Osmium tetroxide	Medium	Medium
Oil red "O"	High	Low

type 1 and type 2 fibers. He suggests that the differences in staining reactions for regular edetic acid-activated ATPase reactions is probably due to enzymic differences actually present in the myofibrils, since he had shown (W. K. Engel, 1969) that the fibril density in the type 1 and type 2 fibers is about the same. He feels that the two reactions are demonstrating different isozymes of ATPase. He also draws attention to the fact that microsomes derived from red muscle of the rabbit have twice as much ATPase activity as the microsomes derived from rabbit white muscle.

Engel goes on to point out that despite this, there is an undoubted difference between the physiological properties of muscles according to the proportions of type 1 and type 2 fibers they possess—red muscle containing mostly type 1 fibers, being a slow twitch muscle; and white muscle containing mostly type 2 fibers, being a fast twitch muscle. The presence of some fibers of the other type in each variety of muscle raises the question of how they function in that particular muscle—in other words, do the red type 1 fibers in the fast twitch muscles behave the same way as the type 2 fibers, or do they retain the physiological property suggested by their histochemical reaction? The problem is further complicated by the fact that some muscles start off in early life by being composed predominantly of one type of fiber and later on develop many fibers of the other type.

Buller *et al.* (1960), connected the nerve of the flexor digitorum

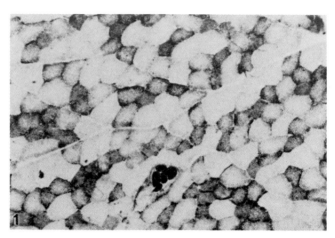

Fig. 1. Normal EDL from control cat showing checkerboard of strongly reacting type 1 fibers, large weakly reacting type 2 fibers, and intermediate fibers. NADH₂ diaphorase; × 110. From Dubowitz (1967).

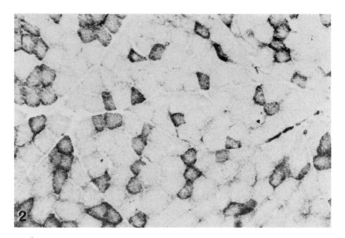

Fig. 2. Cross-innervated R.EDL (kitten K2) showing area with preponderance of large, weakly reacting type 2 fibers, a very small proportion of type 1 fibers, and no intermediate fibers. NADH₂ diaphorase; × 110. From Dubowitz (1967).

longus (a fast muscle) to the soleus (a slow muscle) and found that the nerve altered the physiological properties of the soleus, making it a fast muscle. Dubowitz has repeated and extended this observation (1967). He pointed out that the soleus is comprised entirely of type 1 fibers, giving reactions rich in oxidative enzymes and poor in phosphorylase. The flexor digitorum longus contains both type 1 and type 2 and some intermediate fibers. Using both rabbits and kittens, Dubowitz found that as a result of crossing the innervation of these muscles, the histochemical pattern of the fibers was altered to correspond with the pattern of other muscle.

B. Pathological Muscle

In 1951, Humoller and associates, using a biochemical method, observed that denervation of rat gastrocnemius led to a decrease in its succinic dehydrogenase content. From the work on human muscle biopsies taken from various types of muscular or neuromuscular disorders (Beckett and Bourne, 1958a), this loss of respiratory enzyme did not appear to be very striking in cases where, from the clinical diagnosis (e.g., in polyneuritis, motor neuron disease, and peripheral neuritis) one might expect the muscular innervation to be at fault. Only in one case of such a disease, which was peroneal muscular atrophy, was there considerable fall in succinic dehydrogenase activity, and in this case the picture was complicated by the fact that the patient was a child of $5\frac{1}{2}$ years. The low reaction intensity here might have been due in part to the child's age, since we have observed that in fetuses the succinic dehydrogenase activity in muscle is considerably lower than that in adults, and it is not yet known at what age adult levels of activity are attained.

The general impression we have gained from our work, which covered specimens from a variety of muscular and neuromuscular disorders, including muscular dystrophies, polyneuritis and polymyositis, motor neuron disease, carcinomatous myopathies, and others, is that a decrease in succinic dehydrogenase activity tends to occur in the course of these diseases, but that it is a secondary change following gross structural disorganization of the muscle fibers and their surrounding connective tissue. In other words, where there is slight atrophy or hypertrophy of the muscle fibers, there is no histochemically observable change in the concentration of succinic dehydrogenase in them. However, where there is a gross change in muscle fiber size, often accompanied by a great increase in the amount of interstitial dense connective tissue and/or adipose tissue, as is observed in some muscular dystrophies, there is a marked decrease in the enzyme activity detectable by histochemical means.

In only one case did there seem to be a primary disruption of the succinic dehydrogenase system. This was a sample of muscle from a case of polymyositis in which there was little histological evidence of muscle fiber destruction, but in the center of each fiber there was an area of slight enzyme activity surrounded by a broad outer band of greater reaction intensity (Fig. 32). This suggests a disruption of the respiratory system in the center of each muscle fiber; but a similar picture is not always obtained in cases of this disease, so that it offers no explanation of the pathological processes involved.

In 1956, however, just after Beckett and Bourne had seen (but not published until 1958a) this central loss of succinic dehydrogenase activity, Shy and Magee described a muscular syndrome in patients in which the central cores of the muscle fibers showed loss of mitochondria. It is possible that the case from which Beckett and Bourne obtained their biopsy was a case of central core disease, but since no followup of patient or biopsies are possible at this time, it is impossible to be sure. Subsequently, Dubowitz and Pearse (1960a), W. K. Engel and his colleagues (1961), and A. G. Engel (1966) showed that the cores lacked phosphorylase activity and also ATPase activity. In some recently studied patients, A. G. Engel (1966) found the central core defect was restricted to the type 1 fibers. It is of interest that Golarz de Bourne and Bourne (1972) have found fibers showing this central core deficiency of enzymes in the leg muscles of a chimpanzee with a chronic limp. At first this was thought to be central core disease, but electromyography suggested nerve damage, and there was some evidence that the onset of the disease followed a fist fight with another chimpanzee in which the first animal probably suffered some trauma to the nerve. There was a slow progressive improvement of the limb, but the animal (2 years after the first biopsy) still favors the leg on which he previously limped. A recent biopsy showed no sign of central core, histochemical, or morphological defects (there were not even ultrastructural signs).

Central core disease was originally considered to be a myopathy by Shy and Magee (1956) (and see Shy et al. 1963), but W. K. Engel (1962) has suggested that it might have some connection with an embryonic defect in the innervation of the muscle. Evidence for this suggestion comes from W. K. Engel's (1961) work in which he showed that muscle fibers that had been denervated also showed this central core defect. He has also found the condition in one hundred of biopsies taken from two hundred patients suffering lower motor neuron disease, (W. K. Engel and

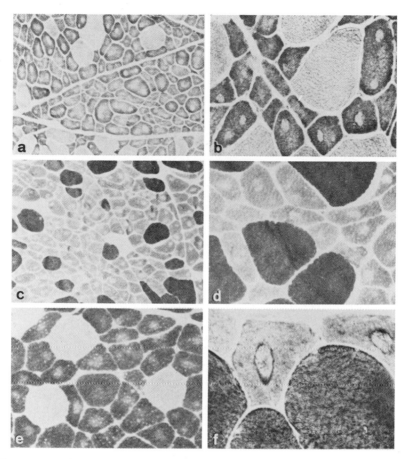

Fig. 3. Central core disease. Some fibers have one or two cores. The fibers with one core are small and belong to type I. (a) and (b) are DPN dehydrogenase reactions. (c) and (d) are pryofibrillar ATPase reactions. (e) and (f) are phosphorylase preparations.

Brooke, 1966). Engel has described muscle fibers which show this central core defect as target fibers. It was of interest that target fibers are usually type 1 fibers, and it is the type 1 fibers that slow the defect in clinical human central core disease.

Resnick and Engel (1967) point out that : "Target fibers also illustrate how one muscle fiber, designated histochemically, may consistently react in a completely different manner from other fiber types in the same muscle at the same time under abnormal conditions. It is not known why the target fibers show a strong preferential occurrence in Type 1 muscle fibers, nor why the center of the fiber is most severely altered."

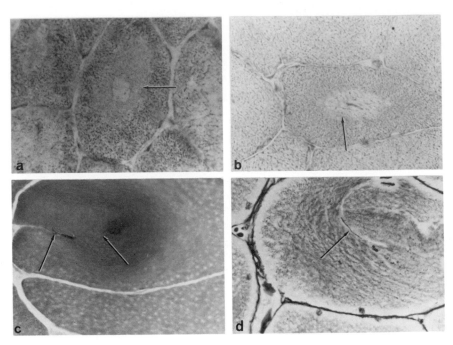

Fig. 4. Muscle fibers showing central core deficiency in chimpanzee with polyneu-ritis, probably of traumatic origin. (a) Three concentric zones in a thigh muscle fiber (*Pan troglodytes*). DPN diaphorase reaction (× 320). (b) Muscle fiber from thigh of chimpanzee (same as a). Similar concentric staining; hexokinase reaction (× 320). (c) Muscle fiber with bizarre structure in an apparent process of split-ting. The fissure can be followed at both arrows. Same animal as above (*Pan troglodytes*). Thigh muscle; hematoxylin and eosin (× 320). (d) The same speci-men as above stained with Wilders' reticulin technique. A vestige of endomysium formation can be seen at arrow. Thigh muscle (× 320).

Engel has pointed out that target bers show three concentric zones. The outermost zone is normal as based on the criteria used; the inter-mediate zone just inside this normal ring appears to have increased oxidative enzyme, myofibrillar ATPase activity, and phosphorylase activ-ity; the central core within this has little or no activity for any of these enzymes. There is some evidence that occurrence of target fibers indi-cates that fibers are being reinnervated, although they may in fact result from denervation—Engel favors the latter.

Another type 1 fiber myopathy is known as rod or membrane myopathy. Originally described by Shy *et al.* (1963), this condition shows the presence of many small rodlike granules in the fibers. They appear actually to be associated with the myofibrils and to be derived

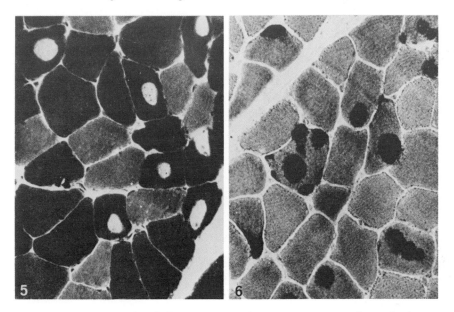

Fig. 5. Myopathy with tubular aggregates; 5 μ cryostat section of muscle showing absence of myosin from the aggregates, which are confined to type II fibers. Myosin ATPase; $\times$ 309. From A. G. E. Pearse and Johnson (1970).

Fig. 6. Myopathy with tubular aggregates; 5 μ cryostat section showing the very strong reaction for ubiquinone or similar compounds in the aggregates. MTT–hydroquinone reaction; $\times$ 309. From A. G. E. Pearse and Johnson (1970).

originally from the region of the Z band. In one case, W. K. Engel (1969) found rods in the type 2 fibers.

Another type of myopathy showed the presence of aggregates of material of a tubular nature in type 2 fibers that gave a strong reaction for DPNH and TPNH tetrazolium reductase activity. Patients from which the biopsies showing this abnormality occurred had only mild or no neuromuscular disease. The aggregates appeared to be in the form of tubules within the muscle fiber that appeared to be derived from the sacroplasmic reticulum. A. G. E. Pearse and Johnson (1970) have shown that these tubular aggregates contain a considerable proportion of lipids. This same condition has also been found in cases of familial hypokalemic periodic paralysis (Odor *et al.*, 1967; Gruner, 1966). A. G. E. Pearse and Johnson (1970) described a case of a patient with a short history of muscular weakness whose biopsy showed tubular aggregates. They found that cytochrome oxidase as well as NADH diaphorase were present in the aggregates. The lipid which was present, showed a high proportion of plasmalogens and unsaturated fatty acids.

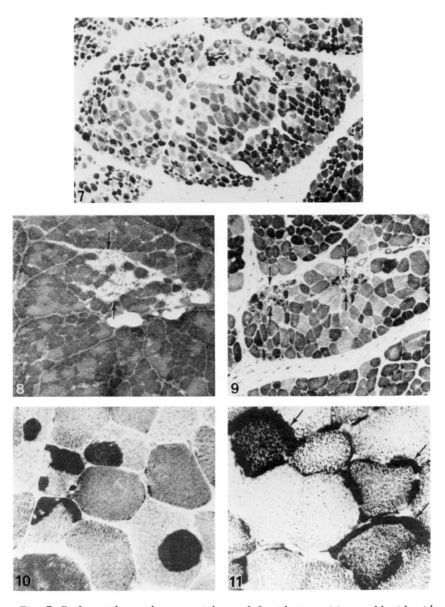

Fig. 7. Preferential atrophy at periphery of fascicle in a 14-year-old girl with untreated dermatomyositis. DPNH dehydrogenase; × 75. From W. K. Engel (1970).

   Fig. 8. Grouping of lightly stained necrotic fibers (arrows) in an otherwise intact area. Childhood X-linked pseudohypertrophic muscular dystrophy. Modified trichrome; × 75. From W. K. Engel (1970).

   Fig. 9. Two groups of darkly stained necrotic and regenerating fibers (arrows)

III. Esterases

The term "esterase" in histochemistry covers simple esterases (enzymes that split esters of short chain fatty acids), lipases (enzymes that split esters of long chain fatty acids) and cholinesterases (which split various choline esters). However, it has become clear in recent years that there exists a whole spectrum of esterases with varying degrees of substrate specificity, so that one cannot clearly define three separate enzyme groups. For the purposes of this present discussion, it is proposed to consider the esterases under two headings: (1) a simple esterases and lipases and (2) cholinesterases, but merely because it is convenient to do so from the point of view of techniques which have been used for the study of muscle. It is very necessary however, to bear in mind the overlap in substrate specificity that exists between the groups.

A. *Esterases and Lipases*

The Gomori Tween technique for the demonstration of lipases was introduced in 1945 (Gomori, 1945). In this technique, a Tween substrate—i.e., a long-chained fatty acid ester of sorbitan or mannitan in which the remaining hydroxyl groups are etherified with ethylene oxide side chains of varying lengths—is split by the enzyme present in sections of acetone-fixed tissue. The free fatty acid so released is precipitated as its calcium salt at the site of enzyme action, then converted to its lead salt, and finally to brown lead sulfide, which is clearly visible under the microscope.

A few years later (Gomori, 1949), a technique for the demonstration of simple esterases was elaborated by Nachlas and Seligman (1949a,b). In their method, the simple esterases of acetone-fixed tissue split β-naph-

---

in an otherwise virtually intact region. Childhood X-linked pseudohypertrophic muscular dystrophy. DPNH dehydrogenase; × 75. From W. K. Engel (1970).

Fig. 10. Tubular aggregates appear as darkly stained masses in five otherwise light (type II) fibers. DPNH dehydrogenase reaction; × 190. From W. K. Engel (1970).

Fig. 11. Accumulations of material darkly stained with oxidative enzyme reactions in subsarcolemmal regions of three fibers (arrows) are abnormal for human muscle (compare with two normal, dark type I fibers in right center of Fig. 10). Electron microscopy showed accumulations to be mitochondrial. Affected fibers are type I (dark) in this and other reactions. Succinate dehydrogenase reactions; × 300. From W. K. Engel (1970).

thol acetate, and the free naphthol released combined with a diazo dye to give a colored end product. The technique suffered somewhat from diffusion artifacts, and was later modified in several ways to improve localization. Gomori (1950, 1952a–c) introduced α-naphthyl acetate and naphthol AS acetate, respectively, as substrates (A. G. E. Pearse, 1953a,b); commercially available diazotates were used; and for some tissues it was found that brief formalin fixation yielded better results than the originally used acetone fixation.

An alternative method for the demonstration of simple esterases was that of Barrnett and Seligman (1951). Their technique made use of sections cut from fresh or lightly formalin fixed tissues. The esterases in the tissues split indoxyl acetate, thereby releasing free indoxyl, which was subsequently oxidized by the oxygen of the air to form insoluble crystals of indigo. This method was criticized by Gomori (1952a) on the grounds that the aerobic oxidation of indoxyl to indigo is a slow process, and that because of this, the localization would be poor. By 1954, however, Holt and others (Holt, 1954) had modified the technique, not only by the introduction of 5-bromoindoxyl acetate or 5, 5'-diindoxyl acetate as substrate, but also by the use of copper ions and a ferroferri-cyanide oxidation–reduction system to hasten the process of conversion of the enzymically released indoxyl derivative to a compound of the indigo type. The localization of esterases demonstrated by this technique was immeasurably improved by these modifications; in fact, some of Holt's (1954) pictures of rat motor end plates rival the best so far obtained with any available histochemical method.

All of these histochemical techniques for the demonstration of lipase and esterase have been used to some slight extent for the study of muscle, usually in conjunction with work on other tissues.

Gomori (1946), Richterich (1952), and Buño and Mariño (1952), all using the Tween technique on acetone-fixed material, noted that the skeletal muscle for several animals, including man, always gave a negative reaction. Nachlas and Seligman (1949b), obtained the same result in human muscle with their original β-naphthol acetate method. Chessick (1953), on the other hand, found a positive reaction for α-naph-thol esterase and naphthol AS esterase in the motor end plates, and possibly also in the muscle spindles, of mouse muscle after acetone or acetone–alcohol fixation. The skeletal muscle of cat, rabbit, rat, and man, however, gave a negative result with these substrates.

Denz (1953), using the β-naphthyl acetate method, observed staining of rat motor end plates that was inhibitable with eserine, and in addition, a diffuse noninhibitable esterase reaction in the muscle substance, whereas with the Gomori Tween technique he could only obtain the diffuse staining of muscle fibers. Both of these procedures were carried out on fresh frozen sections.

Other workers have also described the presence of esterase activity in fresh or lightly formalin-fixed rat muscle. Barrnett (1952), using the original indoxyl acetate method, obtained a positive reaction in both the striated and the smooth muscle of this animal, and Pearson and Defendi (1957) observed high levels of esterase activity in rat motor end plates with either the 5-bromoindoxyl acetate method or the α-naphthyl acetate or naphthol AS acetate techniques. These latter workers also found that reaction intensity obtained with a given technique depended upon pH. The greatest activity with 5-bromoindoxyl acetate as substrate was observed between pH 4.8 and 5.8, whereas the highest degree of activity with α-naphthyl acetate or naphthol AS acetate as substrates could be seen in a higher pH range, i.e., pH 7.3–8.4.

From these studies of the esterase activity of skeletal muscle, it appears that the enzyme(s) is present, particularly in the motor end plates, but that it is inhibited in most animals by acetone fixation and/or paraffin embedding. It also seems that the intensity of reaction obtained depends upon the pH used for a given technique.

B. Cholinesterases

The subject of cholinesterases and motor end plates is dealt with in full by Couteaux elsewhere in this treatise, so we shall confine ourselves to a brief discussion of histochemical techniques for this enzyme and to a few remarks about cholinesterase-positive structure in human muscle and their changes in some pathological conditions.

There are three types of technique available for the demonstration of cholinesterases. The simple esterase techniques described above, and particularly the various naphthyl acetate methods, have been used from time to time, but rarely for the study of muscle. Since these techniques primarily demonstrate simple esterases it is essential to use eserine in conjunction with them in order to differentiate between simple and cholinesterase, and to use DFP (diisopropyl fluorophosphonate) to distinguish between the so-called true and pseudo cholinesterases. It must, however, be remembered that these classifications according to inhibitor sensitivity are purely arbitrary and that the histochemical results for a given site or a given tissue may not correspond with the biochemical findings. A good review of this subject was published by Gomori and Chessick (1953).

A second type of technique which has been used for the demonstration of cholinesterases is Gomori's (1948) myristoylcholine method and its variations. In this technique, a long chain fatty acid is released from the substrate by the action of the cholinesterase and then combines with cobalt, and so is precipitated as a cobalt soap at sites of activity. This cobalt soap is then converted to microscopically visible cobalt sulfide by the action of ammonium sulfide.

The introduction of this technique represented an advance on the use of the simple esterase methods, since it did at least utilize an ester of choline, but Gomori himself realized that it was far from specific and often failed to give a result at sites that were known to contain large amounts of cholinesterase; for example, the electric organ of the electric eel (Gomori, 1952b). This technique has been used by Hard and his co-workers (Hard, 1950; Hard and Hawkins, 1950; Hard and Peterson, 1950) for work on the central nervous system of the dog. It has hardly ever been used for muscle, since it tends to stain muscle nuclei, a possible artifact, and moreover, it is difficult to interpret the structures demonstrated at motor end plates, which may also be nuclei.

The most commonly used type of histochemical method for the demonstration of cholinesterase is that based on the hydrolysis of acetylthiocholine and butyrylthiocholine. The former compound was shown by biochemical means to be split by cholinesterases at a rate greater than that for acetylcholine, and was first employed for histochemical purposes by Koelle and Friedenwald (1949).

The original Koelle and Friedenwald technique was rather cumbersome, and it has since been modified by several workers, including Koelle (1951), Gomori (1952c), Couteaux and Taxi (1952), Cöers (1953a), and Gerebtzoff (1953). The last three modifications involved the use of formalin fixation and acetate buffers at different pH levels to secure better morphological localization. More recently, Bull *et al.* (1957) used alcoholic ammonium sulfide instead of the usual aqueous solution to attain the same end.

The thiocholine techniques are undoubtedly the best for the demonstration of cholinesterases, both from the point of view of specificity and of precise localization. The localization obtained in motor end plates has been sufficiently good to allow studies of the fine morphology of the subneural apparatus of these structures to be carried out.

When the thiocholine technique is carried out, the acetyl or butyryl group is split off from the substrate by the enzyme, and the thiocholine moiety combines with copper ions in the incubating medium, to be precipitated as copper thiocholine or copper thiocholine sulfate (Malmgren and Sylvén, 1955) at sites of enzyme activity. It is then converted to brown copper sulfide during the process of visualization.

1. Cholinesterase in Normal Human Muscle

Cöers (1953b, 1955a) described two types of motor end plate to be found in human adult muscle: (1) the *terminaisons en plaque,* in which the "gutter" of the subneural apparatus is arranged like a continuous rope twisted into a complex pattern, and which have been observed in many different types of animal, including the rat, mouse, lizard, and goat; and (2) the *terminaisons en grappe,* which are composed of small islets of subneural apparatus grouped together but without any apparent connections between them. These latter vary considerably in size, since the number of islets of which they are composed may be three or four or may be thirty to forty, and they seem to be characteristic of human muscle. Although in our previous work (Beckett and Bourne, 1957), we were using Gomori's modification of the thiocholine technique, which differs somewhat from the method used by Cöers, we were able to confirm his observations and also to confirm the presence of cholinesterase-positive motor end plates at the poles of muscle spindles.

In addition, we noticed that of the cholinesterase-positive structures present in muscle, only the *terminaisons en plaque,* or as we have called them, the classic motor end plates, ever gave a reaction when acetylthiocholine was replaced by the butyryl compound. This was true of both normal and pathological samples of muscle.

In addition to the structures just mentioned, which are almost without doubt to be considered as motor end plates, a number of other cholinesterase-positive objects have been seen in both animal and human skeletal muscle.

In 1953, Couteaux described in the muscle of frog, mouse, and fish, structures consisting of typical subneural apparatus "gutters" situated over the ends of muscle fibers at musculotendinous junctions. The gutters projected like parallel fingers into the muscle substance. Between then and 1956, this type of structure was also seen in a variety of other animals by Gerebtzoff and his co-workers (Gerebtzoff, 1956). A little later, we confirmed their existence in human muscle (Beckett and Bourne, 1957) and suggested that they might be stretch receptors.

In the course of our investigations, a range of other structures was seen which we consider to be normal, although they were not always observed in normal human muscle. Possibly earlier workers had missed them because their material had been formalin fixed. It is known that formalin has a varying inhibitory effect on cholinesterase (Taxi, 1952; Couteaux and Taxi, 1952; Beckett and Bourne, 1957, 1958a), and from our preparations it appeared that these previously unreported entities contain less cholinesterase activity than the *terminaisons en plaque* or *ter-*

minaisons en grappe. With less sensitive variations of the technique, therefore, they may well have remained invisible.

In some of our muscle samples, objects were observed that consisted of a system of parallel gutters lying in the same direction as the long axis of the muscle fiber. The gutters were arranged rather closer together at the center of the structure than at its edges, so that the whole looked rather like a cake frill or wheat sheaf. These structures could occur singly on a muscle fiber, or there could be two on adjacent muscle fibers. Sometimes they were seen at points where muscle fibers ended in the middle of a fasciculus. One cannot be certain what their function could be, but their appearance suggests that they might be stretch receptors.

In other muscle specimens, spirals of a continuous or discontinuous gutter were observed winding around single muscle fibers. These again may very well represent sensory nerve endings. Yet other entities were seen, which were probably motor end plates and not sensory endings, and which were composed of islets of gutter structure arranged in a band lying transversely across muscle fibers. This contrasts with the longitudinal arrangements observed by Cöers. There were also some similar bodies of parallel gutters aligned at right angles to the long axis of the muscle fibers.

In addition to cholinesterase present in specific structures in muscle fibers, our observations (Beckett and Bourne, 1957) have indicated that the enzyme is present in the muscle substance. If unfixed frozen sections are incubated in a medium containing acetylthiocholine iodide and are visualized in the normal fashion, it is found that they have an overall deep yellow-brown coloration, suggesting the presence of cholinesterase. This is confirmed by the fact that the depth of color is even; i.e., it is not intensified at all in regions containing motor end plates, and it does not vary in depth according the number of end plates present. Indeed, the coloration is no less intense when there are no end plates at all in the section. If sections are incubated in a medium containing butyrylthiocholine instead of the acetyl compound, then there is virtually no coloration of the muscle fibers.

2. Cholinesterase in Pathological Human Muscle

The most thorough work which has been done on changes in the motor end plate in human pathological muscle is undoubtedly that of Cöers (1955a, b), in which he used the methylene blue and Bielschowsky silver techniques in addition to the thiocholine cholinesterase technique in order to study the changes occurring in some muscular and neuromuscular disorders.

Cöers' general conclusions from the study of forty-odd biopsies and some post mortem material were as follows: In muscular atrophies—e.g., peroneal muscular atrophy of Charcot–Marie–Tooth—no end plates were present in atrophied fibers, but in dystrophies (a heading under which Cöers includes myopathies, myotonic dystrophies, and myotonia congenita) there were good terminal arborizations as demonstrated by methylene blue. In children, there was no change in the form of the subneural apparatus, but in adults there was an increase in the size of end plates.

In addition, Cöers found that in most of the abnormal muscle studied, the size of the end plates in atrophied fibers was reduced, whereas in the hypertrophied fibers seen in cases of dystrophy or partial neurogenic degeneration, there were large extended end plates. In contrast to this, however, Cöers observed that in certain muscular atrophies, and particularly in dystrophia myotonica, the end plates were very extended and complex, even where the muscle fibers were extensively atrophied. Further details of this work have been published by Cöers and Woolf (1959).

Our own work (Beckett and Bourne, 1957) has confirmed some of Cöers observations. The main object of our studies was to see whether or not there was any decrease in the amount of cholinesterase present in the end plate areas of muscle taken from a variety of muscular disorders, including dystrophies of different sorts, motor neuron disease, carcinomatous myopathies, and neuropathies of unknown or obscure etiology. For this reason, we used a thiocholine technique—Gomori's (1952c) modification—which did not employ formalin fixation prior to incubation with the substrate, so that the morphological picture suffered somewhat from diffusion. However, it is possible to say something about the changes in morphology associated with muscle pathology.

First, it must be said that the variety of normal structural forms of human subneural apparatus makes any pathological changes difficult in the extreme to detect with any certainty. For this reason, too, it is a fallacy to carry out experiments (e.g., denervation experiments) on animals and use these results to interpret the picture seen in pathological human muscle.

In the specimens we examined, there appeared to be no loss of cholinesterase activity at motor end plates, but in two cases (one of peroneal muscular atrophy and one of polyneuritis), there was a decrease in the amount of this enzyme present at the musculotendinous junctions. Furthermore, except in one case of peroneal muscular atrophy, there did not seem to be any decrease in the number of cholinesterase positive structures present. In one or two cases, including one of pseudohypertrophic muscular dystrophy, one of facioscapulohumeral muscular dys-

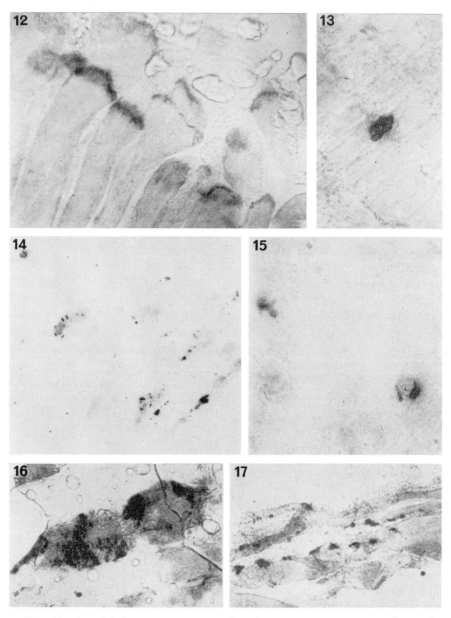

Fig. 12. Acetylcholinesterase at musculotendinous junctions in normal muscle. The structures present here may represent stretch receptors. Magnification approximately × 150.

Fig. 13. A classic motor end plate from a case of motor neuron disease. Magnification approximately × 1000.

trophy, and one of polymyositis, there may have been some breakup of the structure of the subneural apparatus, but it is difficult to be sure of this. In one of the "control" specimens, which was taken from the iliacus during cup arthroplasty of the hip, there seemed to be evidence of disruption of normal end plate structure, since there were present numerous small dots with gutter structure about the size of the nuclei scattered over the muscle fibers.

One of the most striking things observed in our series of specimens was that where there was extreme muscle fiber atrophy accompanied by the replacement of muscle fibers by dense collagenous tissue, such as is seen in cases of facioscapulohumeral dystrophy or familial dystrophy, the remaining fibers were smothered in pieces of end plate structure. This tremendous concentration of end plate material was also seen in cases of pseudohypertrophic muscular dystrophy. where there was disruption of muscle fiber structure and a high degree of filtration by fatty tissue between the remaining muscle fibers. These observations suggest that the presence of the end plate or of some chemical substance such as the enzyme in it has a protective influence on the surrounding muscle tissue and prevents or delays its destruction. That this effect, however, is not due to the cholinesterase is suggested by the fact that in the muscle substance itself, the enzyme is increased in quantity in muscle fibers showing evidence of atrophy and necrosis.

From what has already been said here, it can be seen that no clearcut picture has emerged of the changes that occur in the subneural apparatus of human muscle in the course of muscular or neuromuscular disorders. A great deal more work must be done on normal human muscle in order to try and understand the miscellany of cholinesterase-positive structures present in it before it is safe to argue about what constitutes a normal human end plate and to decide what changes are due to disease processes.

An interesting suggestion was made in two papers by Barker and Ip (1965, 1966) that the axons of the motor neurons can sprout either

---

Fig. 14. Very atrophied muscle fibers in a sea of dense connective tissue from an advanced case of facioscapulohumeral dystrophy showing numerous remaining end plate structures. Magnification approximately × 150.

Fig. 15. Single muscle fiber with its end plate surrounded by dense connective tissue from the same case of facioscapulohumeral dystrophy. Magnification approximately × 300.

Fig. 16. A large end plate on an isolated hypertrophic muscle fiber from a case of pseudohypertrophic dystrophy in which the muscular tissue had been almost entirely replaced by fat. Magnification approximately × 150.

Fig. 17. End plates, apparently normal in form, in muscle taken from a case of familial dystrophy. Magnification approximately × 100.

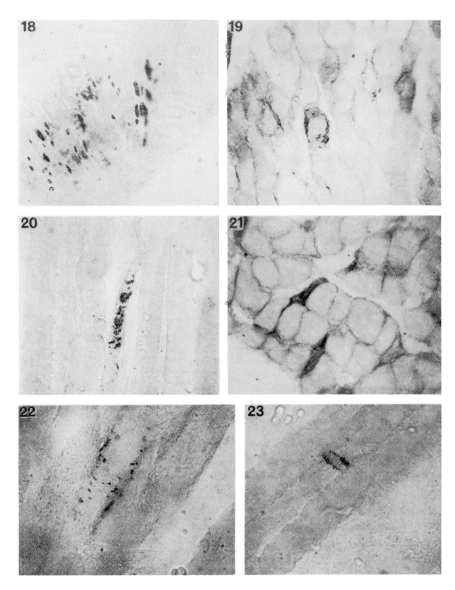

Fig. 18. A group of end plate gutters in muscle from a case of facioscapulohumeral dystrophy, to show well marked transverse lamellae. Magnification approximately × 1000.

Fig. 19. Linear spiral structure with high concentrations of cholinesterase at intervals seen in muscle from a case of thyrotoxic myopathy. This might be a sensory nerve ending, e.g., stretch receptor. Magnification approximately × 150.

Fig. 20. Acetylcholinesterase in an extended dotted motor end plate seen in

terminal or collaterally and that these sprouts are capable of forming new motor end plates. They suggested that older end plates degenerate and are continuously replaced by these newly formed plates. This probably occurs in normal muscles and maybe explains why some times fibers can be found with two end plates. Cöers (1955a) has found the ratio of end plates to muscle fibers to be more than unity, ranging from 1.09/1 to 1.27/1. Cöers (1967) suggests that the axonal branching found in normal muscles could be exaggerated in pathological states.

IV. Phosphatases

A. *Nonspecific Alkaline Phosphatase*

Two types of technique are available for the histochemical demonstration of nonspecific alkaline phosphatase: (1) the Takamatsu–Gomori method in which sodium glycerophosphate in split in the presence of calcium ions to yield calcium phosphate, and the calcium phosphate is then visualized either by the use of silver, or by converting it to cobalt phosphate and thence to black cobalt sulfide; and (2) the diazo dye techniques, first introduced by Menten and associates (1944) and later modified by Manheimer and Seligman (1948), Loveless and Danielli (1949), and Gomori (1951), among others, with the aim of reducing diffusion artifacts. In these diazo dye techniques (except for that of Loveless and Danielli), a naphthyl phosphate is split by alkaline phosphatase, and the released naphthol combines with a diazo dye to give a colored compound.

Although these latter techniques offer some advantages, for example, there is no confusion between end product and preformed calcium, they all suffer to some extent from diffusion artifacts, and there is evidence that the two techniques demonstrate different enzymes.

---

muscle from an amputated leg. This specimen showed signs of disuse atrophy, but the end plate is probably normal.

Fig. 21. Muscle from a case of polymyositis showing the presence of acetylcholinesterase in the substance of the muscle fibers and the increase in concentration in atrophied fibers.

Fig. 22. Acetylcholinesterase in a dotted spiral structure which may represent a normal end plate seen in muscle taken from a case of thyrotoxic myopathy.

Fig. 23. Acetylcholinesterase in a cake-frill, or palisade, type structure. This was seen in muscle from an amputated leg, but is probably normal and may represent some sort of stretch receptor.

1. ALKALINE PHOSPHATASE IN NORMAL MUSCLE

The biochemical work of Kay (1928) indicated that there was no alkaline phosphatase in human muscle, and some of the latter histochemical work (Gomori, 1939; Moog, 1943; Dempsey et al., 1946; Rossi et al., 1954) seemed to indicate that this was also true of muscle taken from a variety of animals, even when, as in the work of Dempsey, Wislocki, and Singer, prolonged incubation periods were used in order to try to obtain a positive result.

Other investigations using one or other of the techniques for alkaline phosphatase, however, showed that although the fibers of skeletal muscle tissue were themselves negative, both the walls of capillaries and the endothelial lining of larger vessels were positive. This was found to be so in the muscle of various animals, including man (Gomori, 1941b), adult man and adult and embryonic mouse (Kabat and Furth, 1941), and adult man and rat (Newman et al., 1950b). Similar observations were made in human muscle by Manheimer and Seligman (1948), in human fetal muscle by Rossi et al. (1954) and also by McKay et al. (1955), and again in adult human muscle by Beckett and Bourne (1958b). These last workers also mentioned that the positive reaction does not extend throughout the whole blood vessel bed of a given area of muscle and suggested that physiological changes in the capillaries, e.g., permeability changes, might find their reflection in the level of alkaline phosphatase activity in their walls.

Zorzoli and Stowell (1947) claimed that alkaline phosphatase was present in the nuclei, cross striations, and the myofibrils of muscle, as well as in the capillaries and surrounding connective tissue, but their incubation periods were so excessively long (up to 70 hr), that grave doubt must exist about the validity of their results.

In recent years, attempts have been made to localize phosphatases in fresh frozen sections in order to eliminate the effects of alcohol or acetone fixation. However, using such a technique, Maengwyn-Davies and Friedenwald (1950) could obtain no reaction in muscle, using glycerophosphate as substrate, even after 48 hr incubation. George et al. (1958) claim that alkaline glycerophosphatase activity may be demonstrated in fresh frozen sections of pigeon breast muscle after prolonged incubation.

2. ALKALINE PHOSPHATASE IN PATHOLOGICAL MUSCLE

In this section, with one exception, we can only discuss the observations made in the course of our own work (Beckett and Bourne, 1958b). These can be regarded as preliminary indications only, but may nevertheless be worthwhile recording. The results obtained, unfortunately, do not show a pattern of change associated with particular clinical diagnoses, so that no general conclusions can be drawn.

In contrast to the picture seen of alkaline phosphatase in normal muscle, a sample taken during cup arthroplasty of the hip for osteoarthritis, which showed signs of degeneration in some areas, contained small stellate cells with alkaline phosphatase in their cytoplasm and nucleoli. The nature of these cells is unknown, but they were not seen in any other specimens of diseased muscle.

Occasionally, among the samples of pathological muscle (i.e., muscle taken from cases of muscular or neuromuscular disorders) where there had been an increase of interstitial connective tissue, e.g., in cases of facioscapulohumeral dystrophy or familial dystrophy, there were a few fine scattered alkaline phosphatase-positive connective tissue fibers. These may have been newly formed fibers, since such fibers have been shown (Bourne, 1943; Fell and Danielli, 1943) to contain this enzyme.

In other specimens, alkaline phosphatase was present in some of the muscle fibers. The occurrence of these enzyme-positive muscle fibers was not correlated in any way with the general histological state of the muscle or with the diagnosis. The samples showing alkaline phosphatase-positive muscle fibers consisted of two out of five cases of facioscapulohumeral dystrophy, one out of two cases of familial dystrophy, one case of dystrophy with periodic paralysis, one case of pseudohypertrophic muscular dystrophy, one case of possible pseudohypertrophic dystrophy, one possible polymyositis, one case of possible thyrotoxic myopathy, and one out of three cases of motor neuron disease.

The enzyme-positive muscle fibers were scattered and were usually in an atrophied state (although this was not so in the case of dystrophy with periodic paralysis), but did not seem to differ greatly morphologically from other atrophied fibers present. In nearly every case, the enzyme-positive muscle fibers were surrounded and invaded by a network of fine reticular fibers which also contained alkaline phosphatase; none of the surrounding connective tissue, however, showed any evidence of a positive reaction.

W. K. Engel and Cunningham (1970) have also found alkaline phosphatase-positive fibers in certain neuromuscular diseases. It is of interest that they identified the abnormal fibers using the α-naphthyl phosphate reaction, whereas the results described by Beckett and Bourne were obtained using the Comori–Takamatsu technique. Engel and Cunningham found the positive reaction to be in the form of scattered black dots or, if the reaction was very strong, to be continuous through the

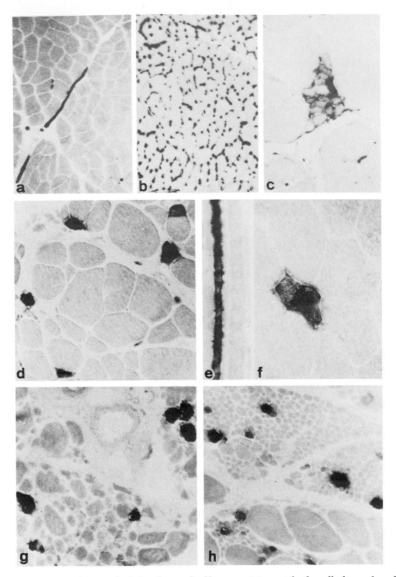

Fig. 24. Human abnormal skeletal muscle fibers positive with the alkaline phosphatase reaction are gray to dark black, as are normal arterioles. (a) Normal human muscle; arterioles but not capillaries or muscle fibers are positive (× 75). (b) Normal rat muscle (gastrocnemius); capillaries and arterioles but not muscle fibers are positive (× 75). (c) Morphologically nonspecific myopathy in a 10-year-old girl; group of positive fibers (× 75). (d) Duchenne muscular dystrophy; five positive fibers, one (upper right) positive only in part of the fiber (× 190). (e) Morphologically nonspecific myopathy in a 4-year-old girl; one positive fiber is stained along its length (× 190). (f) Carrier of Duchenne dystrophy; two positive fibers (× 300). (g) Severe denervation atrophy (amyotrophic lateral sclerosis); six positive fibers (× 190). (h) Infantile spinal muscular atrophy; at least ten positive fibers (H 190). From W. K. Engel and H. Meltzer (1970).

TABLE II

ALKALINE PHOSPHATASE-POSITIVE FIBERS IN VARIOUS MUSCLE DISEASES[a]

Disorders	Number of patients having the following number of alkaline phosphatase-positive fibers per biopsy								
	0	1	2	3	4	5–15	16–50	50–100	100
Normal	43	—	—	—	—	—	—	—	—
Others with minimally abnormal muscle by histochemistry[b]	37	—	—	—	—	—	—	—	—
Duchenne dystrophy carriers	7	1	2	1	2	4	—	—	—
Duchenne dystrophy patients	—	—	—	—	—	1	—	2	3
Myotonic dystrophy, severely abnormal muscle					1	1			
Other active dystrophies		2			1	6	4	3	
Active collagen-vascular myopathy (polymyositis)	1					1		1	2
Denervation,[c] moderate (10–50% of muscle fibers in biopsy abnormal)	2	1	3			3	1	1	
Infantile spinal muscular atrophy with moderate denervation						1	2	1	
Denervation,[c] severe (50% of muscle fibers in biopsy abnormal)	1		1			4	1	4	3
Myotonic dystrophy, slightly abnormal muscle	1	2	1						
Possible defect of long-chain fatty acid utilization with episodic rhabdomyolysis			1						
Denervation,[c] slight (10% of muscle fibers in biopsy abnormal)	16	4							
Severe type II fiber atrophy			2	3					

[a] From W. K. Engel and H. Meltzer (1970).

[b] Includes: benign congenital hypotonia, 3 cases; paramyotonia congenita, 2; ocular neuromuscular disease (limb muscle, 2; old inactive collage–vascular disease, 2; benign myopathy with small fibers having central nuclei, 1; slight type II fiber atrophy, 6; moderate type II fiber atrophy, 11; pyruvate decarboxylase deficiency, 1; nonprogressive nonspecific myopathy, 3; mild thyrotoxic myopathy, 1; metachromatic leukodystrophy, 1.

[c] Includes amyotropic lateral sclerosis and various familial and sporadic peripheral neuropathies.

intermyofibrillar regions. These fibers were generally smaller than the normal fibers; the nuclei were enlarged and usually centrally placed; they also "have irregularity, coarseness and increased staining of the intermyofibrillar network pattern with the reduced di- and triphospho-pyridine nucleotide and malate and lactate dehydrogenase reactions and with the modified trichrome reaction." These fibers were found to contain esterase activity such as is found in fibers suffering from necrosis and phagocytosis, but fibers that are obviously in this condition do not show a positive alkaline phosphatase reaction. Engel and Cunningham have published a table that shows the incidence of alkaline phosphatase posi-tive fibers in muscle biopsied from various human neuromuscular diseases.

B. Acid Phosphatase

Two varieties of technique are available for the histochemical demon-stration of acid phosphatase. The glycerophosphate method, introduced by Gomori (1941a), involves the precipitation of the inorganic phosphate released from the substrate by enzyme action in the form of lead phos-phate and its subsequent conversion to brown lead sulfide. This method is renowned from its waywardness, but, although the efforts of many workers (for instance, Newman *et al.*, 1950a; Gomori, 1950; Goetsch and Reynolds, 1951) have been directed toward improving this state of affairs, no modification has emerged to make the technique reliable.

The diazo dye methods are very similar to those for alkaline phos-phatase, i.e., they involve the enzymic splitting of a naphthyl phosphate, but at an acid pH, followed by the combination of the free naphthol radical so formed with a diazo dye to give a colored reaction product. The first of these techniques was introduced by Seligman and Manheimer (1949). It has since been modified by Friedman and Seligman in 1950, Grogg and Pearse in 1952, and Burton in 1954 (see Rutenburg and Seligman, 1955) in order to make it technically more simple and to decrease the diffusion artifacts of the original method.

All of these diazo dye methods were simultaneous coupling tech-niques; i.e., both the substrate and the dye were present in the mixture in which the sections were incubated. The improved technique introduced by Rutenburg and Seligman (1955) represents a new departure from established practice for two reasons; first, these authors used sodium 6-benzoyl-2-naphthyl phosphate instead of the more usual naphthyl phos-phates, and second, their technique is a post incubation coupling method; i.e., the sections are first incubated with the substrate and then treated with a diazo dye solution afterwards. Using this technique, Rutenberg and Seligman claim that diffusion artifacts are virtually eliminated, but, unfortunately, the colored end product fades after 2 weeks or more.

Rutenburg and Seligman's view concerning the lack of diffusion artifacts produced by this technique is supported by the work of Defendi (1957), in which he observed that 6-benzoyl-2-naphthol has a strong affinity for various tissue components. It would therefore remain at the sites at which it was released by enzyme activity until coupled with the dye in the second stage of the procedure.

1. ACID PHOSPHATASE IN NORMAL MUSCLE

For the study of acid phosphatase in muscle, the Gomori technique seems to have been the one chosen by most investigators. Gomori's work (1941a) led to the conclusion that there was no acid phosphatase in human skeletal muscle taken either at biopsy or within 4 hours postmortem, and this result was supported by observations on rat and monkey muscle by Dempsey *et al.* (1946). Wolf and associates (1943), however, obtained a positive reaction in the skeletal muscle of various adult animals. They found that the sarcolemma and muscle nuclei showed a moderately intense reaction and that the cross striations were as a rule clearly visible. Sometimes, however, the muscle fibers were unreactive.

In addition, Wolf *et al.* demonstrated that the axons, the nuclei, and the cytoplasm of Schwann cells, and the endoneurial cells of peripheral nerves were acid phosphatase positive; that in arteries and veins all nuclei and smooth muscle cells were positive, and that in capillaries only the nuclei gave a reaction. A few years later, Rossi *et al.* (1953, 1954) obtained a positive acid phosphatase reaction in the nuclei of human fetal muscle, and in their later paper they stated that there was enzyme activity in the sacrolemmal nuclei of adult human muscle.

Our own observations (Beckett and Bourne, 1958b) support many of the findings of Wolf *et al.* (1943). In specimens of normal muscle, the most reactive elements were the axons and the so-called neurokeratin of the peripheral nerves, the cell walls of adipose tissue, the fibroblasts of tendon, and also the groups of granules situated at the poles of the nuclei of both muscle fibers and connective tissue. These granules probably represent the Golgi apparatus. The nuclei of connective tissue, of blood vessel walls, and some also of those in muscle fibers contained a moderate amount of the enzyme. The muscle fibers themselves, on the other hand, and the smooth muscle of blood vessel walls were less

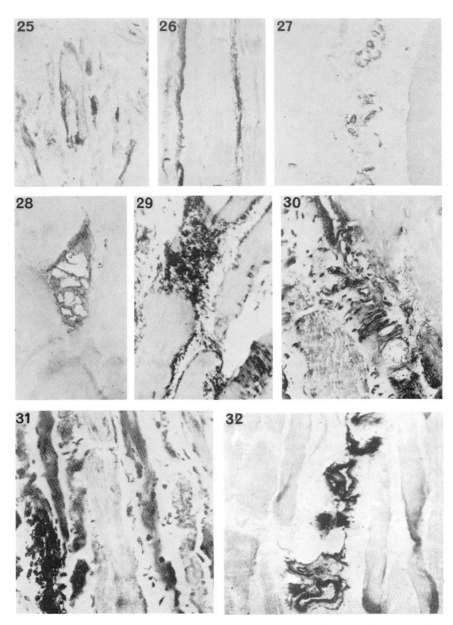

Fig. 25. Alkaline phosphatase in muscle from a case of facioscapulohumeral dystrophy. Several of the few remaining muscle fibers give a positive reaction. Magnification approximately × 150.

Fig. 26. Atrophied, alkaline phosphatase-positive muscle fibers, seen in a case

reactive. The cross striations of voluntary muscle were often visible. As did Wolf *et al.*, we sometimes found that the muscle cells were negative, but this was usually a reflection of the length of incubation used. Capillary walls were occasionally positive, the nuclei being more so than the cytoplasm of the endothelial walls.

Rutenburg and Seligman's (1955) work on formalin-fixed tissue indicated that their method produced a granular reaction for acid phosphatase in skeletal muscle. This reaction had no particular localization. It is difficult to reconcile this observation with the results obtained with the Gomori technique.

2. ACID PHOSPHATASE IN PATHOLOGICAL MUSCLE

Again, for this section we must rely mainly upon our results and must emphasize that these have not been confirmed. In general, there appeared to be little change from normal in the acid phosphatase activity of muscle specimens taken from cases of muscular and neuromuscular disorders. There was a reduced reaction, as judged from the incubation time needed to produce minimal coloration, in three cases of periodic paralysis and one of peroneal muscular atrophy. There was a slight reduction in activity in one case each of polymyositis, motor neuron

---

of thyrotoxic myopathy. Note the fine, enzyme-positive connective tissue fibers surrounding the muscle fibers. Magnification approximately × 150.

Fig. 27. A very much disrupted muscle fiber with alkaline phosphatase in the remnants. This was observed in a case of doubtful diagnosis (pseudohypertrophic muscular dystrophy or polymyositis) in a young boy. Magnification approximately × 150.

Fig. 28. A nonatrophied alkaline phosphatase-positive muscle fiber from a case of muscular dystrophy with periodic paralysis. Magnification approximately × 350.

Fig. 29. Muscle from a case of familial muscular dystrophy, showing a mass of acid phosphatase-positive cells invading a muscle fiber. Magnification approximately × 200.

Fig. 30. Muscle, again from a case of familial muscular dystrophy, showing invasion of muscle fibers by acid phosphatase-positive connective tissue fibers. The very atrophied fibers present in this field contain much acid phosphatase. Magnification approximately × 200.

Fig. 31. Acid phosphatase in muscle from a case of pseudohypertrophic muscular dystrophy. The muscle fiber on the extreme left contains a high concentration of the enzyme and is being invaded by masses of cells. The larger fiber in the center contains pale-staining nuclei, and the fiber to the right of this contains deep-staining nuclei. Magnification approximately × 400.

Fig. 32. 5-Nucleotidase in the walls of blood vessels of muscle from a case of facioscapulohumeral dystrophy. This is a normal distribution pattern for the enzyme.

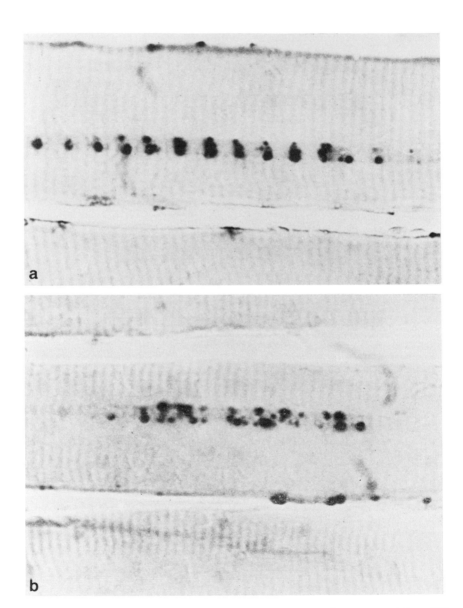

Fig. 33. (a) Progressive muscular dystrophy of the Duchenne type (Case 17); (b) myotonic muscular dystrophy (Case 23). Diazo method (hexazonium pararosaniline) for acid phosphatase. A highly intense reaction for acid phosphatase in granular form is evident. From Monticone *et al.* (1970).

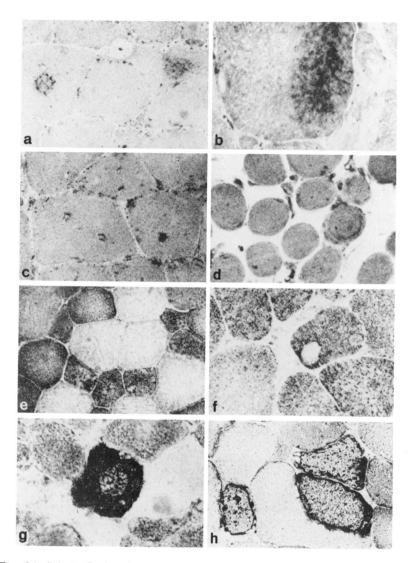

Fig. 34. (a) Acid phosphatase activity occurs in both types of fiber. Duchenne
($\times$ 450). (b) Typical distribution of acid phosphatase in a fiber. Duchenne
($\times$ 450). (c) High acid phosphatase activity situated mainly around the internal
sarcolemmal nuclei. Myotonic dystrophy ($\times$ 450). (d) Low B glucuronidase activity. Only present around sarcolemmal nuclei and macrophages. Duchenne ($\times$ 450).
(e) Four type I fibers with a moth-eaten appearance. Myotonic dystrophy; DPNH
reductase ($\times$ 250). (f) One type I fiber with subsarcolemmal blebs. Myotonic
dystrophy; DPNH reductase ($\times$ 450). (g) Typical sarcoplasmic mass in one type
I fiber. Myotonic dystrophy; succinic dehydrogenase ($\times$ 450). (h) Three type
I fibers with focal enzyme accumulation at the periphery and disorders of internal
distribution. Myotonic dystrophy; DPNH reductase ($\times$ 450).   From Scarlato and
Cornelio (1970).

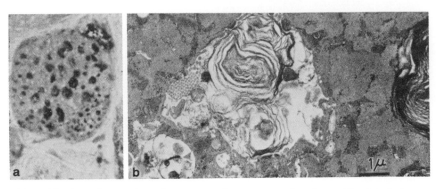

Fig. 35. Autophagy. Autophagic vacuoles are reactive for acid phosphatase as seen in light microscope, × 490 (a) and contain cytoplasmic degradation products as seen in electron microscope, × 16,900 (b). Sporadic late-onset myopathy. From A. G. Engel and MacDonald (1970).

disease, and myopathy of unknown origin. Although we have listed these cases here, it must be said that the change in acid phosphatase activity of a given specimen is not linked with its clinical diagnosis.

A generalized increase in acid phosphatase activity was never observed but it was very noticeable that all atrophied fibers, whether or not they also showed signs of degeneration, contained more acid phosphatase activity than normal. This was very strikingly shown in peroneal muscular atrophy and motor neuron disease, where the atrophied fibers were situated in groups among the more normal muscle fibers.

Groups of inflammatory cells invading degenerating fibers often showed a moderate to strong histochemical reaction for acid phosphatase, although this type of distribution was not as striking as that seen in preparations demonstrating 5-nucleotidase activity. The invading cells were particularly obvious in cases of muscular dystrophy of different types and in cases of polymyositis. Occasionally, too, invading acid phosphatase-positive capillaries were observed.

V. 5-Nucleotidase

5-Nucleotidase was originally studied by biochemical means. In 1934 (cited in Gulland and Jackson, 1938), Reis found 5-nucleotidase in the retinas of various animals and later work by this author (Reis, 1951) and also by Gulland and Jackson (1938), demonstrated that it was present in a variety of tissues taken from animals and also from the human body.

The histochemical techniques for 5-nucleotidase are of two types. The one evolved by Gomori, which uses acetone-fixed material, arose from the study of phosphatase specificity (Gomori, 1949; Newman *et al.,* 1950b). A second type of technique, with which comparatively little work has been carried out, utilizes fresh frozen sections of mammalian tissues (Maengwyn-Davies *et al.,* 1952; Padykula and Herman, 1955a,b). Both types of method are based essentially on the Gomori alkaline phosphatase technique, but with the replacement of glycerophosphate by muscle adenylic acid.

A. 5-Nucleotidase in Normal Muscle

The apparent distribution of 5-nucleotidase in muscular tissue depends upon the technique used. Newman and his co-workers (1950b) were unable to demonstrate 5-nucleotidase in acetone-fixed muscle taken from a range of animals. Later work by the present authors (Beckett and Bourne, 1958c) showed that in normal adult human muscle there may be no 5-nucleotidase activity at all, or there may be a slight reaction for the enzyme. The reaction, when present, is restricted to the nuclei and the intimas of vessels of larger size than capillaries, to the axons and so-called neurokeratin network of peripheral nerves present, to the connective tissue sheaths of muscle spindles, and to occasional fibroblasts. Incubation periods of up to 24 hr are needed, however, to demonstrate 5-nucleotidase in human muscle. This enzyme activity, demonstrated at pH 8.25, is due to a specific 5-nucleotidase and not to alkaline phosphatase, since if the substrate is replaced by sodium glycerophosphate a reaction is never obtained.

The distribution of 5-nucleotidase is curious, since the enzyme is associated with only certain of the larger vessels, and this phenomenon does not appear to be due to technical difficulties. The physiological interpretation of this specific distribution is obscure.

In rat rectus abdominis muscle incubated for 6 and 8 hr at pH 8.0, this variable distribution in vessel walls is most marked. Those the size of arterioles and venules are especially affected, for quite commonly one of an adjacent pair shows an intensely positive reaction while its neighbor remains negative. Capillaries are also positive to a variable degree, and where they are strongly reactive, diffusion to nearby muscle nuclei occurs and gives the latter a spurious positivety. As with human material, the muscle fibers are consistently negative. From the point of view of the amount and distribution of 5-nucleotidase, acetone-fixed mouse muscle is more akin to human than to rat muscle.

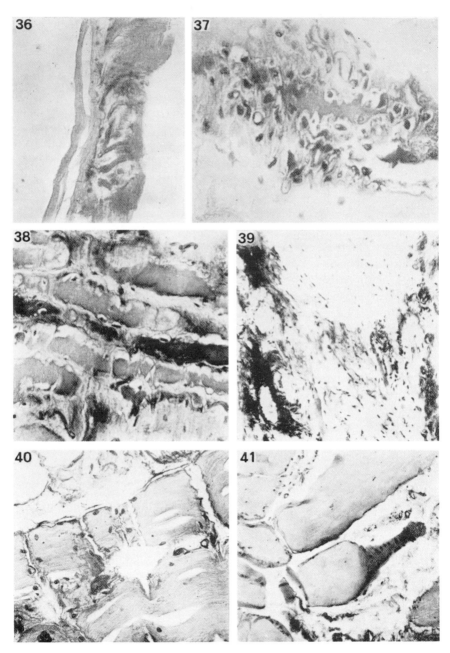

Fig. 36. Muscle from a case of motor neuron disease, showing a 5-nucleotidase positive capillary that has apparently "eaten" its way into the muscle fiber. Magnification approximately × 350.

It does not seem to be particularly easy to demonstrate 5-nucleotidase in fresh frozen sections of mammalian skeletal muscle. Maengwyn-Davies *et al.* (1952) could obtain only a diffuse staining of rat skeletal muscle fibers after 72 hr incubation at pH 8.25. Some muscle nuclei and the adventitia and media of blood vessels also became positive under these conditions. Three years later, Padykula and Herman (1955b) observed no reaction in lingual and leg muscle after incubation for 30 min with muscle adenylic acid at pH 9.4. This latter negative result may in part be due to the high pH used, since we have observed that 5-nucleotidase activity varies considerably with pH (Beckett and Bourne, 1958c).

B. 5-Nucleotidase in Pathological Muscle

Usually, in pathological muscle, the pattern of distribution of 5-nucleotidase activity was essentially the same as in normal samples, i.e., the reaction was absent or was limited to groups of blood vessels and to small areas of connective tissue in their immediate vicinity (Beckett and Bourne, 1958c).

In cases where there had been a considerable reduction in the size of muscle fibers present, with a concomitant increase in the amount of dense collagenous connective tissue (as is, for instance, seen in facioscapulohumeral dystrophy or familial dystrophy), and especially where this was associated with inflammatory changes in and adjacent to the muscle fibers, there was frequently a very great increase in the amount of enzyme present. Where there was less widespread muscle fiber atrophy and necrosis, a smaller increase in the amount of 5-nucleotidase was observed.

Fig. 37. A very atrophied muscle fiber seen in a case of query pseudohypertrophic muscular dystrophy, query polymyositis. The fiber substance contains a high concentration of 5-nucleotidase and so also do the invading cells. Magnification approximately × 450.

Fig. 38. Muscle from a case of familial dystrophy of late onset, showing high 5-nucleotidase activity in atrophied muscle fibers and in the cells and fibers that are destroying them. Magnification approximately × 175.

Fig. 39. 5-Nucleotidase in muscle from a case of facioscapulohumeral dystrophy. Note the enzyme activity in very atrophied muscle fibers, in capillaries, in the fibroblasts of the dense connective tissue, and in some of the connective tissue fibers. Magnification approximately × 100.

Fig. 40. 5-Nucleotidase. Muscle from a case of motor neuron disease to show enzyme-positive capillaries invading a muscle fiber.

Fig. 41. 5-Nucleotidase. A sample of muscle taken from a case of familial dystrophy to show a muscle fiber undergoing atrophy, and at the same time acquiring 5-nucleotidase activity.

Where the activity of this enzyme was increased, it was present in its usual sites, i.e. in the blood vessels, nerves, etc., but the areas of connective tissue showing a positive reaction were considerably enlarged. When there was a lymphocytic infiltration of the connective tissue, the lymphocytes also contained 5-nucleotidase activity. In addition to this increased connective tissue reaction, some muscle fibers were enveloped and appeared to be actively eroded by inflammatory cells, reticular fibers, and capillaries, all of which were enzymically active. At the same time, the muscle fibers themselves accumulated 5-nucleotidase. The occurrence of this increase in 5-nucleotidase activity was not limited to neuromuscular and muscular disorders, since an essentially similar picture was seen in muscle taken from around the hip joint of a case of osteoarthritis of 5 years' standing.

VI. Other Phosphatases

The literature concerning other phosphatases is sparse, particularly in connection with skeletal muscle tissue, and the results obtained in muscle

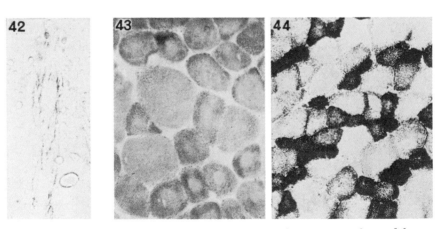

Fig. 42. Succinic dehydrogenase in gastrocnemius from a case of pseudohypertrophic muscular dystrophy. Note the curious woven pattern of the diformazan granules, which reflects the pattern of alignment of the myofibrils. Magnification approximately × 180.

Fig. 43. Succinic dehydrogenase in gastrocnemius from a case of polymyositis. Note the larger fibers with little reaction in them and the smaller fibers with a broad peripheral band of reaction. The central light zone possibly represents a primary biochemical lesion. Magnification approximately × 120.

Fig. 44. Muscle from a rat, showing the checker-board effect of fibers with high and low concentrations of succinic dehydrogenase activity.

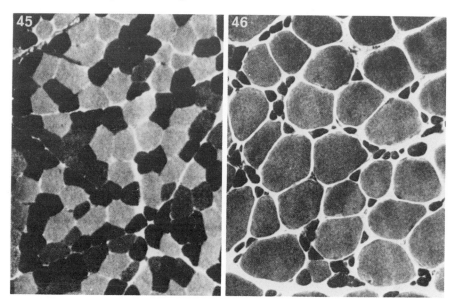

Fig. 45. Normal human skeletal muscle (triceps); 5 μ cryostat section showing distribution of myosin ATPase and distinction between type I and II fibers. × 156. From A. G. E. Pearse and Johnson (1970).

Fig. 46. Skeletal muscle (Werdnig–Hoffmann disease); 5 μ cryostat section showing the survival of type I (pale) fibers confined to restricted areas. × 390. From A. G. E. Pearse and Johnson (1970).

with other phosphate substrates seem to depend primarily upon the technique used. Zorzoli and Stowell (1947) found that an alkaline pH, fructose 1,6-diphosphate gave the same reaction distribution as sodium glycerophosphate in acetone-fixed, paraffin-embedded skeletal muscle.

Three years later, Newman *et al.* (1950b), also using acetone-fixed muscle from several different animals, and carrying out their incubation at pH 9.2, observed that a wide variety of substrates, including glycerophosphate, glucose 1-phosphate, hexose diphosphate, creatine phosphate, yeast adenylic acid, yeast nucleic acid, thiamine pyrophosphate, and barium phytate, gave a typical alkaline phosphatase reaction in muscle; i.e., the endothelium of blood vessels was positive. All of these substrates and adenosine triphosphate and 5-nucleotide gave nuclear staining also, but only after prolonged incubation periods.

Bourne (1954a,b,c) described the reaction of striated muscle of the rat incubated with a variety of substrates at pH 9.0. With glucose 1-phosphate, fructose 1-phosphate, galactose 6-phosphate, and tetrasodium 2-methyl-1:4-naphthohydroquinone phosphate, only the capillaries in skeletal and heart muscle were positive, although the intercalated disks of the latter were also positive. Tetrasodium 3:4-dihydroxy-4-methoxy-chalcone phosphate and tetrasodium 2′,3-dihydroxy-4:4-dimethoxychal-cone diphosphate gave no reaction at all in skeletal muscle, but the nuclei, and to a lesser extent the fibers in heart muscle, were positive. Fructose 1-phosphate gave a positive reaction in capillaries in some muscle fibers and in the connective tissue nuclei. Estrone phosphate, hexestrol phosphate, and diethylstilbestrol phosphate all gave very faint reactions in the fibers of skeletal muscle. In heart muscle, the nuclei were positive and the intercalated disks were positive with the first substrate and negative with others.

When riboflavin phosphate and pyridoxal phosphate were used as substrates with heart muscle, the nuclei and capillaries were positive, the fibers slightly positive, and the sarcosomes strongly positive. The intercalated disks were very positive with the first of these two substrates.

The authors conclude with the statement that "the alkaline phospha-tase reaction provides a means of identifying a particular kind of ab-normal human muscle fiber and of highlighting such fibers in minimally to severely involved muscle."

VII. ATPases

These results indicated, therefore, that if specific phosphatases existed in fresh muscular tissue, only a little adenosine triphosphatase and 5-nu-cleotidase activity could survive acetone fixation, or at any rate, the remainder of the phosphatases were not demonstrable under the experi-mental conditions employed. The only evidence to the contrary was that of Glick (1946), who, using acetone-fixed paraffin-embedded cock-roach muscle, obtained a positive reaction with ATP but not with glycerophosphate.

Later work on the phosphatases of muscle was directed toward the development of methods using fresh frozen sections in order to demon-strate specific enzymes, and particularly to demonstrate ATPase. The early efforts of Maengwyn-Davies and Friedenwald (1950) do not seem to have met with much success, since after 48 hr incubation they obtained a reaction limited to nuclei and blood vessels with glucose 1-phosphate but no reaction at all with hexose diphosphates, β-naphthyl phosphate, or sodium glycerophosphate.

Two years later, the technique of these workers had been improved by the use of varying pHs and the introduction of inhibitors and activators to demonstrate the specificity of the phosphatases (Maengwyn-Davies *et al.*, 1952). Using the modified technique at pH 9.9 with ATP as substrate, they found that after 24 hr incubation, the muscle fibers were blackened and had darker staining nuclei. These nuclei had a blackened membrane and then an unstained region between it and the chromatin. The sarcolemma and the endothelium of blood vessels also showed activity.

At pH 8.25, a pH at which these workers supposed myosin ATPase to be active, the muscle fibers themselves and their sarcolemmas, and also the chromatin of muscle, nuclei were intensely stained after 22 hr incubation with ATP. The endothelium of blood vessels of all sizes and the smooth muscle of the media were diffusely darkened. In addition, the nuclei of the blood vessels were moderately positive, and activity was also observed in the various connective tissue sheaths of peripheral nerves. With creatine phosphate and yeast adenylic acid, these workers could only obtain patchy reactions after very long periods of incubation (48–96 hr).

Further investigations on the technical side by Padykula and Herman (1955a) indicated that the sodium acetate used in the method of Maengwyn-Davies and her colleagues had a strong inhibitory effect on phosphatases, and by using a modified Gomori substrate at pH 9.4, Padykula and Herman obtained a positive reaction with ATP in 15–30 min, a much more satisfactory result than those of the earlier workers.

This positive reaction observed by Padykula and Herman was probably situated in the myofibrils, and there was no sarcoplasmic staining. Blackening of cross striation was sometimes seen, and smooth muscle and endothelium of blood vessels were also positive. In contrast to this, both adenosine diphosphate and muscle adenylic acid gave little reaction under the experimental conditions used.

The energy for muscular contraction comes directly from the hydrolysis of ATP by ATPase. The association of this enzyme with contractile structures is a very fundamental biological fact, a point made originally by Szent-Györgyi in 1953. At that time, he pointed out that contractility is in general dependent upon an interraction between ATP and actomyosin, a mechanism which requires the presence of ATPase. This enzyme has been found in both vertebrate and invertebrate muscle, and even in such elementary organisms as protozoa, it has been found associated with contractile elements. An example of this is Levine's work (1960) on Vorticella in which he found a highly specific ATPase present in cilia and in the myonemes, which are the contractile elements. It is also of interest that sperm tails, which are very motile, contain ATPase.

Stein and Padykula (1962) divided the fibers of muscle into three

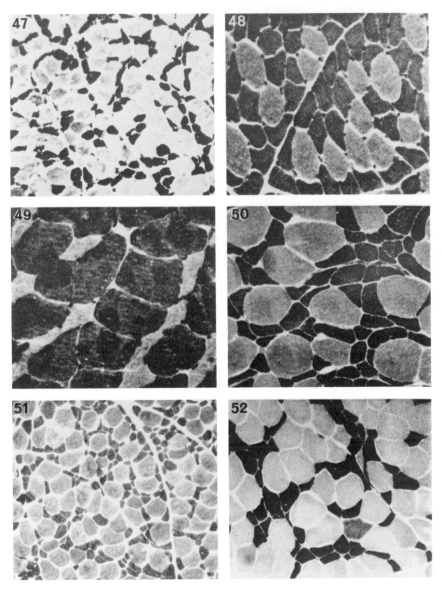

Fig. 47. Prominent preferential atrophy of type II (dark) fibers in guinea pig soleus 27 weeks after total denervation. Regular ATPase; × 130. From W. K. Engel (1970).

Fig. 48. Cat gastrocnemius 4 weeks after sciatic neuroectomy showing early stage of preferential atrophy of type II (dark) fibers. Regular ATPase; × 190. From W. K. Engel (1970).

Fig. 49. Cat gastrocnemius 2 weeks after s Achilles tenotomy showing greater

types based on their SDH activity and designated them A, B, and C fibers. They found the A fibers to have a high ATPase activity which was sensitive to fixation, the type B fibers to have a low activity, and the type C fibers to have a high activity that was resistant to fixation. In 1963, Padykula and Gauthier published a histochemical study of AT-Pase in muscle fibers on the rat diaphragm. Most of the fibers in this muscle were small and red with many subsarcolemmal mitochondria and other small mitochondria aligned at the I band. ATPase activity at pH 7.2 was found to be localized in the mitochondria of both sites and was magnesium dependent. At pH 9.0, a myofibrillar ATPase (myosin ATPase) reacted and was located in the A bands. A cysteine-dependent ATPase was also demonstrated and was especially prominant in the large white (type 2) fibers. This ATPase appeared to be in the region of the sarcoplasmic reticulum. W. K. Engel (1963) has shown a sarcoplasmic reticulum associated ATPase in human skeletal muscle at pH 9.4 and indicated that uridine triphosphate (UTP) was hydrolyzed at identical sites, which indicates either a dual ability on the part of the ATPase or that there is a separate UTPase activity present. Barden and Lazarus (1964) found ATPase activity in mouse muscle only in mitochondria and in the A band of the fibers.

We found ATPase activity in muscle nuclei in our laboratory, especially in denervated muscle; and Engel has also described a nuclear ATPase reaction. Some authors have, however, denied the validity of a nuclear ATPase (e.g., Novikoff, 1958), but it was demonstrated by Sandler and Bourne (1962) that such lack of results was due to a technical error. The presence of ATPase in the endoplasmic reticulum has been confirmed biochemically by Martinosi and Ferretos (1964) and Krespi *et al.* (1964).

Zebe and Falk (1963), of Heidelberg, combining histochemistry and electron microscopy, showed that in insect flight muscles the Z band of the fibers gave a strong ATPase reaction with a higher activity in the M line. They also found a strong mitochondrial ATPase activity, which was localized exclusively in the cristae. They did not mention any association of the reaction with the sarcoplasmic reticulum; however, in heart muscle, ATPase has been found to be associated with the reticu-

---

atrophy of type I (light) fibers. Regular ATPase; × 190. From W. K. Engel (1970).

Fig. 50. Preferential atrophy of type II (dark) fibers; myasthenia gravis. Regular ATPase; × 190. From W. K. Engel (1970).

Fig. 51. Preferential trophy of type II (dark) fibers; subacute sclerosing panencephalitis. Regular ATPase; × 190. From W. K. Engel (1970).

Fig. 52. Preferential atrophy of type I (dark) fibers; myotonic dystrophy. Edetic acid–low pH-activated ATPase; × 75. From W. K. Engel (1970).

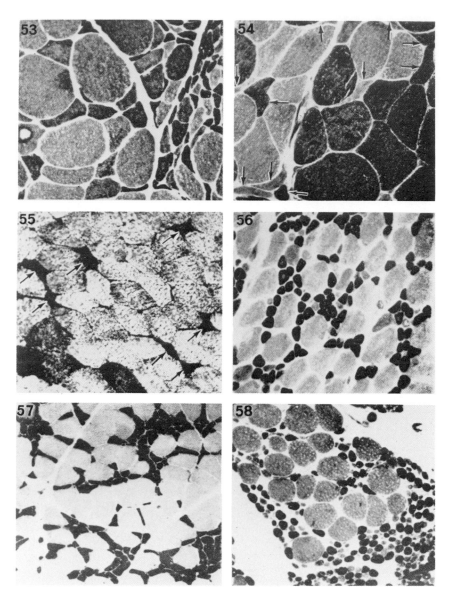

Fig. 53. Mixed atrophy affecting both fiber types; chronic peripheral neuropathy in 14-year-old girl. Regular ATPase; × 190. From W. K. Engel (1970).

Fig. 54. Mixed atrophy affecting both fiber types; amyotrophic lateral sclerosis; atrophic type I fibers (vertical arrows) and type II fibers (horizontal arrows). Regular ATPase; × 190. From W. K. Engel (1970).

Fig. 55. "Scattered atrophy" typical of early denervation. Atrophic fibers (arrows) are angular and darker than normal dark type I fibers (lower left); mild idiopathic

lum at the Z line region, the cell membranes and the intercalated disks. The association of a number of phosphatases with the intercalated disks of heart muscle was first demonstrated by Bourne in 1953.

These results demonstrated that, in addition to the ATPase activity associated with the actomyosin, practically all the other elements of the muscle cell have some activity for this enzyme. It is probable that these are not all the same enzyme, but that at least some of them represent ATPase isozymes. Padykula and Gauthier showed the presence of a fixation-resistant ATPase, which was probably an isozyme; and Sandler and Bourne have demonstrated another that is both fixation and heat resistant.

For some time in our laboratory we have been carrying out studies on the histochemistry of an ATPase that was present in alcohol fixed paraffin embedded tissues. The possibility that we were dealing with a nonspecific phosphatase was one that had to be eliminated, and this was done as follows: The enzymes in a sample of muscle homogenate were spread on starch gel by electrophoresis. The gel was then dehydrated and allowed to stand in 80% alcohol (the equivalent of fixation) for some hours. Dehydration was then completed and the gel cleared in xylol. It was then placed in hot xylol (58°C) in an incubator for 1 hr (the equivalent of embedding). It was then taken out of xylol, rehydrated, and then incubated in ATP substrate solution, the sites of enzyme activity being visualized by the inclusion of alizarin in the incubating medium as originally described by Sandler and Bourne (1961). This technique demonstrates that skeletal muscle has three or four specific ATPase isoenzymes, that most of them show some resistance to the fixing and embedding procedures, and that at least one survives these procedures very well and does not lose its specificity. It is of interest that in this connection, Sandler and Bourne (1962) have obtained evidence that the extraction and purification of a phosphatase enzyme can result in a loss of specificity.

---

familial chronic neuropathy in 14-year-old girl. Quadriceps biopsy; lactate dehydrogenase; × 190. From W. K. Engel (1970).

Fig. 56. Type I fiber hypotrophic with central nuclei. All type I fibers (dark) are small, and all large fibers are type II (light). Central nuclei appear as unstained areas of type I fibers. Case 1, 12-month-old boy. Edetic acid–low pH-activated ATPase; × 480. From W. K. Engel (1970).

Fig. 57. Type I fiber hypotrophic with central nuclei. All type I fibers (dark) are small, and all large fibers are type II (light). Central nuclei appearing as unstained areas are evident. Case 2, 12-year-old boy. Edetic acid–low pH-activated ATPase; × 75. From W. K. Engel (1970).

Fig. 58. Selective hypertrophy of type I (light) fibers in 4-month-old boy with infantile spinal muscular atrophy. Regular ATPase; × 190. From W. K. Engel (1970).

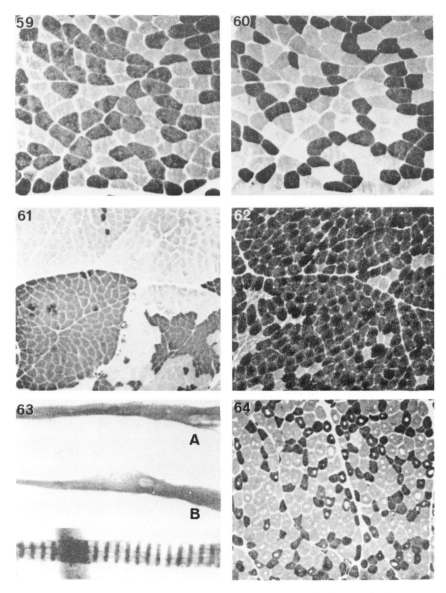

Fig. 59. Serial section of apparently normal biceps brachialis muscle from uninvolved side of 18-year-old boy with idiopathic hemiparesis and focal seizures of one year's duration. Type II fibers are darker. Regular myofibrillar ATPase; × 75. From W. K. Engel (1970).

Fig. 60. Serial section of same muscle as in Fig. 63. Type I fibers darker. Subtle differences are evident among type II fibers. Edetic acid–low pH-activated ATPase; × 75. From W. K. Engel (1970).

Fig. 61. Type grouping is evident as large contiguous collections of each fiber type, in contrast to more evenly mixed mosaic pattern normally seen (Figs. 63

While the energy obtained from the hydrolysis of the terminal phosphate bond of ATP is thought to be the immediate source of energy for contractions of muscle, the restoration of this phosphate bond is essential for the processes that follow contraction. It has been known for many years that creatine phosphate is the high energy phosphate which is capable of restoring the third phosphate bond in ATP to enable relaxation and further contraction to occur. The reaction between ADP and creatine phosphate to reform ATP is the well known Lohman reaction; the enzyme involved is creatine kinase, which is capable of working in either direction:

$$\text{creatine} \cdot \text{PO}_4 + \text{ADP} \rightleftarrows \text{ATP} + \text{creatine}$$

Creatine kinase is therefore a transphosphorylating enzyme.

Whether there is, in addition to creatine kinase, a separate phosphatase for creatine phosphate is not clearly established. In our histochemical studies on alcohol- and acetone-fixed muscle embedded in wax, a strong reaction is given by muscle capillaries when creatine phosphate is used as a substrate, together with a moderate diffuse reaction in the muscle fiber. In dystrophic muscle, although there is an active dephosphorylation of nucleotides by the proliferating endomysium, creatine phosphate is not hydrolyzed at all by the endomysium. We have evidence that the modest hydrolysis of creatine phosphate in our preparations is not due to a nonspecific phosphatase; but whether or not it is due to creatine kinase, we have no information at present.

In our laboratory, we have studied the histochemical results obtained by the use of a large variety of different phosphate esters as substrates for phosphatase studies. It has been suggested that many of the results obtained with these substrates represent nonspecific phosphatase reactions, especially since they are carried out on alcohol- or acetone-fixed

---

and 64). Chronic peripheral neuropathy. Edetic acid–low pH-activated ATPase; × 30. From W. K. Engel (1970).

Fig. 62. Type II fiber predominance, probably the result of unrepresentative sampling, in a 12-year-old boy with chronic infantile spinal muscular atrophy. At left are three groups of atrophic type I fibers; other areas of biopsy contained atrophic type II fibers. Regular ATPase; × 75. From W. K. Engel (1970).

Fig. 63. A, 21-day-old culture from thigh skeletal muscle of 13-day-old chick embryo, grown without innervation. Regular myofibrillar ATPase reaction shows fairly mature cross-striated appearance of outgrown muscle fibers. No innervation into different histochemical fiber types present. B, 26-day-old culture of thigh skeletal muscle from 13-day-old chick embryo. Phosphorylase reaction (dark areas) localized in I band but not in A band or Z line. This demonstrates advanced degree of histochemical maturity of outgrown muscle fibers cultured with innervation.

Fig. 64. Hypokalemic periodic paralysis in a 13-year-old boy. Biopsy taken from a strong muscle not during an attack. Vacuoles are present in both fiber types. Regular ATPase; × 75. From W. K. Engel (1970).

tissues. Studies of the differences of distribution of the reaction with various substrates together with starch gel electrophoresis studies demonstrate, however, that we are dealing with a range of specific phosphatase enzymes. Some of the substrates used were sugar phosphates directly or indirectly related to the glycolytic cycle; these included glycerophosphate, glucose 1- and glucose 6-phosphates, fructose 1-phosphate, and galactose 6-phosphate. Other substrates were coenzymes, such as pyridoxal phosphate, which is a codecarboxylase and acts also as a codeaminase and a transaminase; riboflavin 5′-phosphate (riboflavin mononucleotide), which is a constituent of Warburg's yellow enzyme; cytochrome reductase and amino acid oxidase and thiamine pyrophosphate (cocarboxylase, thiamine diphosphate), which is the coenzyme involved in the decarboxylation of α-keto acids, e.g. pyruvic acid, and α-ketoglutaric acid. It also acts as a coenzyme in the transketolation of the direct oxidative chain of glucose. Di- and triphosphopyridine nucleotides (DPN and TPN) were also used as substrates. These coenzymes are, in fact, codehydrogenases and play an extremely important part in most of the reactions we have been discussing in this chapter.

Using alcohol- or acetone-fixed muscle and paraffin embedding, practically the only structures that gave a reaction when sugar phosphates and glycerophosphate are used as substrates are the muscle capillaries. DPN and TPN and riboflavin phosphate, when used as substrates, show a reaction in the muscle nuclei, especially in the nucleoli; pyridoxal phosphate gives more of a reaction in the fibers, and in some preparations, it seems also to be in the sarcosomes. There is also quite a good nuclear membrane reaction in the muscle nuclei. It is hard to judge the significance of the dephosphorylating activity produced by these substances when used as substrates. Does this mean that *in vivo* specific phosphatases for these coenzymes exist, or are we really looking at the activity of the synthesizing enzyme operating in reverse under the unnatural conditions of a histochemical reaction? If the enzyme that adds the phosphate moieties of the coenzyme is really a transphosphorylase, presumably it could act in either direction; and our studies suggest that most if not all phosphatases, rather than being simple hydrolytic enzymes, are really transphosphatases. In this case, the reactions we get using these substances as substrates should indicate the sites of their synthesis. We know from biochemical studies that at least one DPN synthesizing enzyme is present in the nuclei of cells and not at all in mitochondria. Our studies that demonstrate the presence of dephosphorylating enzyme for DPN in muscle nuclei may, in fact, be demonstrating a DPN synthesizing or at least a phosphorylating enzyme.

Estrone phosphate, hexestrol phosphate, and stilbestrol phosphate have been used as substrates for skeletal muscle, and all give a very strong reaction in the nuclei of the muscle and connective tissue cells. The physiological significance, if any, of phosphorylated steroid hormones is unknown at present, so any attempt at the interpretation of these results is impossible.

Acetyl phosphate was also used as a substrate. This is a high-energy compound and was originally isolated by Lipman (1946) as an intermediate compound in the oxidative decarboxylation of pyruvate by crude extracts of the microorganism *Bacillus delbruckii*. Subsequently, Lipman (1946) found the compound to be present in appreciable amounts in mammalian muscle; and Bourne and Golarz (1959a) have shown histochemically that there is a fixative- and heat-resistant specific acetyl phosphatase in mammalian and human muscle. There is, however, little information as to the role this enzyme or its substrate play in normal muscle metabolism. Bourne and Golarz (1959a) found this enzyme to be present in the sarcoplasm of muscle and not to be associated with any particular organelle. It is an extremely active enzyme, even after fixation and paraffin embedding.

Nucleic acid and a variety of nucleotides were also used as substrates; those used included mono- and triphosphates of adenosine, uridine, guanosine, cytidine, and inosine. Nucleic acid does not appear to be hydrolyzed under our particular histochemical conditions. Most of the other compounds, when used as substrates, produced a slight reaction in the nucleoli and some capillary reaction. Creatine phosphate was also used and gave a similar result together with a faint myofibrillar reaction. The connective tissue of the elements of the endomysium in all these cases gave a completely negative reaction in normal human muscle.

It has been shown by Bourne and Golarz (1959b) and by Golarz *et al.* (1961) in human muscular dystrophy, but not in mouse hereditary dystrophy, duck or chicken dystrophy, or the nutritional (vitamin E deficient) dystrophy of rabbits, that the proliferating endomysium strongly dephosphorylates all the nucleotides used as substrates; it also hydrolyzes acetyl phosphate (a nonnucleotide), DPN, TPN, and, in some cases, thiamine pyrophosphate. It does not hydrolyze riboflavin phosphate, pyridoxal phosphate, or creatine phosphate, none of which are nucleotides or contain nucleotides.

This reaction appears to be associated with the cells, the connective tissue fibers, and the capillaries. ATP and UTP are particularly actively hydrolyzed by the endomysium in active cases of dystrophy. Some other workers, however, have not been able to reproduce our ATPase reactions in dystrophy; but it should be noted that they have been using techniques such as that of Padykula and Herman on fresh or lightly fixed tissue, in which case they get a satisfactory reaction in half an hour or so; it takes 8 or 9 hr for our ATPase to show up. In that time, with the Padykula–Herman technique, the myosin ATPase would have swamped the section with reaction products, and it would not be possible to distinguish anything.

This specific ability of the proliferating endomysium to hydrolyze nucleotides and only two other compounds (acetyl phosphate and thiamine pyrophosphate) may be of some significance; but any interpretation must be made cautiously since, as we have mentioned previously, we do not know if the reactions seen under artificial conditions *in vitro* are also taking place *in vivo*. At this point, however, it might be appropriate to say a word or two about the connective tissue itself. The connective tissue cells apparently have two functions, to produce connective tissue fibers and to produce ground substance. Presumably the macromolecules of collagen are produced by the normal protein synthetic mechanisms of the cell in which activation of amino acids by ATP occurs, and it may be that increased ATPase activity might be related to increase of collagen production. In healing wounds of skin and bone, it has been shown by Bourne (1943) and by Fell and Danielli (1943) that the newly formed collagen (the precollagen) actively dephosphorylates glycerophosphate and that, when the fibers change into mature collagen, the enzyme activity disappears. Fell and Danielli thought this enzymic reaction might have been related to the maturation of the fibers. There is no glycerophosphatase reaction given by the fibers in the connective tissue of muscular dystrophy.

In addition to collagen production, the connective tissue cells are concerned with the production of ground substance. We should consider briefly the reactions involved in this process. An integral part of the ground substance is sulfated mucopolysaccharide, which is usually closely bound to protein to form a mucoprotein complex. It has been mentioned earlier that amino acids are activated by ATP prior to their incorporation into proteins; likewise, acetyl amino sugars become activated as uridine nucleotides. Uridine and guanosine phosphate esters are known to play a part in the activation of sugars in the synthesis of polysaccharides (Strominger, 1961). Adenosine and uridine nucleotides also play an essential part in the sulfation of mucopolysaccharides. Strominger has reviewed the evidence indicating the role of uridine nucleotides in chondroitin sulfate synthesis. A number of synthetic processes involved in mucopolysaccharide synthesis are mediated by DPN and TPN. Xylulose, glucose 6-phosphate, glucose 1-phosphate, uridine diphosphate, glucose, uridine triphosphate, glucuronic acid, ascorbic acid, ATP, DPN, and

TPN are involved in a cycle known as the uridine nucleotide shunt; it is the pathway for the production of glucuronic acid, an essential constituent of some mucopolysaccharides. Any assessment of the factors involved in muscle disease should take into consideration the integrity of this cycle; since, if a defect in the production of mucopolysaccharides for the ground substance develops, proper transport of oxygen and nutrients to the muscle will not take place. Two defects in the uridine nucleotide shunt are already known in human beings. All humans, together with guinea pigs, monkeys, and apes, suffer from a defect of this cycle which makes it impossible for them to synthesize ascorbic acid. Another defect in the cycle is a benign condition known as congenital pentosuria which occurs in some humans and in which large amounts of L-xylulose derived from the free glucuronic acid formed from the uridine nucleotide are excreted in the urine.

The presence of hydrolytic enzymes in the proliferating connective tissue of muscular dystrophy that are active against ATP and other trinucleotides and against coenzymes, such as TPN and DPN, raises the question as to whether this does not indicate some interference with the normal operating of the uridine nucleotide shunt (Bourne and Golarz, 1963). This would result in a failure to produce the mucopolysaccharides that form the basis of the ground substance of connective tissue and, physiologically, a failure of the transport mechanism between blood capillary and muscle fiber. Experimental studies are being made in our laboratory to determine if there is any change in the mucopolysaccharides in dystrophic muscle.*

Muscle Histochemistry in Psychiatric Patients

Meltzer (1968; Meltzer *et al.*, 1972) found that psychotic exacerbations in psychotic patients were accompanied by elevated levels of isozymes of creatine phosphokinase of the "muscle" type. W. K. Engel and Meltzer in 1970 showed that muscle biopsies from twenty-nine patients suffering from similar exacerbations showed "histochemical abnormalities of a myopathic type." The histochemical tests carried out were the NADH-tetrozolium reductase, myofibrillar ATPase (pH 9.4) basophilia with thionine and alkaline phosphatase. The types of abnormality found were positive alkaline phosphatase fibers, necrotic fibers with phagocytosis, abnormal and irregular intermyofibrillar network with trichrome and NADH–tetrazolium reductase staining and reactions and internal nuclei,

---

* This account of ATPase is taken from "Histochemistry of Normal Skeletal Muscle" by Golarz and Bourne (1966).

and "end stage atrophic neurons." Engel and Meltzer sum up these
results as follows:

> If extraneous factors are not responsible for the histochemical abnormali-
> ties in the acutely psychotic patients, the AP positive fibers represent
> morphological evidence of an extra-cerebral organic disease process in some
> acutely psychotic patients. It remains to be determined what relationship,
> if any, these muscle abnormalities have to the mental disturbance in such
> patients; but their high degree of correlation with acute psychosis supports
> the possibility of a significant relationship.

In view of this paper, the results of Golarz de Bourne and Bourne
(1972) of the histopathology and histochemistry of the muscles of iso-

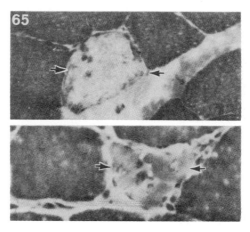

Fig. 65. Muscle fiber undergoing necrosis and phagocytosis. Examples from two
acutely psychotic patients. Modified trichrome stain. From W. K. Engel and Meltzer
(1970).

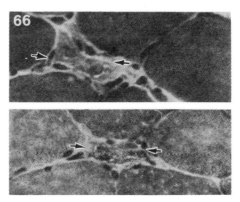

Fig. 66. End-stage atrophic necrosis of muscle fiber. Examples from two acutely
psychotic patients. Modified trichome stain. From W. K. Engel and Meltzer (1970).

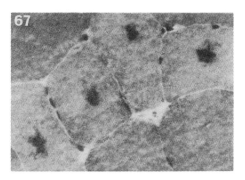

Fig. 67. Clusters of darkly stained rods in the centers of four otherwise normal type 2 muscle fibers. Example from an acutely psychotic patient. Modified trichrome stain. From W. K. Engel and Meltzer (1970).

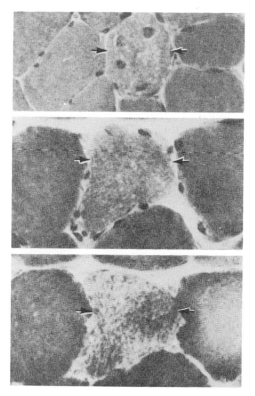

Fig. 68. Moderate architectural abnormality of muscle fiber. Bottom two figures are serial sections; top two were stained with modified trichrome, bottom one treated with NADPH-TR reductase. Examples are from two acutely psychotic patients. From W. K. Engel and Meltzer (1970).

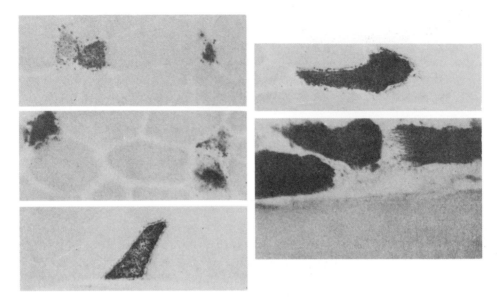

Fig. 69. Examples from five acutely psychotic patients. Alkaline phosphatase-positive muscle fibers are dark; normal fibers are unstained.

lated monkeys are of special interest. Two monkeys were placed in total and separate isolation for a year. During this time, they showed little or no spontaneous activity, being content to sit on their haunches most of the time and getting up only to eat or drink. The muscles of these animals including the diaphragm showed a variety of changes, including extensive myopathy. Among the pathology present there were cellular infiltration, fat infiltration, central nuclei, rowing of nuclei, hyalinization of fibers, type 1 fiber atrophy, and myotube formation. It is not possible to decide whether these changes were due to the physical inactivity or whether there was a psychosomatic factor involved.

VIII. Sulfhydryl and Disulfide Groups

Various methods have been devised for the histochemical detection of sulfhydryl and disulfide groups, and these have been the subject of a review by Gomori (1956). The one that has probably been most used for the study of striated muscle is the dihydroxyl dinaphthyl disulfide (DDD) technique first introduced by Barrnett and Seligman (1952) for the demonstration of sulfhydryl groups, and later modified by the inclusion of thioglycollate reduction (Barrnett and Seligman,

1954) in order to detect disulfide groups as well. In this technique, the substrate combines with sulfhydryl groups and at the same time splits into two parts. One part represents a by-production of the reaction and is removed by washing. The second, which is attached to the protein and which is a naphthol compound, combines with a dye such as tetraazotized diorthoanisidine to give a colored end product. This end product is pinkish mauve where there are few sulfhydryl groups and monocoupling has occurred, but bluish mauve where there are more sulfhydryl groups and dicoupling has occurred.

1. Normal Muscle

According to Barrnett (1953), in skeletal muscle there is an intense reaction for sulfhydryl groups, with no apparent differentiation between A and I bands, and no nuclear reaction. Similar results were obtained by us a few years later (Beckett and Bourne, 1958d). In human, rat, and goat adult skeletal muscle, the reaction intensity for sulfhydryl and disulfide groups was quite strong, and there were no apparent species differences. The picture obtained seemed to indicate that all of the sulfur linkages in muscle were in the sulfhydryl form, since the slides showing sulfhydryl groups were of the same intensity as those showing myofibrils and the sarcoplasm was virtually negative. Cross striations were visible, especially in post mortem specimens. Areas of deeper reaction intensity were sometimes seen near the sides and ends of muscle fibers, but their significance was not clear.

Like Barrnett (1953), the present authors observed a strong reaction for sulfhydryl groups in the smooth muscle of blood vessels and also found that the nuclei of these muscle fibers were less positive than the cytoplasm.

In peripheral nerves, Barrnett (1953) noted a reticulate network which was positive for sulfhydryl groups, and this was later confirmed by us (Beckett and Bourne, 1958d). Barrnett and Seligman (1954), however maintain that this reticulate network showed a reaction for disulfide groups only. The reason for these apparent differences is not obvious. In the connective tissue between the muscle fibers, there is a slight positive reaction (Beckett and Bourne, 1958d), the cellular elements being rather more deeply stained than the connective tissue fibers.

2. Sulfhydryl Groups in Pathological Muscle

The only work on the subject of sulfhydryl groups in pathological muscle appears to be that of the present authors. As judged by histo-

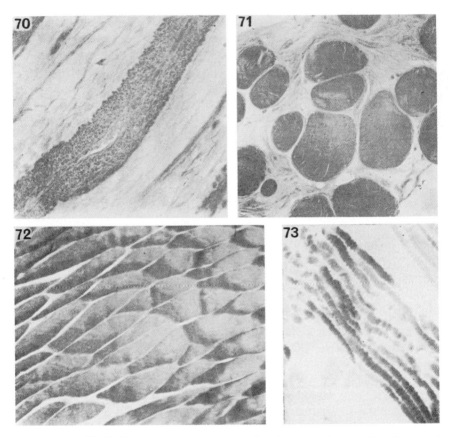

Fig. 70. Sulfhydryl groups in an artery wall. The central area of each smooth muscle fiber, which probably represents the nucleus, is lighter than the periphery. (× 350.)

Fig. 71. Transverse section through muscle fibers from a case of facioscapulohumeral dystrophy, showing that the concentration of sulfhydryl groups is the same in atrophied and hypertrophied fibers. An identical picture is seen in amino group preparations. (× 170.)

Fig. 72. Muscle taken from a case of periodic paralysis showing "shading" of reaction for sulfhydryl groups. Again, amino group preparations were identical in appearance. (× 170.)

Fig. 73. Very atrophied, but otherwise apparently normal, muscle fibers from a case of peroneal muscular atrophy. The concentration of sulfhydryl groups is unchanged, and this was true also of amino agroups. (× 170.)

chemical methods, we could find no change at all in the concentration or distribution of sulfhydryl groups in the range of muscular and neuromuscular disorders examined (Beckett and Bourne, 1958d). This was true even where there was gross change in the size of muscle fibers with or without necrotic changes, and where fibers were being replaced by adipose or dense collagenous tissue.

IX. Protein-Bound Primary Amino Groups

The histochemical technique for the demonstration of these groups was first described by Weiss and associates (1954), and is based upon the use of 3-hydroxy-2-naphthaldehyde as substrate. The aldehyde grouping on this compound reacts with primary amino groups of the protein, and the naphthyl part of the molecule then reacts with tetra-azotized diorthoanisidine to give a color very like that obtained with the technique for sulfhydryl and disulfide groups. As with this latter technique, the color of the end product differs according to whether monocoupling or dicoupling has occurred.

1. NORMAL MUSCLE

The work of Weiss *et al.* (1954) and of the present authors (Beckett and Bourne, 1958d) indicates that the distribution of protein-bound primary amino groups is very similar to that of sulfhydryl groups. In skeletal muscle fibers themselves, the myofibrils have a moderate to strong reaction with fine banding due to cross striations, and there is sometimes a darker reaction at the sides and ends of muscle fibers. The nuclei and sarcoplasm are negative or nearly so. The reaction in the smooth muscle of blood vessel walls is of about the same intensity as that in the striated muscle fibers. The interstitial connective tissue shows little reaction and the cells in it are less positive than in sulfhydryl group preparations. Similarly, the reaction in nerve fibers is much less conspicuous than in the latter preparations.

2. AMINO GROUPS IN PATHOLOGICAL MUSCLE

The picture in muscular and neuromuscular disorders can be summed up quite simply by saying that, as far as our own observations go, there is no change in concentration or distribution of primary protein-bound amino groups as demonstrated by histochemical methods. It must be emphasized, however, that these results are as yet unconfirmed by other workers.

X. Oxidative–Schiff Procedures

If tissue sections are treated with oxidizing agents, such as chromic acid, acidified potassium permanganate, periodic acid, or lead tetraacetate, aldehyde groupings are formed that can then react with Schiff reagent to form a magenta color.

Chromic acid and acidified potassium permanganate were the first known oxidizing agents (Lillie, 1951), but the intensive use of oxidative–Schiff procedures for histochemical work dates from the introduction for this purpose of periodic acid (Hotchkiss, 1948; McManus, 1946). The use of periodic acid and the later introduction of lead tetraacetate (Shimizu and Kumamoto, 1952) for the oxidation of tissue sections have prompted intensive investigation of the chemical reactions involved in oxidative–Schiff methods. As a result, it has become apparent that as well as the 1,2-glycol groups previously known to be oxidized by the reagents mentioned above, other groups, chiefly substituted 1,2-glycol groups are also attacked (e.g., Glegg et al., 1952a; Hale, 1957). Unsubstituted and substituted 1,2-glycol groups are present in a variety of tissue components, including glycogen, mucopolysaccharides, mucoproteins, and some lipids, and all of these substrates may be demonstrated by oxidative–Schiff techniques. Because of this, these techniques have been modified to make them more selective; for instance, saliva or diastase (Bensley, 1939; Lillie and Greco, 1947; among others) have been used to remove glycogen, and lipid solvents (Leblond, 1950) have been used to remove fats. Other workers (Lillie, 1951; Glegg et al., 1952a,b; Casselman, 1954) have directed their attentions toward the possibilities of increasing the selectivity of oxidative–Schiff methods by careful choice of oxidizing agent and the conditions under which it is used.

1. APPLICATION TO PARAFFIN SECTIONS OF MUSCLE

This type of method has been applied to skeletal muscle by Dempsey et al. (1946), who used a chromic acid–Schiff technique on sections of skeletal muscle taken from goat, man, and rat, and also by the present authors (Beckett and Bourne, 1958c) who used periodic acid or lead tetraacetate oxidation followed by Schiff's reagent on sections of normal and pathological human muscle.

The conclusions reached from the two sets of results were similar. In both cases it was observed that stainable material could be present dispersed within the muscle fibers either as fine granules or in large, irregularly distributed aggregates. In human muscle (Beckett and

Bourne, 1958c), it appears that if the specimen is taken at biopsy or within an hour or so of death, the aggregations of stainable material tend to be small, whereas at 12 hr postmortem, the size of the aggregates has increased, and by 48 hr postmortem, they have disappeared. In addition to this randomly distributed material, there is also staining of the cross striations, which Dempsey *et al.* (1946) claim is at the level of the I disks.

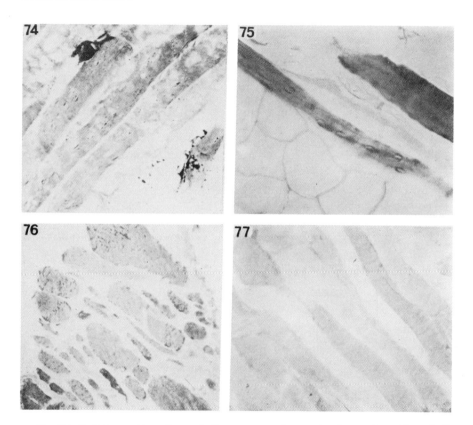

Fig. 74. Gelatine section of muscle from a case of muscular dystrophy with periodic paralysis to show numerous droplets of PAS-positive material situated within the muscle fibers. (× 170.)

Fig. 75. Perinuclear glycogen seen in a case of pseudohypertrophic muscular dystrophy. (× 570.)

Fig. 76. Glycogen in muscle from a case of familial dystrophy to show that atrophy and hypertrophy do not affect the concentration of glycogen in the muscle fibers. (× 170.)

Fig. 77. Muscle taken from a patient with myopathy 48 hr postmortem. Most of the glycogen has disappeared and that which remains is in the cross striations. (× 350.)

The work of the present authors indicates that the amount of stainable material in normal muscle is very variable and that there is no correlation between the amount present and the anatomical site of the muscle concerned. In cases of muscular and neuromuscular disorder, the amount of stainable material is also variable, but the variations are within normal limits and there seems to be no correlation between the amount of stainable material present and the clinical diagnosis involved. Even muscle fibers which show gross atrophy or hypertrophy appear to contain normal concentration of oxidative–Schiff-positive material.

Both Dempsey and his colleagues and we ourselves have observed that diastase (or saliva) will not remove all of this oxidative–Schiff-positive material. Dempsey *et al.* found that the material in irregularly arranged aggregates was removable, whereas that in the cross striations was not. Their interpretation of this was that not all of the positive material was glycogen. In our experience, the picture was not as simple as this. Sometimes diastase would remove both types of stainable material, sometimes only that in the irregular aggregates, and sometimes none at all, even if the sections were all incubated in the same diastase bath together. We agree with Dempsey *et al.* (1946) that the material in aggregates is more readily attacked by diastase or saliva than that in cross striations, but in view of our results we are inclined to think that all of the stainable material is glycogen, but that it is protected to varying extents in different sites by protein associated with it. However, it would be unwise to be too dogmatic on this point, particularly as it is known that diastase preparations (which we have used instead of saliva for most of our work) are contaminated with variable amounts of protelytic enzyme.

Glycogen can accumulate in muscle to a considerable degree in certain diseases. According to Pearce *et al.* (1968), these can be classified into at least eight varieties. The enzyme defect that causes this is not known in all of these conditions. Two diseases in which the defect is known are McArdle's disease and Pompe's disease. In the first of these, absence of muscle phosphorylase has been demonstrated, and in the second, there is a deficiency of acid maltase in the muscle.

In McArdle's disease, the periodic acid–Schiff (PAS) reaction demonstrates considerable increase of glycogen in the subsarcolemmal spaces and also in the interior of the fiber. In Pompe's disease "gross vacuoles was found in the majority of the fibers and greatly increased amounts of PAS positive material were found." Phosphorylase and succinic dehydrogenase activities were found to be normal in the muscle fibers in this disease.

It is of interest that two of us (Golarz de Bourne and Bourne) have found accumulations of glycogen similar to those found in McArdle's disease in the muscles of a baby chimpanzee that had been placed in a whole-body cast from birth for 4 months.

2. Application to Gelatin Sections of Muscle

In 1946, McManus first introduced his periodic acid–Schiff technique, and two years later he suggested (McManus, 1948) that it might be applied to formalin-fixed frozen sections in conjunction with the Baker (1944) Sudan black technique for fats. McManus was of the opinion that glycogen would not be stained in such sections. As far as we are aware, the present authors (Beckett and Bourne, 1958f) are the only authors to have attempted to apply this technique to the study of muscle.

In normal muscle, an irregular diffuse periodic acid–Schiff background staining is obtained. The material that is stained in this is totally different in appearance from that observed in muscle fixed by a suitable method for the preservation of glycogen, and so is almost certainly not this compound. One must qualify this, however, by saying that just occasionally, traces of material are present that might be glycogen that has not been removed during the processes of fixing and embedding.

In addition, both in rat and normal human muscle, there are sometimes droplets of a material that gives a strikingly intense periodic acid–Schiff reaction. In pathological samples of muscle, the number of these droplets is often considerably increased. The number present is not correlated in any way with clinical diagnosis, and it seems that these droplets accumulate at points of mechanical damage in the muscle fibers, e.g., in areas of Nageotte's change, or in areas where the myofibrils are distorted, or again, where there is vacuolation. The presence of more of these droplets in pathological samples of muscle is probably due to the fact that this muscle is more fragile than normal and would therefore be more liable to mechanical damage.

The nature of the strongly periodic acid–Schiff-positive droplets is far from clear. They are certainly not glycogen, since their morphological appearance and distribution is totally dissimilar and they are never removed to the slightest extent by diastase. They do not represent gelatin that has become entrapped in the sections, since this gelatin stains poorly with the periodic acid–Schiff technique. Further tests on the material of which the droplets are composed have shown that it is not metachromatic and does not stain either with hematoxylin and eosin or with Schiff's reagent without prior oxidation and periodic acid. The material is not soluble in cold absolute alcohol, xylol, benzene, or methyl benzoate, but it is extracted during paraffin embedding of formalin or calcium–formalin-fixed material. This extractability suggests that it might be a lipid of some sort, but it is not sudanophilic and does not give a positive Schultz reaction for cholesterol. In fact, the nature of this material remains a mystery.

XI. Fat in Skeletal Muscle

The distribution of fat in skeletal muscle is a problem that has occupied the thoughts of various workers for the last hundred years or more, although from time to time confusion has crept in because of the difficulty of differentiation between true fat droplets and mitochondria.

Henle in 1841 (see Holmgren, 1910) was the first to observe granules between the myofibrils that were less soluble in acetic acid than were the myofibrils themselves, but the nature of the granules which he saw is not clear from the description given. In the next half century or so, the intermyofibrillar granules were the subject of investigation by various workers including Kölliker (1856) and Knoll (see Schaffer, 1893), and it became apparent that the light density of different muscle fibers had a relationship with the amount of sacroplasm present and also with the number of granules in it. No clear line of distinction was drawn, however, between the fat droplets and mitochondria.

Most of this early work was carried out on insect muscle, because of the large size of the myofibrils in its structure, but in later investiga-

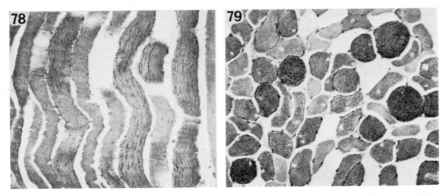

Fig. 78. Muscle from a case of periodic paralysis to show distribution of fat in large streaks in some fibers. (× 170.)

Fig. 79. Muscle from a case of myopathy to show the great variation possible in the number of fat droplets in different muscle fibers. The dark dots at the periphery of the fiber indicate the position of perinuclear fat. (× 170.)

tions human material was used to quite a large extent. In 1889, Walbaum (see Bullard, 1912) studied the occurrence of fats in the skeletal muscle of 119 human bodies, and found "fat" droplets in some muscles of about two thirds of them, but many of these "fat" droplets were not stainable with Sudan III. In the course of his work, Walbaum (see Bell, 1911) also observed that the fat content of the muscles of children, as studied with Sudan III, was not related in any way to nutritive condition.

A few years later, Schaffer (1893), who also studied human muscles (including pectoralis major and gastrocnemius), established that there were wide variations in the numbers and distribution of fat droplets in these muscles, and that there was not always a correlation between the numbers of granules and the degree of optical turbidity of muscle fibers. In addition, Schaffer made one or two observations on the effect of disease on the fat content of muscle. He stated that in chronic disease the "granules," i.e., the mitochondria, were converted to fat, but that few fatty droplets remained, whereas in febrile diseases there are many fatty droplets derived from the "granules."

Bell (1911) studied the effect of nutrition on the fat droplet population of animal muscle and found that starvation reduced the number of fat droplets present, whereas overfeeding increased it. Starvation also tended to remove the differences in turbidity that had existed between different muscle fibers. Like Schaffer before him, Bell found that there was not any constant correlation between the degree of turbidity of muscle fibers and the numbers of fat droplets present.

Bullard (1912) continued to work on this subject and confirmed many of the earlier observations on human muscle. Using diaphragm, pectoral, and eye muscle taken at autopsy, he found that fat droplets occurred constantly and abundantly. When fat was absent, Bullard considered it to be a result of post mortem changes, poor nutrition, or pathological changes. He also thought that post mortem changes were responsible for the occurrence of fat in a granular form.

Bullard differentiated between fat droplets and mitochondria by observing that the latter are not dissolved by absolute alcohol and do not take up fat stains, e.g., Scharlach R. readily. He found no mitochondria in human muscle, a fact he attributed to lack of fresh material. From our own experience in this matter, we think that this failure to demonstrate mitochondria was probably also due in part to the very small size of these bodies in human muscle and to the extreme difficulty of staining them differentially. Both the present authors and Dempsey *et al.* (1964) before us found that mitochondrial stains also demonstrate cross striations with equal clarity.

After the work of Bullard in 1912, there was a considerable lapse

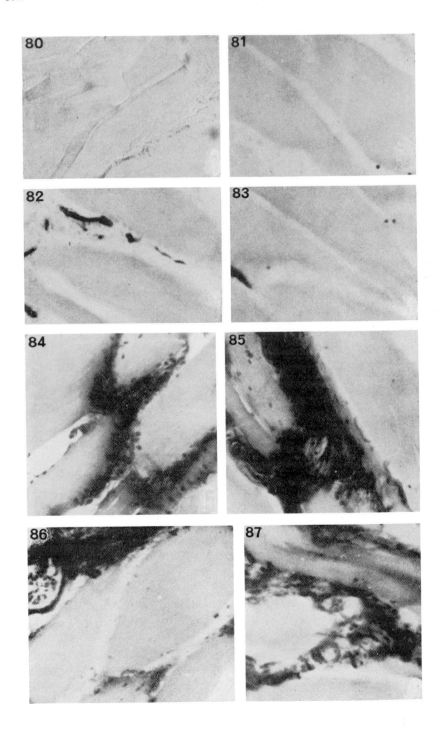

of time before any further observations were carried out on this subject, but in 1946, Dempsey *et al.* carried out a very careful study on skeletal muscle of rats and monkeys using several different techniques. These included Sudan IV and Sudan black staining associated with extraction techniques, the Smith–Dietrich method for phospholipids, and polarizing and fluorescence microscopy.

In the course of their work, they found that both Sudan IV and Sudan black stained minute, discrete fat droplets that were occasionally and irregularly distributed among the striations. These droplets were much more numerous in some fibers than in others and were almost invariably located in the isotropic bands. The droplets were easily removed with acetone or alcohol at room temperature. Sudan black also stained the cross striations, but less intensely than the fat droplets, and the staining was much less easily removed with fat solvents. In addition, the cross striations also gave a positive reaction for phospholipids with the Smith–Dietrich test.

The work of Beckett and Bourne (1958f), using Sudan black staining has confirmed most of these previous observations. In normal human muscle, fat was present in droplets of a very variable size. These were mostly distributed at random, but there were a few situated at the poles of nuclei. In addition, the cross striations were stained, but with a lesser intensity, as has been previously observed by Dempsey *et al.* 1946). The so-called neurokeratin network of the peripheral nerves also contained sudanophilic material.

As far as we could judge, there was no change in the amount of

---

Fig. 80. Human muscle. Normal (fixed in acetone); substrate adenosine triphosphate (ATP); pH 9.0; negative reaction.

Fig. 81. Human muscle. Normal (fixed in acetone); substrate diphosphopyridine nucleotide; pH 9.0; negative reaction.

Fig. 82. Human muscle. Dystrophic; substrate glucose 6-phosphate; pH 9.0; reaction only in capillaries.

Fig. 83. Human muscle. Dystrophic; substrate fructose 6-phosphate; pH 9.0; reaction only in some of capillaries.

Fig. 84. Human muscle. Dystrophic; substrate adenosine triphosphate (ATP); pH. 9.0; intense positive reaction in all elements of proliferating connective tissue; Muscle fibers negative (heat-stable ATPase).

Fig. 85. Human muscle. Dystrophic; substrate adenosine diphosphate (ADP); pH 9.0; reaction similar to ATP.

Fig. 86. Human muscle. Dystrophic; substrate diphosphopyridine nucleotide (DPN); pH 9.0; intense reaction in all elements of proliferating connective tissue; muscle fibers negative.

Fig. 87. Human muscle. Dystrophic; substrate triphosphopyridine nucleotide (TPN); pH 9.0; intense reaction in all elements of connective tissue; muscle fibers negative.

fat within the muscle fibers of any of the samples taken from cases of muscular or neuromuscular disorders, but since the fat content of normal muscle is so very variable, one cannot be certain of the validity of these results. Spectacular accumulations of fat can of course occur between the muscle fibers in diseases such as pseudohypertrophic dystrophy.

ACKNOWLEDGEMENTS

Some of the work by Bourne and Golarz recorded in this chapter was aided by U.S.P.H.S. grant B-2038 (Institute of Neurological Diseases and Stroke), some by grant NGR 11-001-016 from the National Aeronautics and Space Administration and some by Grant RR00165 of the Division of Research Resources, National Institutes of Health.

REFERENCES

Azzone, G. F., and Carafoli, E. (1960). *Exp. Cell Res.* **21**, 447.

Baker, J. R. (1944). *Quart. J. Microsc. Sci.* **85**, 1.

Barden, H., and Lazarus, S. S. (1964). *Lab. Invest.* **13**, 1345.

Barker, D., and Ip, M. C. (1965). *J. Physiol. (London)* **176**, 11.

Barker, D., and Ip, M. C. (1966). *Proc. Roy. Soc., Ser. B* **163**, 538.

Barrnett, R. J. (1952). *Anat. Rec.* **112**, 307.

Barrnett, R. J. (1953). *J. Nat. Cancer Inst.* **13**, 905.

Barrnett, R. J., and Seligman, A. M. (1951). *Science* **114**, 579.

Barrnett, R. J., and Seligman, A. M. (1952). *Science* **116**, 323.

Barrnett, R. J., and Seligman, A. M. (1954). *J. Nat. Cancer Inst.* **14**, 769.

Beckett, E. B., and Bourne, G. H. (1957). *J. Neurol., Neurosurg. Psychiat.* **20**, 191.

Beckett, E. B., and Bourne, G. H. (1958a). *Acta Anat.* **33**, 289.

Beckett, E. B., and Bourne, G. H. (1958b). *Acta Anat.* **35**, 326.

Beckett, E. B., and Bourne, G. H. (1958c). *J. Neuropathol. Exp. Neurol.* **17**, 199.

Beckett, E. B., and Bourne, G. H. (1958d). *J. Histochem. Cytochem.* **6**, 13.

Beckett, E. B., and Bourne, G. H. (1958e). *Acta Anat.* **34**, 235.

Beckett, E. B., and Bourne, G. H. (1958f). *Acta Anat.* **34**, 112.

Bell, E. T. (1911). *Int. Monatsschr. Anat. Physiol.* **28**, 297.

Bensley, C. M. (1939). *Stain Technol.* **14**, 47.

Bourne, G. H. (1943). *Quart. J. Microsc. Sci.* **32**, 1.

Bourne, G. H. (1953). *Nature (London)* **172**, 588.

Bourne, G. H. (1954a). *Quart. J. Microsc. Sci.* **95**, 359.

Bourne, G. H. (1954b). *Acta Anat.* **22**, 289.

Bourne, G. H. (1954c). *J. Physiol. (London)* **124**, 409.

Bourne, G. H., and Golarz, M. N. (1959a). *Nature (London)* **183**, 1741.

Bourne, G. H., and Golarz, M. N. (1959b). *Arch. Biochem. Biophys.* **85**, 109.

Bourne, G. H., and Golarz, M. N. (1963). *J. Histochem. Cytochem.* **11**, 286.

Bull, G., Lawes, M., and Leonard, M. (1957). *Stain Technol.* **32**, 59.

Bullard, M. H. (1912). *Amer. J. Anat.* **14**, 1.

Buller, A. J., Eccles, J. C., and Eccles, R. M. (1960). *J. Physiol.* (*London*) **150**, 399.

Buño, W., and Mariño, R. G. (1952). *Acta Anat.* **16**, 85.

Casselman, W. B. (1954). *Quart. J. Microsc. Sci.* **95**, 323.

Chessick, R. D. (1953). *J. Histochem. Cytochem.* **1**, 471.

Cöers, C. (1953a). *Rev. Belge Pathol. Med. Exp.* **22**, 306.

Cöers, C. (1953b). *Arch. Biol.* **64**, 133.

Cöers, C. (1955a). *Acta Neurol. Psychiat. Belg.* **55**, 2.

Cöers, C. (1955b). *Acta Neurol. Psychiat. Belg.* **55**, 741.

Cöers, C. (1967). *Int. Rev. Cytol.* **22**, 239.

Cöers, C., and Woolf, A. C. (1959). "The Innervation of Muscle." Blackwell, Oxford.

Couteaux, R., and Taxi, J. (1952). *Arch. Anat. Microsc. Morphol. Exp.* **41**, 532.

Defendi, V. (1957). *J. Histochem. Cytochem.* **5**, 1.

Dempsey, E. W., Wislocki, G. B., and Singer, H. (1946). *Anat. Rec.* **96**, 221.

Denz, F. A. (1953). *Brit. J. Exp. Pathol.* **34**, 329.

Dreyfus, J., Schapira, G., and Schapira, F. (1954). *J. Clin. Invest.* **33**, 794.

Dubowitz, V. (1967). *In* "Exploratory Concepts in Muscular Dystrophy and Related Disorders" (A. T. Milhorat, ed.), p. 164. Exerpta Med. Found., Amsterdam.

Dubowitz, V., and Pearse, A. G. E. (1960a). *Histochemie* **2**, 105.

Dubowitz, V., and Pearse, A. G. E. (1960b). *Nature* (*London*) **185**, 701.

Dubowitz, V., and Pearse, A. G. E. (1960c). *Histochemie* **2**, 105.

Engel, A. G. (1966). *Mayo Clin. Proc.* **41**, 713.

Engel, A. G. (1970). *In* "Muscle Diseases" (J. N. Walton, N. Canal, and G. Scarlato, eds.), p. 236. Exerpta Med. Found., Amsterdam.

Engel, W. K. (1963). *Nature* (*London*) **200**, 588.

Engel, W. K. (1967). *In* "Exploratory Concepts in Muscular Dystrophy and Related Disorders" (A. T. Milhorat, ed.), p. 27. Exerpta Med. Found., Amsterdam.

Engel, W. K. (1969). 2nd Conf. Clin. Delineation Birth Defects, Baltimore, May 28, 1969.

Engel, W. K. (1970). *Arch. Neurol.* **22**, 97.

Engel, W. K., and Brooke, M. H. (1966). *Progr. Muskeldystrophie, Myotonie, Myasthenie, Symp., 1965* p. 203.

Engel, W. K., and Cunningham, G. C. (1970). *J. Histochem. Cytochem.* **18**, 55.

Engel, W. K., and Meltzer, H. (1970). *Science* **168**, 273.

Engel, W. K., Foster, J. B., Hughes, B. P., Huxley, H. E., and Mohler, R. (1961). *Brain* **84**, 167.

Engel, W. K., Wanko, T., and Fenichel, G. M. (1964). *Arch. Neurol.* **11**, 22.

Engel, W. K., Vick, N. A., Glueck, C. J., and Levy, R. I. (1970). *N. Engl. J. Med.* **282**, 697.

Fell, H. B., and Danielli, J. F. (1943). *Brit. J. Exp. Pathol.* **24**, 196.

George, J. C., Nair, S. M., and Scaria, K. S. (1958). *Curr. Sci* **27**, 172.

Gerebtzoff, M. A. (1953). *Acta Anat.* **19**, 366.

Gerebtzoff, M. A. (1956). *Extr. Ann. Histochim.* **1**, 26.

Glegg, R. E., Clermont, Y.., and Leblond, C. P. (1952a). *Stain Technol.* **27**, 277.

Glegg, R. E., Clermont, Y.., and Leblond, C. P. (1952b). *J. Nat. Cancer Inst.* **13**, 228.

Glick, D. (1946). *Science* **103**, 599.

Goetsch, J. B., and Reynolds, P. M. (1951). *Stain Technol.* **26**, 145.

Golarz, M. N., and Bourne, G. H. (1966). *In* "Symposium on Progressive Muscular Dystrophy" (E. Kuhn, ed.), Springer-Verlag, Berlin and New York.

Golarz, M. N., Bourne, G. H., and Richardson, H. D. (1961). *J. Histochem. Cytochem.* **9**, 132.

Golarz de Bourne, M. N., and Bourne, G. H. (1972). *Proc. Int. Congr. Primatol., 3rd, 1971* Karger, Basel.

Gomori, G. (1939). *Proc. Soc. Exp. Biol. Med.* **42**, 23.

Gomori, G. (1941a). *Arch. Pathol.* **32**, 189.

Gomori, G. (1941b). *J. Cell. Comp. Physiol.* **17**, 71.

Gomori, G. (1945). *Proc. Soc. Exp. Biol. Med.* **58**, 362.

Gomori, G. (1946). *Arch. Pathol.* **41**, 121.

Gomori, G. (1948). *Proc. Soc. Exp. Biol. Med.* **68**, 354.

Gomori, G. (1949). *Proc. Soc. Exp. Biol. Med.* **72**, 449.

Gomori, G. (1950). *Stain Technol.* **25**, 81.

Gomori, G. (1951). *J. Lab. Clin. Med.* **37**, 526.

Gomori, G. (1952a). "Microscopic Histochemistry," p. 147. Univ. of Chicago Press, Chicago.

Gomori, G. (1952b). *Int. Rev. Cytol.* **1**, 323.

Gomori, G. (1952c). "Microscopic Histochemistry," p. 211. Univ. of Chicago Press, Chicago.

Gomori, G. (1956). *Quart. J. Microsc. Sci.* **97**, 1.

Gomori, G., and Chessick, R. D. (1953). *J. Cell. Comp. Physiol.* **41**, 51.

Gruner, J. E. (1966). *C. R. Soc. Biol.* **157**, 181.

Grutzner, P. (1884). *Zentralbl. Med. Wiss.* **22**, 375.

Gulland, J. M., and Jackson, E. M. (1938). *Biochem. J.* **32**, 597.

Hale, A. J. (1957). *Int. Rev. Cytol.* **6**, 194.

Hard, W. L. (1950). *Anat. Rec.* **106**, 201.

Hard, W. L., and Hawkins, R. C. (1950). *Anat. Rec.* **108**, 577.

Hard, W. L., and Peterson, A. C. (1950). *Anat. Rec.* **108**, 57.

Holmgren, E. (1910). *Arch. Mikrosk. Anat. Entwicklungsmech.* **75**, 240.

Holt, S. J. (1954). *Proc. Roy. Soc., Ser.* **B42**, 160.

Hotchkiss, R. D. (1948). *Arch. Biochem.* **16**, 131.

Humoller, F. L., Griswold, B., and McIntyre, A. R. (1951). *Amer. J. Physiol.* **164**, 742.

Kabat, E. A., and Furth, J. (1941). *Amer. J. Pathol.* **17**, 303.

Kay, H. D. (1928). *Biochem. J.* **22**, 855.

Koelle, G. B. (1951). *J. Pharmacol. Exp. Ther.* **103**, 153.

Koelle, G. B., and Friedenwald, J. S. (1949). *Proc. Soc. Exp. Biol. Med.* **70**, 617.

Kölliker, A. (1856). *Z. Wiss. Zool.* **8**, 311.

Krespi, V., Fozzard, A. A., and Sleator, W. (1964). *Circ. Res.* **15**, 545.

Leblond, C. P. (1950). *Amer. J. Anat.* **86**, 1.

Levine, L. (1960). *Science* **131**, 1377.

Lillie, R. D. (1951). *Stain Technol.* **26**, 123.

Lillie, R. D., and Greco, J. (1947). *Stain Technol.* **22**, 67.

Lipman, F. (1946). *Advan. Enzymol.* **6**, 231.

Loveless, A., and Danielli, J. F. (1949). *Quart. J. Microsc. Sci.* **90**, 57.

McKay, D. G., Adams, E. C., Hertig, A. T., and Danziger, S. (1955). *Anat. Rec.* **122**, 125.

McManus, J. F. A. (1946). *Nature (London)* **158**, 202.

McManus, J. F. A. (1948). *Stain Technol.* **23**, 99.

Maengwyn-Davies, G. D., and Friedenwald, J. S. (1950). *J. Cell. Comp. Physiol.* **36**, 421.

Maengwyn-Davies, G. D., Friedenwald, J. S., and White, R. T. (1952). *J. Cell. Comp. Physiol.* **39**, 95.

Malaty, H. A., and Bourne, G. H. (1953). *Nature (London)* **171**, 295.

Malmgren, H., and Sylvén, B. (1955). *J. Histochem. Cytochem.* **3**, 441.

Manheimer, L. H., and Seligman, A. M. (1948). *J. Nat. Cancer Inst.* **9**, 181.

Martinosi, A., and Ferretos, R. (1964). *J. Biol. Chem.* **239**, 659.

Meltzer, H. (1968). *Science* **159**, 1368.

Meltzer, H., Elkin, L., and Moline, R. (1972). *Science* (in press).

Menten, M. L., Junge, J., and Green, M. H. (1944). *J. Biol. Chem.* **153**, 471.

Monticone, G. F., Gobella, G., and Bergamici, D. D. (1970). In "Muscle Diseases" (J. N. Walton, N. Canal, and G. Scarlato, eds.), Excerpta Med. Found., Amsterdam.

Moog, F. (1943). *J. Cell. Comp. Physiol.* **22**, 223.

Nachlas, M. M., and Seligman, A. M. (1949a). *J. Nat. Cancer Inst.* **9**, 415.

Nachlas, M. M., and Seligman, A. M. (1949b). *Anat. Rec.* **105**, 677.

Newman, W., Kabat, E. A., and Wolf, A. (1950a). *Amer. J. Pathol.* **26**, 489.

Newman, W., Feigin, I., Wolf, A., and Kabat, E. (1950b). *Amer. J. Pathol.* **26**, 257.

Novikoff, A. B. (1958). *J. Histochem.* **6**, 61.

Odor, D. L., Patel, A. N., and Pearce, L. A. (1967). *J. Neuropathol. Exp. Neurol.* **26**, 98.

Ogata, T. (1958). *Acta Med. Okayama* **12**, 216, 228, and 233.

Padykula, H. A. (1952). *Amer. J. Anat.* **91**, 107.

Padykula, H. A., and Gauthier, G. F. (1963). *J. Cell Biol.* **18**, 87.

Padykula, H. A., and Herman, E. (1955a). *J. Histochem. Cytochem.* **5**, 161.

Padykula, H. A., and Herman, E. (1955b). *J. Histochem. Cytochem.* **5**, 170.

Pearce, G. W., Adamson, D. G., and Salter, R. H. (1968). In "Research in Muscular Dystrophy," p. 171. Pitman, London.

Pearse, A. G. E., and Johnson, M. (1970). In "Muscle Diseases" (J. N. Walton, N. Canal, and G. Scarlato, eds.), p. 25. Excerpta Med. Found., Amsterdam.

Pearse, E. (1953a). "Histochemistry," p. 262. Churchill, London.

Pearse, E. (1953b). "Histochemistry," p. 264. Churchill, London.

Pearson, B., and Defendi, V. (1957). *J. Histochem. Cytochem.* **5**, 72.

Ranvier, L. (1880). "Leçons d'anatomie générale sur le système musculaire." Delabayes, Paris.

Reis, J. L. (1951). *Biochem. J.* **48**, 548.

Resnick, J. S., and Engel, W. K. (1967). In "Exploratory Concepts in Muscular Dystrophy and Related Disorders" (A. T. Milhorat, ed.), p. 255. Exerpta Med. Found., Amsterdam.

Richterich, R. (1952). *Acta Anat.* **14**, 342.

Rosa, C. G., and Velardo, J. T. (1954). *J. Histochem. Cytochem.* **2**, 110.

Rossi, F., Pescetto, G., and Reale, E. (1953). *Z. Anat. Entwicklungsgesch.* **117**, 36.

Rossi, F., Reale, E., and Pescetto, G. (1954). *Extr. C. R. Ass. Anat.* **74**, 1.

Rutenburg, A. M., and Seligman, A. M. (1955). *J. Histochem. Cytochem.* **3**, 455.

Sandler, M., and Bourne, G. H. (1961). *Exp. Cell Res.* **24**, 174.

Sandler, M., and Bourne, G. H. (1962). *Nature (London)* **194**, 389.

Scarlato, G., and Cornelio, F. (1970). *In* "Muscle Diseases" (J. N. Walton, N. Canal, and G. Scarlato, eds.), Excerpta Med. Found. Amsterdam.

Schaffer, J. (1893). *Sitzungsber. Akad. Wiss. Wien.* Abt. III, **102**, 7.

Seligman, A. M., and Manheimer, L. H. (1949). *J. Nat. Cancer Inst.* **9**, 427.

Shimizu, N., and Kumamoto, T. (1952). *Stain Technol.* **27**, 97.

Shy, G. M., and Magee, R. R. (1956). *Brain* **79**, 610.

Shy, G. M., McLean, L., Di Giacomo, N., and Saunders, S. (1963). *Excerpta Med.* p. 357.

Stefano, D. D., and Lorenzi, D. D. (1678). See Ranvier (1880).

Stein, J. M., and Padykula, H. A. (1962). *Amer. J. Anat.* **110**, 103.

Strominger, G. L. (1961). *Tex. Rep. Biol. Med.* **19**, 217.

Szent-Györgyi, A. (1953). "Chemical Physiology of Contraction in Body and Heart Muscle." Academic Press, New York.

Taxi, J. (1952). *J. Physiol.* (*Paris*) **44**, 595.

Weiss, L. P., Tsou, K. C., and Seligman, A. M. (1954). *J. Histochem. Cytochem.* **2**, 29.

Wolf, A., Kabat, E. A., and Newman, W. (1943). *Amer. J. Pathol.* **19**, 423.

Zebe, E., anf Falk, H. L. (1963). *Exp. Cell Res.* **31**, 340.

Zorzoli, A., and Stowell, R. E. (1947). *Anat. Rec.* **97**, 495.

6

MYOPATHY, THE PATHOLOGICAL CHANGES IN INTRINSIC DISEASES OF MUSCLES

F. D. BOSANQUET, P. M. DANIEL, AND H. B. PARRY
Revised by G. H. Bourne and M. Nelly Golarz de Bourne

I. Introduction

The greater part of the human body is composed of muscle tissue, and imperfect working of this tissue causes the most severe and disastrous effect on the day-to-day life of the patient, yet we know less about the pathology of the muscles than about that of any other organ. However, there appears to be a growing interest in the subject, for within the last few years several monographs have appeared (Adams et al., 1953; Greenfield et al., 1957; Walton and Adams, 1958) that supplement earlier work (Erb, 1891; Pick, 1900; Durante, 1902; Steinert, 1909; Marinesco, 1910; Jendrassik, 1911; von Meyenburg, 1929; S. Wohlfahrt and Wohlfart, 1935; Slauck, 1936), though our knowledge of this obscure field still remains slight. A number of studies, though concerned more with the clinical features than with the pathological changes in myopathy, should be consulted by all those interested in the subject (Meryon, 1852, 1864; Duchenne, 1868, 1872; Gowers, 1879, 1902; Erb, 1884, 1894; Batten, 1910a,b; Spiller, 1913; Bing, 1926; Rouquès, 1931; Curschmann, 1936; Hurwitz, 1936; Sjövall, 1936; Wilson, 1940; Bell, 1943, 1947; Shank et al., 1944; Levison, 1951; Welander, 1951; Becker, 1953; Schoen and Tischendorf, 1954; Walton and Nattrass, 1954; Bouman, 1955). More recently, there have been two valuable volumes on muscle diseases, "Exploratory Concepts in Muscular Dystrophy," edited by A. T. Milhorat (1967), and "Muscle Diseases," edited by J. N. Walton and colleagues (1970). See also a result of a congress of British workers in muscle diseases entitled "Research in Muscular Dystrophy," edited by the Research Committee of the Muscular Dystrophy Group (1968), and Pearce, G. W. (1964). "Disorders of Voluntary Muscle." An earlier work was "Muscular Dystrophy in Man and Animals," edited by G. H. Bourne and M. N. Golarz (1963a).

Genetic studies have thrown considerable light on the problem of myopathy, especially those of Tyler in America (Tyler and Wintrobe, 1950; Tyler and Stephens, 1950, 1951; Stephens and Tyler, 1951; Tyler, 1951, 1954) and Stevenson in Ireland (Stevenson, 1953, 1955; Stevenson et al., 1955). See also Chapter 8 by Kloepfer and Walton in this volume.

In this paper, "myopathy" is used to describe any disease that is primary to muscle and not of neural origin. Knowledge of the pathology of muscle when the first edition of this treatise was published was limited to the results obtained by light microscopy. The development of histochemistry and electron microscopy has revolutionized our knowledge of the nature of many of these diseases and has greatly assisted in their diagnosis and classification.

II. Varieties of Change in Pathological Muscle

A. *Changes in Muscle Fibers*

Normal skeletal muscle fibers are polyhedral in cross section and appear of relatively equal diameter. They have peripherally situated subsarcolemmal nuclei. The extrinsic ocular muscles, however, have in general small and rounded muscle fibers with some central nuclei; occasional fibers of large diameter are present (Cooper *et al.*, 1955). Occasional small muscle fibers, which have no pathological significance, are seen in all muscles; they are the extracapsular parts of the intrafusal muscle fibers of the muscle spindles.

Abnormal muscle fibers are usually rounded in cross section and may be either larger (hypertrophied) or smaller (atrophied) than normal. They may also show a variety of changes in the cytoplasm, while the nuclei may be increased in number, unusual in appearance, and or centrally placed within the muscle fiber. An abnormally great variation in the caliber of the muscle fibers together with obviously atrophied fibers is usually the most striking change in a severely diseased muscle, and the distribution of such fibers may give the clue to the disease. Atrophying muscle fibers may simply lose substance so that they appear as merely thin normal fibers (Fig. 2), or the cytoplasm may show degenerative changes, such as breaking up into short segments. The cytoplasm may disappear from the sarcolemmal tube, which is left empty and shrunken except for a row of nuclei. The sarcolemma may disappear so that isolated nuclei or small masses of nuclei alone remain.

Hypertrophy of the muscle fibers is less easy to recognize with certainty than atrophy, since normal muscle fibers may appear enlarged if adjacent fibers are atrophied; when all the fibers in a field are enlarged, hypertrophy may also be difficult to recognize. Genuinely enlarged fibers may be otherwise normal, as in compensatory hypertrophy and some dystrophies, but enlargement of muscle fibers is often only part of a degenerative change as, for example, when a segment of a muscle fiber is swollen and the sarcoplasm is glassy and without striations (Fig. 20). Such swollen fibers commonly fragment, forming at first isolated segments, with an empty sarcolemmal sheath stretched between them, (Fig. 13) and then later becoming irregular masses of sarcoplasm (Fig. 17), with either peripheral nuclei or nuclei in their substance.

Measurement of muscle fiber diameter requires accurate transverse sections. As muscle fiber sizes vary greatly with age, nutritional status (Hammond, 1932), and in different skeletal muscles, it is essential to know the normal variation in fiber size for the particular muscle before interpreting any changes as being pathological. Data on normal human fiber sizes are scanty, but some information is available (Halban, 1894; Feinstein et al., 1955; Sissons, 1956; Greenfield et al., 1957). In muscles concerned with delicate movements, e.g., the lumbricals, the fiber sizes are small, 18μ (Feinstein et al., 1955), while in large muscles concerned with coarse movements, e.g., tibialis anterior, the fibers are large, 56 μ (Feinstein et al., 1955). However, the range of size is considerable, for Greenfield et al. (1957) in specimens of vastus lateralis from normal volunteers found a range of fiber sizes from 25 to 90 μ. We would regard any considerable number of fibers over 100 μ as beyond the range of normal. In the present state of knowledge, the actual measurement of muscle fibers, particularly in biopsy material, is of limited value.

The cytoplasm of muscle fibers may undergo a number of pathological changes. The striations may be lost and the cytoplasm may become hyaline (Fig. 20). The cytoplasm may show fine granules (granular degeneration, Fig. 16) or larger floccules (floccular degeneration, Figs. 18 and 20), becoming at the same time intensely eosinophilic. Vacuolation may occur, especially in much swollen fibers. Floccular degeneration may be segmental and is accompanied by an ingrowth of nuclei, some of which are sarcolemmal nuclei and some, invading phagocytes. Portions of the sarcolemmal tubes may thus come to be filled with nuclei and the remains of sarcoplasm (Figs. 14 and 21).

A pathological basophilia of the cytoplasm (sometimes difficult to distinguish from artifactual staining) also occurs. Isolated necrotic muscle fibers are often intensely basophilic, staining purple throughout with hematoxylin. A milder degree of basophilia is seen in thin regenerating fibers (Fig. 19) and in the multinucleated sarcoplasmal masses ("muscle buds") containing numbers of vesicular nuclei, which are also found in regenerating muscle, although it is not certain that the presence of these muscle buds always denotes regeneration.

An increase, or at least an apparent increase, in the number of sarcolemmal nuclei is seen in almost every abnormal muscle and is commonly the first change discernable. The nuclei tend to be more rounded and vesicular than normal and prominent nucleoli may appear (Fig. 16). The nuclei may form peripheral chains (Fig. 15), but often they invade the sarcoplasm, becoming centrally placed in rows or long chains, especially in dystrophia myotonica (Figs. 8–11). Centrally placed nuclei in adult muscles other than the extrinsic eye muscles are usually abnormal. Nuclei of abnormal muscle fibers may be pyknotic and densely staining. Large clumps of darkly staining nuclei may be seen in and among atrophic fibers (Figs. 2 and 6).

Other unusual appearances seen in muscle fibers are ringed fibers and split fibers. Ringed fibers (*ringbinden* or spiral annulets) consist of muscle fibers encircled by cross-striated muscle fibrils as a fiber is encircled by a ring lying within the sarcolemmal sheath. The arrangement of these cross-striated encircled fibrils is thought to be spiral (Greenfield *et al.*, 1957). Their significance is unknown, but they may increase with age, though we have seen them in muscles, including the eye muscles (Daniel, 1946), of normal people of all ages. Heidenhain (1918) thought they were specifically associated with dystrophia myotonica, but though common in this disease, they cannot be regarded as diagnostic (G. Wohlfart, 1951). Longitudinal splitting of muscle fibers is seen only in pathological muscles; the complete muscle fiber, the sarcolemma split, and in longitudinal section actual division of a fiber may be seen, as in cardiac muscle. In cross section, the appearance is of two fibers lying within a single endomyosial sheath.

B. Changes in the Interstitial Tissue

The interstitial tissue shows changes in many varieties of muscle disease. The changes, like those in the muscle fiber, are not specific for any particular disease but appear rather to reflect the rapidity or chronicity of the change in the muscle tissue. Interstitial fat is increased in most cases of prolonged muscle wasting, but it is most marked in the Duchenne type of muscular dystrophy (Figs. 1, 2) and in thyrotoxic myopathy. Normal muscle contains virtually no fat between the muscle fibers (Volume II, Chapter 4), but as muscle fibers are lost, rows of fat cells may be found separating the remaining fibers. Thus, isolated groups of fibers surrounded by collagen (thickened endomysium) are found lying in adipose tissue that occupies the site of the original muscle (Fig. 1). In this adipose tissue lie the structures that have not degenerated—the blood vessels, nerves, and muscle spindles.

Collagen also is increased in almost all chronic myopathies. The relative amount of adipose and fibrous tissue varies from case to case, but in general, collagen is the main constituent increased after prolonged denervation; and also in the very chronic Landouzy–Dejerine type of muscular dystrophy, fibrous tissue may form the main bulk of what was a muscle, and the original vascular bed, nerves, and muscle spindles lie embedded in collagen (Figs. 3–6). It was formerly thought that muscle fibers were actually transformed into either fat or collagen, but this view is not now accepted, and though the process is not fully understood, it is generally believed that there is replacement rather than metaplasia. The well-marked fibrous capsule of normal muscle spindles and the small intrafusal muscle fibers, which often have central nuclei, have been erroneously described and illustrated as pathological structures (Ingram and Stewart, 1934; Bevans, 1945).

An increase in the numbers of cells in the interstitial tissue (Figs. 12 and 19) may be found in any active muscle disease. In most cases, mononuclear cells predominate, while phagocytes, lymphocytes, plasma cells, and fibroblasts, as well as free sarcolemmal nuclei from disintegrated fibers, may all be present. Both neutrophilic and eosinophilic polymorphonuclear cells may be present, but they are seldom the predominating cell; it should be mentioned, however, that a noticeable number of polymorphonuclear cells may be present even in a dystrophic muscle of many years standing.

The endothelial nuclei of the capillaries are often enlarged and may on occasion be difficult to distinguish from sarcolemmal nuclei. A cellular reaction may be found in an actively degenerating muscle in muscular dystrophy, yet, in general, the infiltration is greater in polymyositis. It is to be noted that a quite marked cellular reaction may also be present in a muscle after rapid denervation (as after poliomyelitis) and even in the muscles of bedridden patients.

Lymphorrhages are focal collections of small round cells (lymphocytes) lying between muscle fibers that appear normal, and are seen classically in myasthenia gravis, but they may also be found in polymyositis, rheumatoid conditions, and in the ocular muscles in exophthalmic ophthalmoplegia, where the foci of round cells can reach an enormous size.

C. Artifactual Appearances

Variation in the staining of fibers may cause considerable difficulty in sections of muscle tissue. Whether artifactual or due to genuine differences in the fibers, the following variations are found in sections of normal muscle. The staining may simulate either basophilia or eosinophilia, and furthermore, the same fibers may show an abnormal staining in several sections. The artifactual nature of the staining is fairly easily recognized if it occurs in a block of fibers or perhaps at one side of the section. When, however, scattered fibers stain abnormally, they may be difficult to assess, and the only safe rule is to ignore all variations in staining unless there are other unequivocal pathological changes. This also applies to an apparent loss of cross striations, a common finding in normal muscle fibers, and one that should not be accepted without other, and certain, evidence of abnormality. Tissue fixed in extreme contraction or too rapidly (notably with osmic acid or alcohol) may show fragmentation of the myofibrils into irregular, darkly staining bands (Nageotte's contraction bands), or into longer segments with condensed contraction bands at the ends. Bad cutting and brittle material may also cause multiple fragmentation, but this is usually obvious. Longitudinal splitting within the sarcolemmal sheath, vacuolation, and shrinkage of the fibers away from the endomysium are difficult to assess, as they often accompany pathological changes (Figs. 11 and 14), though they are often artifactual and should therefore be discounted. The scattered dropping out of fibers, sometimes seen in transverse sections, may simulate early fatty infiltration.

III. Distinction between Neuropathic and Myopathic Changes in Muscle

Degeneration of a motor nerve is, as is well known, followed by degeneration of the muscle fibers supplied (Fig. 7). It is therefore necessary in all cases of muscular wasting to distinguish between lesions of neuropathic origin and those that are myopathic. This may be simple or very difficult. See Fig. 1.

The histological characteristics of a neuropathic muscle lesion are dependent upon the fact that only those muscle fibers atrophy that have lost their nerve supply, and second, that compared with myopathic conditions, the lesion is a relatively pure atrophy with little other change. Thus, the typical histological finding is groups of small muscle fibers lying beside normal sized ones. This is best seen in a transverse section where the atrophic fibers appear as small scattered groups (i.e., motor units), or there may be larger groups consisting of whole fasciculi (Fig. 7), or again an entire muscle may be atrophied, as after poliomyelitis. The atrophy is generally quite obvious, the affected muscle fibers often being only 5 or 10 μ across with an increase in nuclei, while the unaffected fibers, apart from sometimes being rounded, show no abnormal-

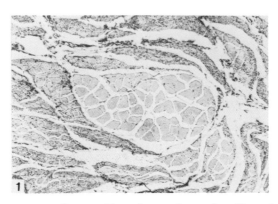

Fig. 1. Motor neuron disease. Typical neural atrophy. Note that the atrophic muscle fibers are in large groups and that the intact fibers in center appear almost normal. Transverse section, sternomastoid; iron hematoxylin and van Gieson; × 80. From A. G. Engel and MacDonald (1970).

ity; the staining reaction of all fibers is usually normal. The nuclei of the small fibers lie close together on account of the shrinkage of the cytoplasm. When degeneration is far enough advanced, nuclei within a sarcolemmal sheath without any sarcoplasm and free sarcolemmal nuclei are present among the atrophic fibers; these nuclei tend to be hyperchromatic. The unaffected muscle fibers remain normal, apart from a tendency to become rounded and possibly to develop some compensatory hypertrophy. Little is known as to the extent of this hypertrophy, but it is seldom striking, and it varies a good deal from case to case. In time, the collagen around groups of atrophied muscle fibers increases, and may in fact demarcate islands of such fibers; this is in contrast to myopathic lesions where the collagen tends to encase each individual fiber.

In longitudinal section, groups of long, very thin muscle fibers, and also fibers represented only by a nucleated sarcolemmal sheath, are found beside normal fibers. Short segments of fibers that have fragmented are also seen, often with cross striations still preserved, and sometimes capped at the ends with hyperchromatic nuclei. The number of nuclei in the degenerating bundles of muscle fibers appears to be greatly increased, and there are frequently clumps of massed nuclei on and between the fibers. It is uncertain whether this apparent increase is due entirely to the approximation of nuclei from loss of the intervening sarcoplasm. There is remarkably little interstitial infiltration with cells, except occasionally when a large amount of muscle tissue is degenerating. An increase of adipose tissue, although often present, is seldom conspicuous. Degenerating muscle fibers may have central nuclei, but swollen segments, floccular degeneration, and phagocytosis of the muscle fibers, though they may be present, are rare.

This is the characteristic picture of a muscle with a partial denervation of some standing. The time taken for a diagnosable lesion to develop varies much, but is at least several weeks, and often, in slowly progressive diseases, many months. Adams *et al.* (1953), after experimental nerve section in animals, found an increase of nuclei within 3 weeks and just visible atrophy in a month, but with unexplained variation in the rate of degeneration even within a single muscle. In humans, after acute denervation, there may be obvious degeneration in 4–6 weeks, but in chronic diseases, it may not be discernable until long after clinical weakness is apparent (9 months or more). No signs of regeneration of muscle fibers occur after complete nerve section. De Giacomo *et al.* (1970), in a study of neurogenic diseases, showed groups of enlarged fibers of some histochemical type. They have also seen angulated type I and type II fibers.

Peripheral nerves included in a muscle biopsy may show evidence of degeneration, such as fibrosis and loss of myelinated fibers; fat staining of frozen sections may show myelin sheaths breaking down in the more acute cases, and silver staining may show degenerating axons. Intravital staining of nerve fibers with methylene blue may also demonstrate abnormal nerve fibers and end plates (Cöers and Woolf, 1959). In practice, however, only a minority of routine biopsies afford this information, partly because nerves to muscles are mixed (motor and sensory) and partly because histological changes must be unequivocal to be significant. Greenfield *et al.* (1957) in a series of 121 biopsies found significant changes in the nerves of only two. We have found the presence of obviously normal nerve trunks more valuable in excluding a neural lesion than the converse.

Anderson *et al.* (1967) point out that the loss of size in muscle fibers in denervation is due to the decrease of myofilaments resulting in a decrease of size of the myofibrils. There was also fragmentation of Z lines and eventually a loss of the pattern of the sarcomeres. There are also changes in the sarcoplasmic reticulum and mitochondria. "The neuromuscular junctions showed retraction of the terminal axonal expansion leading to its disappearance. This was associated with a marked alteration of post-synaptic membrane arrangements that are not unlike those observed in Vincristine myopathy." In summarizing their findings, the authors point out that it is very difficult to differentiate clearly between neuropathy and myopathy based on a single histological examination. "For example, sarcoplasmic changes in neurectomy-induced denervation are apparent on the ultrastructural level, and in the clinical studies relatively few cases of neuropathy showed only segmental atrophy. On the other hand, late stages of drug induced myositis may result in segmental atrophy, thereby mimicking denervation atrophy. Reliable interpretation of muscle biopsy findings thus requires consideration of the entire histologic background. Increased understanding of the pathogenesis of neuromuscular diseases, as well as the appreciation of factors such as site of muscle biopsy and stage of the disease at the time of biopsy can result from continued clinical and experimental studies with the evaluation of sequential morphologic changes."

The muscle changes following denervation vary in different species according to the muscle component affected first and whether the changes are degenerative or proliferative.

In 1961, W. K. Engel drew attention to the fact that a type of muscle fiber, called by him a "target fiber," was diagnostic of a neurogenic myopathy. In 1966, W. K. Engel and Brooke identified target fibers in approximately 50% of the muscle biopsies taken from more than two hundred patients known to have disease of the lower motor neuron. See also Brooks and W. K. Engel (1969).

Target fibers are described as being usually of normal size, but very small target fibers may be found. In phosphotungstic acid–hematoxylin stained sections, target fibers may be seen to have three concentric zones. Histochemical studies show that the central zone contains no oxidative enzymes, the intermediate zone gives a very strong reaction for these enzymes, and the peripheral zone gives a moderate reaction (a normal reaction). ATPase reactions show the same variations. The absence of oxidative enzymes from the center of the target fibers is confirmed by the electron microscope, which shows no mitochondria; it also has no properly formed myofibrils and practically no membranous elements. There are no cross striations in the rudiments of the myofibrils. Electron-dense material of unknown character, though it may be derived from the Z bands, is found in this central zone. In the intermediate zone, all the muscle fiber elements are present, but they show some alteration; there is some breakdown of the Z bands. There is an increased number of mitochondria in this zone, and many glycogen granules are present. The muscle fiber elements appear to be unaffected in the outer zone except for a spreading of the Z bands. Target fibers are most commonly found in type I fibers. They do not appear to be suffering from phagocytosis or to cause invasion by inflammatory cells; in fact, there is some evidence that they represent a regenerative stage following reinnervation (Dubowitz, 1970). The significance of these fibers is still uncertain since W. K. Engel et al. (1966b) found them in muscle which had been tenotomised.

Central core disease also shows muscle with many targetlike fibers and was considered to be a myopathy by Shy and Magee (1956); W. K. Engel (1962) has suggested that it may result from some defect in the innervation in the muscle of the fetus. Resnick and Engel (1967) point out, however, that "core" fibers show only two zones with most histological and histochemical techniques, but that the phosphorylase and periodic acid–Schiff techniques have shown an intermediate zone. They also indicate, on the other hand, that fibers with two zones (targetoid fibers) are sometimes found in patients who have denervated muscles.

Golarz and Bourne (1962) have pointed out that muscle atrophy following denervation is most rapid up to about 4 weeks (see also Hník, 1960) and then becomes much slower. They also found that denervated muscles in rats showed a considerable loss of weight by 4 weeks, compared with control muscle. The muscle/body weight ratio of the denervated side compared with the nondenervated side was about one-quarter. The changes were similar to those recorded by a number of authors (Tower, 1939; Adams *et al.*, 1953; Sunderland and Ray, 1950). This included apparent increase in the number of subsarcolemmal nuclei with vesiculation and great prominence of the nuclei, some central migration of the nuclei, and decrease in diameter in some muscle fibers. However, at this stage there was very little proliferation or replacement of muscle fibers by connective tissue. In agreement with other authors, they saw no signs of mitosis.

Histochemical studies showed that denervation enormously increased the histochemical reaction for a wide variety of phosphatases in the capillaries of the muscle. In addition, the subsarcolemmal nuclei showed a general increase in their general phosphatase activity, especially on the part of the nucleoli that were enlarged. It is noteworthy that there was a slight increase in the nucleolar phosphatase reaction in the subsarcolemmal nuclei in the muscles of the contralateral (nondenervated limb). Animals receiving intraperitoneal injections of nicotinamide-adenosine 3-monophosphate and procaine–adenosine 3-monophosphate showed an increase in the phosphatase reactions in the nucleoli but not in the capillaries.

In a myopathic lesion, the distinction between normal and abnormal muscle fibers is much less sharp, as, in contrast with a partially denervated muscle, each muscle fiber is affected as a single unit although the whole muscle is abnormal. Therefore, atrophic fibers are not found in localized groups as after denervation, but muscle fibers of all sizes may be intermixed, while few fibers are normal. Often many or all of the muscle fibers are rounded in cross sections, and enlargement of fibers is often conspicuous (Fig. 3); this enlargement, either absolute or relative, frequently persists until a late stage in the disease. Degenerative changes other than atrophy are more common in myopathic than in neuropathic muscles; the muscle fibers show changes in staining reaction and loss of cross striations more frequently. Floccular degeneration of part of a fiber and phagocytosis of the sarcoplasm with segmental disruption of the fiber are seen more often, particularly in the more acute myopathies. In an actively progressing lesion, an interstitial infiltration of cells may be found, while an increase in endomysial collagen and infiltration of fat between the individual fibers is a more characteristic feature of a myopathy than a neuropathy.

These are the distinguishing features in typical, fairly well advanced cases of denervation and myopathy, but there are many modifying factors that may make it difficult or impossible to distinguish between the two in a particular biopsy, e.g., when the condition is too far advanced or not advanced enough. The earliest change after denervation is an apparent increase in the number of sarcolemmal nuclei, which at this stage are more vesicular than usual; this change is not specific and may be seen in any abnormal muscle. In the final stages of both a neuropathic atrophy and a myopathy, a few muscle fibers only may remain, surrounded by collagen or adipose tissue; it may not be clear by which process the lesion originated. The characteristic pattern of neural atrophy is not present after complete denervation, as all the fibers show rounding and shrinkage, and furthermore, a few scattered fibers may remain larger than the others for a considerable time. Segmental degeneration may be present in a neuropathy as well as in a myopathy.

It must be emphasized that the histological changes in muscle fibers are not specific to a particular etiology and that the histological appearance often shows little correlation with the clinical state. Age, normal variations, and disuse must all be considered, and it should be remembered that active degeneration may be found in muscle from patients who have been in bed awhile.

It is therefore obvious that a positive diagnosis can only be made in a proportion of muscle biopsies. This proportion is increased if the muscle selected shows well-marked atrophy clinically, if a good-sized specimen is taken in such a way that accurate transverse and longitudinal sections may be made, and also by careful processing.

IV. The Muscular Dystrophies

Muscular dystrophy is the name given to a number of syndromes that have one feature in common—a progressive weakness of the voluntary (and sometimes cardiac) muscles, without evidence of inflammation in the muscles or of involvement of the nervous system. The muscular dystrophies are extremely chronic, usually progressing slowly for many years, and the patients often have a family history of similar disease.

It is not clear whether muscular dystrophy is really one disease, as Erb (1891, 1894) thought when he gave the name dystrophia muscularia progressiva, or a number of related diseases. However, since the clinically different syndromes seem to be constantly inherited (see Section I), it appears that there are some genuine differences between the different types of diseases. Unfortunately, a number of cases are met with which do not fit clearly into the classic types but show some features of each.

The cause of muscular dystrophy of any type is quite unknown, but the diseases are thought to be essentially "degenerative" and possibly due to inborn errors of metabolism or endocrine disturbances. Much work has been done on the biochemical aspects of these conditions, but this has thrown little light on the pathological changes in the muscles. An increase in creatinine excretion and a decrease in creatine excretion is regularly found, but this abnormality occurs in any condition in which muscle tissue is being broken down.

Adams *et al.* (1953) suggest that the lack of evidence of regeneration is the fundamental characteristic of muscular dystrophies. Walton and Adams (1956), however, consider that regeneration of muscle fibers may occur. More recent studies (Denny-Brown, 1962; Gilbert and Hazard, 1965; McArdle, 1967) have established that regeneration, including the formation of myotubes, as a characteristic feature of dystrophic muscle. Little work has been done on the innervation of dystrophic muscles.

A. Duchenne Type of Muscular Dystrophy*

Duchenne type of muscular dystrophy characteristically begins in boys under 5 years of age, being inherited as a sex-linked recessive (Tyler, 1950; Stevenson, 1953). The muscles of the pelvic girdle and the quadriceps femoris are usually involved first, while enlargement of the calf muscles is very characteristic. The condition is almost invariably progres-

---

* Synonyms: pseudohypertrophic paralysis, Duchenne (1868); pseudohypertrophic muscular dystrophy; progressive muscular dystrophy of childhood, Tyler and Wintrobe (1950); Duchenne type rapidly progessive muscular dystrophy of young boys, Stevenson (1953); and severe generalized familial muscular dystrophy, Adams *et al.* (1953).

sive, and most cases die of intercurrent disease (usually pulmonary infection as a result of weakness of the respiratory muscles) before they are 20. The facial muscles may or may not be involved.

The histological appearances vary greatly, depending on the stage that the disease has reached and the degree of involvement of the particular muscle examined. Examination of many muscles obtained at autopsy gives the only adequate idea of the extent of the disease, and of the severity of the pathological changes in the muscles; thus, biopsy specimens may be misleading.

A muscle taken from an early case may show little change beyond rounding of the muscle fibers and a slight increase in the sarcolemmal nuclei. There may, however, be occasional fibers of small diameter and also a few fibers of abnormally large diameter, over 100 μ, but otherwise normal. The cause of the enlargement of the whole muscle is not clear, since often so few of the individual muscle fibers are enlarged, but there may already be an increase in the amount of the interstitial fat. Small focal areas of necrosis may be seen in a muscle fiber, but this is only an occasional finding.

When the terminal stage of the disease has been reached necropsy provides an astonishing sight. The bulk of the musculature of the body, sometimes without much loss of size, appears to have been sculptured in fat. The architecture of the various muscles is quite clear, but of ordinary brown muscle tissue, hardly a trace is seen. The intercostal muscles, the diaphragm, the tongue and muscles of mastication, the extrinsic eye muscle, and perhaps the lumbricals of hands and feet alone probably resemble normal muscles, though they are very pale.

Microscopically, a cross section of one of these muscles shows that the muscle tissue has disappeared to an astounding degree. Only occasional rounded muscle fibers are seen (Fig. 1), some of which may be unusually large and others small. Whole microscopic fields may have to be searched before a single muscle fiber is found in the expanses of fibrofatty tissue in which the muscle spindles stand out clearly; for the intrafusal muscle fibers of these organs seem to be relatively well preserved, although we know little of their functional state or innervation.

Clearly, there are very many intermediate states to be seen between the stage in the disease when there is a little simple rounding of muscle fibers with slight variations in fiber size and the terminal stage of the disease when virtually no muscle fibers are to be seen. See Figs. 2 and 3.

Kakulas (1970) has pointed out that greater recognition is now being given to the fact that there is a cardiac involvement in cases of progressive muscular dystrophy (Perloff et al., 1967) and that a third of typical

Duchenne patients show mental retardation. Rosman and Kakulas (1966) have found microscopic heterotopia and pachygyria in such patients. Electron microscope studies of Duchenne dystrophy have demonstrated loss of mitochondria from the fibers at an early stage, together with loss of sharp definition of the Z bands and the presence of large bodies that contain the pigment lipofuscin. Splitting was found in some fibers and necrosis in others. Shortening of the sarcomeres followed dissolution of the myofilaments.

It is now well established that muscle regeneration occurs in Duchenne dystrophy simultaneously with degradation of muscle. Evidence of regeneration can be seen in the presence of misformed myofibrils that contain an increased amount of ribosomal material and in the presence of myotubes. Kakulas and his colleagues (1968) have found that dystrophic muscle in tissue culture shows abnormalities of growth and differentiation compared with normal controls.

In general, it can be said that the electron microscope has demonstrated no structural change that is specific for Duchenne dystrophy, and Kakulas points out that this directs attention to the metabolic energy pathways. He says, "In comon with other inherited diseases it is likely that the abnormal gene in muscular dystrophy manifests biochemically as an enzyme deficiency. The light microscopic lesions are clearly not specific and simply reflect a cycle of continuing necrosis and intrinsically abnormal regeneration. . . . This causes eventual loss of fibers and fat and connective tissue replacement occurs. From this it is evident that the basic (biochemical) lesion must pre-dispose to focal muscle fiber necrosis. . . . It may be ventured that the inherited biochemical disorder in the Duchenne form of progressive muscular dystrophy manifests in the region of oxidative (electron transfer) metabolism of the muscle fiber."

Démos (1961) suggested that human myopathy may be due to a defect in the microcirculation rather than being due to a primary defect in the muscles themselves. Démos *et al.* (1970) have extended this view by providing additional data derived from a study of muscle blood flow, slowing of the development of a myopathy by the use of a vasodilator and by demonstration of a platelet enzyme in the blood of Duchenne dystrophy patients and carriers; this enzyme has an electrophoretic migration different from that of the same enzyme in the platelets of normal persons.

Hathaway *et al.* (1970) have pointed out that the earliest changes seen in the muscles of Duchenne patients are "small foci of grouped muscle fibers undergoing necrosis or regeneration, all fibers in the group being in about the same stage; less often there are single scattered

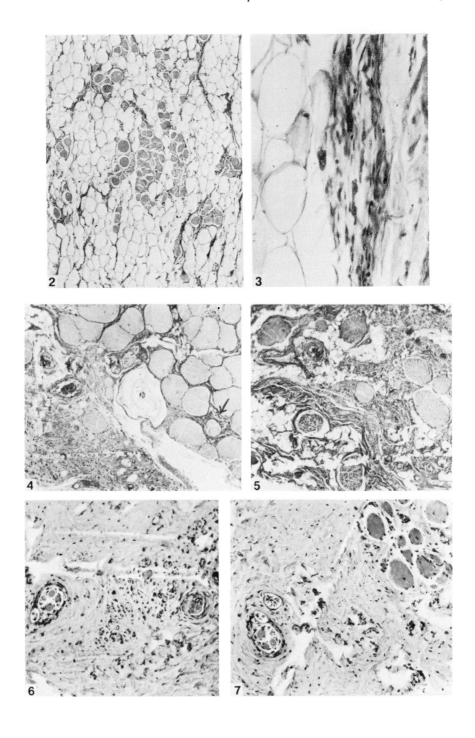

necrotic or regenerating muscle fibers." The fibers surrounding these areas are usually normal. Engel regards these focal abnormalities as "virtually diagnostic of that kind of dystrophy." Hathaway and his colleagues suggest that the occurrence of these focal areas could best be explained by assuming that the blood supply to these groups of fibers has been affected. "The size of the foci suggests that they represent the critical territory of a terminal arteriole or perhaps a vessel distal to the arteriole . . . alterations of the walls of small blood vessels (thickening and mononuclear infiltrates) and narrowing of their lumens are found in the early phases of the disease."

Hathaway and his colleagues caused intermittant blocking of the microcirculation to the skeletal muscle of a rabbit. The animals were killed at intervals of 1 week to 3 months and the following pathology was found: focal grouped necrotic muscle fibers; focal grouped regenerating muscle fibers; single, scattered necrotic or regenerating muscle fibers; newly healed groups of fibers; thickened, sometimes occluded arterioles, occasionally with perivascular cellular reaction, and later areas with marked increase of endomysial connective tissue surrounding muscle

---

Fig. 2. Duchenne type of muscular dystrophy; from a boy aged 14. Scattered groups of rounded muscle fibers of all sizes embedded in adipose tissue. Transverse section, soleus; iron hematoxylin and van Gieson; × 65.

Fig. 3. Same case as Fig. 2. Longitudinal section of a small bundle of atrophic muscle fibers in adipose tissue; showing well preserved cross striations, clumps of sarcolemmal nuclei, and an increase of collagen around the muscle fibers. Plantaris; iron hematoxylin and van Gieson; × 350.

Fig. 4. Facioscapulohumeral muscular dystrophy in a man aged 21 with a family history of the disease. Severe atrophy on the left; tiny muscle fibers are seen and also sarcolemmal nuclei in empty sheaths, all embedded in dense fibrous tissue. On the right, very large rounded muscle fibers. One fiber (arrow), undergoing active phagocytosis. There is an increase in endomysial collagen. Normal-looking muscle spindle in center with nerve above. Transverse section, vastus medialis; hematoxylin and eosin; × 65.

Fig. 5. Same case as Fig. 4. Another field of the same muscle, showing more severe atrophy and fatty infiltration. Scattered, large muscle fibers and fibers in various stages of atrophy remain. There are two normal-looking nerve trunks. Transverse section; hematoxylin and eosin; × 65.

Fig. 6. Facioscapulohumeral muscular dystrophy from girl aged 15 with family history, showing muscle virtually replaced by collagen. Between the two muscle spindles (right and left), one recognizable muscle fiber and the remains of many atrophic fibers are seen. Transverse section, biceps brachii; iron hematoxylin and van Gieson; × 140.

Fig. 7. Adjacent field to Fig. 6. Severe atrophy. A single bundle of rounded muscle fibers of very unequal size on the right. Normal-looking muscle spindle on the left, with sarcolemmal nuclei in intervening dense fibrous tissue. Transverse section biceps brachii; iron hematoxylin and van Gieson; × 140.

fibers that varied in size and sometimes contained central nuclei. The authors pointed out that these changes were very similar to those seen in Duchenne dystrophy. It may also be of significance that they found that in the areas where the muscle fibers were affected, the capillary vessels were devoid of alkaline phosphatase activity.

It is of interest, in view of these studies, that Bourne and Golarz, some ten years ago (1963b), suggested that failure of transport of nutrients and gases from the blood vessels to the muscle fiber constituted the primary lesion in muscular dystrophy. They did not suggest, however, that this was due to a defect in the blood vessels but to a defect in mucopolysaccharides that exist in the connective tissue between the blood vessel and the muscle fiber and through which substances must diffuse to reach one from the other. They indicated an almost complete absence of metachromatic staining in the connective tissue of dystrophic muscle (except for mast cells) and pointed out that metachromosia is related to the presence of sulphated mucopolysaccharides.

As has been pointed out elsewhere in this treatise, Duchenne dystrophy in an X-linked inherited disease. Daughters transmit the disease, but sons develop it. Mölbert, in 1960, showed, with the electron microscope, that in carriers there was a homogenization of the myofibrils and a degeneration of sarcosomes. Beckmann et al. (1970) have shown that even in cases where enzymic and electromyography and biopsy have given negative results in Duchenne carriers, ultrastructural studies can result in identification. In these cases, homogenization of the myofibrils is constant; other characteristic features are fatty degeneration of the mitochondria and accumulation of glycogen. There is also an increase of striated collagen fibrils between the muscle fibers. The mothers of boys with the disease always show some signs of muscular degeneration, including loss of mitochondria and thinning of myofibrils.

B. Facioscapulohumeral Type of Muscular Dystrophy*

The onset of this variety of muscular dystrophy is usually later than it is in that described above as the Duchenne type, though it may be diagnosed at any age. Males and females are about equally affected. Tyler considers that the disease is inherited as a simple somatic Mendelian

* Synonyms: atrophic progressive myopathy of Landouzy-Dejerine (1885a); juvenile form of progressive muscular atrophy, Erb (1884, 1894); progressive muscular dystrophy, facioscapulohumeral type, Tyler and Wintrobe (1950); autosomal limb-girdle muscular dystrophy, Stevenson (1953); and mild restricted muscular dystrophy, Adams et al. (1953).

dominant (Tyler, 1951); others believe that a family history is not in-variable. Weakness and wasting of the shoulder girdle, especially the lower fibers of the pectoralis major and trapezius, is often the earliest feature of the disease, followed by involvement of the muscles of facial expression, though Batten (1910a) points out that these latter may be affected at birth. There is seldom enlargement of the affected muscles in the facioscapulohumeral type of dystrophy, which thus differs mark-edly from the Duchenne type, where hypertrophy of the leg muscles is common and where the disease is virtually confined to males (entirely confined to boys, according to Stevenson, 1953). The inheritance is very different (Tyler, 1951).

Facioscapulohumeral muscular dystrophy is an astonishingly chronic disease. Landouzy and Dejerine (1885b) described an autopsy on a case whose symptoms began in the face at the age of 3 years and who died of phthisis at the age of 24 at a relatively early stage of the muscle disease. The central and peripheral nervous systems were normal. They described a simple atrophy of muscle fibers, which retained their stria-tions and showed no increase in nuclei. The connective tissue was in-creased and there was some increase in fat. In 1886, these authors re-ported another autopsy (scapulohumeral type) in a man of 66, who had suffered from the disease for over 40 years. There was an intact nervous system with atrophy of the limb-girdle muscles. However, in this case, they reported a marked increase of the sarcolemmal nuclei and hypertrophy of muscle fibers in certain muscles. In the very atro-phied muscles, they found an increase of interstitial fat without much sclerosis. The third classic case upon whom any autopsy was performed was a patient described by Landouzy and Dejerine in 1884 and 1885 (1885a), who did not die until 1902, aged 45, having been studied by these acute observers for nearly 30 years. Landouzy and Lortat-Jacob (1909) performed a very careful dissection of this case. They found that the muscles of the limbs were much atrophied, pale, yellow-gray in color, and very fibrosed. Some were mere rigid cords. The muscles of the thenar eminence were degenerated and resembled tendons, al-though the hypothenar and lumbrical muscles were not so badly degen-erated. The muscles of the neck and trunk were also much atrophied. The supra- and infraspinati and subscapulares appeared normal in color. Only a trace of the facial muscles was found.

Histologically, they found that, in the case of the most severely affected muscles, they could not be sure that striations were present, the muscle fibers appearing granular with only doubtful striations. In less severely affected muscles, they found proliferation of nuclei in chains. The interstitial fibrous tissue was much increased, and they thought the muscle fibers were compressed. There was much adipose tissue in the interstices. The central nervous system and peripheral nerves did not show degeneration. We have been unable to find any other autopsy reports in the literature in which a systematic study of the muscles has been made, although Denny-Brown performed an autopsy on a case and examined some of the muscles. This case is referred to in Adams *et al.* (1953).

In general, in this disease, the histological appearances show the changes of myopathy. That is to say, the atrophic muscle fibers are scattered at random, or if they are in groups, these are irregular, in contrast to a neural degeneration. Muscle fibers of large diameter, some of which are always present except in the very last stages of the disease, are even more irregularly scattered. All the muscle fibers tend to be rounded when seen in cross section (Figs. 4, 5, and 7). The interstitial collagen increased around individual muscle fibers until in time most of the shrunken muscle mass is replaced by connective tissue in which the remaining muscle fibers lie either singly (Fig. 4) or in small groups (Fig. 7). This mass of connective tissue also contains the vascular bed of the replaced muscle, the nerve trunks, and the muscle spindles, now very prominent owing to the disappearance of so many extrafusal muscle fibers (Figs. 4, 6, and 7).

The great increase in the collagenous tissue has been stressed because in this type of muscular dystrophy it is very characteristic. In this condition, there is always some, and often much, increase in the adipose tissue of the affected muscles, but the extreme degree of fatty change such as is seen in the Duchenne type of dystrophy is not so striking a feature of this disease. See Figs. 4–7.

The earliest evidence of the disease is a rounding of the muscle fibers and irregularity of caliber, with little if any increase in interstitial tissue. Both large and small fibers are seen at an early stage. The large muscle fibers (over 100 μ) may be homogeneous and swollen over short distances of their length. These are presumably undergoing some form of degeneration. However, fibers of well over 100 μ diameter are seen that are well striated and, apart from their increased girth, show no abnormality. The significance of these large normal-looking muscle fibers is not clear, but it seems possible that, for some time, a work hypertrophy might occur in some fibers of a muscle which is much used. Scattered foci of necrosis may be seen in some muscle fibers, and an increase in sarcolemmal nuclei in others; granular degeneration and phagocytosis (Figs. 4 and 5) also occur during the early stages of the disease, and during this phase there may be some interstitial infiltration with cells, including polymorphs. These changes are found in any actively degen-

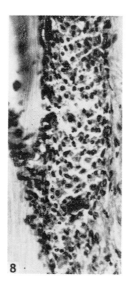

Fig. 8. Facioscapulohumeral muscular dystrophy; same case as Figs. 3 and 4. Active phagocytosis in fiber of muscle which had not shown recent clinical deterioration. Longitudinal section, vastus medialis; hematoxylin and eosin; × 350.

erating muscle and in dystrophic muscle may be present many years after the onset of the disease. The differential diagnosis from mild polymyositis may be difficult. The vessels and nerves in these cases show no pathological change, as far as is known, other than those secondary to loss of muscle fibers.

Ionasescu *et al.* (1970) claim that biochemical studies indicate that in fascioscapulohumeral dystrophy there is uncoupling of phosphorylation and respiration, and this has led to the adaptation of the respiratory process to low energy requirements and is probably the reason why the disease is so protracted. See Fig. 8.

Electron microscope studies by Mataglia and his colleagues (1969) have shown that an early change in this disease is dilatation of the sarcoplasmic reticulum and the loss of the I bands of the sarcomeres.

C. Dystrophia Myotonica*

The characteristic feature of this disease is muscular weakness accompanied by myotonia. The onset is seen most commonly in adolescence or early adult life, but it may be at any age. The disease is usually

* Synonyms: myotonic dystrophy, myotonia atrophica.

inherited as a dominant Mendelian characteristic, affecting both sexes (Maas, 1937; Ravin and Waring, 1939; Thomasen, 1948). Weakness and myotonia are followed, after a varying period of years, by atrophy of the affected muscles.

This disease differs from the other muscular dystrophies in a number of ways. It is genetically distinct. The age of onset tends to be later, and it seems probable that a number of distal myopathies of late onset are in fact examples of this disease. The distal muscles, particularly of the hands and forearms, are affected first, together with those of the face (ptosis being common) and the sternomastoids; later, there may be dysphagia and dysarthria, which are rare in other dystrophies. Myotonia usually precedes the weakness, though it may be overlooked. Maas and Paterson (1947) believe dystrophia myotonica to be a generalized disease; certainly it is accompanied by endocrine disorders. Premature baldness and testicular atrophy occur in a high proportion of cases. Thyroid or pituitary deficiency is suggested by a low basal metabolic rate and enophthalmos, while a low blood sugar curve is occasionally found. Blood cholesterol has been normal where recorded. General asthenia, suggesting adrenal deficiency, is the rule, but the serum sodium and potassium level has been normal in the few cases in which it has been recorded. Hypoparathyroidism, once suggested as a cause of the disease, has not been substantiated, and blood calcium is normal. Acrocyanosis is often present. Premature cataract is very frequent—nearly 90% of Thomasen's (1948) cases—as was first noted by Greenfield (1911). This may be found as the only abnormal feature in the relatives of a patient suffering from the overt disease. Hormonal studies have not contributed much so far, though Liversedge and Newman (1956) have found that cortisone and corticotropin reduce the duration of the myotonia. Gonadotropic hormones are decreased when there is clinical gonadal atrophy. Creatinuria is usually slight, corresponding to the slow rate of muscular atrophy.

Myotonia, a hallmark of the disease, is usually found in the hands and forearms, though it may also be present in muscles that are not weak or wasted; for instance, it is common in the tongue, which is rarely wasted. Myotonia is made worse by cold or fatigue and may, in fact, only be elicited after cooling. The electromyogram is characteristic, consisting of prolonged bursts of potentials during and after a single contraction. Atrophy may be long delayed; however, in time the weak muscles become atrophied and are then paler macroscopically than normal muscles.

Microscopically, the changes, once established, are essentially those of other chronic myopathies, that is to say, rounding of the muscle fibers in cross section with atrophy of fibers singly or in groups, in either case forming an irregular pattern when seen in cross section (Figs. 8 and 9). In addition to the small atrophied fibers, large fibers (100–150 μ in diameter) may be present, and indeed may be common in the early stages. These may appear normal apart from their size or may have central nuclei (Fig. 8). Any of the various types of muscle fiber degeneration may be present, but in so chronic a disease, atrophy with empty sarcolemmal tubes is usually more prominent than acute degeneration. Splitting of muscle fibers is said to be rare (Adams *et al.*, 1953), but we have seen it in several cases, and G. Wohlfart (1951) speaks as though it is common. In fact, he describes (and illustrates) capillaries apparently within the muscle fibers as a result of splitting. There is a generalized increase in cellularity, but on the whole, interstitial infiltration of cells is slight. The degree of increase of collagen and fat corresponds with the normal amount of muscle fiber atrophy, and until the atrophy is severe, the interestitial overgrowth is only moderate in amount. A great excess of fat between the muscle fibers is seldom seen except in the final stages of a severe atrophy.

A striking feature histologically, and one much stressed by writers on the subject, is the presence of long rows of central nuclei (Figs. 11 and 12) within the muscle fibers (Adie and Greenfield, 1923). These nuclei are more rounded than usual and often contain prominent nucleoli. Such rows of nuclei in otherwise normal-looking muscle fibers have been called diagnostic (Adams *et al.*, 1953). This view should be accepted with caution, particularly in biopsy specimens, as rows of central nuclei of considerable length may be found in other conditions, but it is true that really long chains (10, 20, or more) of central nuclei are often a striking feature in dystrophia myotonica and that they may also be present in clinically unaffected muscles in cases of this disease. The peripheral muscle nuclei may also be arranged in chains, but this is not unusual in other conditions (Fig. 14). G. Wohlfart (1951) has observed that a thick peripheral layer of clear sarcoplasm without myofibrils is present in many muscle fibers due to loss of myofibrils without a corresponding shrinkage of the sarcoplasm, and he believes that this is characteristic of the disease. He also agrees with others (e.g., Heidenhain, 1918) that *ringbinden* are common, though he is careful to point out that these may be found in normal muscles. According to W. K. Engel *et al.* (1968), nuclear chains occur most frequently in type 1 fibers, and when these are very small, the appearance is that of a fetal myotube.

Owing to the extreme sensitivity of the muscle to mechanical stimulation, biopsy material, unless it is allowed to lose its contractility before

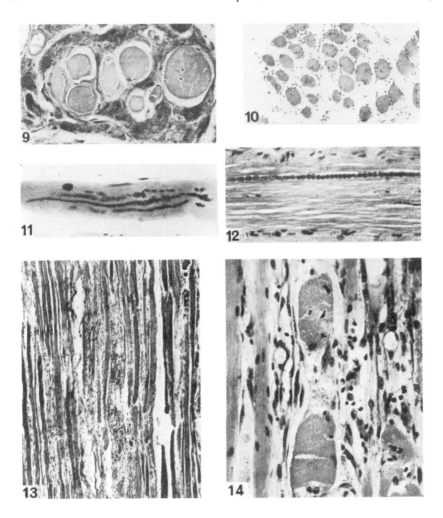

Fig. 9. Dystrophia myotonica in man aged 34 with family history. The muscle fibers are rounded with central nuclei, and there is unusually severe fibrosis. Transverse section, peroneus longus; iron hematoxylin and van Gieson; × 350. From A. G. Engel and MacDonald (1970).

Fig. 10. Dystophia myotonica in man aged 66 with family history. Note uneven size and rounding of muscle fibers with central nuclei but no fibrosis. Transverse section, tongue; hematoxylin and eosin; × 80. From A. G. Engel and MacDonald (1970).

Fig. 11. Dystrophia myotonica in man aged 66. Single muscle fiber showing multiple chains of muscle nuclei. Longitudinal section, tongue; hematoxylin and eosin; × 350. From A. G. Engel and MacDonald (1970).

Fig. 12. Dystrophia myotonica in woman aged 48. There is a very long chain of central nuclei. Longitudinal section; biceps brachii; hematoxylin and eosin; × 350.

Fig. 13. Polymyositis in man aged 61, who died 6 weeks from onset of disease.

fixation, often shows fragmentation and contraction bands, and thus, the possibility of artifacts must be considered. A number of autopsies have been performed on patients with dystrophia myotonica by Adie and Greenfield (1923) (who give a very good and full account of the pathological changes in the muscles), Rouquès (1931), Bielschowsky *et al.* (1933), Black and Ravin (1947), G. Wohlfart (1951), and others. Dystrophic changes in the myocardium have not been reported. The endocrine organs have been studied by some authors (Black and Ravin, 1947; Thomasen, 1948).

Electron microscope studies of muscle in myotonic dystrophy have shown that annular myofibrils are present, frequently associated with sarcoplasmic masses, periphically situated. According to Lapresle and Fardeau (1968), "Occasionally peripheral masses may show a special type of spatial disorganization of the sarcomeres, which are stacked, without order, at different spatial levels and lack an identifiable Z-line at their extremities." These changes are, however, not exclusive to myotonic dystrophy.

Fardeau (1970) has shown an accumulation of mitochondria and fatty pigments near one pole of the nucleus. He also found retangular inclusions in the mitochondrial cristae and some were lying free in the cytoplasm. Fardeau has indicated that abnormal inclusions in mitochondria can be seen in many dystrophies but that they are most numerous and most pleomorphic in myotonic dystrophy.

Aloise and Margreth (1967) reported swelling vacuolation in the sarcoplasmic reticulum in myotonic dystrophy, and this had also been recorded in 1964 by Pearce for progressive muscular dystrophy.

Fardeau (1970) has also described honeycomblike tubular complexes which he believes to be formed from the transverse T system in both myotonic dystrophy and in Duchenne's dystrophy. They are nonspecific structures. Fardeau (1970) has the following to say about nuclear structure in dystrophies:

> Deeply indented, "grimacing" nuclei were demonstrated in myotonic dystrophy by Gruner (1963). The deformed nuclei show numerous cytoplasmic inclusions where bundles of filaments, microtubules or membrane profiles can be found. All intermediate stages can be demonstrated down to pyknotic residues limited by a double membrane. In the present series,

---

Shows general disorganization of the muscle, with necrosis of many fibers and interstitial cellular infiltration. Longitudinal section, pectoralis minor; hematoxylin and eosin; × 65.

Fig. 14. Polymyositis, same case as Fig. 13. Severe acute degeneration. Segmental swelling and fragmentation of muscle fiber. Longitudinal section, pectorialis minor; hematoxylin and eosin; × 350. From Fardeau (1970).

such degenerate nuclei were especially common in myotonic dystrophy. Occasionally they were found in other dystrophies and in polymyositis. Lee and Altschul (1963) found the same appearances in experimental denervations.

Sarcoplasmic masses and ring fibers were also seen in a case of myotonic dystrophy studied by Giacanelli and Perciaccante (1970). In two cases, they found giant muscle cells that showed metachromatic staining with toluidine blue. The authors thought these might have been regeneration cells, especially as they stained well with reagents that indicated the presence of RNA.

Radu and his colleagues (1970) state that dystrophia myotonica shows lesions of the sarcoplasmic reticulum, mitochondria, and myofibrils. There is type I fiber atrophy. In addition to the uncoupling of respiration and oxidative phosphorylation, myotonic muscle has a reduced uptake of calcium, and there is a substantial increase in ATPase activity. These characters are, according to these authors, specific for myotonia. The other histopathological changes seen in the muscle in this condition are simply due to the dystrophic process. There is some evidence that type I fibers are more susceptible to dystrophic processes, and it is of interest that the ratio of type I to type II fibers varies in different directions in the two types of myotonia; there is a definite preferential atrophy of type I fibers in myotonia dystrophica.

Schröder (1970) has described infoldings of the sarcolemma in cases of myotonic dystrophy. These folds were originally shown by Cöers and Woolf (1959) to be positive for acetylcholinesterase. Schröder believes these infoldings represent the remains of denervated motor end plates. They were not present in the one case of myotonia congenita he examined.

D. Myotonia Congenita*

This rare congenital and hereditary disease was first described by Thomsen (1876), who himself suffered from it. It is in most cases non-progressive, and the muscles remain strong; creatine and creatinine excretion are normal. Myotonia is widespread, and a myotonic electromyogram is obtained from many muscles. Some authors (Maas and Paterson, 1939, 1950) consider that this disease is a variant of dystrophia myotonica. Paramyotonia (Eulenburg, 1886), in which the myotonia is

---

* Synonyms: Thomsen's disease, myotonia hereditaria.

only observable after chilling, appears to be only a variant of Thomsen's disease (Thomasen, 1948).

Little work has been done on the pathology of Thomsen's disease. The myotonic muscles are enlarged and strong, the lower limbs characteristically showing most change, but the whole musculature may be hypertrophied. Histologically, the chief abnormal finding (in biopsy material) has been enlarged muscle fibers, with a diameter often well over 100 μ. There may also be some tendency to rounding in cross section and a few central nuclei (G. Wohlfart, 1951; Greenfield *et al.*, 1957). Post mortems have been reported by Erb (1886) and Dejerine and Sottas (1895).

Two outstanding findings in this condition are a relative uncoupling of oxidative phosphorylation and a relative increase of type II fibers. An accumulation of mitochondrial (tubular) aggregates has been seen in this condition. The aggregates were found to contain lipid, NADH diaphorase, and cytochrome oxidase, and to give a positive reaction with MTT hydroquinone, which demonstrates the presence of ubiquinone or some other redox compound (Radu *et al.*, 1970).

E. Other Muscular Dystrophies

1. Ocular Dystrophy[*]

This condition has been described in Kiloh and Nevin (1951) and the literature fully reviewed. The disease appears to be confined at least in the early stages, to the extrinsic ocular muscles and perhaps the orbicularis oculi. There is said to be great variation in the size of the muscle fibers, with some hypertrophied fibers and a variable amount of fat and connective tissue between the muscle fibers; the nerves are normal. The syndrome may be a variant of the Landouzy–Dejerine type of musclar dystrophy.

Ferber and his colleagues (1970) stated that half the patients they saw with ocular dystrophy had dystrophic processes in the skeletal muscles. In the exophthalmic ophthalmoplegia occurring during the course of thyrotoxicosis, changes have been found in the extraocular muscle that indicate a primary muscle affection is present. There are degenerative processes in the muscle fibers, increase of lymphocytes, proliferation of endomysium, and fatty infiltration (Naffziger, 1933; Dobyns, 1950). The extraocular muscles are also affected in myosthenia gravis.

---

[*] Synonym: Progressive dystrophy of the external ocular muscles.

However, interpretation of biopsies from eye muscles requires care, for normally the muscle fibers vary greatly in size, are rounded, and often have central nuclei (Cooper and Daniel, 1949; Cooper et al., 1955).

2. DYSTROPHY OF LATE ONSET

A number of cases of muscular dystrophy of late onset that are difficult to classify have been reported. Nevin (1936) and Welander (1951) have reviewed the pathology of these cases. Welander (1951) studied 249 cases of a condition which she called distal late hereditary myopathy. The age of onset varied from 20 to 77 years. Muscle biopsies were obtained from twenty-six patients, and autopsies were performed on three who had shown symptoms for from 9 to 16 years. The histological picture seen was typical of a primary myopathy with, in early cases, rounding and variation in size of muscle fibers and increase of sarcolemmal nuclei with some central nuclei. Later, there was great increase in the interstitial fibrous and fatty tissue.

Barnes (1932) was able to find details of 283 descendants of a man born in about 1749 who died in 1836 and who suffered from a form of myopathy in the latter part of his life. The disease in this family was always of late onset. As young adults, those later to be affected were unusually strong and tended to excel at games. Many achieved a ripe old age. The muscles first affected were usually the proximal limb muscles, in contrast with Welander's (1951) cases in which the distal limb muscles were most commonly affected. Unhappily, muscle for histological examination was obtained from only one member of this remarkable family. The illustrations show the changes characteristic of any muscular dystrophy—rounded swollen muscle fibers, excess of nuclei, and in one field, excess of interstitial fat and fibrous tissue. No evidence was seen of phagocytosis or lymphocytic infiltration.

Although the number of cases of late myopathy reported is comparatively small, a considerable number of muscle biopsies reach the pathologist from cases presenting as muscle diseases arising in middle-aged and elderly patients. Some of these are neural degenerations that are clinically atypical. Of those that are true myopathies, neither the clinical nor pathological classification is yet certain, nor do the two coincide. Clinically, after separation of those subsequently found to have carcinoma, there are a group of distal myopathies and a group with a proximal syndrome. Either of these may or may not have a family history of muscle disease and may run a longer or shorter course. Pathologically, some cases are histologically similar to a myositis, while a few are both clinically and pathologically similar to the dystrophies of young people.

V. Myopathies of Infants

A. Amyotonia Congenita*

This disease has more names than pathology. It is here used to signify a congenital or infantile disease of muscles, as opposed to Werdnig–Hoffman's disease, which is an infantile spinal muscular atrophy in which the muscle changes are secondary to those in the motor nerves. The two conditions have been much confused, but there now appears no doubt that there is both a spinal muscular atrophy (Werdnig–Hoffman's disease) and an infantile myopathy (for discussion of this, see the clinical section of this treatise). Werdnig–Hoffman's disease is progressive and fatal, whereas, as the name benign congenital myopathy suggests, in true amyotonia congenita, there is a tendency to recover. There are therefore extremely few necropsy reports on cases of myopathy, i.e., true amyotonia congenita. Spiller (1905), Councilman and Dunn (1911), Lerebouillet and Baudouin (1909), and Menges (1931) have reported such cases, as has also Turner (1949). Turner reported the post mortem findings in a case that was unusual in showing muscular wasting. Histological examination revealed only slight changes in the less affected muscles, but in the more severely atrophied ones, there were scattered foci of necrosis and invasion of nuclei, while some fibers otherwise normal had chains of central nuclei. Dr. Greenfield reported these changes as those of a chronic myopathy.

Biopsies have usually shown no abnormality, according to Walton (1957), although others have found small muscle fibers in biopsies from "flabby babies," but some at least of these are no doubt cases of Werdnig–Hoffman's disease. At present, there is no diagnostic pathology for this condition.

Cöers and Pelc (1954) and Woolf and Till (1955), using intravital staining of nerve fibers with methylene blue in muscle biopsies, have suggested that the motor end plates may be immature for the age of the infant, which is interesting, as others have suggested that the lesion is primarily a delay in development of the muscle.

B. Congenital Nonprogressive Myopathy

A form of amyotonia congenita has been described by Shy and Magee (1956) as a new congenital nonprogressive myopathy. In this familial

* Synonyms: Oppenheim's disease; myatonia congenita; benign congenital myopathy.

disease, the proximal muscles of the limbs were most severely involved and wasting was not prominent. The pathological changes in the muscles are most striking and appear to be diagnostic. The muscle fibers have a central core of fibrils running almost the whole length. Fibrils in this central core have a striking difference in staining reaction from the peripheral fibrils; with Gomori trichrome, the peripheral fibrils stain red, while the central core stains blue. The core also gives a more strongly positive periodic acid–Schiff reaction than the peripheral fibrils. Some very large muscle fibers, up to 240 μ in diameter, were found, and these had a special tendency to show central nuclei in chains. There was a moderate increase in interstitial fat, but no increase in collagen. The nerves appeared normal.

C. Arthrogryposis Multiplex Congenita*

As its name suggests, this disease consists of congenital deformities of the limbs with rigidity of the joints. It appears that this state may be due to at least two pathological conditions, one a disease of the nervous system, the other a primary disorder of the muscles. Adams et al. (1953) in humans, and Whittem (1957) in calves, found evidence of neural disease, and a myopathic change was seen by Gilmour (1946). More recently, Banker et al. (1957) found severe myopathic changes in two infants, so that the disease may clearly also be due to a congenital myopathy.

VI. Polymyositis and Related Conditions

Polymyositis is the name now generally accepted for another idiopathic disease of muscle, and it does not here refer to the myositis caused by bacteria, viruses, or parasites. There is general agreement that dermatomyositis, first described as a clinical entity by Unvericht in 1887, is a variant of this disease with a skin eruption. The skin lesion may indeed overshadow the muscle disease, but primary degeneration of skeletal muscle appears to be a basic part of the pathology, and the condition is therefore a true myopathy whose etiology is quite unknown. In this account, we have confined the use of the term polymyositis to the pure

* Synonyms: Amyoplasia congenita; multiple congenital articular rigidities; myodystrophica congenita deformans; myodystrophia foetalis deformans; congenital contractures of the extremities.

muscle disease and to dermatomyositis. The menopausal dystrophy of Shy and McEachern (1951) and some of the myopathies of late onset appear to fall within this group and have the characteristic pathology. It must be emphasized, however, that the name is misleading, as the disease is not primarily inflammatory. Acute cases, running a fatal course of a few months, or even weeks, may have myoglobinuria, but myoglobinuria may occur in any very severe muscle degeneration and of itself does not alter the diagnosis. A number of cases have been associated with malignant disease. This and other pathological and clinical features have been reviewed by Dowling (1955) and are fully treated by Walton and Adams (1958) in their monograph on the subject.

Affected muscles are macroscopically pale, and they may be a striking gray-white in color or have circumscribed areas of pallor that are often slightly swollen, or again pale flecks are sometimes present throughout the muscle; in the most acute cases, there may be small hemorrhages. Any muscle can be affected, but the trunk, neck, and proximal limb muscles are usually most severely involved; the respiratory muscles, tongue, and muscles of deglutition also sometimes have lesions.

Histologically, the muscles show an acute myopathy. The most striking changes are a degeneration of the muscle fibers, which are rounded, vary markedly in size, show genuine changes in staining reaction, and have greatly increased numbers of nuclei (Figs. 13, 14, and 15). Fibers may be necrotic throughout their length, but segmental degeneration is very typical, short lengths of fiber being swollen and without striations (hyaline or cloudy degeneration) (Fig. 21). Fragmentation, often with areas of empty sarcolemmal sheath stretched between the fragmented ends of sarcoplasm, is seen (Fig. 14). Floccular degeneration (Fig. 19) is common, and segments in which the sarcolemmal sheath is filled with nuclei (Fig. 15) may be numerous. The finding of these last two changes is in fact almost a necessity for making a diagnosis. In the interstitial tissue lie both free sarcolemmal nuclei and many types of cell (Figs. 13 and 20)—macrophages, plasma cells, lymphocytes, (the last two are most common) and a variable number of polymorphs, including eosinophiles. The infiltrated cells may be present diffusely throughout the tissue or in collections between the muscle fibers and around vessels. Large collections of lymphocytes may be found, and curiously enough, these are more often seen in the less degenerated areas of muscle. The muscle fibers are of all sizes, some are obviously swollen (up to 200 μ in diameter) but many are shrunken, and some, having lost all their cytoplasm, are represented only by chains of nuclei. Mixed with the degenerating fibers are thin regenerating fibers (Fig. 20), basophilic and containing vesicular nuclei, and sarcoplasmal giant cells (muscle buds), which are also probably regenerating fibers. Good illustrations of these various changes will be found in Kinney and Maher (1940) and Walton and Adams (1958).

Muscle spindles may be involved in the general degeneration. Blood vessels and nerves are normal (reports of neuromyositis have not been substantiated). Infiltration with fat is not a feature. There may be some fibrosis, but we have not found it to be very marked even in chronic cases.

This overall picture is very characteristic when present. Unfortunately, the histology is extremely variable. Not all the muscles are affected in any one case, and even a single muscle may have areas that are normal and areas of the severest damage. There may be little evidence of interstitial reaction, or of regeneration, and the muscle fibers may show only hyaline change and a mild increase in nuclei. Again, there may be only collections of round cells between normal fibers. Neither of these last two appearances are diagnostic and the "interstitial nodular myositis" is more characteristic of other conditions, such as rheumatoid arthritis and myasthenia gravis.

The heart may show small foci of myocardial degeneration with lymphocytic infiltration, or a patchy fibrosis. The kidney tubules in those cases with myoglobinuria may contain myoglobin. In dermatomyositis, the skin shows atrophy of the epidermis with thickening and condensation of the underlying collagen, sometimes with vascular proliferation and focal infiltration with cells.

The more chronic cases of polymyositis cannot always be distinguished from muscular dystrophy (in which, incidentally, skin changes have many times been reported). In our experience, increase in collagen has not been very great in polymyositis, and severe fibrosis would favor a diagnosis of dystrophy. Muscle fibers with segments of acute necrosis and interstitial infiltration may be found in muscular dystrophy (Fig. 21), but in dystrophy there is usually less acute degeneration and no regeneration.

Some authors do not distinguish between polymyositis and scleroderma, considering them both to be varieties of the group of conditions that have been called collagen diseases. In scleroderma, however, visceral lesions are the rule (Dowling, 1955), particularly atrophy of the plain muscle of the gastrointestinal tract and arterial lesions especially affecting the kidneys. This is not so in typical polymyositis. A review of the literature on the pathological changes in scleroderma will be found in Dowling (1955), who also considers that the skin lesions in the two conditions are different. The muscle changes in scleroderma may no doubt sometimes be identical with those in polymyositis, but in our admittedly limited experience, the muscle lesions in scleroderma have been more purely degenerative than in polymyositis, with little cellular reaction and no muscle buds present. Furthermore, the histological appearances of polymyositis may be produced experimentally in a number of unrelated ways (e.g., by poisons) and are probably simply the picture of any acute muscle degeneration. Therefore, it would seem to us best to classify the conditions separately for the present.

Polymyositis, at least, appears to be a primary muscle fiber degeneration and not a degeneration secondary to disease of the interstitial collagen, so that there seems little reason to call it a collagen disease. Pearson (1967) has listed the characteristics of a muscle biopsy in both polymyositis and dermatomyositis as

1. Primary focal or extensive degeneration of muscle fibers, sometimes with vacuolization

2. Evidence of regeneration as demonstrated by sarcoplasmic basophilia and the presence of large vesicular nuclei and prominent nucleoli

3. Necrosis of a part of the whole or one or more fibers with phagocytosis of their substance

4. Interstitial infiltrates of chronic inflammatory cells, sometimes focal and sometimes diffuse and often with a perivascular component

5. Significant variation in fiber size, especially in cases of several months' duration

6. Interstitial fibrosis; this characteristic is not agreed with by some other authors

See Figs. 13–21.

Bedivan (1970) has described an association of what she described as chronic pseudomyopathic polymyositis with chronic adrenal disease. Muscle biopsies from her case showed interstitial lymphocytic inflammatory nodules and some fibrosis. The small blood vessels in the muscle showed thickening of the walls and occlusive proliferations of the endothelium. This patient had originally complained, among other symptoms, of myalgia and weakness. She had then been given steroid therapy and some months later returned with a marked asthenia, which was said to clear up with further adrenal corticoid treatment. After a 6 year "course," presumably of treatment with steroids, the patient developed generalized muscle atrophy. After further treatment, the patient was again improved. It should be pointed out here that steroid hormones can themselves cause myopathic changes.

De Vivo and Engel (1970) have described a remarkable recovery of a patient suffering from what was described as recurrent polyneurop-

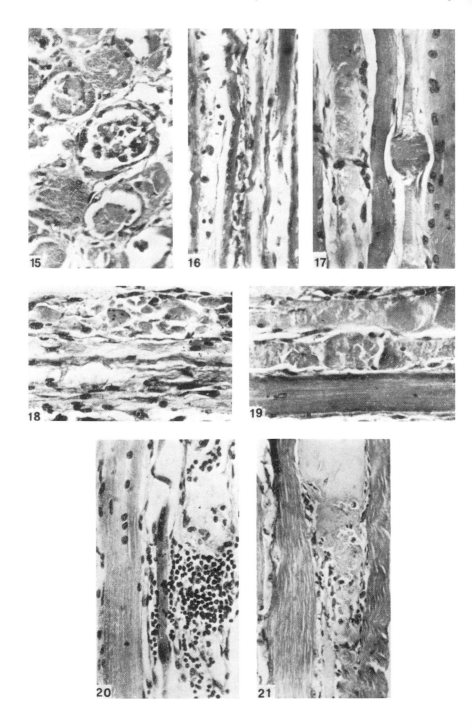

athy after steroid treatment. Muscle biopsies in this patient showed marked variation in fiber size which occurred in both type I and type II fibers. In the biopsy, some small, angular fibers were present, which stained intensely for oxidative enzyme activity, suggesting that they were denervated fibers.

VII. Muscle Diseases Associated with Systemic and Metabolic Disorders

A. Myopathy Associated with Malignant Disease

Patients suffering from carcinoma frequently develop muscular weakness, and when the clinical signs are severe enough to suggest muscle disease, these cases form, clinically, part of the heterogenous group of "late myopathies." Denny-Brown (1948), Henson *et al.* (1954), and Heathfield and Williams (1954) have reported somewhat indefinite pathological changes. There is also an undoubted association of dermatomyositis and carcinoma (Domzalski and Morgan, 1955; Walton and Adams, 1958). The changes found in malignant disease usually consist of simple atrophy of muscle fibers and an increase in nuclei. Sometimes a more acute degeneration of muscle fibers with some cellular infiltration resembling a mild polymyositis is seen. Carcinomatous neuropathy is

---

Fig. 15. Acute polymyositis, showing disintegration of muscle fibers. In the center is a muscle fiber undergoing active phagocytosis. Transverse section, pectoralis major; hematoxylin and eosin; × 350.

Fig. 16. Acute polymyositis, showing atrophic muscle fibers. The central dark fiber, with a chain of nuclei, may be regenerating. Longitudinal section, pectoralis major; hematoxylin and eosin; × 350.

Fig. 17. Acute polymyositis. Segmental swelling and fragmentation. Muscle fiber on the left shows granular degeneration, that on the right has vesicular nuclei. Longitudinal section, pectoralis minor; iron hematoxylin and van Gieson; × 350.

Fig. 18. Acute polymyositis. Disintegrating muscle fiber (right); complete loss of fibers (left). Longitudinal section, pectoralis minor; iron hematoxylin and van Gieson; × 350.

Fig. 19. Acute polymyositis, showing floccular degeneration in two muscle fibers. Longitudinal section, pectoralis major; hematoxylin and eosin; × 350.

Fig. 20. Acute polymyositis, showing mixed cellular infiltration, including polymorphonuclear leukocytes and a thin, basophilic, regenerating muscle fiber (dark). Longitudinal section, pectoralis major; hematoxylin and eosin; × 350.

Fig. 21. Chronic polymyositis in man aged 38. Muscle fiber showing hyaline degeneration (above) and floccular degeneration and phagocytosis (below). Longitudinal section; hematoxylin and eosin; × 350.

well recognized, and we believe that the muscle lesions seen may be partly accounted for by a scattered neural degeneration.

Little is known of the changes produced in human muscle by prolonged bed rest, cachexia, and senility, but that considerable changes occur in the muscles of such cases is certain. Until we know more of these matters, the significance of the changes seen in the muscles of patients with carcinoma must remain in doubt. Rebeiz (1970) has studied muscle changes in a number of cancer patients and makes the following correlations.

1. A case of malignant thymoma and a case of gastric carcinoma showed changes resembling polymyositis. The pathology noted were segmental muscle fiber necrosis, phagocytosis, regeneration, and perivascular infiltration with lymphocytes and plasma cells.

2. In cases of lymphomas, leukemias, pancreatic carcinomas, glioblastoma, glioma, breast carcinoma, and seminoma, there was no well defined pathology, but there was a significant decrease in fiber size.

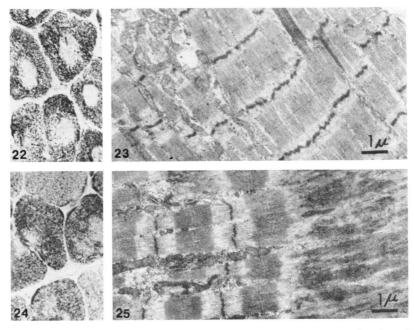

Figs. 22–25. Mitochondrial loss in central core disease (Figs. 22 and 23) and in polymyositis (Figs. 24 and 25). In both cases Z disk streaming occurs in region where the mitochondria are absent.

3. In a number of cases of bronchogenic carcinomas, mammary carcinomas, and carcinomas of stomach, urinary bladder, prostate, parotid gland and tonsils, two types of muscle atrophy were seen: (a) Individual and groups of atrophic fibers varying greatly in diameter and shape are distributed at random. (b) Groups of closely packed fibers scattered among normal or hypertrophied fibers indicate a condition reminiscent of neurogenic atrophy.

The etiology and pathogenesis of these changes is not clear. Rebeiz makes the following comment.

> As far as the generalized muscle cell atrophy is concerned, the chronic debilitating state that occurs in malignant disease is the result of increased catabolism which leads to negative protein balance. This in turn results in increased demand on the protein stores of the body. One of the important stores is the muscle system.

This could easily explain the generalized atrophy of the fibers but does not explain the changes which resemble neurogenic atrophy. However, Rebeiz points out that some cases may, in fact, be due to "dying back of axons," which causes loss of motor neurons. In other cases he says, "in some cases it is neurogenic in origin and is independent of classical motor neuropathy."

B. Periodic Paralysis

The attacks of paralysis, from which the disease is named, are accompanied in most cases by a fall in serum potassium, associated with a decrease in urinary excretion of potassium (i.e., a retention of potassium in the body as a whole). The relationship to potassium metabolism is not, however, a simple one, as serum potassium may not be lowered during attacks (Tyler *et al.*, 1951) and may even be raised (Bull *et al.*, 1953); Conn *et al.* (1957) have shown that sodium retention may be the precipitating factor. Whether periodic paralysis should be considered as one disease or several, the pathology is the same in all, and vacuoles are formed within the muscle fibers (Goldflam, 1897; Tyler *et al.*, 1951; Conn *et al.*, 1957). The fibers may be distended by droplets that resemble in lesser degree those seen in von Gierke's disease; they do not, however, contain glycogen.

The droplets are probably autophagic vacuoles and the membranes from these are probably derived from the T tubule networks of the sarcoplasmic reticulum (A. G. Engel and Macdonald, 1970). This condition pathologically is the one that most commonly shows mitochondrial (tubular) aggregates (Pearse and Johnson, 1970). W. K. Engel *et al.,* 1970), although they found the most abundant aggregates in the muscles of patients with periodic paralysis, found that only a minority of these patients showed tubular aggregates in their muscles. The aggregates are limited to type II muscle fibers. The presence of characteristic double-walled or multivesicular tubes in hypokalemic periodic paralysis was also described by Gruner (1966) and by Odor *et al.* (1967). It was originally thought by W. K. Engel (1964), who first described the condition, and by others that these aggregates were mitochondrial in origin, but later studies, including those by W. K. Engel *et al.,* showed that these tubular aggregates are derived from the sarcoplasmic reticulum.

C. Glycogen Storage Disease

A condition that may cause difficulty in diagnosis because of its remarkable clinical similarity to amyotonia congenita is a form of glycogen storage disease in which there is profound muscular weakness due to infiltration of the skeletal muscle fibers with glycogen. The diagnosis can be made readily by muscle biopsy, but without the aid of this device, it may be impossible to differentiate between the two diseases. Humphreys and Kato (1934) illustrated the pathological changes of the striated muscle fibers in this form of von Gierke's disease. The muscles involved show marked pallor, and histologically, the muscle fibers show a curious form of vacuolation. In an advanced case, the vacuoles, which are filled with glycogen, replace all the sarcoplasm in the fiber, so that only a swollen sarcolemmal sheath is seen. In the lesser degrees of involvement, there may merely be small vacuoles filled with glycogen in the middle of the muscle fibers. The remaining nuclei in badly affected fibers may be pyknotic. R. Günther (1939) found that a case, diagnosed clinically as one of amyotonia congenita, had deposition of glycogen in vacuoles in the skeletal muscles. Di Sant'Agnese *et al.* (1950) found that when glycogen storage disease affects the heart, the skeletal muscles are also often affected. Krivit *et al.* (1953), in an interesting paper on three infants (siblings) all with a flaccid weakness of the skeletal muscles, describe the post mortem findings on two of the children and illustrate the pathological changes in the muscle fibers. Clinically, the three children resembled cases of amyotonia congenita, and these authors stress the necessity for muscle biopsy in the diagnosis of the muscular atrophies of infancy.

D. Polyarteritis Nodosa*

Polyarteritis nodosa is not primarily a muscle disease but is included here since the muscles are probably more commonly involved than any other tissue, and muscle biopsies from suspected cases are often presented for diagnosis. The diagnostic finding is arteritis of an intramuscular arteriole. The vessel wall and perivascular space are infiltrated by acute and chronic inflammatory cells. Fibrinoid necrosis may be demonstrable, and there may be thrombosis. The effects upon the muscle are essentially due to ischemia, which causes either a direct degeneration of muscle fibers or a secondary degeneration owing to the involvement of nerves. Multiple small infarcts are found, with hyaline eosinophilic muscle fibers and also groups of small atrophic muscle fibers resulting from neural atrophy. Later, the dead fibers are removed and a patchy fibrosis of the muscle is seen. Perivascular or interstitial collections of cells may be the only finding, but this alone is not diagnostic. Cöers (1970) has published a photograph showing a grouping of type I fibers (replacing the normal checkerboard distribution) in muscle biopsy taken from a case of periarteritis nodosa with polyneuritis.

E. Thyroid Disease

1. Hyperthyroidism

According to Kissel et al. (1970) "hyperthyroidism affects the motor end plate in a complex manner." Muscular weakness, wasting, and fascicular twitchings occur commonly in thyrotoxicosis, and this muscular weakness may precede other symptoms, though it is not necessarily proportional to the basal metabolic rate. If the muscular symptoms are pronounced enough, they may constitute a myopathy (chronic thyrotoxic myopathy).

Kissel and his colleagues (1970) have described degenerative changes in the muscle fibers and round cell infiltration in this condition. The pathology has however been little studied, but as long ago as 1898, Askanazy reported an infiltration of fat between the muscle fibers and atrophy of the fibers themselves. Adams et al. (1953) found the same changes and we have seen a muscle biopsy that showed a histological picture very similar to their illustrations. The marked increase in fat is interesting, as subcutaneous fat is reduced in thyrotoxicosis. The condi-

* Synonym: Periarteritis nodosa.

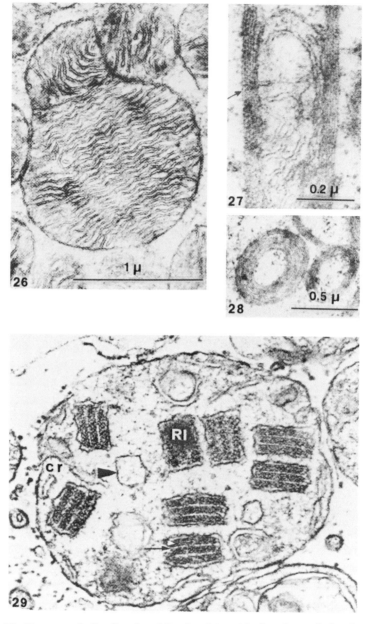

Fig. 26. Hypermetabolic disorder. Mitochondria with densely packed cristae that have zigzag arrangement. From W. K. Engel *et al.* (1970).

Figs. 27 and 28. Hypermetabolic disorder. Mitochondria with concentrically arranged peripheral layers of membranes (see above).

Fig. 29. Megaconial myopathy with rectangular mitochondrial inclusions (R1). White arrowhead indicated osmiophilic rodlike component of inclusion. Black arrow points to 80 Å interspace. Black arrowhead shows membrane-limited rectangle, which appears empty. CR = cristae ($\times$ 70,000) (see above).

tion is known as chronic thyrotoxic myopathy, to distinguish it from acute thyrotoxic myopathy, a very rare clinical syndrome. There is some evidence that thyrotoxicosis and myosthenia gravis are related (see Section VIII). See Figs. 26–28.

2. Hypothyroidism

Enlarged muscles with slow movements may be found in hypothyroidism both in adults and children. In myxedematous adults, the condition is called Hoffman's syndrome (Thomasen, 1948), and in cretins, Debré and Semelaigne's syndrome (1935). In cretins, enlargement of the tongue is a usual feature. In spite of the clinical enlargement of the muscles, little is known of the pathology; hypertrophy of the muscle fibers has not been adequately substantiated. Kissel *et al.* (1965) indicated that there was neuromuscular involvement in hypothyroidism. Other authors have described muscle changes (Scarlato and Spinner, 1967). The accumulation of mucoid (mucopolysaccharide) in the endo- and perimysium and in the endo- and perineurium of peripheral nerves has been reported in cases of myxedema (Avancini and Caccia, 1970). Light microscope study of muscle from hypothyroid myopathy (Godet-Guillain and Fordeau, 1970) does not demonstrate extensive lesions. The diameter of the fibers is variable; a few fibers show vacuoles and some sarcoplasmic masses; there was also a modest increase of the numbers of nuclei; the motor end plates, though variable in size, appeared normal in structure. See Figs. 31, cf. Figs. 32–38.

Under the electron microscope, considerable structural damage was evident in muscle from these cases. Changes were found in myofibrils, mitochondria, and sarcoplasmic reticulum; the peripheral zone of the muscle fibers was particularly affected. "In this area, annular myofibrils could be seen together with laterally placed sarcoplasmic masses, sometimes sequestered between complex membraneous structures." The myofibrils were severely damaged in the affected fibers, many of them being in the form of thin bundles of myofilaments. The Z bands appeared to be especially affected; they were "widened, sometimes with 'streaming' of Z material over the neighbouring sarcomeres; here and there rods were found whose density, repetitive periodicity of 150 Å along their length, and transverse structure were entirely compatible with those described in nemaline myopathy."

Inside the mitochondria, rectangular bodies formed in cristae; dense amorphous inclusions were also found inside these organelles. There was a tendency for the mitochondria to be aggregated around the poles of the subsarcolemmal nuclei. The sarcoplasmic reticulum "showed either

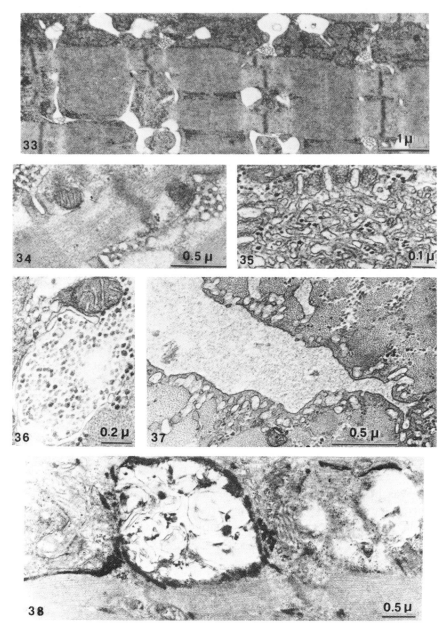

Fig. 33. Dilated anastomosing, and proliferating T tubules. Myxedema myopathy. × 12,800. From A. G. Engel and Macdonald, (1970).

Fig. 34. Proliferating Tubules. Primary hypokalemic paralysis. × 32,000. From A. E. Engel and Macdonald (1970).

case was shown clearly by Russell (1953), though much earlier, Buzzard (1905, 1910) had noted that in addition to lymphorrhages, there were slight degenerative changes in the muscle fibers in this disease, though well marked atrophy of muscles was rare.

Russell (1953) described a necrosis of muscle fibers, with swelling of the fibers and eosinophilia of the sarcoplasm, followed by the development of an inflammatory exudate in and around the fiber. In examples of muscle lesions where lymphorrhages were most prominent, a progressive atrophy of muscle fibers was seen, the nuclei often becoming central. In other specimens, there was simple atrophy of single muscle fibers and groups of fibers. Walton and Adams (1958) think that there is a similarity between the pathological changes described by Russell (1953) and those found in cases of polymyositis. Rowland *et al.* (1956) review some twenty-six postmortem reports on cases of myasthenia gravis, and Woolf *et al.* (1956) describe changes in the intramuscular nerve endings in one case.

Rennie (1908) first suggested there was a relationship between myasthenia gravis and thyrotoxicosis. Subsequently, Cohen and King (1932) also referred to similarities between these two diseases. One of the most striking was the hypertrophy of lymphatic tissue. Occasionally, lymphocytosis can be seen in both diseases, and lymphorrhages are invariably present in both of them. Gunn *et al.* (1964) have indicated that germinal centers may be present in the thymus in both thyrotoxicosis and myasthenia gravis. Despite these similarities, Simpson (1968) states, that myasthenia is only a rare complication of thyroid disease. Simpson also suggested (1960, 1964) that myasthenia gravis was an autoimmune disease. He says, "It has been accepted for a long time that a constitutional or genetic defect may predispose to the development of Graves disease. It was, therefore, suggested that a gene may have a variable expression, producing either thyroid disorder, myasthenia gravis, or both. It could be responsible for altering immunological responses by action on the thymus."

What, precisely, is the function of the muscle antibody in myasthenia gravis is not known, although, according to Simpson, there is "increasing

Fig. 35. Disordered proliferation of T tubules. Dermatomyositis. × 56,800. From A. G. Engel and Macdonald (1970).

Fig. 36. Glycogen sequestration by proliferating T tubule. Pompe's disease. × 43,200. From A. G. Engel and Macdonald (1970).

Fig. 37. Proliferating T tubules open into and merge with membranous boundary of vacuole. Primary hypokalemic periodic paralysis. × 36,600. From A. G. Engel and Macdonald (1970).

Fig. 38. Horseradish peroxidase-filled T networks surround autophagic vacuole. Experimental chloroquine myopathy. × 18,800.

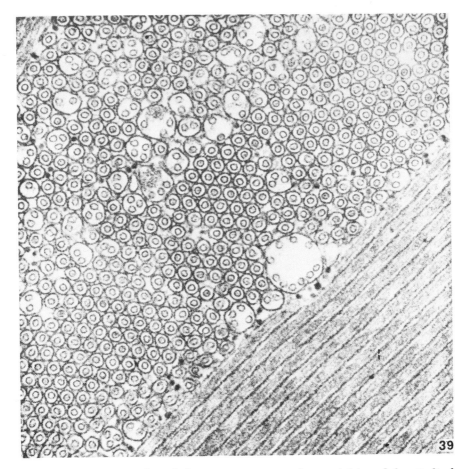

Fig. 39. Detail of the tubules in cross section (upper left) and longitudinal section (lower right). Most have a single concentric inner tubule; some have several inner tubules and a small amount of amorphous material. $\times$ 100,000. From W. K. Engel *et al.* (1970).

recognition of abnormal immunological reaction in thyrotoxicosis." Originally, it was suggested that the antibody might affect the motor end plates, and in fact, some ultrastructural changes have been reported from these organs.

Downes (1968) has discussed in some detail the possible autoimmune nature of myasthenia gravis. He believes one of the most important links is the high incidence of thymic abnormalities in the disease. Also, most autoimmune diseases show accumulation of mononuclear cells in the affected organs, and the lymphorrhages of myasthenia gravis are examples of this accumulation. In this disease, both nonspecific and organ-specific autoantibodies are found in higher proportion is myasthenic patients than in normal people. Downes has also found antibodies to skeletal muscle in 104 out of 203 myasthenia gravis patients examined by him.

Cöers and Demstedt (1959) and Demstedt (1962) have indicated that the lesion in myasthenia gravis lies in the terminal aborization of the motor axon, and they have found that "The myasthenic dysplasia consists of a peculiar elongation and lack of arborization of the terminal part of the motor axon which establishes synaptic contacts with the muscle. The sub-neural apparatus of Couteaux on the post synaptic side of the junction appears to have adapted itself to the geometry of the elongated nerve endings without disclosing obvious changes." There are usually little or no dystrophic changes in the affected muscle. The conclusion is that myasthenia results from defective neuromuscular transmission.

IX. Myoglobinuria

Myoglobinuria occurs either as a spontaneous, paroxysmal disease or "symptomatically" as a result of the rapid destruction of a large quantity of muscle from any cause. Biörck (1949) describes the various conditions in which myoglobinuria is found. In the "crush syndrome," myoglobinuria follows extensive trauma to muscle and leads to anuria (Bywaters, 1944), and a similar condition may result from extensive infarction of muscle (Bywaters and Stead, 1945). It is well known that myoglobinuria may be seen in severe dermatomyositis and that it occurs occasionally in cases of muscular dystrophy (Acheson and McAlpine, 1953).

Idiopathic paroxysmal myoglobinuria is a rare condition in man; it is sometimes familial. In Germany outbreaks of paroxysmal myoglobinuria (Haff disease), thought to be due to eating poisoned fish, have occurred, and the condition has been reviewed by H. Günther (1940). Paroxysmal myoglobinuria is a well-known condition in horses, developing after good feeding and unusual exertion; march "hemoglobinuria" in man may possibly be a related condition. The pathology of idiopathic paroxysmal myoglobinuria has been little studied. Schaar *et al.* (1949) performed a necropsy on a case, and Elek and Andersen (1953) reported a muscle biopsy and discussed the literature. Vuia (1970) described the muscle pathology of a patient who died in

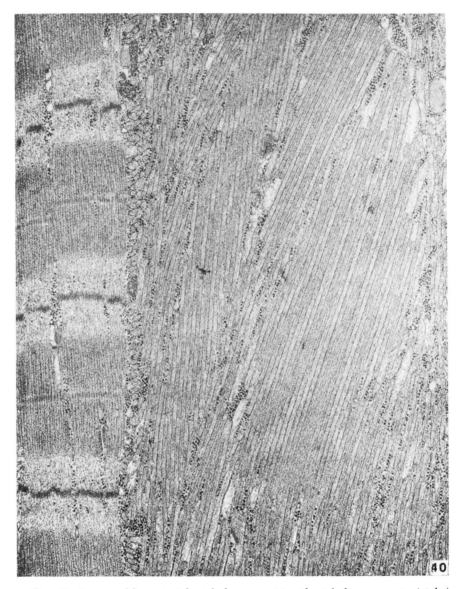

Fig. 40. A mass of long straight tubules comprising the tubular aggregate (right) adjacent to the normal part of the muscle fiber. × 25,000. From W. K. Engel *et al.* (1970).

the course of developing an attack of paralytic paroxysmal myo-globinuria. In the skeletal muscle, he found large eosinphilic fibers with a homogeneous expression accompanied by a loss of myofibrils. Some fibers showed a loss of material from the center of the fiber. The periphery of such fibers appeared swollen, and the sarcoplasm in this region showed radially arranged spicules. Fat droplets were found in the sarcoplasm of the affected fibers. There was no glycogen staining, and only a slight positive mucopolysaccharide reaction. There was a strong metochromatic reaction in the affected fibers. The pathological changes in general appear to consist of acute degeneration and necrosis of the muscle fibers followed by regenerative changes, so that it much resembles polymyositis.

X. Nemaline Myopathy

This condition was first described by Shy *et al.* (1963) in a 4-year-old girl. She had suffered for some time from a "Familial, congenital, non-progressive weakness of the proximal muscles of the extremities" (Gonatas *et al.*, 1966). The muscle fibers were found to contain rod or threadlike bodies, but the fibers were otherwise normal. Ultrastructural studies of these bodies showed them to be fibrillary in nature and to bear transverse bands regularly spaced 145 Å apart. This gave the name "nemaline" to the condition (Greek *nema*, a thread). The nemaline bodies were found to be larger and to occur more frequently near the nucleus and the sarcolemma. Most of them were approximately 5 μ long and 1 μ wide and in some cases had I band filaments 75–100 Å thick inserted into their edges. Under the electron microscope, the nemaline bodies had the same density of the material as the Z disks, and some of these disks showed fusiform expansions. This suggests that the bodies are derived from the Z disks, and Gonatas *et al.* state, "The demonstration of nemaline bodies arising from the Z band gives supporting evidence that an excess of a protein arising from the Z band is the characteristic hallmark of this disease."

Nienhuis and his colleagues (1967) have also described this disease. It is probable that the nemaline bodies, like the Z band, have a high content of tropomyosin B. Nienhuis *et al.* therefore assumed they were rich in tyrosine. According to Fardeau (1969), "using two different methods of fixation (osmium tetroxide and glutaraldehyde–osmium) allowed the comparison of the transverse rod-structure with the two different patterns presented by the Z-bridges on transverse sections: the quadratic lattice of the rods, 180 Å each side, differs from the 'oblique' and larger Z network observed after osmium tetroxide, and resembles the transverse network obtained after aldehyde fixation. Consequently, the hypothesis

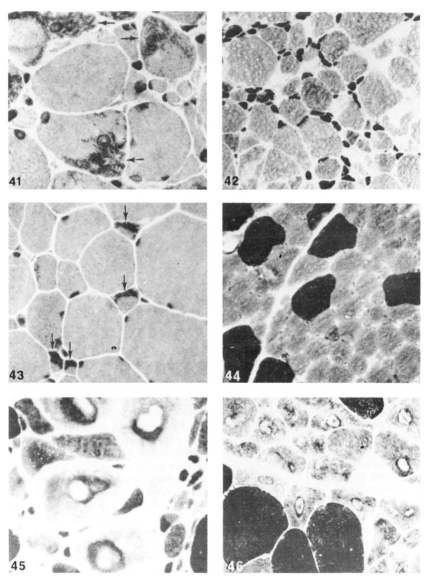

Fig. 41. Rod (nemaline) myopathy (case 1). Rods (dark particles, arrows) are only in large fibers. Modified trichrome; × 480. From W. K. Engel (1970b).

Fig. 42. Rod (nemaline) myopathy (case 1). Only type II fibers (dark) are small. Regular ATPase; × 90. From W. K. Engel (1970b).

Fig. 43. Rod (nemaline) myopathy (case 2). Rods (dark particles, arrows) are only in small fibers. Modified trichrome; × 480. From W. K. Engel (1970b).

was put forward that rods are formed only by one of the two morphological components of the Z band. Furthermore, selective extractions showed that the rods and the Z band are not equally extracted after actin-dissolving solution" (A. G. Engel and Gomen, 1967). Shafiq *et al.* (1967) and Hudgson *et al.* (1967) have also reported cases of nemaline myopathy, but Fardeau points out that there is growing evidence that the formation of nemaline rods is a nonspecific activity. According to W. K. Engel and Resnick (1966) and other authors, they have been seen in muscle in inflammatory or late onset myopathies in toxic myopathy (Newcastle and Humphrey, 1965). W. K. Engel has seen them after tenotomy (W. K. Engel *et al.*, 1966a). Cornag and Gonatas (1967) have seen them in rhabdomyosarcoma, and Fawcett (1968) has even seen them in the normal myocardium. Fardeau (1970) has also seen rods associated with other Z band alterations in chloraquine neuromyopathy. See also W. K. Engel (1970ab) and W. K. Engel *et al.* (1970). See Figs. 41–44; cf. Figs. 45–48.

XI. Myopathy from Local Trauma

Hathaway *et al.* (1969) have described myopathic changes in guinea pig muscle following local trauma. They draw attention to the fact that the process of "needling" muscles will produce focal inflammatory myopathic changes and should be avoided in taking muscle biopsies. In their study, carried out with guinea pigs, they found that the use of a percussion hammer on muscle, combined with vigorous massage, can produce focal pathology. They found the following changes at various times after this treatment.

2 minutes	Hemorrhage between groups and individual muscle fibers
6 hours	In addition to the hemorrhage, there were scattered muscle fibers undergoing necrosis and phagocytosis. Neutrophiles were seen both intra- and extracellularly
9 hours	The same as 6 hours, but all the changes were increased

Fig. 44. Rod (nemaline) myopathy (case 2). Type II fibers (dark) are only large. Regular ATPase; × 190. From W. K. Engel (1970b).

Fig. 45. Four target fibers having broad darkly stained intermediate zones and pale centers only in type I (light) fibers. Peripheral neuropathy. Regular ATPase; × 190. From W. K. Engel (1970b).

Fig. 46. Central core disease. Cores (unstained) are only in type I (light) fibers. Some core fibers have a thin dark intermediate zone. Phosphorylase; × 190. From W. K. Engel (1970b).

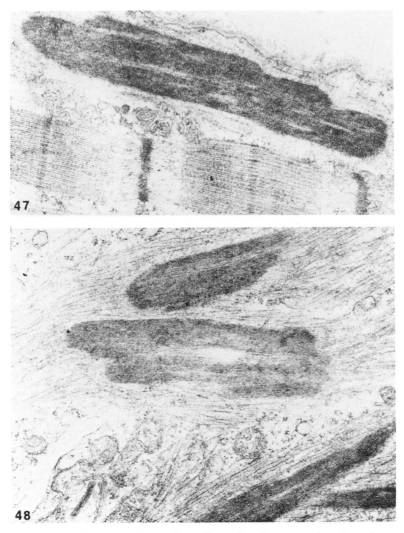

Fig. 47. Nemaline myopathy. Subsarcolemmal rods 3 μ in length with regular transverse striations at intervals of 160 Å. Glutaraldehyde–osmium tetroxide fixative; uranyl acetate and lead citrate stain; × 31,500. From Fardeau (1970).

Fig. 48. Hyperthyromyopathy. Numerous rods in a peripheral zone of myofibrillary disorganization with 150 Å periodicity. Glutaraldehyde–osmium tetroxide fixative; uranyl acetate and lead citrate stain; × 25,500. From Fardeau (1970).

42 hours	Hemorrhage was less obvious, but there were many more necrotic and phagocytosed muscle fibers. There was infiltration around muscle fibers and fiber groups
90 hours	Same as 42 hours, but there was now a reduced number of necrotic

fibers. There was, however, a great increase in mononuclear cell infiltration

1 week The hemorrhage had disappeared. The number of necrotic fibers and the cellular reaction were decreased. "Groups of small, round muscle fibers were present in the middle of collections of mononuclear cells with an increase of endomyseal connective tissue and central nuclei"

2 weeks Except for two small foci of mononuclear infiltration, the muscle appeared normal. The authors point out that the cause of these changes is unknown and suggest that they may be due either to direct damage of the fibers or may have resulted from a temporary loss of vascular supply resulting from the damage to the vessels which caused the hemorrhage.

The authors described the case of a 38-year-old woman suffering from myalgia who, on biopsy, showed some necrotic fibers. Enzyme and electromyographic tests were normal; however, the patient not infrequently received very vigorous massages, and the authors speculate as to the possibility that the necrotic fibers may have been due to this fact. See Figs. 49 and 50.

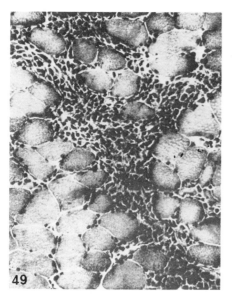 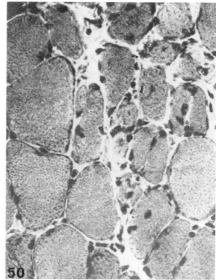

Fig. 49. Biopsy of left gastrocnemius muscle from guinea pig killed 90 hours after trauma. Several necrotic fibers and cellular infiltration of monocytic cells are present. × 300. From Hathaway *et al.* (1969).

Fig. 50. Biopsy of left gastrocnemius muscle from guinea pig killed 1 week after trauma. Note relative diminution of cellular infiltrate, appearance of groups of small round fibers, and increase in central nuclei. × 750. From Hathaway *et al.* (1969).

XII. Myopathy in Animals

Degeneration of the skeletal muscles in animals is not rare. The common muscular degenerations, however, are acquired conditions related to dietetic and environmental factors. In the veterinary literature, these are described as muscular dystrophies, though they bear no resemblance to the conditions known as muscular dystrophy in the human. The animal myopathies which do resemble muscular dystrophies in the human are exceedingly rare.

Michelson *et al.* (1955) have described as dystrophia muscularis a hereditary primary myopathy in the house mouse, and this disease has many of the features of a human muscular dystrophy. The disease develops early in life, beginning in the hindlimbs and running a slowly progressive course to involve the trunk and forelimbs. At necropsy, the nervous system has been found to be normal, while the histological appearance of the muscles is strikingly similar to that seen in a human muscular dystrophy. There is rounding and atrophy of muscle fibers, which have some central nuclei, no evidence of regeneration, and an increase of fibrous interstitial tissue. The illustrations in this paper might well be taken from a human case.

One example of a muscular dystrophy in animals has been extensively studied. This is the congenital myotonia of goats, which was fully described by Kolb in 1938; he considered that the condition was essentially the same as myotonia congenita (Thomsen's disease). Unfortunately, the pathological changes in these animals were not fully studied, but the illustrations show rounding of muscle fibers and central nuclei, suggestive of a dystrophy. A number of physiological studies have been made on these myotonic goats by G. L. Brown and Harvey (1939), who concluded that the myotonia was in every way similar to myotonia in man and that the primary dysfunction was in the muscle fiber.

Innes (1951), in a paper in which he reviews what little is known of myopathies in animals, describes a case of possible muscular dystrophy affecting the gastrocnemii of a dog, where the histological changes were typical of a true muscular dystrophy in man. Ziegler (1929) described a somewhat similar condition in a dog with generalized disease. Bosanquet *et al.* (1956) have described a myopathy occurring sporadically in adult sheep under good nutritional conditions. The condition appears to be fairly common, as they found forty-five cases. Affected muscles were distributed all over the body, and histologically, the changes resembled a polymyositis, though the chronicity of the condition

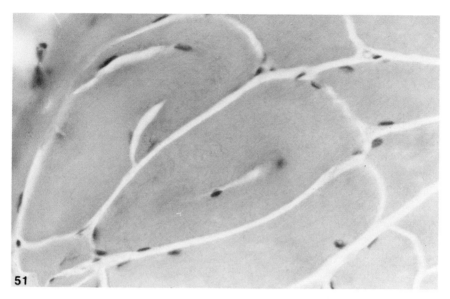

Fig. 51. Muscle from thigh, chimpanzee "Todd," suffering from a polyneuritis, showing splitting of fibers.

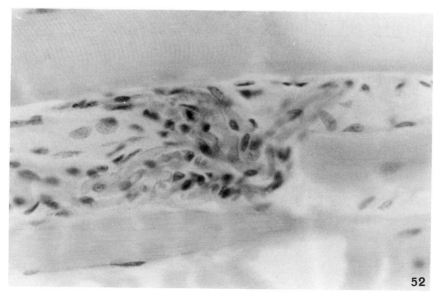

Fig. 52. Preparation from diaphragm of rhesus monkey (*Macaca mulatta*) showing evidence of muscle destruction and regeneration (myotubes).

Fig. 53. Preparation from diaphragm of rhesus monkey showing fat and cellular and connective tissue invasion.

and the possibility of an inherited factor are thought to be more suggestive of a dystrophy.

It was pointed out earlier that one of several etiological factors may induce the condition of arthrogryposis multiplex congenita in the human. This is probably true in animals also, for Whittem (1957) has suggested that in calves it has a neurogenic etiology, while one form of the condition in sheep is thought to be myogenic (Middleton, 1934).

Paralytic myoglobinuria of horses is a well known condition. It is a paroxysmal disease that follows severe exertion after rest. At necropsy, the muscles are salmon pink in color and histologically show acute hyaline degeneration and necrosis of muscle fibers (Carlström, 1931; Minett, 1935). Myoglobinuria is also known to occur in various animals suffering from severe "dystrophy," where it is apparently secondary to the muscle degeneration and comparable with "symptomatic" myoglobinuria in human polymyositis. Myoglobinuria that follows transportation or driving may however possibly be related to the paroxysmal disease of horses. The pathology in all these conditions is that of a nonspecific acute degeneration of muscle.

A number of animal diseases are described in the literature as muscular dystrophies, but these do not resemble, either pathologically or clini-

Fig. 54. Preparation from thigh muscle of rhesus monkey showing cellular infiltration.

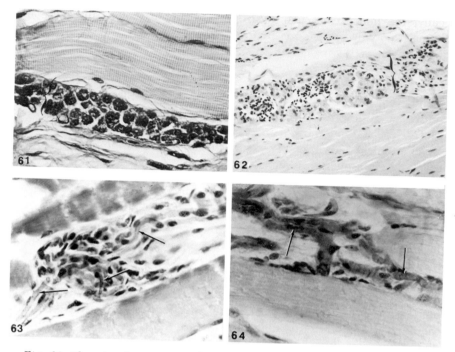

Fig. 61. Floccular degeneration of a muscle fiber. The floccular masses stain intensely with PAS. *Macaca mulatta;* PAS stain; × 400.

Fig. 62. From the same lesion as Fig. 59, flocculation with cellular infiltration and phagocytosis of a muscle fiber. *Macaca mulatta;* hematoxylin and eosin; × 160.

Fig. 63. Muscle regeneration in an isolated animal; evidence of myotube formation, their cross striations (arrows) could be observed under the microscope by slightly varying the focus. *Macaca mulatta;* diaphragm muscle; hematoxylin and eosin; × 320.

Fig. 64. PAS-positive myotubes in an area of muscle regeneration. *Macaca mulatta;* diaphragm muscle; PAS stain, × 480.

changes were found were the galago (*Galago crassicaudatis*), Celebes black ape (*Cynopithecus niger*), chimpanzee (*Pan troglodytes*), gorilla (*Gorilla gorilla*) (see Figs. 75–78), Java macaque (*Macaca fascicularis*), drill (*Mandrillus leucophaeus*), orangutan (*Pongo pygmaeus*), pigtail (*Macaca nemestrina*), rhesus macaque (*Macaca mulatta*) (see Figs. 52–64 and 68–73), squirrel monkey (*Saimiri boliviensis nigriceps* and *Saimiri sciureus*), potto (*Perodicticus potto*), slow loris (*Nycticebus coucang*), spider monkey (*Ateles geoffroyi*), owl monkey (*Aotus trivirgatus*), and mandrill (*Mandrillus sphinx*). The majority of these animals were wild born; a few were born in captivity. The rhesus monkey was the only species of which enough specimens were available to establish that animals born and raised under optimum laboratory conditions show a lesser degree of pathology than wild-born imported animals. Practically all wild-born animals seem to show some muscle pathology.

In the rhesus monkey, individual muscle samples from these animals have shown at least one of the following pathological entities discussed below. Necrosis of individual and groups of muscle fiber with and without regeneration of the fibers was found in several animals. The regeneration process was evident from the formation of myotubes, formed of myoblasts with large, pale, globular nuclei. In some of the preparations, faint cross striations could be seen in some of these myotubes. In several specimens, necrotic fibers were being heavily phagocytized. In some specimens, fine reticular fibers were seen extending over the degenerating fibers—the first stage of connective tissue replacement.

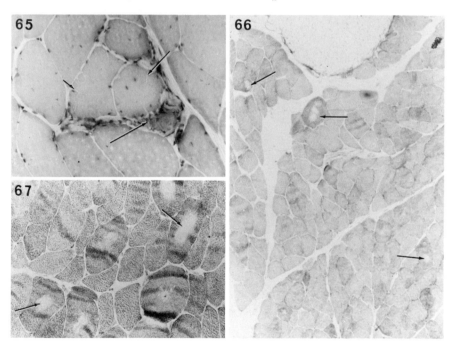

Fig. 65. Central muscle nuclei; fiber necrosis and "splitting" can be observed in several fibers. Gomori's technique for alkaline phosphatase. *Pan troglodytes;* deltoid muscle; substrate adenosine diphosphate; × 400.

Fig. 66. Several targetoid fibers (arrows) from the leg which showed muscular weakness in a chimpanzee with peripheral polyneuritis. *Pan troglodytes;* thigh muscle; reaction for DPN diaphorase; × 75.6.

Fig. 67. Several muscle fibers show zonal differential staining described in the literature as targetoid fibers. Same specimen as above. *Pan troglodytes;* thigh muscle; reaction for DPN diaphorase; × 128.

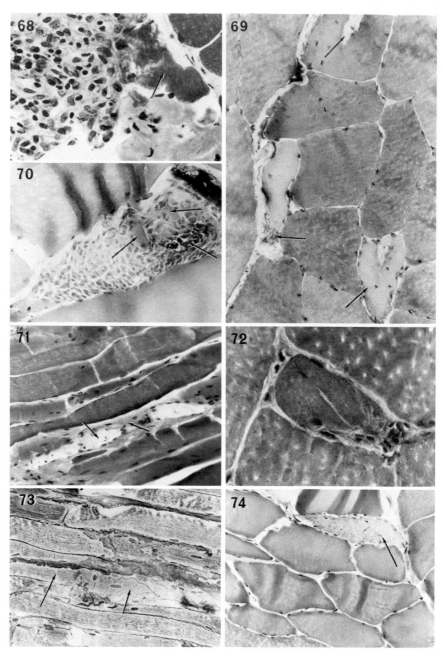

Fig. 68. One muscle fiber with a "bursting" of necrosis and phagocytosis. *Macaca mulatta;* diaphragm muscle; hematoxylin and eosin × 320.

Some specimens showed muscle fibers with vacuoles in the center of the fiber, and others, narrow splits extending from the periphery of the fiber to the center. In some cases, it appeared that such splitting may have been preceded by the development of a central vacuole. In other specimens, these splits were widened and invaded by connective tissue. In some cases, the vacuolation of the fiber was accompanied by flocculation and invasion by phagocytes. Several specimens showed great variation in size and shape of individual muscle fibers. In some animals, nuclear changes were seen; central migration of nuclei and rowing of nuclei were also present.

In one specimen, a number of fibers showed a concentration of lactic dehydrogenase activity in the center of the fiber, indicating that there may be a concentration of mitochondria in this area. It is of interest that W. K. Engel and Meltzer (1970) have shown the accumulation of darkly staining rod-shaped particles in the center of muscle fibers in some psychotic patients, though they make no claim that these are mitochondria.

One rhesus monkey showed an infection of muscle fibers with sarcosporidia. These have been seen before in many animals, including monkeys. Sarcosporidia were also seen in a skeletal muscle biopsy of one of our gorillas.

Fig. 69. A muscle bundle still maintaining some normal architecture. While one of the fibers is undergoing hyalinization and phagocytosis (arrow), two other fibers are undergoing invasion by the proliferating endomysium; migrated nuclei are observed in the three affected fibers. Gomori's technique for alkaline phosphatases was applied, which shows a positive reaction in the capillaries and nuclei present. The horseshoe appearance of the connective tissue invasion is not common in the author's experience. *Macaca mulatta;* lumbar muscle; substrate adenosine 5'-triphosphate; × 224.

Fig. 70. A single muscle fiber undergoing near total destruction, at center necrosis (lower arrow). Above it are remnants of the muscle fiber (arrows), which can be seen also at the lower left angle. Artificially produced contraction bands at the top fiber and ATP-positive blood vessel to the right. *Macaca mulatta;* lumbar muscle; Gomori's technique for alkaline phosphatase; × 224.

Fig. 71. Centrally located nuclei, some atrophic fibers and, in one fiber, simultaneous hyaline and floccular degeneration (arrows). *Macaca mulatta;* diaphragm muscle; hematoxylin and eosin; × 128.

Fig. 72. Splitting of a fiber and phagocytosis. *Macaca mulatta;* lumbar muscle; Gomori's alkaline phosphatase plus hematoxylin and eosin; × 480.

Fig. 73. In the same area as the section in Fig. 70, reticular fibers can be seen at right arrow in the site of connective tissue invasion of a muscle fiber still partly intact (left arrow). *Macaca mulatta;* diaphragm muscle; Wilder's reticulum stain; × 192.

Fig. 74. Floccular degeneration in a single fiber in the same muscle as above. *Macaca mulatta;* lumbar muscle; hematoxylin and eosin; × 160.

One of the gorillas we studied extensively was a 7-year-old female. At death, she had marked atrophy of the limbs on the left side. The clinical diagnosis of her condition was that of infantile spastic hemiplegia and intractable cerebral seizures. The condition had continued progressively over the last 5 years. The animal died during surgical procedures unrelated to her condition. Most of the muscles observed on the affected side showed extensive changes, and even those fasciculi that were architecturally normal presented fibers with central nuclei. The most common changes observed were small fibers with rounded contours and complete loss of normal polygonal architecture. There was vacuolar degeneration without cellular infiltration or phagocytosis. The contents of the fibers were in various stages of disintegration, some with myoplasm contracted away from the sarcolemma and even some empty sarcolemmal sheaths could be observed.

In muscle taken from thirty-three chimpanzees, the whole range of pathological changes could be seen, though not all in any one animal, and mostly in isolated foci. With the exception of one, these animals showed no clinical symptoms of muscle disease. In one animal, which showed no clinical symptoms, the deltoid muscle showed great variation in muscle fiber diameter, and there were a large number of swollen fibers interspersed with small angular fibers. In some of the fascicles, the swollen fibers showed the beginnings of formation of fissures. Sporadic necrotic muscle fibers were also found in this specimen, but central migration of muscle fiber nuclei was found in nearly all the fibers.

One of the chimpanzees studied showed weakness of the hind legs, especially on one side. This became progressively worse over some months and then slowly began to improve. The vastus lateralis and the gastrocnemius muscle were especially studied in this animal. The overall architecture of the muscle showed little change. With standard histological techniques, some swollen fibers and a rare angular necrotic fiber were found, but many of the fibers showed a zonal differential staining; some fibers were deeply fissioned with endomysial penetration. Histochemical studies for oxidative enzymes showed the reaction to be restricted to the periphery of the fibers, with the center giving little or no reaction. There was obviously loss of normal structure in the center of these fibers, some of which also had a centrally located nucleus. (Figs. 51, 65, 66, and 67.) This kind of structural change in the muscle fiber, giving it a target or targetoid appearance, is considered a sign of denervation in some neuromyopathies. There is also a rare human muscle disease called central core disease, which presents similar zonal irregularities in the individual muscle fibers. This disease is characterized by a nonprogressive muscular weakness. Sixty days after the original biopsy, electro-

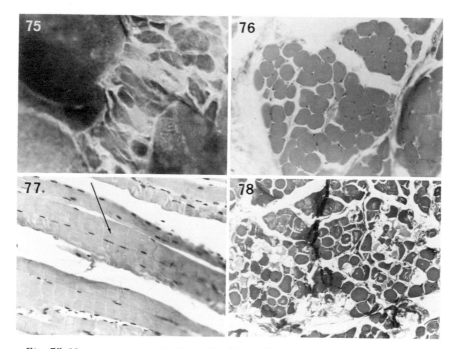

Fig. 75. Necrosis in a muscle fiber. *Gorilla gorilla;* gastrocnemius muscle; trichrome stain; × 1000.

Fig. 76. In a cross section, all the muscle fibers show complete loss of the usual polizonal architecture. *Gorilla gorilla;* thigh muscle; hematoxylin and eosin; × 160.

Fig. 77. Centrally located nuclei in an otherwise normal muscle, in longitudinal section. *Gorilla gorilla;* arm muscle; trichrome stain; × 160.

Fig. 78. Extensive vacuolar degeneration without cellular infiltration or phagocytosis. Empty sheaths can be seen, some with remnants of myoplasm and nuclei, others at different stages of loss of contents. *Gorilla gorilla;* thigh muscle; hematoxylin and eosin; × 63.

myographic studies suggested that the condition was peripheral polyneuritis of unknown etiology, although there was some evidence that it might have had a traumatic origin. Subsequently, the animal improved clinically and biopsies taken at the time of the electromyography showed no target or targetoid fibers. No necrosis, cellular infiltration, or connective tissue proliferation were found in the biopsies of this animal.

We have no explanation for the appearance of fissures and splitting of the muscle fibers in the muscles of two of the chimpanzees studied. Some publications in recent years describe fiber splitting as a result of hypertrophy following periodic strenuous work; however, there was no record of any unusual muscular exercise in our animals.

The changes found in squirrel monkey muscles were so extensive in some cases that it is difficult to accept that many of these affected animals could function competently. It is of interest that squirrel monkeys readily reach a stage of chronic stress in which cortisol levels in the blood would produce Cushing's disease in man (G. M. Brown *et al.*, 1970). Myopathic changes are known to occur in humans with Cushing's disease, and perhaps the muscle changes found so commonly in this species of monkey may be a steroid myopathy.

One instance of these nonclinical myopathic changes is of interest and in general cannot be accounted for. A few years ago, M. N. Golarz de Bourne and G. H. Bourne made a study of muscle biopsies from a group of young, healthy medical students (unpublished). One of these students showed extensive muscles changes, including a number of those described in this chapter, but there were no clinical symptoms. It was not established whether this student was related to any patient with hereditary muscular disease, but we saw him again some years later, and although he was fatter, he seemed to be in normal health.

ACKNOWLEDGEMENTS

The work by G. Bourne and M. N. Golarz de Bourne described in this chapter has been carried out under NASA grant number NGR 11-001-016 and grant number RR 00165 of the division of Research Resources National Institutes of Health.

Figures 65–78 are taken from an article by M. Golarz de Bourne and G. Bourne (1972). *In* "The Pathology of Simian Primates" (R. Fiennes, ed.), Vol. I. Karger, Basel.

REFERENCES

Acheson, D., and McAlpine, D. (1953). *Lancet* **2**, 372.

Adams, R. D., Denny-Brown, D., and Pearson, C. M. (1953). "Diseases of Muscle, A Study in Pathology." Cassell, London.

Adie, W. J., and Greenfield, J. G. (1923). *Brain* **46**, 73.

Aloise, M., and Margreth, A. (1967). *In* "Exploratory Concepts in Muscular Dystrophy" (A. T. Milhorat, ed.), p. 305. Excerpta Med. Found., Amsterdam.

Alström, I. (1948). *Skand. Veterinaertidskr.* **38**, 593.

Anderson, P. J., Slotwiner, P., and Song, S. K. (1967). *J. Neurol.* **26**, 15.

Askanazy, M. (1898). *Deut. Arch. Klin. Med.* **61**, 118.

Avancini, G., and Caccia, M. R. (1970). *In* "Muscle Diseases" (J. N. Walton, N. Canal, and G. Scarlato, eds.), p. 521. Excerpta Med. Found., Amsterdam.

Banker, B. Q., Victor, M., and Adams, R. D. (1957). *Brain* **80**, 319.

Barnes, S. (1932). *Brain* **55**, 1.

Batten, F. E. (1910a). *In* "A System of Medicine" (C. Allbutt and H. D. Rolleston, eds.), Vol. 7, p. 31. Macmillan, New York.

Batten, F. E. (1910b). *Quart. J. Med.* **3**, 313.

Becker, P. E. (1953). "Dystrophia Musculorum Progressiva." Thieme, Stuttgart.

Beckman, R., Klohe, W. D., and Freund Mölbert, E. R. G. (1970). In "Muscle Diseases" (J. N. Walton, N. Canal, G. Scarlato, eds.), p. 438. Excerpta Med. Found., Amsterdam.

Bedivan, M. (1970). In "Muscle Diseases" (J. N. Walton, N. Canal, and G. Scarlato, eds.), p. 499. Excerpta Med. Found., Amsterdam.

Bell, J. (1943). In "The Treasury of Human Inheritance" (R. A. Fisher, ed.), Vol. 4, Part 4, p. 283. Cambridge Univ. Press, London and New York.

Bell, J. (1947). In "The Treasury of Human Inheritance" (R. A. Fisher, ed.), Vol. 4, Part 5, p. 343. Cambridge Univ. Press, London and New York.

Bevans, M. (1945). *Arch. Pathol.* **40**, 225.

Bielschowsky, M., Maas, O., and Ostertag, B. (1933). In "Volume Jubilaire Marinesco," p. 71. Marvan, Bucarest.

Bing, R. (1926). In "Handbuch der inneren Medizin" (G. von Bergmann and R. Staehelin, eds.), 2nd ed., Vol. 5, Part 2, p. 1154. Springer-Verlag, Berlin and New York.

Biörck, G. (1949). *Acta Med. Scand.* **133**, Suppl. 226, 24.

Black, W. C., and Ravin, A. (1947). *Arch. Pathol.* **44**, 176.

Blaxter, K. L., and McGill, R. F. (1955). *Vet. Rev. Annotations* **1**, 91.

Bosanquet, F. D., Daniel, P. M., and Parry, H. B. (1956). *Lancet* **2**, 737.

Bouman, H. D. (1955). *Amer. J. Phys. Med.* **34**, 1–324.

Bourne, G. H., and Golarz, M. N. (1963a). "Muscular Dystrophy in Man and Animals." Karger, Basel.

Bourne, G. H., and Golarz, M. N. (1963b). *J. Histochem. Cytochem.* **11**, 286.

Brain, W. R., and Turnbull, H. M. (1938). *Quart. J. Med.* [NS] **7**, 293.

Brooke, M. H., and Engel, W. K. (1969). *Neurology* **19**, 591.

Brown, G. L., and Harvey, A. M. (1939). *Brain* **62**, 341.

Brown, G. M., Grota, T. J., Penney, D. P., and Reichlin, S. (1970). *Can. Psychiat. Ass. J.* **15**, 425.

Bull, G. M., Carter, A. B., and Lowe, K. G. (1953). *Lancet* **2**, 60.

Burch, F. E. (1929). *Minn. Med.* **12**, 668.

Buzzard, E. F. (1905). *Brain* **28**, 438.

Buzzard, E. F. (1910). In "A System of Medicine" (C. Allbutt and H. D. Rolleston, eds.), Vol. 7, p. 50. Macmillan, New York.

Bywaters, E. G. L. (1944). *J. Amer. Med. Ass.* **124**, 1103.

Bywaters, E. G. L., and Stead, J. K. (1945). *Clin. Ser.* **5**, 195.

Carlström, B. (1931). *Skand. Arch. Physiol.* **61**, 161; **62**, 1.

Cöers, C. (1970). In "Muscle Diseases" (J. N. Walton, N. Canal, and G. Scarlato, eds.), p. 365. Excerpta Med. Found., Amsterdam.

Cöers, C., and Demstedt, J. E. (1959). *Neurology* **9**, 238.

Cöers, C., and Pelc, S. (1954). *Acta Neurol. Psychiat. Belg.* **54**, 166.

Cöers, C., and Woolf, A. L. (1959). "The Innervation of Muscle." Blackwell, Oxford.

Cohen, S. J., and King, F. H. (1932). *Arch. Neurol. Psychiat.* **28**, 1338.

Conn, J. W., Louis, L. H., Fajans, S. S., Streeten, D. H. P., and Johnson, R. D. (1957). *Lancet* **1**, 802.

Cooper, S., and Daniel, P. M. (1949). *Brain* **72**, 1.

Cooper, S., Daniel, P. M., and Whitteridge, D. (1955). *Brain* **78**, 564.

Cornag, J. F., and Gonatas, N. K. (1967). *J. Ultrastruct. Res.* **20**, 433.

Councilman, W. T., and Dunn, C. H. (1911). *Amer. J. Dis. Child.* **2**, 340.

Curschmann, H. (1936). *In* "Handbuch der Neurologie" (O. Bumke and O. Foerster, eds.), Vol. 16, p. 431. Springer-Verlag, Berlin and New York.

Daniel, P. (1946). *J. Anat.* **80**, 189.

Debré, R., and Semelaigne, G. (1935). *Amer. J. Dis. Child.* **50**, 1351.

De Giacomo, P., Buscaino, G. A., and Perniola, T. (1970). *In* "Muscle Diseases" (J. N. Walton, N. Canal, and G. Scarlato, eds.), p. 61. Excerpta Med. Found., Amsterdam.

Dejerine, J., and Sottas, J. (1895). *Rev. Med. (Paris)* **15**, 241.

Démos, J. (1961). *Ref. Fr. Etud. Clin. Biol.* **6**, 876.

Démos, J., Place, T., and Chereau, H. (1970). *In* "Muscle Diseases" (J. N. Walton, N. Canal, R. Scarlato, eds.), p. 408. Excerpta Med. Found., Amsterdam.

Demstedt, J. E. (1962). *In* "Neurochemistry" (K. A. C. Elliott, I. H. Page, and J. H. Quastel, eds.). Thomas, Springfield, Illinois.

Denny-Brown, D. Personal communication.

Denny-Brown, D. (1948). *J. Neurol., Neurosurg. Psychiat.* **11**, 73.

Denny-Brown, D. (1962). *Rev. Can. Biol.* **21**, 507.

De Vivo, D. C., and Engel, W. K. (1970). *J. Neurol., Neurosurg. Psychiat.* **33**, 62.

Di Sant'Agnese, P. A., Andersen, D. H., and Mason, H. H. (1950). *Pediatrics* **6**, 607.

Dobyns, B. M. (1950). *J. Clin. Endocrinol.*, **10**, 202.

Domzalski, C. A., and Morgan, V. C. (1955). *Amer. J. Med.* **19**, 370.

Dowling, G. B. (1955). *Brit. J. Dermatol.* **67**, 275.

Downes, J. M. (1968). *In* "Research in Muscular Dystrophy" (Research Committee of The Muscular Dystrophy Group, eds.), p. 93. Pitman, London.

Dubowitz, V. (1970). *In* "Muscle Diseases" (J. N. Walton, N. Canal, and G. Scarlato, eds.), p. 568. Excerpta Med. Found., Amsterdam.

Duchenne, G. B. (1868). *Arch. Gen. Med.* [4] **11**, 5, 179, 305, 421, and 552.

Duchenne, G. B. (1872). "De l'electrisation localisée et son application a la pathologie et à la thérapeutique." Baillière, Paris.

Durante, G. (1902). *In* "Manuel d'histologie pathologique" (V. Cornil and L. Ranvier, eds.), 3rd ed., Vol. 2, p. 1. Alcan, Paris.

Elek, S. D., and Anderson, H. F. (1953). *Brit. Med. J.* **2**, 533.

Engel, A. G., and Gomen, M. R. (1967). *J. Neuropathol. Exp. Neurol.* **26**, 601.

Engel, A. G., and MacDonald, R. D. (1970). *In* "Muscle Diseases" (J. N. Walton, N. Canal, and G. Scarlato, eds.), p. 71. Excerpta Med. Found., Amsterdam.

Engel, W. K. (1961). *Nature (London)* **191**, 389.

Engel, W. K. (1962). *Neurology* **12**, 778.

Engel, W. K. (1964). *J. Histochem. Cytochem.* **12**, 46.

Engel, W. K. (1970a). *Science* **168**, 273.

Engel, W. K. (1970b). *Arch. Neurol.* **22**, 97.

Engel, W. K., and Brooke, M. H. (1966). *In* "Progressive Muskeldystrophie, Myotonie, Myasthenie" (E. Kuhn, ed.), p. 203. Springer-Verlag, Berlin and New York.

Engel, W. K., and Meltzer, H. (1970). *Science* **68**, 273.

Engel, W. K., and Resnick, J. S. (1966). *Neurology* **16**, 305.

Engel, W. K., McFarlin, D. E., Daws, G., and Wochner, R. D. (1966a). *J. Amer. Med. Ass.* **195**, 754 and 837.

Engel, W. K., Brooke, M. H., and Nelson, P. G. (1966b). *Ann. N.Y. Acad. Sci.* **138**, 160.

Engel, W. K., Gold, G. N., and Karpati, G. (1968). *Arch. Neurol.* 18, 435.

Engel, W. K., Bishop, D. W., and Cunningham, G. C. (1970). *J. Ultrastruct. Res.* 31, 507.

Erb, W. H. (1884). *Duet. Arch. Klin. Med* 34, 467.

Erb, W. H. (1886). "Die Thomsen'sche Krankheit (Myotonia congenita)." Vogel, Leipzig.

Erb, W. H. (1891). *Deut. Z. Nervenheilk.* 1, 13 and 173.

Erb, W. H. (1894). *In* "Clinical Lectures on Subjects Connected with Medicine and Surgery" (by various German authors), 3rd ser., p. 231. New Sydenham Society, London.

Eulenburg, A. (1886). *Neurol. Zentralbl.* 5, 265.

Fardeau, M. (1969). *Acta Neuropathol.* 13, 250.

Fardeau, M. (1970). *In* "Muscle Diseases" (J. N. Walton, N. Canal, and G. Scarlato, eds.), p. 98. Excerpta Med. Found., Amsterdam.

Fawcett, D. W. (1968). *J. Cell Biol.* 36, 266.

Feinstein, B., Lindegard, B., Nyman, E., and Wohlfart, G. (1955). *Acta Anat.* 23, 127.

Frauchiger, E., and Fankhauser, R. (1957). "Vergleichende Neuropathologie des Menschen und der Tiere." Springer-Verlag, Berlin and New York.

Giacanelli, M., and Perciaccante, G. (1970). *In* "Muscle Diseases" (J. N. Walton, N. Canal, and G. Scarlato, eds.), p. 65. Excerpta Med. Found., Amsterdam.

Gilbert, R. K., and Hazard, J. B. (1965). *J. Pathol. Bacteriol.* 89, 503.

Gilmour, J. R. (1946). *J. Pathol. Bacteriol.* 58, 675.

Godet-Guillain, J., and Fardeau, M. (1970). *In* "Muscle Diseases" (J. N. Walton, N. Canal, and G. Scarlato, eds.), p. 512. Excerpta Med. Found., Amsterdam.

Golarz, M. N., and Bourne, G. H. (1962). *Exp. Cell Res.* 25, 691.

Golarz de Bourne, M. N., and Bourne, G. H. (1971). *In* "Pathology of Primates" (R. N. Fiennes, ed.), p. 520. Karger, Basel.

Goldflam, S. (1897). *Duet. Z. Nervenheilk.* 11, 242.

Gonatas, N. K., Shy, G. M., and Godfrey, E. H. (1966). *N. Engl. J. Med.* 274, 535.

Gowers, W. R. (1879). Pseudo-hypertrophic Muscular Paralysis." Churchill, London.

Gowers, W. R. (1902). *Brit. Med. J.* 2, 89.

Greenfield, J. G. (1911). *Rev. Neurol. Psychiat.* 9, 169.

Greenfield, J. G., Shy, G. M., Alvord, E. C., and Berg, L. (1957). "An Atlas of Muscle Pathology in Neuromuscular Diseases." Livingstone, Edinburgh.

Gruner, J. E. (1963). *C. R. Soc. Biol.* 157, 181.

Gruner, J. E. (1966). *C. R. Soc. Biol.* 160, 193.

Gunn, A., Michie, D., and Irvine, W. J. (1964). *Lancet* 2, 776.

Günther, H. (1940). *Ergeb. Inn. Med. Kinderheilk.* 58, 331.

Günther, R. (1939). *Virchows Arch. Pathol. Anat. Physiol.* 304, 87.

Halban, J. (1894). *Anat. Hefte, Arb. Anat. Inst., Wiesbaden* 3, 267.

Hammond, J. (1932). "Growth and the Development of Mutton Qualities in the Sheep." Oliver & Boyd, Edinburgh.

Hartley, W. J., and Dodd, D. C. (1957). *N. Z. Vet. J.* 5, 61.

Hathaway, P. W., Dahl, D. S., and Engel, W. K. (1969). *Arch. Neurol.* 21, 355.

Hathaway, P. W., Engel, W. K., and Zellweger, J. (1970). *Arch. Neurol.* 22, 365.

Heathfield, K. W. G., and Williams, J. R. B. (1954). *Brain* 77, 122.

Heidenhain, M. (1918). *Beitr. Pathol. Anat. Allg. Pathol.* 64, 198.

Henson, R. A., Russell, D. S., and Wilkinson, M. (1954). *Brain* **77**, 82.

Hník, P. (1960). *Experientia* **16**, 326.

Hudgson, P., Pearce, G. W., and Walton, J. N. (1967). *Brain* **90**, 565.

Humphreys, E. M., and Kato, K. (1934). *Amer. J. Pathol.* **10**, 589.

Hurwitz, S. (1936). *Arch. Neurol. Psychiat.* **36**, 1294.

Ingram, J. T., and Stewart, M. J. (1934). *Brit. J. Dermatol. Syph.* **46**, 53.

Innes, J. R. M. (1951). *Brit. Vet. J.* **107**, 131

Ionasescu, V., Luca, N., and Vuia, O. (1970). *In* "Muscle Diseases" (J. N. Walton, N. Canal, and G. Scarlato, eds.), p. 246. Excerpta Med Found., Amsterdam.

Jendrassik, E. (1911). *In* Handbuch der Neurologie" (M. Lewandowsky, ed.), Vol. 2, Part 1, p. 321. Springer-Verlag, Berlin and New York.

Kakulas, B. A. (1970). *In* "Muscle Diseases" (J. N. Walton, N. Canal, and G. Scarlato, eds.), p. 377. Excerpta Med. Found., Amsterdam.

Kakulas, B. A., Papadimitrou, J. M., Knight, J. O., and Mataglia, F. L. (1968). *Proc. Aust. Ass. Neurol.* **5**, 79.

Kiloh, L. G., and Nevin, S. (1951). *Brain* **74**, 115.

Kinney, T. D., and Maher, M. M. (1940). *Amer. J. Pathol.* **16**, 561.

Kissel, P., Hartemann, P., and Duc, M. (1965). "Les syndromes myothyroïdiens." Masson, Paris.

Kissel, P., Schmitt, J., Duc, M., and Duc, M. L. (1970). *In* "Muscle Diseases" (J. N. Walton, N. Canal, and G. Scarlato, eds.), p. 464. Excerpta Med. Found., Amsterdam.

Kolb, L. C. (1938). *Bull. Johns Hopkins Hosp.* **63**, 221.

Krivit, W., Polglase, W. J., Gunn, F. D., and Tyler, F. H. (1953). *Pediatrics* **12**, 165.

Landouzy, L., and Dejerine, J. (1884). *C. R. Acad. Sci.* **98**, 53.

Landouzy, L., and Dejerine, J. (1885a). "De la myopathie atrophique progressive; myopathie héréditaire sans neuropathie, débutant d'ordinaire dans l'enfance par la face." Alcan, Paris.

Landouzy, L., and Dejerine, J. (1885b). *Rev. Med.* (*Paris*) **5**, 81 and 253.

Landouzy, L., and Dejerine, J. (1886). *Rev. Med.* (*Paris*) **6**, 977.

Landouzy, L., and Lortat-Jacob, L. (1909). *Presse Med.* **17**, 145.

Lapresle, J., and Fardeau, M. (1968). *Acta Neuropathol.* **10**, 105.

Lee, J. C., and Altschul, R. (1963). *Z. Zellforsch. Mikrosk. Anat.* **61**, 168.

Lerebouillet, G., and Baoudouin, A. (1909). *Bull. Mem. Soc. Med. Hop. Paris* [3] **27**, 1162.

Levison, H. (1951). *Acta Psychiat. Neurol. Scand., Suppl.* **76**, 1.

Liversedge, L. A., and Newman, M. J. D. (1956). *Brain* **79**, 395.

Maas, O. (1937). *Brain* **60**, 498.

Maas, O., and Paterson, A. S. (1939). *Brain* **62**, 198.

Maas, O., and Paterson, A. S. (1947). *Monatsschr. Psychiat. Neurol.* **113**, 79.

Maas, O., and Paterson, A. S. (1950). *Brain* **73**, 318.

McArdle, B. (1967). *In* "Exploratory Concepts in Muscular Dystrophy" (A. T. Milhorat, ed.), p. 334. Excerpta Med. Found., Amsterdam.

Marinesco, G. (1910). In "Nouveau triaté de médecine et de thérapeutique" (P. Brouardel, A. Gilbert, and L. Thoinot, eds.), Part 38, p. 1. Baillière, Paris.

Mataglia, F. L., Papadimitrou, J. M., and Kakulas, B. A. (1969). *Proc. Aust. Ass. Neurol.* **6**, 95.

Mendell, J. R., Engel, W. K., and Derrer, E. C. (1971). *Science* **172**, 1143.

Menges, O. (1931). *Deut. Z. Nervenheilk.* **121**, 240.

Meryon, E. (1852). *Med. Chir. Trans.* **35**, 73.

Meryon, E. (1864). "Practical and Pathological Researches on the Various Forms of Paralysis." Churchill, London.

Michelson, A. M., Russell, D. S., and Harman, P. J. (1955). *Proc. Nat. Acad. Sci. U.S.* **41**, 1079.

Middleton, D. S. (1934). *Edinburgh Med. J.* [N. S.] **41**, 401.

Milhorat, A. T., ed. (1967). "Exploratory Concepts in Muscular Dystrophy." Excerpta Med. Found., Amsterdam.

Minett, F. C. (1935). *Proc. Roy. Soc. Med.* **28**, 672.

Mölbert, E. (1960). *Naturwissenschaften* **47**, 186.

Montecone, F. G., Gabella, G., and Bergamini, L. (1970). *In* "Muscle Diseases" (J. N. Walton, N. Canal, and G. Scarlato, eds.), p. 44. Excerpta Med. Found., Amsterdam.

Mulvany, J. H. (1944). *Amer. J. Ophthalmol.* **27**, 589, 693, and 820.

Naffziger, H. C. (1933). *Arch. Ophthalmol.* **9**, 1.

Nevin, S. (1936). *Quart. J. Med.* [N. S.] **5**, 51.

Newcastle, N. B., and Humphrey, I. G. (1965). *Arch. Neurol.* **12**, 570.

Nienhuis, A. W., Coleman, R. F., Brown, W. J., Munsat, T. L., and Pearson, C. M. (1967). *Amer. J. Clin. Pathol.* **48**, 1.

Odor, D. L., Patel, A. N., and Pearce, L. A. (1967). *J. Neuropathol. Exp. Neurol.* **26**, 98.

Pearce, G. W. (1964). "Disorders of Voluntary Muscle," p. 220. Churchill, London.

Pearse, A. G. E., and Johnson, M. (1970). *In* "Muscle Diseases" (J. N. Walton, N. Canal, and G. Scarlato, eds.), p. 25. Excerpta Med. Found., Amsterdam.

Pearson, C. M. (1967). *In* "Exploratory Concepts in Muscular Dystrophy and Related Disorders" (A. T. Milhorat, ed.), p. 13. Excerpta Med. Found., Amsterdam.

Perloff, I. T., Roberts, W. C., and DeLyon, A. C. (1967). *Amer. J. Med.* **42**, 179.

Pick, F. (1900). *Deut. Z. Nervenheilk.* **17**, 1.

Price, H. M. (1967). *In* "Exploratory Concepts in Muscular Dystrophy" (A. Milhorat, ed.), p. 52. Excerpta Med. Found., Amsterdam.

Radu, T., Penefunda, G., Blücher, G., Radu, A., Darko, Z., and Gödri, I. (1970). *In* "Muscle Diseases" (J. N. Walton, N. Canal, and G. Scarlato, eds.), p. 322. Excerpta Med. Found., Amsterdam.

Ravin, A., and Waring, J. J. (1939). *Amer. J. Med. Sci.* **197**, 593.

Rebeiz, J. J. (1970). *In* "Muscle Diseases" (J. N. Walton, N. Canal, and G. Scarlato, eds.), p. 383. Excerpta Med. Found., Amsterdam.

Rennie, G. E. (1908). *Rev. Neurol. Psychiat.* **6**, 229.

Research Committee of The Muscular Dystrophy Group, eds. (1968). "Research in Muscular Dystrophy." Pitman, London.

Resnick, J. S., and Engel, W. K. (1967). *In* "Exploratory Concepts in Muscular Dystrophy and Related Subjects" (A. T. Milhorat, ed.), p. 255. Excerpta Med. Found., Amsterdam.

Rouquès, L. (1931). "La myotonie atrophique." Legrand, Paris.

Rowland, L. P., Hoefer, P. F. A., Aranow, H., Jr., and Merrit, H. H. (1956). *Neurology* **6**, 307.

Russell, D. S. (1953). *J. Pathol. Bacteriol.* **65**, 279.

Scarlato, G., and Cornelio, F. (1970). *In* "Muscle Diseases" (J. N. Walton, N. Canal, and G. Scarlato, eds.), p. 33. Excerpta Med. Found., Amsterdam.

Scarlato, G., and Cornelio, F. (1967). *Acta Neurol.* **22**, 177.

Schaar, F. E., La Brie, T. W., and Gleason, D. F. (1949). *J. Lab. Clin. Med.*
34, 1744.

Schoen, R., and Tischendorf, W. (1954). *In* "Handbuch der inneren Medizin"
(G. von Bergmann, W. Frey, and H. Schwiegk, eds.), 4th ed., Vol. 6, Part
1, Sect. 21, p. 886. Springer-Verlag, Berlin and New York.

Schröder, J. M. (1970). *In* "Muscle Diseases" (J. N. Walton, N. Canal, and G.
Scarlato, eds.), p. 109. Excerpta Med. Found., Amsterdam.

Shafiq, S. A., Milhorat, A. T., and Gorycki, M. A. (1967). *Neurology* 17, 934.

Shank, R. E., Gilder, H., and Hoagland, C. L. (1944). *Arch. Neurol. Psychiat.*
52, 431.

Shy, G. M., and McEachern, D. (1951). *J. Neurol., Neurosurg. Psychiat.* 14, 101.

Shy, G. M., and Magee, K. R. (1956). *Brain* 79, 610.

Shy, G. M., Engel, W. K., Somers, J. E., and Wonko, T. (1963). *Brain* 86, 793.

Simpson, J. A. (1960). *Scot. Med. J.* 5, 419.

Simpson, J. A. (1964). *J. Neurol., Neurosurg. Psychiat.* 27, 485.

Simpson, J. A. (1968). *In* "Research in Muscular Dystrophy" (Research Committee
of The Muscular Dystrophy Group, eds.), p. 31. Pitman, London.

Sissons, H. A. (1956). *Proc. Int. Congr. Neuropathol., 2nd, 1955* p. 443.

Sjövall, B. (1936). *Acta Psychiat. Neurol. Scand.* Suppl. 10, Part 1.

Slauck, A. (1936). *In* "Handbuch der Neurologie" O. Bumke and O. Foerster,
eds.), Vol. 16, p. 412. Springer-Verlag, Berlin and New York.

Spiller, W. G. (1905). *Univ. Pa. Med. Bull.* 17, 342.

Spiller, W. G. (1913). *Brain* 36, 75.

Steinert, H. (1909). *Deut. Z. Nervenheilk.* 37, 58.

Stephens, F. E., and Tyler, F. H. (1951). *Amer. J. Hum. Genet.* 3, 11.

Stevenson, A. C. (1953). *Ann. Eugen. (London)* 18, 50.

Stevenson, A. C. (1955). *Ann. Hum. Genet.* 19, 159.

Stevenson, A. C., Cheeseman, E. A., and Huth, M. C. (1955). *Ann. Hum. Genet.*
19, 165.

Sunderland, S., and Ray, L. J. (1950). *J. Neurol., Neurosurg. Psychiat.* 13, 159.

Thomasen, E. (1948). "Myotonia." Universitetsforlaget, Aarhus.

Thomsen, J. (1876). *Arch. Psychiat. Nervenkrankh.* 6, 702.

Tower, S. S. (1939). *Physiol. Rev.* 19, 1.

Turner, J. W. A. (1949). *Brain*, 72, 25.

Tyler, F. H. (1950). *AMA Arch. Neurol. Psychiat.* 63, 425.

Tyler, F. H. (1951). *Proc. 1st and 2nd Med. Conf. Muscular Dystrophy Ass. Amer.,*
p. 46.

Tyler, F. H. (1954). *Res. Publ., Ass. Res. Nerv. Ment. Dis.* 33, 283.

Tyler, F. H., and Stephens, F. E. (1950). *Ann. Intern. Med.* 32, 640.

Tyler, F. H., and Stephens, F. E. (1951). *Ann. Intern. Med.* 35, 169.

Tyler, F. H., and Wintrobe, M. M. (1950). *Ann. Intern. Med.* 32, 72.

Tyler, F. H., Stephens, F. E., Gunn, F. D., and Perkoff, G. T. (1951). *J. Clin.
Invest.* 30, 492.

Unvericht, H. (1887). *Z. Klin. Med.* 12, 533.

von Meyenburg, H. (1929). *In* "Handbuch der speziellen pathologischen Anatomie
und Histologie" (F. Henke and O. Lubarsch, eds.), Vol. 9, Part 1, p. 299.
Springer-Verlag, Berlin and New York.

Vuia, O. (1970). *In* "Muscle Diseases" (J. N. Walton, N. Canal, and G. Scarlato,
eds.), p. 412. Excerpta Med. Found., Amsterdam.

Walton, J. N. (1957). *Proc. Roy. Soc. Med.* 50, 301.

Walton, J. N., and Adams, R. D. (1956). *J. Pathol. Bacteriol.* **72,** 273.

Walton, J. N., and Adams, R. D. (1958). "Polymyositis." Livingstone, Edinburgh.

Walton, J. N., and Nattrass, F. J. (1954). *Brain* **77,** 169.

Walton, J. N., Canal, N., and Scarlato, G., eds. (1970). "Muscle Diseases." Excerpta Med. Found., Amsterdam.

Welander, L. (1951). *Acta Med. Scand.* **141,** Suppl. 265.

Whittem, J. H. (1957). *J. Pathol. Bacteriol.* **73,** 375.

Wilson, S. A. K. (1940). "Neurology." Arnold, London.

Wohlfart, G. (1951). *J. Neuropathol. Exp. Neurol.* **10,** 109.

Wohlfart, S., and Wohlfart, G. (1935). *Acta Med. Scand., Suppl.* **63.**

Woolf, A. L., and Till, K. (1955). *Proc. Roy. Soc. Med.* **48,** 189.

Woolf, A. L., Bagnall, H. J., Bauwens, P., and Bickerstaff, E. R. (1956). *J. Pathol. Bacteriol.* **71,** 173.

Zeigler, M. (1929). *In* "Handbuch der speziellen pathologischen Anatomie der Haustiere" (E. Joest, ed.), Vol. 5, p. 383. Schoetz, Berlin.

CLINICAL ASPECTS OF SOME DISEASES OF MUSCLE

RONALD A. HENSON

I. Introduction

The fifteen years that have passed since the publication of the first edition of this treatise have seen a substantial increase in knowledge of all aspects of muscle disease, and many new types of myopathy have been identified and described. It remains true that on a global view the common affections of muscle are due to trauma, infection, infestation, and circulatory disorder. Cachexia from serious systemic disease, nutritional problems, or malignant disease are frequent causes of generalized muscular atrophy, while localized wasting and weakness occur in muscles around injured or diseased joints. It is a matter of universal experience that inactivity, frank disuse, and increasing age lead to muscular enfeeblement and atrophy. Any condition interfering with the functions of the lower motor neuron may produce muscular atrophy and paralysis. The pathology of some of these conditions (e.g., infection, trauma, and circulatory deficiency) is well understood, and treatment is effective, but the mechanisms involved in others (e.g., cachectic states) are far from clear. These elementary facts are recalled to bring into proportion those rarer disease states, such as muscular dystrophy, the metabolic myopathies, and the myasthenias, which generally spring to mind when the subject of muscle disease arises.

There is no need to discuss the common, well understood myopathies mentioned above. Neural atrophies are also excluded from consideration for reasons of space and convenience. The conditions described have been selected on grounds of clinical and theoretical interest. In most, the muscular system is diffusely affected, sometimes as a manifestation of generalized disease processes. No attempt has been made to cover the whole field. A comprehensive classification of neuromuscular disorders was published in 1968 by the Research Group on Neuromuscular Disorders of the World Federation of Neurology, and a reduced form of this classification is contained in Table I.

TABLE I

CLASSIFICATION OF DISEASES OF MUSCLE

I. Neural muscular atrophies (deriving from disease of the anterior horn cells, motor nerve roots and peripheral nerves)

II. Disorders of neuromuscular transmission—myasthenia gravis, other forms of myasthenia; poisoning with anticholinesterase components, depolarizing drugs, botulism, tick paralysis

III. Disorders of muscle

    A. Genetically determined myopathies

        1. The muscular dystrophies

        2. Obscure congenital myopathies of unknown etiology, e.g., central core disease, nemaline myopathy, benign congenital myopathy

        3. Myotonic disorders, dystrophia myotonica, myotonia congenita, paramyotonia congenita

        4. Glycogen storage diseases

        5. Familial periodic paralysis and related syndromes

    B. Muscle damage by external agents

        1. Physical

        2. Toxic, Haff disease, snake (*Enhydrina schistosa*) bite

        3. Drugs (steroids, chloroquine)

    C. Inflammatory

        1. Infections (viral and bacterial)

        2. Infestations (trichinella, cysticercosis)

        3. Polymyositis or dermatomyositis (possibly manifestations of autoimmune disease)

        4. Of unknown etiology (sarcoidosis, polymyalgia rheumatica)

    D. Muscle disorder associated with endocrine or metabolic disease

        1. Thyrotoxicosis, myxoedema

        2. Hypopituitarism, acromegaly

        3. Cushing's syndrome

        4. Addison's disease

        5. Hyperparathyroidism

        6. Myopathy in metabolic bone disease

        7. Alcoholic myopathy

    E. Myopathies associated with malignant disease

    F. Other myopathies of unknown etiology

        1. Paroxysmal myoglobinuria

        2. Amyloidosis

        3. Disuse atrophy, cachectic myopathy

    F. Tumors of muscle

II. Some General Clinical Features

Muscular development varies among healthy individuals according to age, sex, race, and other factors. The general state of health and nutrition are reflected in the condition of the muscles. The symptoms of the forms of muscle disease under consideration are largely variations on a theme of muscular weakness. In the muscular dystrophies and certain other important myopathies, weakness is commonly predominant in the limb girdles and trunk. Muscular atrophy due to diesease of the lower motor neuron, on the other hand, generally begins in the distal muscle groups, as for example in motor neuron disease and peripheral neuropathy. There are important exceptions in both types of case. Certainly, subacute or chronic proximal muscular atrophy leads to the initial suspicion that the condition is myopathic.

The relationship between the degree of atrophy and weakness varies in different muscular affections. Where there is widespread wasting in an ambulant patient with malignant disease or some chronic infective process without a specific neural lesion, general reduction in power is found; but in the earlier stages, at least, power is often surprisingly good, considering the degree of atrophy. Muscular wasting and weakness proceed equally in most primary myopathies and in many diseases of the lower motor neuron, but this is not an absolute rule. In the muscular dystrophies, the situation is complicated by the occurrence of pseudo-hypertrophy, a term used to denote muscular enlargement without increased power. Pseudohypertrophy is a characteristic feature of the Duchenne type of muscular dystrophy. Muscular hypertrophy, increased muscle bulk and power, is a rare phenomenon encountered in hypertrophia musculorum vera and myotonia congenita.

Myotonia and myasthenia are two classic symptoms of some muscle diseases. Myotonia is a condition in which the affected muscles continue to contract after voluntary contraction has ceased. It occurs in response to mechanically stimulated contraction, as by percussion or electrical stimulation. Voluntary contraction itself may be slowed down. Myotonia is usually worse in states of cold or fatigue. It may be diminished by repeated movement of the affected muscles, though this is not always the case. The function of the muscles involved is naturally interfered with; in the hands, there is difficulty in releasing any held object; when the legs are affected, falling is a common complaint; while implication of the tongue causes dysarthria. The term myasthenia is used to denote a state of abnormal muscle fatiguability occurring in response to exercise. Muscular fatigue is a normal response to prolonged exertion, but this natural tendency is increased when muscles are weakened from any cause.

Tetany is a state of muscular spasm associated with hypocalcemia and alkalosis, the commonest cause being hyperventilation. The muscles of the extremities and larynx are predominantly involved. While the spasms can occur spontaneously, they are mainly precipitated by movement or by mechanical stimulation of muscle or nerve. In tetanus, muscle spasm is continuous, but in generalized cases, severe episodic spasms occur. Muscular cramps may be caused by lesions affecting the anterior horn cells of the spinal cord, e.g., motor neuron disease and poliomyelitis, various forms of peripheral neuropathy, biochemical disorders such as tetany and salt depletion, certain intoxications, and vascular disease. Cramps also occur in some metabolic myopathies, either spontaneously or on exertion. When cramps constitute an isolated symptom and no physical signs can be made out, they rarely derive from neuromuscular disease. Cramps occur as a matter of everyday experience when the untrained undertake violent exercise, especially in cold weather.

The term fasciculation is used to denote contraction of fibers of single motor units, a phenomenon visible as spontaneous movements in muscles when the anterior horn cells are diseased. However, fasciculation also appears as a benign condition, called myokymia, in normal persons, and distinction from the more serious form demands clinical expertise. The term myokymia is also used to indicate a rare type of spontaneous muscular contraction, slower in tempo than fasciculation and linked with various myopathies. Myoidema is seen when muscles are percussed in debilitated individuals; a ridgelike contraction develops in response to the mechanical stimulus. The appearance can be confused with myotonia, but myoidema is electrically silent, in sharp contrast with myotonic states.

The tendon reflexes are diminished *pari passu* with weakness in the muscular dystrophies and many other myopathies, but they are preserved and even exalted in myopathy associated with osteomalacia. Tone is reduced in muscle disease, and hypotonia is one of the important signs of myopathy in infants. Sensation is not disturbed in myopathy, but muscular tenderness may be present. Pain is a common feature of ischemic, traumatic, and inflammatory diseases of muscle and also occurs in some metabolic myopathies; it is rarely experienced by sufferers from muscular dystrophy, save as the result of abnormal posture or fatigue of weakened muscles.

The differential diagnosis between neurogenic and myogenic weakness is not usually difficult when the whole clinical picture is considered. Sometimes it is hard to distinguish chronic forms of spinal atrophy from myopathy. However, electrodiagnostic tests and muscle biopsy are commonly used to establish the site of the disease process, and these investigations should always be made in doubtful cases. Histochemical and biochemical studies are needed for the diagnosis of several myopathies as the following text illustrates.

III. The Muscular Dystrophies

A. Introduction

The muscular dystrophies form a group of disorders characterized by progressive muscular wasting and weakness. Genetically determined, they are thought to be due to a primary degenerative affection of muscle fibers, though recent electrophysiological findings have raised the question of neural lesions, but the cause or causes remain unknown. The distribution of weakness is peculiar. Affected muscles, either in the same patient or in different forms of the disease, may show muscle enlargement with increased or diminished power as well as atrophy and weakness.

B. Classification of Muscular Dystrophies

As the etiology of muscular dystrophy is obscure, classification must be on a clinical and genetic basis. The clinical classification of genetically determined disease states is admittedly unsatisfactory, and excessive characterization can lead to confusion. In this account, the categories listed by Walton and Gardner-Medwin (1969) will be adopted as an acceptable rational classification in the present state of knowledge.

C. Clinical Features

1. GENERAL FEATURES

The patient with muscular dystrophy presents with complaints stemming from muscle weakness, though the term weakness is rarely mentioned by patients or parents. Thus, parents may say that a child has never been able to walk, that his gait is clumsy, or that he tends to fall without cause. Specific defects may be mentioned, e.g., difficulty in rising from a chair, in mounting stairs, or in lifting the arms. Dragging of the legs may be noted. In adults, the range of complaints is wider, but the underlying pattern is the same. Weakness usually results from affection of the muscles of limb girdles and trunk. The onset is insidious and signs of disease are usually obvious by the time the patient comes for examination. Examination shows muscular wasting and weakness, commonly proximal, with or without muscle enlargement. Tendon reflexes are generally reduced in the affected parts. There are no signs of nervous lesion. Skeletal deformities, e.g., increased lumbar lordosis or spinal kyphosis, may be present. Detailed reviews of the clinical aspects of the muscular dystrophies have been made by Walton and Nattrass (1954) and Walton and Gardner-Medwin (1969).

2. X-Linked Muscular Dystrophy

a. Duchene Type. This form of muscular dystrophy was described by Duchenne (1868), but he did not recognize the paralysis as myogenic. Although muscular enlargement (pseudohypertrophy) frequently occurs in this variety, the term pseudohypertrophic muscular dystrophy is best avoided, as a similar increase in muscle bulk occurs in other forms, and it is not invariably present in patients who otherwise appear to belong to this group (Stevenson, 1953; Walton and Nattrass, 1954). This is an affection of young children, occurring almost exclusively in males; rare examples have been reported in girls with Turner's syndrome. It is inherited as a sex-linked recessive. The onset may be remarked by parents in the first year of life, and before the age of 4 in one-half of all cases (Walton and Nattrass, 1954). The muscular weakness and wasting begin symmetrically in muscles of the pelvic girdle and thighs, especially the iliopsoas, quadriceps, gluteus maximus, and sacrospinalis. The child walks with a waddling gait and has difficulty in running, climbing stairs, or rising from a chair. If asked to rise from the ground, he "climbs up himself" in characteristic fashion. The anterior tibial group and peronei are involved later. The calf muscles, hamstrings, and abductors of the thighs are relatively spared, and this accounts for the muscular contractures which are liable to develop. After a time, the shoulder girdles and upper trunk muscles become affected and, later still, the arms and forearms. The calf muscles are commonly enlarged and firm. This enlargement may be shared by the quadriceps, hamstrings, glutei, deltoid, or other muscles in the shoulder girdle and upper limb. As already noted, the enlarged calf muscles retain their power remarkably well in comparison with other muscle groups. The enlargement may disappear later, or the muscles may become less firm to the touch.

This complaint is inexorably and, for a muscular dystrophy, rather rapidly progressive. Contractures appear; immobility leads to severe osteoporosis and gross skeletal deformities. Remissions or periods of arrest do not occur. Death results from respiratory infection or cardiac involvement usually in adolescence. A few patients survive into the third decade, in a state of helplessness.

b. BENIGN X-LINKED TYPE (BECKER AND KEINER). This is a benign X-linked recessive variety of muscular dystrophy (Becker and Keiner, 1955). Formerly, examples of this disease were included with the Duchenne and other types. The condition differs from the severe Duchenne form in age at onset (5–25 years), mode of inheritance, and rate of progression. While the pattern of weakness is similar in both complaints, patients with the benign form are crippled over a period of many years and may survive into late adult life. Cardiac involvement is uncommon.

3. FACIOSCAPULOHUMERAL TYPE

This variety was first described by Landouzy and Déjèrine (1884). Onset is commonly early in the second decade, but it may be earlier or later. It is rarely possible to determine the precise length of the history in this insidious complaint. Males and females are affected, and inheritance is usually autosomal dominant. Muscles of the face are involved at an early stage and the lips tend to be everted, forming the so-called tapir mouth. The patient cannot whistle. Attempts to smile produce a sneer, due to weakness of the zygomatici. Later, eye closure and wrinkling of the brow become enfeebled. The presenting symptoms, however, are referable to an atrophic palsy of the shoulder girdle, arms, and trunk. Spinati, rhomboids, serratus anterior, pectoralis major, deltoid, biceps, and triceps are affected early. The disturbance may be asymmetrical. Trunk weakness is usually prominent, and this leads to an exaggerated lordotic stance. The complaint is slowly progressive, though there may be long periods of arrest. After a few years, the pelvic girdle and lower limbs become affected. Muscular enlargement is rare. Generally speaking, the older the patient at the time of onset, the more favorable the prognosis. Most patients remain mobile for many years, being able to work or look after their homes. Complete incapacity may ultimately ensue, but it is unusual. Taking the group as a whole, there seems to be no serious reduction in the expectation of life. Abortive forms are encountered, usually in the families of known sufferers. In such cases the weakness tends to be localized to a few muscles, and the condition is nonprogressive.

4. OTHER LIMB–GIRDLE MUSCULAR DYSTROPHIES

This group contains all those patients with limb girdle dystrophy in whom there is no facial weakness, the Duchenne type being also excluded. Tyler and Wintrobe (1950), Adams *et al.* (1953), and Stevenson

(1953), among others, would include all these cases with the facioscapulohumeral group. Walton and Nattrass (1954) and Walton (1963), on the other hand, regard limb–girdle dystrophy as a definable, separate type. In this heterogeneous group, somewhat differing clinical pictures are found. For example, weakness may begin in the pelvifemoral group of muscles or in the scapulohumeral region. In some families, muscular enlargement, with or without increased power, is prominent. Onset is commonly from the second to the fifth decade. Either males or females may be affected. Weakness is progressive, but the rate is variable and long periods of arrest may occur. Severe disability usually results in from 15 to 30 years. Taking the group as a whole, the expectation of life is shortened, but again the later the onset, the longer the period of useful survival. Contractures and skeletal changes occur late or not at all.

5. Rarer Forms of Muscular Dystrophy

There are several rarer forms of muscular dystrophy of clinical or theoretical interest.

a. Congenital Muscular Dystrophy (Synonym Congenital Myopathy). Batten (1909–1910) and Turner (1940, 1949) described a benign form of congenital muscular dystrophy. This is one of the causes of the amyotonia congenita syndrome (see Section IV). More severe examples have been recorded, some with death in the early months of life (Greenfield *et al.*, 1958; Short, 1963; Wharton, 1965).

b. Distal Form of Muscular Dystrophy. It has been emphasized that muscular dystrophies commonly originate in the limb girdles and trunk, but there is a form, rare outside Scandinavia, which begins in the distal parts of the limbs and progresses centripetally. Welander (1951, 1957) recorded a large series of cases from Sweden. Inherited as a dominant character, the disease usually begins between the ages of 40 and 60 and affects both men and women.

c. Ocular Muscular Dystrophy. Kiloh and Nevin (1951) have given a detailed account of this condition, which for many years had been believed to be due to some chronic affection of the oculomotor nuclei. As a result of myogenic atrophy the external ocular muscles become weakened; ptosis and paralysis of external ocular movements follow. Later, the muscles of the face and shoulder girdle are likely to become involved, but this is not invariable. Ocular myopathy occurs in a variety of hereditary and acquired neurological disorders from which the dystrophic condition must be distinguished.

d. LOCALIZED FORMS OF MUSCULAR DYSTROPHY. Sometimes the dystrophic process may be very localized. For example, a form in which the quadriceps alone is affected has been described (Bramwell, 1922; Denny-Brown, 1939).

e. MUSCULAR DYSTROPHY WITH HYPERTROPHY (HYPERTROPHIA MUSCULORUM VERA). This is a very rare form, but it is important because of the peculiar way in which it develops. Barnes (1932) described a family in which muscular enlargement with increased power (hypertrophy) was followed by weakness and ultimate wasting. The occurrence of a similar phenomenon on a less dramatic scale in other forms of dystrophy has been mentioned. Cramps may occur in these exceptionally strong muscles. The rare phenomenon malignant hyperpyrexia under anesthesia has been observed in relatives of sufferers from this complaint.

D. Muscular Dystrophy and Myotonia

The clinical features of myotonia have been previously described. Myotonia is associated with muscular dystrophy in the condition known as dystrophia myotonica and also occurs in myotonia congenita, a much rarer complaint. It is also a feature of muscle phosphorylase deficiency and adynamia episodica hereditaria (see Section VI).

1. MUSCULAR DYSTROPHY WITH MYOTONIA (DYSTROPHIA MYOTONICA)

In this hereditary disease, described by Steinert (1909) and reviewed by Caughey and Myrianthopoulos (1963), muscular dystrophy is linked with myotonia. There are numerous associated abnormalities, including cataract, gonadal atrophy, premature baldness, cardiac conduction defects, hypotension and syncope, and defective esophageal contraction (Lee and Hughes, 1964; Gleeson *et al.*, 1967). Mental deficit is apparent in some affected children; progressive dementia is encountered in adults (Thomasen, 1948), and air encephalography reveals enlarged ventricles.

Dystrophia myotonica occurs in both males and females. Symptoms commonly arise beteen 15 and 50 years. Age at onset is often difficult to define, as patients are wont to conceal and even deny their disabilities; however, Lynas (1957) and Klein (1958) showed that symptoms did not begin until after the age of 30 in one-half of all cases; personal experience covers patients presenting in the neurological clinic from 3 to 81 years. The characteristic features are the myotonia, which is usually limited to the hand, forearm and tongue, and muscular atrophy and weakness. The wasting and weakness are commonly conspicuous in the facial muscles, sternomastoids, muscles of the shoulder girdles, forearms and hands, trunk, quadriceps, and the muscles of the legs below the knees. Ptosis is present in about half of all cases. Emaciation, impotence, testicular atrophy, amenorrhea, baldness, excessive perspiration, and mental abnormalities are other features. Affection of the pharyngeal and esophageal muscles may lead to dysphagia. Dysarthria is a common feature, being due to myotonia in the tongue muscles. The diagnosis can usually be made at sight from the characteristic expressionless face, hollowed temples and cheeks, drooping eyelids, and baldness. Muscular enlargement is rare. The condition is progressive, leading to severe disability, generally within 20 years. The patient usually succumbs in late middle age from respiratory infection, due to weak coughing and esophageal spillover, or cardiac complications. Patients constitute poor anesthetic risks and only essential surgical procedures should be advised (Caughey and Myrianthopoulos, 1963).

Dystrophia myotonica has been a field of continuing investigation since Steinert's original paper, but the cause remains unknown. Physiological studies indicate that the myotonia is due to muscle disorder, but histopathological studies of muscle spindles have raised the question of neural involvement. Quinine, diphenylhydantoin, and procaine amide diminish myotonis, and the latter is commonly used in treating patients with significant disability from this element.

2. Myotonia Congenita (Thomsen's Disease)

This condition was described by Thomsen (1876), who himself suffered from the disease. The myotonia is usually present from birth, and symptoms are commonly noted in childhood. Males and females are affected. The myotonia, which may be generalized or localized, is associated with exceptional muscle development and increased muscular power. There is no muscular wasting, and the reflexes are normal. The basic symptom is an interference with movement through tonic muscular spasm. Any attempt at forceful movement from rest, or a change from a state of easy activity to strong muscular contraction, produces painless cramp of the muscle group involved. Emotional disturbance, cold, or fatigue tend to bring on the myotonia. The patient is able to "loosen up," so to speak, by a series of laborious movements of the affected part. During the spasm, the muscles are prominent to the eye and hard to the touch. Life is not shortened. Some patients with myotonia congenita develop muscular wasting and weakness earlier or later in life, and this is one of the facts that had led some to suppose that the condition is simply another manifestation of the dystrophia myotonica syndrome. The severity of the myotonia tends to decrease as the patient grows older.

3. PARAMYOTONIA CONGENITA

In this form of myotonia the symptom only occurs on exposure to cold. It is doubtful whether the complaint represents an independent disease state, for all types of myotonia are aggravated by cold. Some cases are probably examples of adynamia episodica hereditaria (Shy, 1961; McArdle, 1969).

4. MYOTONIA ACQUISITA

This term has been used to describe cases of myotonia coming on in adult life in the absence of a family history of the symptom.

E. Differential Diagnosis of Muscular Dystrophy

As muscular dystrophy is progressive and incurable, diagnostic certainty is required. First, the site of the lesion must be established beyond doubt. Neural muscular atrophy caused by motor neuron disease and other spinal cord disorders, peroneal atrophy, or chronic types of peripheral neuropathy can generally be differentiated on clinical grounds alone. However, there is a chronic form of spinal muscular atrophy that produces a clinical picture closely resembling muscular dystrophy (Kugelberg and Welander, 1956); electrodiagnostic studies and muscle biopsy are needed to elucidate the position. In infants and small children, it is possible to confuse bilateral hip disease and muscle disorder.

Second, once the diagnosis of myopathy has been made the nature of the disease process must be established. Many patients with muscular dystrophy fall clearly into the categories described above, and there may be a family history of the complaint. Sporadic or unusual cases demand special caution, for there are a number of congenital and endocrine–metabolic myopathies that can resemble muscle dystrophy clinically. There is particular danger in diagnosing muscular dystrophy in adults with a negative family history before detailed electrodiagnostic studies, muscle biopsy, and appropriate biochemical tests have been made. Serum levels of several enzymes are raised in the Duchenne type of muscular dystrophy, but estimation of serum creatine kinase has proved the most delicate index (Ebashi *et al.*, 1959). Values of this enzyme can be raised several hundred times in preclinical and active stages of the disease (Pennington, 1969). These high levels decline as the disease progresses (Pearce *et al.*, 1964). Last, polymyositis may resemble muscular dystrophy in children and adults. Polymyositis (see Section V) is not hereditary, signs of system disorder are often present, and muscle biopsy appearances are generally distinctive.

F. Treatment of Muscular Dystrophy

There is no known treatment that modifies the course of muscular dystrophy. Patients are encouraged to use their limbs to the best of their ability, but short of fatigue. Physical methods can prevent the development of contractures. Obesity becomes a problem as patients grow less mobile, and strict dietary control is needed. Respiratory, cardiac, and orthopedic complications are treated as they arise. Understanding psychological management of patient and family is important. Education should not be neglected, and advice on appropriate employment should be provided. If the severe Duchenne type is excluded, many sufferers are able to make continuing useful contributions to the economy and life of the community. Carrier detection is important in prevention of muscular dystrophy, especially the X-linked forms. Walton and Gardner-Medwin (1969) and Dreyfus *et al.* (1970) have provided recent authoritative reviews of the subject. Bundey *et al.* (1970) report that clinical examination, slit-lamp examination of the eyes, and electromyography are the most useful ways of detecting heterozygotes for dystrophia myotonica. Measurement of serum immunoglobulins can also be helpful.

IV. Muscular Disorders in Infancy and Early Childhood

The term amyotonia congenita was applied by Oppenheim (1900) to a condition marked by hypotonia and muscular weakness occurring in infancy. It soon became clear that the complaint was a syndrome and not a nosological entity. The features of the clinical picture are extreme flaccidity and lack of movement generally noticed in the first days of life. The child is so limp as to slip through the hands; it may be placed in extraordinary postures. The whole problem has been clarified in recent years by Brandt (1950), Sandifer (1955), Walton (1956, 1957a), Dodge (1960), and Tizard (1949). Hypotonia and poverty of movement are found in neonates who are premature or have suffered brain injury at birth. Other causes include serious systemic disease, organic brain lesions other than trauma, skeletal and ligamentous abnormalities, and a whole range of disorders of the central nervous system, peripheral nerves, and muscle. The largest single causes are cerebral palsy, mental defect, and benign congenital hypotonia according to Paine (1963). Walton (1957b) introduced the embracing term benign congenital hypotonia to describe cases in which partial or complete recovery is the rule.

There are several nonprogressive benign types of congenital myopathy. Turner (1940, 1949) and Turner and Lees (1962) published the record of a family of whom several members presented with infantile hypotonia and slow development of motor activities such as walking. Although the patients were affected by permanent muscular wasting and weakness, the complaint, histologically a form of muscular dystrophy, proved to be nonprogressive. Other benign forms include central core disease (Shy and Magee, 1956), nemaline myopathy (Shy *et al.*, 1963), and three types characterized by mitochondrial abnormalities. On the other hand, examples of progressive and severe congenital muscular dystrophy have been described (Zellweger *et al.*, 1967).

There are many other forms of muscle disease that can present in infancy or early childhood, including several types of metabolic myopathy and muscular dystrophy. These are described elsewhere in the text. The important neural atrophies in this age group are those caused by infantile spinal muscular atrophy (Werdnig–Hoffman Disease), the spinal muscular atrophy of Wohlfart, Kugelberg, and Welander, and peripheral neuropathy. Special investigations are often required to site the lesion with assurance.

It will be evident that the cause of muscular hypotonia and weakness in these young patients can often be established without recourse to detailed tests. However, when the underlying lesions involve motor neurons, peripheral nerves or muscle appropriate tests must be made for purposes of prognosis, genetic counseling, and treatment in the few conditions in which amelioration or relief can be achieved. Estimation of serum enzymes, electrodiagnosis, and histological and histochemical studies on muscle biopsy specimens may all be needed. Obviously, these tests are restricted to essentials. In some patients, the family history will yield the diagnosis.

V. Inflammatory Diseases of Musule

A. *Specific Diseases*

Bacterial, viral, fungal, and parasitic infections of muscle will not be considered here, they are mentioned for completeness.

B. *Nonspecific Diseases*

1. POLYMYOSITIS AND DERMATOMYOSITIS

The term polymyositis was introduced to denote a group of non-bacterial inflammatory disorders of muscle. Pearson (1969) has provided an authoritative review of the subject. In polymyositis, there is diffuse involvement of muscles, while in dermatomyositis the skin is also affected. About half of all cases are associated with malignant disease or connective tissue disorder, and other organs, such as the heart, kidneys, liver, spleen, and joints, may be involved. Both polymyositis and dermatomyositis can appear from early childhood to old age, but most cases are encountered in persons over 30. Females are more frequently affected than males. In some patients, the complaint follows an infective illness or treatment with penicillin or sulfonamides. The rash of dermatomyositis can be precipitated by exposure to sunlight, and photosensitization is common in the established complaint.

Classification of polymyositis is difficult because of our ignorance of causation and on account of the variable clinical and pathological manifestations. In this situation, Pearson (1969) has advanced the classification shown in Table II, which he commends as an aid to prognosis and treatment.

a. CLINICAL FEATURES. The clinical picture is extremely varied, but the pattern and tempo of muscular paralysis are similar whether skin

TABLE II
CLASSIFICATION OF POLYMYOSITIS

Type	I	Polymyositis in adults
Type	II	Typical dermatomyositis in adults
Type	III	Typical dermatomyositis (occasionally polymyositis) with malignancy
Type	IV	Childhood dermatomyositis
Type	V	Acute myolysis
Type	VI	Polymyositis in Sjögren's syndrome

lesions are present or not. The pelvic girdles are affected first in subacute and chronic forms, the weakness later spreading to involve the shoulder girdles and trunk. Dysphagia and dysarthria afflict more than half of all patients and the neck muscles are peculiarly susceptible to the process. Distal muscles retain reasonable power until the later stages of the disease. In acute cases, severe widespread paralysis develops; death commonly ensued within weeks before the introduction of treatment with corticosteroids. At the other extreme, the process continues insidiously, with or without periods of arrest or partial remission, for many years. Most cases fall between these extremes, and the pattern is one of subacute or chronic progression of weakness with episodes of remission and relapse; general deterioration over the years is the rule. Muscular pain and tenderness are common in the acute type, but rare in more chronic forms.

There is a diversity of skin lesions in dermatomyositis. The characteristic rash is an erythema of the face and upper part of the body, sometimes with a violet hued discoloration about the eyes, but seborrheic and urticarial eruptions are also encountered. Hyperkeratosis of the extremities and desquamation of the skin are other features, while hyperemia about the base of the finger nails is visible on careful inspection. Subcutaneous tissues and muscles may become edematous.

Transient arthritic or rheumatic symptoms occurred in between a third and a half of Pearson's (1969) cases. Constitutional symptoms, such as general malaise, anorexia, loss of weight, and fever, are present in acute examples of polymyositis and during active stages of more chronic forms. Raynaud's phenomenon is a frequent concomitant.

b. LABORATORY INVESTIGATIONS. The erythrocyte sedimentation rate is raised and there may be polymorphonuclear leukocytosis in the blood during active phases. About half of all patients show elevated α_2 and γ serum globulin values. Tests for circulating rheumatoid factor may also be positive. While serial serum enzyme studies are useful in assessing activity of muscle involvement and response to treatment, they have limited diagnostic value, as the changes found are not specific. Transaminase, aldolase, and creatine phosphokinase levels are the most sensitive indicators (Pearson, 1966; Rose and Walton, 1966). Electrodiagnostic studies are helpful in identifying cases of polymyositis, but again, the abnormalities found lack specificity. Muscle biopsy reveals a wide range of abnormalities, depending on the type of case and point in the illness when the examination is made. Degeneration of muscle fibers and varying degrees of inflammatory infiltration and interstitial fibrosis are seen. Regenerating fibers are a common finding. In acute cases, massive destruction of fibers and intense inflammatory cellular infiltrates are found. When the process is chronic, interstitial fibrosis is prominent. Muscle biopsy constitutes the most important single diagnostic investigation in patients without skin lesions and is rarely omitted in cases where the diagnosis is visually apparent. However, interpretation of the abnormalities found is a matter for the expert in myopathology. Valuable guides are afforded by Adams *et al.* (1962) and Adams (1969).

c. DIAGNOSIS. The identification of dermatomyositis is a simple matter. On the other hand, certain recognition of subacute and chronic cases of polymyositis demands clinical, electrophysiological, and pathological expertise. In general, the problem posed can be stated as the diagnosis of polymyositis from other acquired myopathies. Success is important in view of the value of corticosteroids in treatment. Total diagnosis requires multisystemic clinical and laboratory investigation with a careful search for associated malignant disease and connective tissue disorder.

d. ETIOLOGY. This remains unknown. The association with malignant disease and connective tissue disorder in many, and with infection or drug therapy in a few, suggests that immunological mechanisms are involved, but appropriate laboratory studies on sera of patients with polymyositis have been negative or inconclusive. Kakulas (1966a,b) showed that muscle fiber necrosis can be caused experimentally by sensitized lymphocytes. More recently, the same worker (Kakulas, 1970) has studied the effects of concentrated peripheral blood lymphocytes from patients with polymyositis and other conditions on cultures of human fetal skeletal muscle. While the results are not conclusive, they suggest that cell-mediated mechanisms may operate in the production of muscle lesions in polymyositis in man. This is clearly a field in which research will be intensified during the next few years. Electron microscopy of muscle specimens from one patient with polymyositis showed structures morphologically similar to the myxoviruses (Chou, 1967), a finding that indicates that virus infection may be responsible for some cases of polymyositis. This is another area in which further inquiry must be made.

e. TREATMENT AND PROGNOSIS. The introduction of corticosteroid treatment has brought distinct benefit to sufferers from polymyositis and certainly modifies the course of the disease or diseases. It is particularly effective in dermatomyositis without associated malignant disease and in acute or subacute forms of polymyositis. Corticosteroids are suppressive, not curative, and maintenance doses are needed for an indefinite period. Pearson (1969) gives a full account of the subject.

Prognosis naturally varies in the different types of polymyositis, and the presence of malignant disease is decisive in many older patients. Rose and Walton (1966) surveyed eighty-nine patients; the mortality for the whole series was 30%, but this descended to 16% when cases with malignant tumors were excluded. No deaths occurred in patients under 30, and over this age mortality increased in direct proportion to patient age. Collagen vascular disease (connective tissue disorder) and malignant disease were important causes of death. From this and other reviews, it is clear that patients with polymyositis are now far less likely to succumb from the direct results of muscular paralysis, and prognosis is much better in the young than the old. However, significant degrees of crippling are common.

2. Myositis in Connective Tissue Disorders

Chronic inflammatory changes are often seen in the muscles of patients with rheumatoid arthritis, rheumatic fever, disseminated lupus erythematosus, and scleroderma. While these findings are of pathological importance, they are not usually associated with remarkable muscle weakness; however, occasional patients with rheumatoid arthritis manifest notable weakness due to this myopathic process. Muscular weakness in both rheumatoid arthritis and disseminated lupus erythematosus is more likely to derive from peripheral neuropathy or to be related to arthropathy. The extremities and face are mainly affected in scleroderma, but weakness of the bulbar musculature affects some sufferers. Muscular weakness in scleroderma is multifactorial.

In both polyarteritis nodosa and Wegener's granuloma, muscular lesions are secondary to the characteristic arterial changes found in both conditions. Wegener's granuloma is particularly liable to cause substantial areas of muscle infarction. Ischemic neuritis is a classic feature of polyarteritis nodosa, and this process naturally leads to neurogenic atrophy.

Polymyalgia rheumatica usually affects elderly women, though the complaint is not limited to females (Paulley and Hughes, 1960). Temporal arteritis is present in some instances. The complaint is characterized by muscular pain and weakness, often limited to the shoulder girdles, but sometimes more widely disseminated from the outset. A raised erythrocyte sedimentation rate is found, and fever may be present. Polymyalgia rheumatica responds to treatment with corticosteroids.

3. SARCOIDOSIS

Muscular lesions are commonly found in patients dying with sarcoidosis; the changes are of the type found in other organs in this disease, namely noncaseating tubercles. Rarely, sarcoidosis presents with insidious, progressive muscular wasting and weakness in a middle aged or elderly patient. The clinical state is similar to that found in muscular dystrophy, polymyositis, and other chronic myopathies. Diagnosis is accomplished by muscle biopsy. The condition is not influenced by treatment.

VI. Metabolic Myopathies

A. *Introduction*

Strictly speaking, muscle metabolism is disturbed in all forms of muscle disease. However, the term "metabolic myopathy" is employed to denote a series of disease states in which muscular disorder derives, or is thought to derive, from primary metabolic defect within voluntary muscle or secondarily from metabolic upset elsewhere in the body. Diseases of the endocrine glands are commonly responsible for the secondary type of metabolic myopathy. This area of neurological practice includes several important complaints that are remarkably responsive to treatment.

B. *Muscular Disorders Associated with Disease of the Thyroid Gland*

The muscle disorders associated with thyrotoxicosis include exophthalmic ophthalmoplegia, acute and chronic thyrotoxic myopathy, myasthenia gravis, and periodic paralysis. These disorders may coexist in the same patient; for example, a few of our cases of exophthalmic ophthalmoplegia also have myasthenia gravis or thyrotoxic myopathy. Muscular symptoms also occur in hypothyroidism. The whole subject of muscular disorders and thyroid disease has been reviewed by Millikan and Haines (1953) and McArdle (1969).

1. HYPERTHYROIDISM

a. THYROTOXIC MYOPATHY. Muscular weakness and atrophy have been known features of thyrotoxicosis for many years (Bathurst, 1895). While it is unusual for these myopathic symptoms to constitute leading complaints in thyrotoxic patients, they are present to some degree in about 60% of all sufferers (Satoyoshi *et al.*, 1963). Remarkable weakness is commoner in older patients and those with a long history of thyroid overactivity. Weakness is experienced in the proximal and axial musculature. The rate of onset is variable, but considerable disability may occur within a few weeks. Fasciculation may be seen, so that a suspicion of motor neuron disease is aroused. In one rare form, the external ocular, bulbar, and cervical muscles are predominantly affected. Electromyographic studies have shown a high incidence of myopathic abnormalities in thyrotoxic patients with demonstrable weakness and in those without. Histological changes found in muscle biopsy specimens from affected individuals correlate poorly with the clinical situation and are often sparse. Various biochemical abnormalities have been described (Satoyoshi and Kinoshita, 1970). For example, the total water and intracellular water contents of muscle are decreased, and the intracellular potassium is also reduced with a reciprocal increase in sodium. These changes could be due to increased membrane permeability caused by excess thyroid hormone. Recently, Peter *et al.* (1970) observed an increased yield of mitochondria from fractionated skeletal muscle of thyrotoxic patients; they concluded that this accounted at least in part for the hypermetabolism of hyperthyroidism. Respiratory control and oxidative phosphorylation were normal. There was also an increased yield of sarcotubular vesicles, which might explain the rapid contraction and relaxation of muscle in this disease. The diagnosis of thyrotoxic myopathy is simple, for signs of the underlying disease process are always present. Muscular weakness responds rapidly to antithyroid treatment.

b. ACUTE THYROTOXIC MYOPATHY. It is doubtful whether this rare and life-threatening complaint constitutes an entity. Characteristically, a patient with acute thyrotoxicosis develops rapidly progressive, widespread muscular paralysis, which includes the bulbar and respiratory muscles. Acute thyrotoxic myopathy probably represents an acute manifestation of thyrotoxic myopathy, but in some instances, it represents an association of acute thyrotoxicosis and myasthenia gravis. Treatment consists in urgent antithyroid medication and symptomatic supportive measures. Cholinergic drugs are naturally beneficial when there is a myasthenic element.

c. THYROTOXIC PERIODIC PARALYSIS. The great majority of reported examples of hypokalemic periodic paralysis (see Section VI,E,3) with thyrotoxicosis derives from Japan (Satoyoshi and Kinoshita, 1970). Abnormal membrane permeability due to excess thyroid hormone may explain the association, for this could lead to a shift of potassium and water with paralysis. The muscular disorder yields to successful treatment of thyrotoxicosis. Japanese writers believe a high rice diet, rich in carbohydrate and low in potassium, may be partially responsible. Apparently, there is a declining incidence at the present time, and this could be causally linked with altered dietary habits in modern Japan.

d. THYROTOXICOSIS AND MYASTHENIA GRAVIS. Approximately 5% of myasthenic patients develop thyrotoxicosis at some point in time (Millikan and Haines, 1953). The endocrine disorder may precede, accompany, or follow the neuromuscular disease. Hypothyroidism is less frequently related to myasthenia. Treatment of the thyroid disorder and myasthenia are independent.

e. EXOPHTHALMIC OPHTHALMOPLEGIA. This is the commonest overt myopathy associated with thyroid disease in our material. The relationship is indirect. Approximately one-half of all patients are thyrotoxic at the time of onset of ocular symptoms, while a few are myxoedematous. About 90% suffer from overt thyroid disease, usually hyperthyroidism, at some point in time. In times past, severe examples were often encountered in patients who had recently been treated surgically or with early antithyroid preparations for thyrotoxicosis; this is now rare experience in our practice. Symptoms are due to edema of the orbital contents, including the muscles, which may be greatly swollen. There is also increased intraorbital fat. Microscopically, a cellular infiltration of the involved tissues is seen. In the later stages the muscles tend to become shrunken and fibrosed. The cause of the condition remains uncertain. Excess thyrotrophin and exophthalmos-producing substance are inconstantly present in the blood. Pinchera *et al.* (1965) and McKenzie (1966) have remarked on the presence of long-acting thyroid stimulator (LATS) in blood from these patients, but this substance is not always present and the significance of the finding is not fully understood. The condition is commonly but not always bilateral. As the name implies, the symptoms are exophthalmos and paralysis of the external ocular muscles, with resultant diplopia. The exophthalmos varies greatly in severity; it may be gross, with chemosis and extreme edema of the lids. In such cases, eye closure may be precluded, so that there is grave danger of corneal ulceration and ultimate panophthalmitis. Papilloedema may develop, going on to optic atrophy. The degree of exophthalmos and severity of muscle weakness are usually related, but not invariably. Muscle paralysis without exophthalmos occurs when the infiltrative element is pathologically predominant. Weakness commonly involves the abductors and elevators of the globe first, diplopia being produced in the appropriate planes. Eventually, total external ophthalmoplegia may supervene. The illness is subacute, the initial history being measured in weeks rather than months, and generally self limiting. The exophthalmos is accompanied by much local discomfort in most patients. Treatment is unsatisfactory. If there is thyroid disorder, it must be treated. Thyroid oblation by ^{131}I, oral steroids, different hormones, and irradiation of the pituitary gland or orbits produce no constant benefit. Local measures are directed toward preservation of sight; if this sense is threatened by reason of increasing exophthalmos, orbital decompression is required, and tarsorrhaphy may be useful as an interim measure to prevent corneal ulceration. Inability to close the eyelids is a clear indication for early surgical treatment of this sort. Combined ophthalmic and neurosurgical skills should ensure preservation of vision. The prognosis for full recovery is bad; some residual degree of exophthalmos and ophthalmoplegia is the rule. Once the complaint is inactive, various operations can be performed on the external ocular muscles to overcome residual diplopia.

2. HYPOTHYROIDISM

Debré–Semelaigne syndrome occurs in children who are cretins or suffer from juvenile myxoedema. It is characterized by muscular enlargement with weakness and slowness of movement. When tendon reflexes are elicited, muscular contraction and relaxation are excessively slow. Painful spasms or cramps occur and myoidema can often be demonstrated. The myopathy of myxoedematous adults is qualitatively similar but usually less severe in every respect. Aloujouanine and Nick (1945) described the rare atrophic form of hypothyroid myopathy with clinical features common to other acquired myopathies. Light microscopy shows no consistent pathological changes, but ultrastructural studies reveal varied and extensive abnormalities. Signs of thyroid deficiency are always evident in patients with either type of hypothyroid muscle disease. Treatment with *l*-thyroxine is highly beneficial, though not always completely restorative. True myotonia is rare in myxoedema and when present probably reflects aggravation of mild preexistent myotonia of hereditary type (McArdle, 1969). Cramps, myoidema, and prolonged tendon reflexes in uncomplicated hypothyroid myopathy combine to produce so-called pseudomyotonia.

C. Diseases of the Pituitary and Adrenal Glands

Increased muscle bulk and strength may occur in the early stages of acromegaly, but weakness and fatigability are common complaints as the disease progresses. Recent findings indicate that the proximal muscles of acromegalics are often the seat of a patchy myopathic process, which is probably the basis for the muscular weakness but which can be asymptomatic (Mastaglia *et al.*, 1970). It is likely that muscle changes are related to persistent excess of circulating growth hormone. Generalized muscular weakness and wasting are features of pituitary failure from any cause. These symptoms can be relieved by appropriate replacement therapy.

Cushing's syndrome is also accompanied by weakness, which is particularly evident in the proximal musculature (Muller and Kugelberg, 1959). A similar situation obtains in patients on long-term corticosteroid treatment. Steroid myopathy is commoner when halogenated steroids are employed, and these preparations are best avoided unless specifically indicated (Yates, 1970). Difficult diagnostic situations arise when patients with complaints that cause muscular weakness in themselves are treated with steroids; such complaints include polymyositis, scleroderma, polyarteritis nodosa, and disseminated lupus erythematosus. Steroid myopathy possibly derives from abnormal carbohydrate metabolism. The best treatment is prophylactic, for myopathy is unlikely to appear if the daily steroid dose is restricted to 10 mg of prednisone or its equivalent (Yates, 1970). Once myopathy has developed, recovery occurs over a period of months if the daily dose can be reduced. Treatment of Cushing's syndrome is less rewarding in terms of restored muscle power.

Muscular weakness is a common complaint in adrenal failure (see Section VI,F,2) from any cause and may be the presenting symptom; treatment of the underlying metabolic defect restores muscle power. Primary aldosteronism causes attacks of intermittent paralysis.

D. Osteomalacia and Hyperparathyroidism

Patients with hyperparathyroidism and osteomalacia sometimes manifest a peculiar myopathic syndrome characterized in its complete form by fatigability, wasting, and weakness of the proximal musculature, especially affecting the hips and thighs, hypotonia, increased tendon reflexes, cramps and muscular pains, discomfort on movement, and bone tenderness. This condition was first described by Vicale (1949); Lemann and Donatelli (1964) remarked on the severe muscular weakness that accompanies parathyroid crisis, while Henson (1966) found the syndrome in eleven out of thirty-four patients with parathyroid tumors. Prineas, Mason, and Henson (1965) noted that muscular symptoms were not directly related to hypercalcemia in two patients studied, a finding confirmed and extended by R. Smith and Stern (1967). The causal situation appears to be as follows: a similar syndrome is seen in patients with hypercalcemia and osteomalacia; patients with osteomalacia, which can derive from several causes, are normocalcemic, while those with hyperparathyroidism may be both hypercalcemic and osteomalacic; in osteomalacic individuals and most patients with hyperparathyroidism, the cause of the muscular disability is related to defective vitamin D metabolism, but in a few patients with uncomplicated hyperparathyroidism, and especially parathyroid crisis, or hypercalcemia from other causes, disordered calcium metabolism is responsible. The role of calcitonin is unknown. Osteomalacia, in Britain most often due to intestinal malabsorption, is an important treatable cause of muscle disease. Clinical suspicion is particularly engendered by the exaggerated reflexes and muscular cramps, but other indications of hyperparathyroidism or osteomalacia are usually demonstrable if sought. Important laboratory tests include estimation of serum alkaline phosphatase, commonly raised in osteomalacia, and serum calcium studies. Skeletal survey with X-rays reveals the presence of ostemalacia, but bone biopsy may be needed to clinch the diagnosis. Treatment with vitamin D relieves the myopathy in patients with osteomalacia, whether linked with hyperparathyroidism or not, but this must be carefully controlled by serial estimations of serum calcium and alkaline phosphatase. When hypercalcemia appears to be the sole cause, appropriate steps are taken to restore serum levels to normal, and these steps naturally vary according to the underlying pathology.

E. Periodic Paralysis

Recurrent attacks of flaccid muscular weakness or paralysis occur in inherited and acquired disease states characterized by low serum potassium levels. Hyperkalemia may also be associated with muscular weakness. This section deals with the hereditary group.

1. Familial Hypokalemic Periodic Paralysis

This is a hereditary condition. Although it is usually familial, sporadic cases are encountered with identical clinical and biochemical features.

The onset is commonly in childhood or early adult life. The patient suffers recurrent attacks of flaccid muscular weakness. These episodes of weakness are liable to be present on waking; they are often provoked by a large meal with a high carbohydrate content, exposure to cold, high fluid intake, or vigorous exercise. The weakness affects the limbs and trunk, generally sparing the bulbar and respiratory muscles; it may be mild, or there may be complete paralysis. The distribution can be asymmetrical, and one limb may be affected alone. Even in severe attacks some distal movement of the limbs is usually possible, but tone is reduced according to the severity of the attack, and tendon reflexes are diminished or abolished. Sensation is unaffected. The somatic weakness may lead to difficulty in micturition or defecation. Episodes vary in severity and duration on the same patient. Recovery is often complete in a few hours, but sometimes several days elapse before there is full restoration of power. Death has been recorded in an attack, but this is excessively rare. Permanent muscular atrophy and weakness deriving from muscle damage form an unusual but well recognized complication. In patients thus afflicted, the continuing clinical picture resembles that of muscular dystrophy. Periodic paralysis must be differentiated from hysteria. In severe attacks, the distinction is easy; the abolition of tendon reflexes and loss of muscular response to electrical stimuli are helpful points. When the weakness is mild, the problem becomes more difficult, even when the diagnosis of periodic paralysis has been previously made. In susceptible persons, attacks may be induced by the administration of insulin and glucose, adrenalin, or fluorohydrocortisone.

Biemond and Daniels (1934) noted that the plasma potassium was lowered coincidentally with development of paralysis, and Aitken et al. (1937) showed that attacks could be terminated by giving potassium chloride. Many workers agree that there is positive potassium balance during attacks and that potassium moves into muscles (Allott and McArdle, 1938; Grob et al., 1957; Zierler and Andres, 1957). The excess potassium leaves muscle cells as recovery occurs. In patients susceptible to periodic paralysis, weakness begins to develop when the plasma potassium falls to 3 mEq/liter, becoming more marked with further falls in the level. There must be some peculiarity common to sufferers from the complaint, as experimentally produced hypokalemia in normal subjects does not produce paralysis with plasma levels lower than those encountered in periodic paralysis. In other hypokalemic disease states severe depletion is required before paralysis occurs. Thus, although potassium metabolism is important in this disease, other factors must operate; various theories have been advanced as to their nature, but none has been proved to date. McArdle (1969) has reviewed the state of knowledge. Even the mechanism of paralysis remains to be clarified (Buchthal, 1970). Muscle biopsy reveals a peculiar vacuolation of muscle fibers, and ultrastructural observations show that these vacuoles are the fruit of dilatation of the sarcoplasmic reticulum (Shy *et al.*, 1961; Engel, 1966). This vacuolar appearance is widespread in the muscles of patients with permanent muscular weakness complicating periodic paralysis. A similar change is seen in the muscles of patients with adynamia episodica hereditaria.

The course of periodic paralysis is variable. Prognosis is certainly worse in patients with permanent muscular atrophy, although crippling is partial. In general, the attacks become less severe with increasing years and may disappear altogether. Acute attacks are treated by oral administration of potassium chloride; intravenous injection of potassium is rarely called for. Prophylactic treatment is by no means satisfactory. Some patients are helped by daily medication with potassium chloride, while others respond to chlorothiazide or dichlorphenamide. Provocative situations, including undue exertion, high carbohydrate intake, and exposure to cold, should be avoided.

2. NORMOKALEMIC PERIODIC PARALYSIS

Tyler *et al.* (1951) reported a family in whom attacks of periodic paralysis occurred without any significant change in the serum potassium. The attacks did not respond to the administration of potassium salts. Poskanzer and Kerr (1961) described normokalemic attacks lasting for days or weeks, the paralysis being made worse by administration of potassium but improved by sodium chloride; complete control was achieved by daily medication with acetazolamide and 9-α-fluorohydrocortisone.

3. HYPERKALEMIC PERIODIC PARALYSIS (ADYNAMIA EPISODICA HEREDITARIA)

Gamstorp (1956) described this syndrome and gave it the name adynamia episodica hereditaria. Onset is often in infancy or childhood. Attacks are brief, usually lasting an hour or so, and can occur several times a week. The severity varies, but in most episodes, the patient is unable to stand or walk. Paralysis characteristically develops at rest after exercise, and, peculiarly, developing attacks can be postponed by exercise. Other provocative factors include cold, infections, and general anesthesia. Administration of potassium induces attacks. Some patients develop a permanent myopathy clinically identical with that seen in hyperkalemic periodic paralysis, and histopathological examination shows similar changes in affected muscles. Myotonia is found either clinically or electromyographically in most cases (McArdle, 1969; Gamstorp, 1970). Myotonic lid lag is found in some patients, and this may be accompanied by persistent furrowing of the forehead. Drager *et al.* (1958) and Layzer *et al.* (1967a) believe that hyperkalemic periodic paralysis and paramyotonia congenita are different manifestations of the same disease.

The episodic weakness that occurs in this complaint is accompanied by elevated plasma potassium readings, commonly 5 mEq/liter or above; these are levels that are not accompanied by weakness in normal individuals. Some degree of weakness can occur with values below 5 mEq/liter, but severe paralysis is seen with higher readings. McArdle (1969) has reviewed current knowledge of the disordered electrolyte distribution in these patients. The nature of the underlying enzymic defects is unknown. Buchthal (1970) states that the paralysis is due to a reversible depolarization of varying degree in different fibers, leading to block in some and hyperexcitability in others.

Adynamia episodica usually causes most disability in childhood and adolescence, but some patients remain liable to temporarily crippling attacks well into adult life. As in the hyperkalemic form, there is a general tendency to improvement with increasing age. Myotonia is more persistent. The permanent myopathy can be arrested by successful control of the intermittent paralysis. Treatment with diuretics that promote the excretion of potassium as well as sodium is beneficial. Dichlorphenamide is generally employed in Britain, but chlorothiazide or hydrochlorothiazide is also effective. Again, exposure to provocative situations should be avoided as far as possible.

4. DIFFERENTIAL DIAGNOSIS OF PERIODIC PARALYSIS

Three forms of periodic paralysis have been described. When the disease occurs in a family, the pattern of electrolyte abnormality is the same in affected individuals. However, occasional examples of periodic paralysis occurring with high, normal, or low potassium plasma levels in the same patient are encountered (Pearson, 1964, authors' personal observation). As Aström (1970) has remarked, the three forms are sufficiently similar to suggest that they are related; indeed, they may be differing manifestations of the same fundamental metabolic deficit. Nevertheless, separation of the three types is needed if treatment is to be effective, and it is necessary to measure the plasma potassium in an attack for certain diagnosis, except in young children of fully investigated families. Attacks can be safely induced for metabolic studies by one of the methods previously noted. An electrocardiographic record made during an attack can be helpful.

F. Disturbances of Potassium Metabolism

1. HYPOKALEMIC DISEASE STATES

As stated above, flaccid muscular weakness or paralysis may occur in any disease state where the plasma potassium is lowered, for example, certain types of kidney disease (especially renal tubular acidosis), primary aldosteronism commonly due to adrenal cortical tumor, and remarkable diarrhea and vomiting from any cause. Patients with renal tubular acidosis and adrenal cortical tumor may present with episodic paralysis, but clinical and laboratory investigation differentiate these conditions from the periodic paralysis described above. Persons addicted to aperients or purgatives occasionally develop hypokalemia and muscular weakness from loss of potassium by the bowel. In all these hypokalemic conditions, treatment is that of the underlying disease.

2. HYPERKALEMIC DISEASE STATES

Hyperkalemia occurs in salt-losing nephritis, acute renal cortical ischemia (as in the crush syndrome, mismatched blood transfusions, or after abortion), and in adrenal failure. The electrocardiogram becomes abnormal as the serum potassium rises. Occasionally, potassium poisoning, usually due to renal disease, causes the rapid onset of a flaccid paralysis which may be of ascending type. In these patients the tendon reflexes are lost, and respiratory failure and bulbar palsy may supervene. The muscles respond to electrical or even mechanical stimulation. Death results from cardiac arrest. Commonly, however, patients show nothing more than general muscular enfeeblement, perhaps with reduction of reflexes; indeed, death from cardiac arrest occurs without the appearance of remarkable muscle weakness. From the practical aspect the following points emerge: (1) hyperkalemia due to renal disease must be considered a rare cause of acute flaccid paralysis; (2) adrenal failure may present with muscular weakness as a prominent symptom. Treatment consists in restoring the electrolyte balance by methods appropriate to the underlying disease process.

3. PERIODIC PARALYSIS ASSOCIATED WITH THYROTOXICOSIS

This subject has been dealt with in Section VI,B,c above.

G. Glycogen Storage Diseases

Voluntary muscle is affected in several members of this complex group of rare, hereditary, metabolic defects. There are three varieties in which muscular involvement is known to be clinically important.

1. ACID MALTASE DEFICIENCY [TYPE II GLYCOGENOSIS (CORI, 1957); POMPE'S DISEASE]

Acid maltase is a lysosomal enzyme widely distributed in the body. When it is deficient, glycogen accumulates in many organs, but especially in the heart and skeletal muscles. Symptoms of heart failure appear in early infancy, and the muscles are weak and hypotonic. The patient usually dies before the age of 1 year. Muscle biopsy reveals the accumulation of glycogen. Recently, Zellweger *et al.* (1965) and Hudgson *et al.* (1968) have shown that the disease can exist in milder form with survival into adult life. The clinical picture is similar to that of muscular dystrophy and the disability can be slight. Again, diagnosis is made at muscle biopsy. There is no useful treatment for this complaint. Prognosis in the milder form depends on the presence or degree of cardiomyopathy.

2. MUSCLE PHOSPHORYLASE DEFICIENCY (TYPE V GLYCOGENOSIS; McARDLE'S DISEASE)

This rare disease is important as the first hereditary myopathy in which the metabolic defect was defined (McArdle, 1951). Because of the absence of phosphorylase, glycogen cannot be degraded to lactic acid in muscles, where it consequently accumulates. The fundamental enzymic defect was discovered by Schmid and Mahler (1959) and Mommaerts *et al.* (1959). It appears that the missing phosphorylase isoenzyme is specific to muscle (McArdle, 1969). The absence of phosphorylase can be shown histochemically and by studies on muscle homogenates; but the usual screening test is to measure blood lactate after ischemic exercise; in McArdle's disease, the lactate does not rise. McArdle's disease is marked by muscular pain, stiffness, and weakness on slight or moderate exertion. Symptoms are relieved by rest and their duration is proportionate to the degree of exercise undertaken. About one-third of all sufferers have continuing muscular weakness, but this is usually mild. Approximately one-half manifest myoglobinuria after exertion. Cardiac symptoms are also a feature of the complaint. The condition is apparently nonprogressive when neither permanent weakness nor myoglobinuria is present. Cardiac problems may develop with increasing age because of the large muscle blood flow that accompanies exercise. Treatment is generally disappointing (McArdle, 1969), and patients should try to live within the limits of the disability imposed upon them.

3. PHOSPHOFRUCTOKINASE DEFICIENCY (TYPE VIII GLYCOGENOSIS)

The symptoms of this glycogenosis are similar to those of McArdle's disease. There is absence of phosphofructokinase both from muscle and red blood cells (Tarui *et al.*, 1965; Layzer *et al.*, 1967b).

H. Hypermetabolic Myopathy

Only a single case of this unusual disease has been described (Luft *et al.*, 1962), but the report raises important theoretical implications. Briefly, a woman aged 35 showed signs of hypermetabolism from the age of 7, and her muscles were wasted and weak. Ultrastructural studies showed mitochondria in muscle to be increased in number and some in size. Biochemical investigation indicated that the mitochondria had a high respiratory rate that was not adequately controlled by the usual mechanisms. As a result of this failure, the individual was unable to adapt respiration to the demands of energy. Abnormal mitochondria have been seen in other types of muscle disease (Shy *et al.*, 1966; Price *et al.*, 1967). The significance of these observations is unknown, and the role of mitochondria in myopathy is uncertain. Several workers believe that the changes described have no specific nosological importance; that is, they form a secondary manifestation (Aström, 1970).

I. Myoglobinuria

Myoglobin is never found in the urine under normal conditions. Myoglobinuria occurs when there has been acute muscle damage from a variety of causes, including traumatic and ischemic lesions of muscle, polymyositis, metabolic disorders, such as phosphorylase deficiency, and certain toxic conditions. Among the toxic states are carbon monoxide poisoning, in which there can be acute extensive muscle necrosis, alcoholism, and sea snake (*Enhydrina schistosa*) bite. In all these situations, myoglobin escapes from muscle into the blood stream; when the blood level reaches a threshold, the pigment is excreted in the urine, and there is danger of kidney damage. Rowland *et al.* (1964) and Aström (1970) have reviewed the subject. The term "paroxysmal myoglobinuria" is reserved for a group of disorders represented by muscular pains, cramps, and paralysis, which are followed within hours by myoglobinuria. Korein *et al.* (1959) reviewed the subject and differentiated two types. First, paralytic paroxysmal myoglobinuria usually affects young adults. Attacks follow exertion or trauma and may lead to a permanent degree of incapacity. There is a positive family history in about one-third of all cases. Mortality is not high, and the disease can persist for many years. The second variety is idiopathic or toxic myoglobinuria, which usually begins in childhood and is commonly linked with an acute infection. Attacks are less frequent but more severe than in the first type, and mortality is higher. The pathogenesis of myoglobinuria is unknown. Current theories have been reviewed by Aström (1970). Treatment is prophylactic and symptomatic. Expert nephrological supervision is required.

J. Alcoholic Myopathy

Hed *et al.* (1955, 1962) first recognized the association of myopathy with alcoholism. Previously muscular symptoms in alcoholics had been ascribed to the peripheral neuropathy which afflicts some heavy drinkers. Alcoholic myopathy exists in acute and chronic forms, but affected persons are always chronic alcoholics. The acute form occurs after a bout of heavy drinking, with sudden onset of pain, cramps, muscular tenderness, and weakness, mainly involving the lower limbs. Myoglobinuria is present in severe cases and death can result from renal failure and hyperkalemia. Varying degrees of muscle necrosis are seen at histopathological examination according to the severity of the case. On the biochemical side, Perkoff *et al.* (1966) found low levels of phosphorylase activity in the muscles of alcoholic patients who also showed a poor lactate response to ischemic exercise. Ekbom *et al.* (1964) described the chronic form. Here, the clinical picture is one of slowly increasing weakness, especially in the lower limbs. There is no pain. Muscle biopsy reveals minor pathological changes. Aström (1970) regards the two forms as clinically and pathogenetically distinct, although they may coexist or overlap. The alcoholic myopathies respond to abstinence from the drug.

VII. Muscular Disorders Associated With Malignant Disease

The term carcinomatous myopathy was introduced by Henson *et al.* (1954) to indicate muscle disease arising in patients with cancer, with the implication that there is a causal relationship between the two events. Muscular disorders are also encountered in the reticuloses. There is a wide range of myopathies linked with malignant disease. Some of these are associated with well defined metabolic upsets, but the majority are ill understood. In this description, no account will be taken of the muscular atrophies secondary to neural lesions.

A. *Myopathy in Malignant Cachexia*

In this common condition, widespread muscular wasting and weakness occur, although muscular power is usually good considering the degree of atrophy present. Diminution of tendon reflexes is found when wasting is extreme. Similar clinical findings are commonplace in patients with other forms of chronic debilitating disease. Decubitus plays some part in the production of this syndrome and is particularly responsible for changes in the proximal musculature, which is little employed by a patient in bed, and indeed throughout the lower limbs. However, cachetic myopathy can be extreme in persons who remain ambulant, and there are obviously other mechanisms at work. Hildebrand and Cöers (1967) demonstrated changes in the peripheral nerves which could be responsible for some of the muscular atrophy found; they thought the neural changes were probably nutritional in origin. Histological studies on muscle from these patients reveal variation of fiber size and scattered degenerating and necrotic fibers. There is little correlation between the degree of atrophy and the microscopic changes described. Marin and Denny-Brown (1962) found muscle cells to be affected by a peculiar, granular pigmentary degeneration in some patients examined. More recently, Rebeiz (1970) has described group fiber atrophy similar to that seen in patients with lesions of anterior horn cells or peripheral nerves. He concluded that this type of atrophy was probably due to loss of anterior horn cells, demonstrated in his material, from carcinotoxic activity.

This common myopathy still requires elucidation. The cause or causes are presumably linked with the biochemical behavior of malignant growths. Costa and Holland (1965) have reviewed metabolic disturbances associated with malignant disease that could have systemic effects.

B. Proximal Myopathy

Some patients with malignant disease, especially bronchial carcinoma, present with muscular wasting and weakness affecting the proximal and axial musculature. The complaint is not linked with cachexia, and the growth may be concealed. About one-half of all men with acquired myopathy over 50 years of age ultimately prove to have cancer, usually of the bronchus (Shy and Silverstein, 1965). The condition is much rarer in females. Electrodiagnostic studies simply reveal the changes expected in primary muscle disease and histopathological abnormalities are limited to isolated, scattered, fiber necrosis. The cause of this myopathy is likely to be metabolic derangement from tumor biochemical activity.

C. Subacute Necrotizing Myopathy

This rare complaint is characterized by rapidly progressive muscular paralysis with diffuse fiber necrosis and no inflammatory infiltration (B. Smith, 1969; Urich and Wilkinson, 1970).

D. Myasthenia Gravis

The association of myasthenia gravis and thymic tumor is well known, but there is so far no evidence of significant links with other types of new growth.

E. Myasthenic Syndrome

This complaint is commonly found in persons with bronchial carcinoma, although it has been described with other cancers and in reticuloses. The clinical picture is one of muscular weakness and fatigability involving limb girdles and trunk; the lower limbs are usually affected first. External ophthalmoplegia and bulbar paresis are rarer complaints. While there is persistent weakness, this is increased by exertion (Henson, 1953; Henson *et al.*, 1954). Eaton and Lambert (1957) noticed a temporary increase in muscle power after brief exercise, a reversed myasthenic effect. The same workers and colleagues at the Mayo Clinic detailed the electrodiagnostic changes common but not specific to the syndrome. Elmquist and Lambert (1968) and Grob and Namba (1970) have elegantly demonstrated that the immediate cause of the neuromuscular defect is decreased transmitter release. Other clinical features, evidently not due to the myopathic element, include dryness of the mouth, impotence, and paraesthesiae in the extremities.

This remarkable syndrome is important, as it may constitute the first manifestation of malignant disease. Patients with the complaint are sensitive to muscle relaxants (Croft, 1958; Wise and MacDermot, 1962). Treatment is limited to exhibition of cholinergic drugs and guanidine, but the response is often partial.

F. *Dermatomyositis*

The association of dermatomyositis with malignant disease was noted in the nineteenth century. While the overall incidence of malignant disease in patients with dermatomyositis is about 20%, the figure rises to 50% in males over 40 (Arundell *et al.*, 1960) and even higher over 50 (Shy, 1962). The associated growth is usually a carcinoma; in our material, cancer of the bronchus and prostate have occurred most frequently in men; in women, cancer of the breast, ovary, and bronchus has been implicated. It is rare for dermatomyositis to be linked with lymphomas or leukemias, but we have examples of both. Clinical features of the muscle disease and the significance of the association have been discussed above. Dermatomyositis with malignant disease is treated along the usual lines, but the response is far less satisfactory than in uncomplicated cases. The myopathy can precede symptoms or diagnosis of the associated cancer, but it may also develop in persons with established neoplastic disease.

G. *Metabolic Myopathies*

Watson (1966) states that malignant disease is the commonest cause of hypercalcemia, and this biochemical disorder is particularly associated with benign and malignant tumors of the parathyroid gland, myeloma, and oat cell bronchial carcinoma. It also occurs with other growths, commonly when there are bony metastases. The clinical picture of hypercalcemic myopathy has been described above; the syndrome is rare in malignant disease states despite the relative frequency of hypercalcemia. Muscular weakness also occurs in dilutional and depletional hyponatremia caused by secretion of an antidiuretic hormonelike substance by oat cell bronchial cancer (Ross, 1963). This complication usually develops late in the course of the disease. Cushing's syndrome is manifest in patients with oat cell carcinoma when the growth secretes a corticotrophin-like substance. In these circumstances, the myopathy previously described can be present and may provide the leading complaints.

VIII. Myasthenia Gravis

Myasthenia gravis is generally classified as an affection of muscle, although the immediate defect is one of neuromuscular transmission. This peculiar and fascinating disease was probably first described by Thomas Willis in 1672, but Campbell and Bramwell (1900) provided the earliest detailed account of the clinical features. The overall female:male ratio in affected persons is generally quoted as 2:1, but below the age of 30, the female preponderance is greater. While myasthenia gravis may occur from early childhood to extreme old age, the incidence falls sharply after the age of 50.

A thymic tumor is present in about 10% of cases, and germinal centers are found in the medulla of the thymus in 70–80%. Apart from this association with thymic abnormality, myasthenia gravis is significantly linked with a group of hematological, endocrine, and connective tissue diseases. The more important of these are thyrotoxicosis and myxoedema, rheumatoid arthritis, disseminated lupus erythematosus, and Addisonian or aplastic anemia. Pure red cell aplasia forms a rare but important complication when there is a thymic tumor. Muscle pathology was reviewed by Russell (1953); the classic lesion is the lymphorrhage, but other changes are seen. Modifications in the terminal arborization of motor nerve endings were described by Cöers and Desmedt (1959) and Bickerstaff and Woolf (1960).

Certain indications of immunological disorder have been reported by many workers, including Strauss *et al.* (1960), White and Marshall (1961), and Miller *et al.* (1962), and these have been reviewed by Simpson (1969, 1970). Vetters (1965) noted a bound fluorescence in patients with thymomas, and this seems to be the most consistent finding. Grob and Namba (1970) remark that the defect in myasthenia gravis does not appear to be attributable to circulating globulins and publish experimental work that suggests that lymphocytes are involved.

The nature of the defect of neuromuscular transmission in myasthenia gravis has been the subject of prolonged debate (Simpson, 1969). Elmquist *et al.* (1964) and Elmquist (1965) have provided evidence of decreased transmitter (acetylcholine) release at motor nerve terminals, and Grob and Namba (1970) conclude that it seems likely the transmitter plays an important role in bringing about or aggravating the neuromuscular defect. The weight of current opinion favors a presynaptic abnormality.

In spite of intensive research, the etiology of the disease remains unknown. Simpson (1960) proposed an autoimmune hypothesis on several grounds, including the asociation with thymic abnormality. The characteristic clinical feature is abnormal muscular fatigability occurring above a ground of varying persistent weakness. While any muscle or group of muscles can be involved, the external ocular, facial, and shoulder girdle muscles are most commonly affected, and weakness of the muscles subserving mastication, swallowing, and speech is frequently present. In severe cases, the whole voluntary musculature is involved, but the upper limbs are more affected than the lower. However, severe fatigability and weakness can occur in one area—e.g., the external ocular or bulbar musculature—with little weakness elsewhere. Permanent weakness and atrophy occur in a proportion of cases, commonly those of longstanding. The course and severity of the complaint are remarkably variable, but overall there is reduced life expectancy and much chronic disability. Myasthenia is most active during the first 5 years, and the majority of deaths directly caused by the disease occur within 5–10 years of onset; however, there is no absolute rule. Prognosis is difficult to determine in the individual case. The variable course of the complaint is best illustrated by three personal cases: (1) A woman aged 40 died in 9 months from onset, the last 6 months of her life being spent in hospital while every known form of treatment was employed. (2) A man aged 18 suffered classic myasthenic symptoms that persisted for about 2 months in all, and there has been no recurrence throughout 16 years' observation without treatment. (3) Another man, aged 71, developed myasthenia when 17, but remains fully employed though still demonstrating myasthenia on clinical examination.

Diagnosis rests on the demonstration of weakness that responds to the intravenous injection of edrophonium chloride or to intramuscular administration of neostigmine methyl sulfate. In long-standing cases, and in patients with permanent weakness and atrophy, the response to these drugs may be reduced or abolished. Myasthenia gravis must be distinguished both from the paraneoplastic myasthenic syndrome and polymyositis, in which a distinct myasthenic element is occasionally found.

Treatment can be medical or surgical. Myasthenia may be controlled by cholinergic drugs, of which the chief are neostigmine bromide and pyridostigmine. Immunosuppressive drugs, especially azothiaprine, have been used in recent years, and steroids, and cytotoxic preparations have been employed in patients with thymomas. Surgical treatment consists in thymectomy (Eaton and Clagett, 1955; Simpson, 1956; Henson *et al.*, 1965). Thymectomy is particularly beneficial in female sufferers under the age of 30 with a history of less than 2 years; the operation has proved far less effective in patients with thymomas. Simpson (1969) concludes that surgically treated patients without thymomas show a much lower mortality and have a greater chance of complete remission or substantial improvement when compared with a medically treated group. Assisted methods of respiration and feeding are life saving in persons with respiratory embarrassment and dysphagia. While the treatment of myasthenia gravis is by no means satisfactory, the outlook has changed radically since Walker's (1934) initial observation on the effect of prostigmine and the subsequent introduction of thymectomy by Blalock *et al.* (1941).

While myasthenic patients usually present no major problems during pregnancy or labor, they are delivered of myasthenic infants in about 10% of cases. Neonatal myasthenia is a self-limiting complaint that clears within 3 months of birth, but cholinergic drugs are required to overcome feeding difficulties.

REFERENCES

Adams, R. D. (1969). *In* "Disorders of Voluntary Muscle" (J. N. Walton, ed.), 2nd ed., p. 143. Churchill, London.
Adams, R. D., Denny-Brown, D., and Pearson, C. M. (1953). "Diseases of Muscle. A Study in Pathology," 1st ed. Cassell, London.
Adams, R. D., Denny-Brown, D., and Pearson, C. M. (1962). "Diseases of Muscle. A Study in Pathology," 2nd ed. Harper (Hoeber), New York.
Aitken, R. S., Allott, E. N., Castleden, L. L. N., and Walker, M. (1937). *Clin. Sci.* 3, 47.
Allott, E. N., and McArdle, B. (1938). *Clin. Sci.* 3, 329.
Aloujouanine, T., and Nick, J. (1945). *Paris Med.* 129, 346.
Arundell, F. D., Wilkinson, R. D., and Hasrick, J. R. (1960). *Arch. Dermatol.* 82, 772.
Aström, K. E. (1970). *Acta Neurol. Scand, Suppl.* 43, 177.

Barnes, S. (1932). *Brain* **55**, 1.

Bathurst, L. W. (1895). *Lancet* **2**, 529.

Batten, F. E. (1909–1910). *Quart. J. Med.* **3**, 313.

Becker, P. E., and Keiner, F. (1955). *Arch. Psychiat. Nervenkr.* **193**, 427.

Bickerstaff, E. R., and Woolf, A. L. (1960). *Brain* **83**, 10.

Biemond, A., and Daniels, A. P. (1934). *Brain* **57**, 91.

Blalock, A., Harvey, A. M., Ford, F. R., and Lilienthal, J. L. (1941). *J. Amer. Med. Ass.* **117**, 1529.

Bramwell, E. (1922). *Proc. Roy. Soc. Med.* **16**, 1.

Brandt, S. (1950). "Werdnig-Hoffman's Infantile Progressive Muscular Atrophy." Munksgaard, Copenhagen.

Buchthal, F. (1970). *Acta Neurol. Scand., Suppl.* **43**, 129.

Bundey, S., Carter, C. O., and Soothill, J. F. (1970). *J. Neurol., Neurosurg. Psychiat.* **33**, 279.

Campbell, H., and Bramwell, E. (1900). *Brain* **23**, 277.

Caughey, J. E., and Myrianthopoulos, N. C. (1963). "Dystrophia Myotonica and Related Disorders." Thomas, Springfield, Illinois.

Chou, S. M. (1967). *Science* **158**, 1453.

Cöers, C., and Desmedt, J. E. (1959). *Acta Neurol. Psychiat. Belg.* **59**, 539.

Cori, G. T. (1957). *Mod. Probl. Paediat.* **3**, 344.

Costa, G., and Holland, J. F. (1965). *In* "Remote Effects of Cancer on the Nervous System" (R. L. Brain and F. H. Norris, eds.), Contemporary Neurology Symposia, Vol. 1, p. 125. Grune & Stratton, New York.

Croft, P. B. (1958). *Brit. Med. J.* **1**, 181.

Denny-Brown, D. (1939). *Proc. Roy. Soc. Med.* **32**, 876.

Dodge, P. R. (1960). *Res. Publ., Ass. Res. Nerv. Ment. Dis.* **38**, 497.

Drager, G. A., Hammill, J. F., and Shy, G. M. (1958). *Arch. Neurol. Psychiat.* **80**, 1.

Dreyfus, J. C., Schapira, G., Schapira, F., and Démos, J. (1970). *In* "Muscle Diseases" (J. N. Walton, N. Canal, and G. Scarlato, eds.), p. 417. Excerpta Med. Found., Amsterdam.

Duchenne, G. B. (1868). *Arch. Gen. Med.* **11**, 5, 179, 305, 421, and 552.

Eaton, L. M., and Clagett, O. T. (1955). *Amer. J. Med.* **19**, 703.

Eaton, L. M., and Lambert, E. H. (1957). *J. Amer. Med. Ass* **163**, 1117.

Ebashi, S., Toyokura, Y., Momoi, H., and Sugita, H. (1959). *J. Biochem. (Tokyo)* **46**, 103.

Ekbom, K., Kirstein, L., Hed, R., and Aström, K. E. (1964). *Arch. Neurol.* **10**, 449.

Elmquist, D. (1965). *Acta Physiol. Scand., Suppl.* **249**.

Elmquist, D., and Lambert, E. H. (1968). *Mayo Clin. Proc.* **43**, 689.

Elmquist, D., Hofmann, W. W., Kugelberg, J., and Quastel, D. M. J. (1964). *J. Physiol. (London)* **174**, 417.

Engel, A. G. (1966). *Mayo Clin. Proc.* **41**, 797.

Gamstorp, I. (1956). *Acta Paediat. (Stockholm), Suppl.* **108**.

Gamstorp, I. (1970). *Acta Neurol. Scand., Suppl.* **43**, 109.

Gleeson, J. A., Swann, J. C., Hughes, D. T. D., and Lee, F. I. (1967). *Brit. J. Radiol.* **40**, 96.

Greenfield, J. G., Cornman, T., and Shy, G. M. (1958). *Brain* **81**, 461.

Grob, D., and Namba, T. (1970). *In* "Muscle Diseases" (J. N. Walton, N. Canal, and G. Scarlato, eds.), p. 167. Excerpta Med. Found., Amsterdam.

Grob, D., Liljestrand, A., and Johns, R. J. (1957). *Amer. J. Med.* **23**, 356.

Hed, R., Larsson, H., and Wahlgren, I. (1955). *Acta Med. Scand.* **152**, 459.

Hed, R., Lundmark, C., Fahlgren, H., and Orell, S. (1962). *Acta Med. Scand.* **171**, 585.

Henson, R. A. (1953). *Proc. Roy. Soc. Med.* **46**, 859.

Henson, R. A. (1966). *J. Roy. Coll. Physicians, London* **1**, 41.

Henson, R. A., Russell, D. S., and Wilkinson, M. (1954). *Brain* **77**, 82.

Henson, R. A., Stern, G. M., and Thompson, V. C. (1965). *Brain* **88**, 11.

Hildebrand, J., and Cöers, C. (1967). *Brain* **90**, 67.

Hudgson, P., Gardner-Medwin, D., Worsfold, M., Pennington, R. J. T., and Walton, J. N. (1968). *Brain* **91**, 435.

Kakulas, B. A. (1966a). *Nature (London)* **210**, 1115.

Kakulas, B. A. (1966b). *J. Pathol. Bacteriol.* **91**, 495.

Kakulas, B. A. (1970). *In* "Muscle Diseases" (J. N. Walton, N. Canal, and G. Scarlato, eds.), p. 377. Excerpta Med. Found., Amsterdam.

Kiloh, L. G., and Nevin, S. (1951). *Brain* **74**, 115.

Klein, D. (1958). *J. Genet. Hum., Suppl.* **7**.

Korein, J., Coddon, D. R., and Mowrey, F. H. (1959). *Neurology* **9**, 767.

Kugelberg, E., and Welander, L. (1956). *AMA Arch. Neurol. Psychiat.* **75**, 500.

Landouzy, L., and Déjèrine, L. (1884). *C. R. Acad. Sci.* **98**, 53.

Layzer, R. B., Lovelace, R. E., and Rowland, L. P. (1967a). *Arch. Neurol.* **16**, 455.

Layzer, R. B., Rowland, L. P., and Ranney, H. M. (1967b). *Arch. Neurol.* **17**, 512.

Lee, F. I., and Hughes, D. T. D. (1964). *Brain* **87**, 521.

Lemann, J., Jr., and Donatelli, A. A. (1964). *Ann. Intern. Med.* **60**, 477.

Luft, R., Ikkos, D., Palmieri, G., Ernste, L., and Afzelius, B. (1962). *J. Clin. Invest.* **41**, 1776.

Lynas, M. (1957). *Ann. Hum. Genet.* **21**, 318.

McArdle, B. (1951). *Clin. Sci.* **10**, 13.

McArdle, B. (1969). *In* "Disorders of Voluntary Muscle" (J. N. Walton, ed.), 2nd ed., p. 607. Churchill, London.

McKenzie, J. M. (1966). *Amer. Thyroid Ass., Chicago* p. 44.

Marin, O., and Denny-Brown, D. (1962). *Amer. J. Pathol.* **41**, 23.

Mastaglia, F. L., Barwick, D. D., and Hall, R. (1970). *Lancet* **2**, 907.

Miller, J. F. A. P., Marshall, A. H. E., and White, R. G. (1962). *Advan. Immunol.* **2**, 111.

Millikan, C. H., and Haines, S. F. (1953). *AMA Arch. Intern. Med.* **92**, 5.

Mommaerts, W. F. A. M., Illingworth, B., Pearson, C. M., Guillory, R. J., and Seraydarian, K. (1959). *Proc. Nat. Acad. Sci. U.S.* **46**, 791.

Muller, R., and Kugelberg, E. (1959). *J. Neurol., Neursurg. Psychiat.* **22**, 314.

Oppenheim, H. (1900). *Monatsschr. Psychiat. Neurol.* **8**, 232.

Paine, R. A. (1963). *Develop. Med. Child. Neurol.* **5**, 115.

Paulley, J. W., and Hughes, J. P. (1960). *Brit. Med. J.* **2**, 1562.

Pearce, J. M. S., Pennington, R. J., and Walton, J. N. (1964). *J. Neurol., Neurosurg. Psychiat.* **27**, 96.

Pearson, C. M. (1964). *Brain* **87**, 341.

Pearson, C. M. (1966). *Annu. Rev. Med.* **17**, 63.

Pearson, C. M. (1969). *In* "Disorders of Voluntary Muscle" (J. N. Walton, ed.), 2nd ed., p. 501. Churchill, London.

Pennington, R. J. T. (1969). *In* "Disorders of Voluntary Muscle" (J. N. Walton, ed.), 2nd ed., p. 385. Churchill, London.

Perkoff, G. T., Hardy, P., and Welez-Garcia, E. (1966). *N. Engl. J. Med.* **274**, 1277.

Peter, J. B., Worsfold, M., and Stempel, K. (1970). *In* "Muscle Diseases" (J. N. Walton, N. Canal, and G. Scarlato, eds.), p. 506. Excerpta Med. Found., Amsterdam.

Pinchera, A., Pinchera, M. G., and Stanbury, J. B. (1965). *J. Clin. Endocrinol. Metab.* **25**, 189.

Poskanzer, D. C., and Kerr, D. N. S. (1961). *Amer. J. Med.* **31**, 328.

Price, H. M., Gordon, G. B., Munsat, T. L., and Pearson, C. M. (1967). *J. Neuropathol. Exp. Neurol.* **26**, 475.

Prineas, J. W., Mason, A. S., and Henson, R. A. (1965). *Brit. J. Med.* **1**, 1034.

Rebeiz, J. J. (1970). *In* "Muscle Diseases" (J. N. Walton, N. Canal, and G. Scarlato, eds.), p. 383. Excerpta Med. Found., Amsterdam.

Rose, A. L., and Walton, J. N. (1966). *Brain* **89**, 747.

Ross, E. J. (1963). *Quart. J. Med.* **32**, 297.

Rowland, L. P., Fahn, S., Hirschberg, E., and Harter, P. H. (1964). *Arch. Neurol.* **10**, 537.

Russell, D. S. (1953). *J. Pathol. Bacteriol.* **65**, 279.

Sandifer, P. H. (1955). *Proc. Roy. Soc. Med.* **48**, 186.

Satoyoshi, E., and Kinoshita, M. (1970). *In* "Muscle Diseases" (J. N. Walton, N. Canal, and G. Scarlato, eds.), p. 455. Excerpta Med. Found., Amsterdam.

Satoyoshi, E., Murakami, K., Kowa, H., Kinoshita, M., Noguchi, K., Hoshina, S., Nishiyama, Y., and Ito, K. (1963). *Neurology* **13**, 645.

Schmid, R., and Mahler, R. F. (1959). *J. Clin. Invest.* **38**, 1044.

Short, J. K. (1963). *Neurology* **13**, 526.

Shy, G. M. (1961). *Res. Publ., Ass. Res. Nerv. Ment. Dis.* **38**, 274.

Shy, G. M. (1962). *World Neurol.* **3**, 149.

Shy, G. M., and Magee, K. R. (1956). *Brain* **79**, 610.

Shy, G. M., and Silverstein, I. (1965). *Brain* **88**, 515.

Shy, G. M., Wanko, T., Rolley, P. T., and Engel, A. G. (1961). *Exp. Neurol.* **3**, 53.

Shy, G. M., Engel, W. K., Somers, J. E., and Wanko, T. (1963). *Brain* **86**, 793.

Shy, G. M., Gonatas, N. K., and Perez, M. (1966). *Brain* **89**, 133.

Simpson, J. A. (1956). *Proc. Roy. Soc. Med.* **49**, 795.

Simpson, J. A. (1960). *Scot. Med. J.* **5**, 419.

Simpson, J. A. (1969). *In* "Disorders of Voluntary Muscle" (J. N. Walton, ed.), 2nd ed., p. 541. Churchill, London.

Simpson, J. A. (1970). *In* "Muscle Diseases" (J. N. Walton, N. Canal, and G. Scarlato, eds.), p. 14. Excerpta Med. Found., Amsterdam.

Smith, B. (1969). *J. Pathol.* **97**, 207.

Smith, R., and Stern, G. (1967). *Brain* **90**, 593.

Steinert, H. (1909). *Deut. Z. Nervenheilk.* **37**, 58.

Stevenson, A. C. (1953). *Ann. Eugen., London* **18**, 50.

Strauss, A. J. L., Smith, C. W., Cage, G. W., Van der Geld, H. W. R., McFarlin, D. E., and Barlow, M. (1960). *Proc. Soc. Exp. Biol. Med.* **105**, 184.

Tarui, S., Okuno, G., Okura, Y., Tanaka, T., Suda, M., and Nishikawa, M. (1965). *Biochem. Biophys. Res. Commun.* **19**, 517.

Thomasen, E. (1948). "Thomsen's Disease, Paramyotonia Dystrophia Myotonica." Universitetsforlaget, Aarhus.
Thomsen, J. (1876). *Arch. Psychiat. Nervenkrankh.* **6**, 702.
Tizard, J. P. M. (1949). *Proc. Roy. Soc. Med.* **42**, 80.
Turner, J. W. A. (1940). *Brain* **63**, 163.
Turner, J. W. A. (1949). *Brain* **72**, 25.
Turner, J. W. A., and Lees, F. (1962). *Brain* **85**, 733.
Tyler, F. H., and Wintrobe, M. M. (1950). *Ann. Intern. Med.* **32**, 72.
Tyler, F. H., Stephens, F. E., Gunn, F. D., and Perkoff, G. T. (1951). *J. Clin. Invest.* **30**, 492.
Urich, H., and Wilkinson, M. (1970). *J. Neurol., Neurosurg. Psychiat.* **33**, 398.
Vetters, J. M. (1965). *Immunology* **9**, 93.
Vicale, C. T. (1949). *Trans. Amer. Neurol. Ass.* **74**, 143.
Walker, M. B. (1934). *Lancet* **1**, 1200.
Walton, J. N. (1956). *Proc. Roy. Soc. Med.* **49**, 107.
Walton, J. N. (1957a). *J. Neurol., Neurosurg. Psychiat.* **20**, 144.
Walton, J. N. (1957b). *Proc. Roy. Soc. Med.* **50**, 301.
Walton, J. N. (1963). *In* "Muscular Dystrophy in Man and Animals" (G. H. Bourne and M. N. Golarz, eds.), p. 264. Hafner, New York.
Walton, J. N., and Gardner-Medwin, D. (1969). *In* "Disorders of Voluntary Muscle" (J. N. Walton, ed.), 2nd ed., p. 455. Churchill, London.
Walton, J. N., and Nattrass, F. J. (1954). *Brain* **77**, 169.
Watson, L. (1966). *Australas. Ann. Med.* **15**, 359.
Welander, L. (1951). *Acta Med. Scand., Suppl.,* **265**.
Welander, L. (1957). *Acta Genet. Med. Gemellol.* **7**, 321.
Wharton, B. A. (1965). *Lancet* **1**, 603.
White, R. G., and Marshall, A. H. E. (1961). *Lancet* **1**, 1030.
Wise, R. P., and MacDermot, V. (1962). *J. Neurol., Neurosurg. Psychiat.* **25**, 31.
Yates, D. A. H. (1970). *In* "Muscle Diseases" (J. N. Walton, N. Canal, and G. Scarlato, eds.), p. 482. Excerpta Med. Found., Amsterdam.
Zellweger, H., Illingworth, B., McCormick, W. F., and Jun-Bi Tu (1965). *Ann. Paediat.* **205**, 413.
Zellweger, H., Afifi, A., McCormick, W. F., and Mergner, W. (1967). *Amer. J. Dis. Child.* **114**, 591.
Zierler, K. L., and Andres, R. (1957). *J. Clin. Invest.* **36**, 730.

8

GENETIC ASPECTS OF MUSCULAR AND NEUROMUSCULAR DISEASES

H. WARNER KLOEPFER AND JOHN N. WALTON

I. Basis for the Selection of 139 Genetically Determined
Disorders of the Nervous and Neuromuscular Systems

From a search of the literature, genetic information was located for 139 of the inherited disease entities in the comprehensive classification prepared by the Research Group on Neuromuscular Disorders of the World Federation of Neurology (1968). Neither this list nor the genetic information for any one disorder is complete, just as there was no claim that this comprehensive classification was complete. McKusick (1971) has also attempted to list and classify these disorders, and we acknowledge the valuable help we have obtained from his work.

Names used to identify each of the disorders refer to publications that are most likely to be known to investigators in the field of neuromuscular disease in the same way that identifying names or eponyms used in the comprehensive classification were not necessarily those of the authors who first described the disorders. It should be emphasized that nongenetic factors may also be associated with a specific genetic entity. For example, peripheral neuropathy induced by isoniazid requires both a particular genotype and a drug. Some major genes may require for their expression one or more additional genes, but since no secondary genes have been precisely defined for the disorders mentioned in this chapter, only the known or presumed effects of major genes will be included.

It must be stressed that the emphasis of this chapter is genetic and not clinicopathological. Thus, among the entities listed are some that many neurologists and neuropathologists do not as yet accept as being independent disease entities, as the clinical and investigative evidence reported to date is often inadequate to establish the pathogenesis and nosology of the disorder described. Some indeed may eventually prove to be no more than variants of other well recognised diseases and may not, therefore, prove to be due to the effects of a number of different mutant genes. This is, however, for future research to decide, and no attempt is made in this chapter to do more than simply to list and identify the many disorders of this type that have been described. As we have as yet no reliable means, other than clinicopathological description, of identifying individual genotypes, we do not propose to discuss controversy concerning classification as argued between the "lumpers," who regard many of these disorders as being variants of a small number of genotypes, and the "splitters," who believe that minor differences in clinical expression probably imply basic differences in genotype. For reasons of space, we are including only brief (often telegraphically brief and clinically inadequate) comments upon the clinical features and pathology of many of the disorders described in an attempt to make our commentary as comprehensive as possible and thus the more valuable as a reference source. More detailed clinical descriptions and classifications appear elsewhere in this volume. We have also included comments on a number of inherited disorders of the nervous system that do not affect primarily the lower motor neurons and voluntary muscles (the neuromuscular apparatus), because in many of these disorders (particularly the group of hereditary ataxias), peripheral nerve involvement has been shown by modern electrophysiological and pathological techniques to be more common than was realized even a few years ago.

II. Genetic Terminology and the Risk of Affection in Relatives According to Pattern of Transmission

To visualize more clearly the transmission and effects of the genetic disorders cited in this chapter, we should be reminded that each person was once a fertilized egg with a single nucleus containing some 100,000 genes distributed on 23 pairs of chromosomes. One member of each pair of chromosomes came originally from each parent. After 100 trillion copies of this original nucleus were made, each gene was distributed in the nucleus of all cells of the body, although the clinical results of a mutant gene can be due to damage limited to muscle cells or to neurons that innervate muscle cells. Genes located on any chromosome that is not an X or Y chromosome are said to be autosomal.

A. Autosomal Dominant Pattern of Transmission

When an autosomal gene changes to a mutant gene which causes a neuromuscular disorder and when the mutant gene typically expresses its effect in the presence of a normal contrasting allele (the gene in the same location on the other member of the pair of chromosomes) the new mutation is said to be dominant and the contrasting pair of genes are said to be heterozygous. Except for two (see numbers 45 and 84 in Section V) out of 75 dominant genes listed in this chapter, only heterozygotes have been described because two individuals affected by the same dominant gene seldom have children who could be homozygous. Among the 139 genetic entities listed, 54% are dominant.

When a heterozygote has offspring, half are expected to receive the mutant gene and half are expected to receive the normal contrasting allele. The half that receive the mutant gene will be expected to develop muscular weakness at the appropriate age if the genotype is 100% penetrant. Each mutant gene has a characteristic age range of onset and sometimes also a characteristic age range for duration of affection.

Penetrance is the percentage of individuals with a mutant genotype who become affected after they have lived beyond the latest age for the first appearance of clinical symptoms. Not only may a certain percentage of persons with a given mutant gene not become affected, but also the clinical expression of the genotype may range from very slight to very severe neuromuscular involvement. For most of the disorders due to dominant genes, as listed in Section V of this chapter, penetrance is not 100%. The data are insufficient in most instances to state the exact penetrance, but it may be assumed that penetrance is less than 100% unless a penetrance figure is given. Penetrance often varies in the two sexes to cause a deviation in the relative proportion of affected males and females. A sex ratio of approximately 1:1 may be assumed unless stated otherwise.

An example will be given to show how penetrance affects risk figures. If an autosomal dominant gene is 50% penetrant, the risk of an affected offspring would be 1:4. This value may be obtained by multiplying $\frac{1}{2}$ (chance that the chromosome with the dominant gene will be passed to a gamete) by $\frac{1}{2}$ (chance that the genotype will become affected when the penetrance is 50%). Since about one-half of all rare dominant mutations encountered in a series of cases may be new ones, a parent of an affected child would not be expected to be affected more than half of the time if the gene were 100% penetrant. When a dominant gene is both rare and has a low penetrance, even fewer affected parents will be expected to be found. Another reason a parent may not be found to be affected in a retrospective investigation is due to the fact that offspring are often more likely to be produced by mildly affected heterozygotes (with low penetrance) than by those more severely affected.

B. *Autosomal Recessive Pattern of Transmission*

When an autosomal gene changes to a mutant gene that typically is not expressed in the presence of its normal contrasting allele, the new mutation is said to be recessive. Heterozygotes may pass on a recessive gene to half of their offspring generation after generation without knowing that they are heterozygous carriers. When two such heterozygotes have children, one fourth of them will be expected to receive the mutant gene from each parent. They will be homozygous, and at the appropriate age, they will be expected to develop muscular weakness if the genotype is 100% penetrant. Although typically affection is limited to the homozygote in certain genetic entities, mild clinical manifestations may sometimes be observed in the heterozygous carrier. Of the 139 entities listed in this chapter, 37% are recessive. In contrast to disorders of dominant inheritance, those that are recessive often cause more severe affection, are more frequently 100% penetrant, and often have an earlier age of onset for clinical symptoms. Parents are (statistically) more closely related (consanguinity) than is characteristic for random matings, and when secondary cases occur, they usually are limited to the siblings, or possibly to cousins. However, there are exceptions to these generalizations.

An individual with a recessive disorder seldom has an affected offspring or an affected parent, because the random chance of marrying a carrier with the same rare recessive gene is about 1 in 100. The chance that the heterozygous parent will pass the mutant gene to an offspring is 1:2, which makes the risk that an affected person will have an affected child about 1:200. This risk is about 200 times greater than the random risk of affection. Not only are both normal parents of offspring with recessive conditions heterozygous carriers, but so too are half the siblings of such parents, and one-quarter of the siblings of grandparents will be expected to be heterozygous carriers. The prevalence of carriers among those individuals who marry cousin relatives explains why secondary cases are more likely to be found among cousins for rare recessive disorders than among relatives in previous generations.

C. Recessive X-Chromosomal (Sex-Linked or X-Linked) Pattern of Transmission

Since there are 22 pairs of autosomes and only 1 pair of sex chromosomes, it is no surprise that 93% of the genetic entities listed in this chapter were found to be located on an autosome. When a gene on the X chromosome changes to a mutant gene that is not expressed typically in the female heterozygote, the mutation is a recessive X-chromosomal gene. Since the male has only one X chromosome, such a mutant gene is always expressed with the result that these affections are virtually limited to males. Affected males seldom have children, and when they do, they rarely choose a mate who is a carrier of the same rare gene. Consequently there is little opportunity for female homozygotes to be born. As in the autosomal heterozygotes, females with a recessive X-chromosomal gene on one member of the X chromosome pair typically show no symptoms or possibly only mild ones. However, techniques are now becoming available for the identification of some such heterozygotes (female carriers) as, for instance, in the X-linked Duchenne type muscular dystrophy. Among the list of 139 genetic disorders, about 7% were recessive X-chromosomal. If all chromosomes were the same length and carried the same number of genes, one might expect the X chromosome to have 4.3% of all genes. The fact that the X chromosome is longer than average means that more than 4.3% of the genes can be located on it.

When secondary cases are found in a kindred with this type of mutant gene, they usually will be male siblings, maternal uncles, sons of maternal aunts, or more distant males in the maternal line. An affected male has no greater risk of having an affected offspring than a male picked at random, because in each instance all sons get their X chromosomes from their mothers. All daughters of affected males will be heterozygous carriers with a 1:4 chance of having an affected male child. Female siblings of affected males have a 1:8 chance of having an affected child.

D. Dominant X-Chromosomal Pattern of Transmission

When a gene on the X chromosome changes to a mutant gene that is expressed typically in the female heterozygote, the mutation is a dominant X-chromosomal gene. Only one such entity was found among the 139 listed in this chapter. A male with such a gene will not only be expected to become affected at the appropriate age if the gene is 100% penetrant, but he will also pass the gene on to all his daughters and to none of his sons, all daughters will be expected to become affected. Affected heterozygous daughters will pass the mutant gene to half their offspring of both sexes, but females usually become less severely affected, due to the influence of the normal contrasting allele.

E. Sporadic Pattern of Transmission

Two of the 139 mutant entities in this chapter were found to be sporadic. If a dominant mutation is 100% penetrant and if the effects of the mutation are so severe that affected individuals are never able to reproduce, each affected person will be the result of a new mutation and each case will typically be isolated within families. Reduced penetrance or something less than lethal expression in a given parent might make it possible for more than one offspring to be affected.

III. Linkage

Genetic linkage occurs when genes for two different traits are located on the same pair of chromosomes. Linkage studies give information that may be useful in identifying a particular genetic eentity that otherwise may be confused with another genetic disorder. Only in a few instances have investigators of neuromuscular and muscular disease looked for linkage relations. Typically, correlations between traits within families are not detected except by applying an appropriate statistical procedure because genes may be in repulsion phase in one sibship and in coupling phase in another. This means that the correlation between traits will be positive in one sibship and equally negative in another, causing no correlation between the traits among individuals when information is pooled from various families.

IV. Incidence

Incidence information is important in evaluating the meaning of risk figures for affection within families in relation to random risk. Although incidence values are included in Section V when known, it should be emphasized that such figures relate only to the particular population studied by a given investigator. Incidence figures typically vary greatly from one geographic region to another and from one racial group to another. On the other hand, penetrance figures usually do not vary so much from different population studies for the same gene.

V. Brief Descriptions of Genetically-Determined Disorders of the Nervous and Neuromuscular System

A. *Disorders Principally Involving the Anterior Horn Cells of the Spinal Cord*

1. RECESSIVE TYPE OF WERDNIG (1891)–HOFFMANN (1893) DISEASE (INFANTILE SPINAL MUSCULAR ATROPHY)

Brandt (1949) reported 112 cases in 70 kindreds in Denmark. The disorder was verified in 51 cases in which age of onset was from birth to 12 months, and death usually resulted from respiratory infection within a few months or years after progressive paralysis of skeletal muscles.

Autosomal recessive inheritance associated with an eightfold increase in parental consanguinity was suggested in all kindreds except two in which transmission was compatible with an autosomal dominant gene. Incidence was 4.4×10^{-6}. Oppenheim's amyotonia congenita was the initial diagnosis in 37 of the 112 cases reported. Concordance in monozygotic twins has been reported. Tizard (1969) reviews the recent literature.

2. RECESSIVE TYPE OF SPINAL MUSCULAR ATROPHY (KUGELBERG AND WELANDER, 1956)

Spira (1963) reported 2 sibships with 7 affected in which all 4 parents were second cousins with common grandparents. Consanguinity of parents and affection limited to siblings has been reported frequently. In a review of 60 cases reported between 1942 and 1965, Smith and Patel (1965) noted a male:female sex ratio of 2:1 with an age of onset from 2–18 years (average 9.4) and duration of 2–40 years (average 18). A family history of other affected members was reported in 70% of all index cases. Not only were autosomal recessive and autosomal dominant patterns of inheritance reported in the original families of Kugelberg and Welander (1956), but in addition, both patterns of inheritance have been reported by other authors, though the recessive type is much the more common; no clinical criteria are available that distinguish one genetic type from the other. The clinical features may overlap with those of Werdnig–Hoffmann disease, or else both entities may be variants of a single gene, since cases typical of the Werdnig–Hoffmann and Kugelberg–Welander types have been reported in the same sibship (Lovelace *et al.*, 1966; Gardner-Medwin *et al.*, 1967).

3. DOMINANT TYPE OF SPINAL MUSCULAR ATROPHY (KUGELBERG AND WELANDER, 1956)

Many kindreds showing transmission by an autosomal dominant gene with reduced penetrance (particularly in the female) have been reported including 7 affected individuals in 3 generations (Tsukagoshi *et al.*, 1966) and 20 affected in 4 generations. The clinical features of such cases appear to be virtually indistinguishable from those observed in patients with the recessively inherited disorder.

4. RECESSIVE TYPE OF AMYOTROPHY WITH ARTHROGRYPOSIS MULTIPLEX CONGENITA (OTTO, 1841)

Drachman and Banker (1961) presented the pathology of a case and reviewed twenty reports from the literature. Features included severe joint deformities, micrognathia, neurogenic muscular atrophy with fatty and fibrous replacement of muscle, degeneration and loss of motor neurons in the spinal cord. Von Frischknecht *et al.* (1960) reported 3 siblings with features simulating amyotrophic lateral sclerosis with lack of Betz cells in the gigantopyramidal layer of the motor cortex. Abdominal musculature was absent.

5. DOMINANT TYPE OF SCAPULOPERONEAL MUSCULAR ATROPHY (KAESER, 1964)

Davidenkow (1939) reported 2 kindreds with 12 affected individuals in 3 generations. He believed that the syndrome was first described by Wohlfahrt (1926) who reported 10 affected persons in 4 generations. Features include atrophy of the lower legs, talipes equinovarus, footdrop, bilateral winging of the scapulae, and sometimes bulbar involvement occurring later. Both motor cranial nerve nuclei and spinal anterior horn cells were involved. The age of onset was from the late teens up to 45 years. Males were more severely affected than females.

6. DOMINANT TYPE OF AMYOTROPHIC LATERAL SCLEROSIS (KURLAND AND MULDER, 1955)

Kurland and Mulder (1955) reviewed earlier pedigrees and reported 5 kindreds with 36 affected individuals over 3 and 4 generations. Similar kindreds were reported by Bonduelle *et al.* (1959), W. K. Engel *et al.* (1959), Green (1960), and Eldridge *et al.* (1969). Clinical features, which included weakness, wasting and fasciculation of muscles, bulbar involvement, and corticospinal tract dysfunction, were similar to those noted in the much commoner sporadic form of motor neuron disease. Kurland (1957) estimated a frequency of 1.4 per 100,000 for all forms of amyotrophic lateral sclerosis (motor neuron disease) in the United States.

7. DOMINANT TYPE OF GUAM MOTOR NEURON DISEASE (KOERNER, 1952)

There is some uncertainty as to whether the gene responsible for this type of amyotrophic lateral sclerosis (ALS) is different from that responsible for inherited motor neuron disease elsewhere. The clinical features are similar, but in certain families associated features of Parkinsonism and dementia may occur. The possibility has even been raised recently that the condition could be acquired and due to a slow virus infection, rather than being genetically determined. Koerner (1952) found an incidence of 13 cases per 10,000 population in Guam based on a study of 46 cases. The age of onset was from 26 to 69 years, with only 7 cases outside the range of 30–50 years. The average age of onset was 42 years. Because of reduced penetrance and late onset, only 40% of the cases gave a history of having affected relatives. Espinosa *et al.* (1962) concluded that the Guam type was a different entity from the disorder typically observed in the United States. Since a dominant gene for the Parkinsonism–dementia complex (PD) (Hirano *et al.*, 1961) often segregates in the same families as does the gene for amyotrophic lateral sclerosis, it has been suggested that these two disorders may be due to differing expressions of a single gene, or alternatively, that a third gene accounts for the combined disorder, being different from the two dominant genes that account for ALS and PD and that may segregate independently.

8. Recessive Type of Amyotrophic Lateral Sclerosis (Refsum and Skillicorn, 1954)

The number of reports suggesting recessive transmission of ALS are very limited (McKusick, 1968b) compared to those of the dominant type. The fact that the recessively inherited cases show an early age of onset at 3–5 years (Refsum and Skillicorn, 1954) suggests a possible relationship with the Kugelberg–Welander syndrome in which signs of corticospinal tract dysfunction occasionally occur (Gardner-Medwin *et al.*, 1967).

9. Dominant Type of Progressive Muscular Atrophy (Aran, 1850; Duchenne, 1868)

Brown (1960) reported a family with 26 affected individuals in 4 generations. Muscular wasting usually began in the hands at 50–70 years. Progressive limb and trunk paralysis then led to death from inanition and respiratory infection, usually after a course of 3–5 years, but one patient survived for 30 years. It is not certain whether this disorder is different from dominantly inherited ALS in which similar muscular atrophy is accompanied by corticospinal tract dysfunction.

10. Dominant Type of Spinal Muscular Atrophy with Bulbar Paralysis (Fazio; Londe, 1894)

Gomez *et al.* (1962) described a case, verified by autopsy, whose father and two first cousins had had dysphagia. He stated that the cases of Fazio were a mother and her $4\frac{1}{2}$-year-old son. Very few families of this type have been described.

11. Recessive Type of Spinal Muscular Atrophy with Bulbar Paralysis (Fazio; Londe, 1894)

Londe (1894) reported two male siblings, aged 5 and 6 years, whose parents were first cousins. The brother and sister reported by Marinesco (1915) were aged 8 and 12 years. The recessive type cannot be distinguished from the dominant type clinically.

12. Recessive X-Chromosomal Type of Progressive Bulbar Palsy (Takikawa, 1953)

Takikawa (1953) investigated 2 affected brothers in 1 family and a mother and daughter in another family. Tsukagoshi *et al.* (1965) reported 8 cases in 4 families. Kennedy *et al.* (1968) reported 3 affected brothers and a maternal first cousin male in one kindred and 8 maternally related males over 4 generations in a second kindred. The age of onset was from 27 to 50 years. Features included initial weakness of shoulder and girdle muscles, which spread over 2 years to involve facial and tongue muscles and to give dysphagia and dysarthria. Gynecomastia was common, progression slow, and longevity usually normal. Several female siblings (carriers?) of those affected complained of muscle cramps.

13. Dominant Type of Amyotrophy with Disease of Creutzfeldt (1920) and Jacob (1921)

Worster-Drought *et al.* (1940) reported 12 cases in 3 generations with an onset at 40–60 years and a duration of up to 13 years. Symptoms included dysarthria, striatal tremor and rigidity, dementia, and wasting of the limbs, with widespread fasciculation. This condition is nowadays more often referred to as corticostriatonigral degeneration and is probably different from the sporadic form of transmissible myoclonic dementia with spongiform encephalopathy generally referred to as Jakob-Creutzfeldt disease.

14. Recessive Type of Distal Muscular Atrophy with Mental Deficiency and Choreic Movements (Asano et al., 1960)

Asano *et al.* (1960) reported 7 cases in 3 sibships in which all parents were related. Features included delay in walking, severe atrophy of the peroneal muscles by the age of 2 or 3 years, atrophy of distal muscles in the upper extremities, impaired sensation in the feet, dysarthria, choreic movements of facial and upper extremity muscles, optic atrophy, scoliosis, incontinence, and mental deficiency.

B. Inherited Disorders of the Central Nervous System with Possible Associated Involvement of the Lower Motor Neurons or Muscle

The group of disorders classified as the hereditary ataxias remain heterogeneous and confusing. Except in a few isolated diseases, entities previously included in this group (e.g., Refsum's disease) in which specific enzyme defects have been discovered, presumably resulting from the effects of a single mutant gene, there is still no evidence to indicate whether the many clinical syndromes that have been described are each due to single discrete genes or whether this very large group of diseases results from a much smaller group of genes showing wide variation in expression from one family to another, possibly due to the modifying effects of allelic genes or to other as yet unknown factors that could even in some cases be environmental. Within this group of diseases, there are disorders affecting not only corticospinal, cerebellar, and sensory tracts in the central nervous system, but also the optic nerves and retina, other cranial nerves, the extrapyramidal system, the skeleton, anterior horn cells in the spinal cord, the peripheral nerves, and perhaps even the voluntary muscles, in a series of bewildering combinations. Many of these disorders do not specifically involve the peripheral neuromuscular apparatus, but in others, and then only in some families, varying degrees of muscular atrophy have sometimes been noted, though often no evidence is available to indicate the nature or site of the lesion responsible. This is the justification for including brief consideration of this group of disorders in the present chapter, though our coverage is by no means complete, as we have excluded some conditions in which we are satisfied that there is no current evidence available to indicate neuromuscular affection; on the other hand, we have included some others in which such signs are inconstant, if indeed they ever occur at all.

15. Recessive Type of Amyotrophy with Friedreich's (1863) Ataxia

Heck (1964) described a large kindred with neural and extraneural findings. McKusick (1968b) suggested that the diagnosis of Friedreich's ataxia should be limited to cases showing an autosomal recessive type of inheritance. These cases are the most numerous and are characterized by ataxia, nystagmus, pes equinovarus, scoliosis, and sensory tract disorders. Except for the typical preadolescent age of onset, the symptoms are indistinguishable from those noted in the other genetic types.

16. DOMINANT TYPE OF AMYOTROPHY SIMULATING FRIEDREICH'S ATAXIA

Sylvester (1958) reported a kindred with 7 affected individuals in 2 generations in which optic atrophy and nerve deafness were additional features. Presenting symptoms included fatigue, impaired vision, and muscular weakness. The age of onset was $2\frac{1}{2}$–9 years, and the duration of the disease from $2\frac{1}{2}$ months to 4 years.

17. RECESSIVE X-CHROMOSOMAL DISORDER SIMULATING FRIEDREICH'S ATAXIA (E. V. TURNER AND ROBERTS, 1938)

E. V. Turner and Roberts (1938) described 9 affected males noted in 2 generations, and Brandenberg (1910) reported 4 affected males in 3 generations. Typically, the age of onset was 5 years and the duration 20 years.

18. DOMINANT TYPE OF AMYOTROPHY WITH ATAXIA AND AREFLEXIA (ROUSSY AND LEVY, 1934)

Spillane (1940) reported a kindred with 21 affected individuals in 6 generations. The presenting features in infancy and childhood included delayed walking, bilateral pes cavus, absent tendon jerks, clumsiness and static tremor of the hands, and slight wasting of the muscles of the thenar and hypothenar eminences and in the legs below the knee. Yudell *et al.* (1965) investigated 11 cases in 4 generations and emphasized that the tremor was useful in the differential diagnosis from peroneal muscular atrophy. Variation in expressivity was evident in that 5 cases showed the complete syndrome with slowed motor nerve conduction, 5 cases were younger and lacked the tremor, and 1 patient, aged 44, had no signs other than tremor.

19. DOMINANT TYPE OF SPASTIC PARAPLEGIA (SCHWARZ AND LIU, 1956)

Seeligmuller (1876) first described this condition. Schwarz and Liu (1956) reported a pedigree with 22 affected individuals in 6 generations and Hohmann (1957) another with 7 cases in 2 generations. Aagenaes (1959) reported 14 cases in 4 generations and described bilateral cortico-spinal tract degeneration. The age of onset is usually between 25 and 35 years. Of the various genes causing spastic paraplegia, between 10 and 30% show a dominant pattern of inheritance.

20. Dominant Type of Spastic Paraplegia with Amyotrophy of the Hands (Silver, 1966)

Strumpell (1880) first described this condition. Silver (1966) reported 2 pedigrees with a total of 11 affected individuals in 3 generations. Nystagmus and dysarthria were additional features.

21. Dominant Types of Spastic Paraplegia with Extrapyramidal Signs (Dick and Stevenson, 1953)

Dick and Stevenson (1953) described extrapyramidal signs in 4 out of 7 cases of spastic paraplegia, beginning at 9–10 years of age and occurring in 3 generations.

22. Recessive Type of Spastic Paraplegia (Bell and Carmichael, 1939)

Bell and Carmichael (1939) found evidence suggesting recessive inheritance in 49 out of 74 published pedigrees.

23. Recessive X-Chromosomal Type of Spastic Paraplegia (Johnston and McKusick, 1962)

Clinical features in the pedigree described by Blumel *et al.* (1957) were similar to those noted in the family of Johnston and McKusick (1962) in which onset was early, progression slow, and survival long; eventually there was involvement of the cerebellum, cerebral cortex, and optic nerves.

24. Recessive Types of Spastic Quadriplegia with Oligophrenia and Ichthyosis (Sjögren and Larrson, 1957)

Sjögren and Larrson (1957) traced 28 cases to a 600-year-old mutation that caused pigmentary degeneration of the retina in about half of the cases. Skin changes simulated the features of congenital ichthyosiform erythroderma. The gene frequency of heterozygous carriers in North Sweden was 1.3% of the population.

25. Dominant Type of Dystonic Paraplegia with Progressive Amyotrophy (Gilman and Horenstein, 1964)

Gilman and Horenstein (1964) reported a kindred with 12 cases in 3 generations. Features included mental retardation, pseudobulbar palsy, paraplegia, dystonia, urinary and faecal incontinence, and progressive muscular wasting indicative of anterior horn cell involvement.

26. Dominant Type of Cerebellar Ataxia (Menzel, 1891)

Greenfield (1954) recognized 13 families of cerebellar ataxia with olivopontocerebellar atrophy and degeneration of ascending spinal tracts.

27. Dominant Type of Cerebellar Ataxia (Holmes, 1907)

Clinical features of the Holmes type of cerebellar ataxia are similar to the Mengel type except for a much later age of onset and the more frequent association of mental dysfunction.

28. Dominant Type of Dyssynergia Cerebellaris Myoclonica (Hunt, 1921)

Gilbert *et al.* (1963) reported a kindred with 8 cases noted in 4 generations. Asymmetrical cerebellar signs are associated with myoclonic jerking of the limbs but the progressive dementia of progressive myoclonic epilepsy does not occur.

29. Dominant Type of Myoclonus, Cerebellar Ataxia, and Nerve Deafness (May and White, 1968)

May and White (1968) described 6 cases in 4 generations with various combinations of myoclonus, cerebellar ataxia, and deafness. Hearing loss was the presenting feature at the age of 4 years in one case, while in another case myoclonus and cerebellar symptoms began at the age of $14\frac{1}{2}$ years.

30. Dominant Juvenile Type of Paralysis Agitans (Hunt, 1917)

In a father and daughter, Clark (Ford, 1966) reported typical Parkinsonism with tremor, masklike facies, bradykinesia, dysarthria, and rigidity. There was a loss of large cells in the lenticular nuclei. The onset typically was in the teens with slow progression and long duration.

31. Recessive Type of Paralysis Agitans with Pallidopyramidal Involvement (Hunt, 1917)

Davison (1954) described 5 affected individuals in 3 families. In one family, the parents of an affected son and daughter had an uncle–niece relationship. In another family, the parents of an affected son and daughter were first cousins. Pathological features included pallor of pallidal segments, thinning of ansa lenticularis, early demyelination of the pyramids, and crossed pyramidal tracts. The typical tremor and rigidity of paralysis agitans began at the age of 13 years in one of Davison's (1954) cases which was reported previously by Hunt (1921).

32. Dominant Childhood Type of Acute Intermittent Cerebellar Ataxia with Intention Tremor (Hill and Sherman, 1968)

Hill and Sherman (1968) reported a Negro pedigree in Louisiana with 35 affected persons in 5 generations. The age of onset was as early as 2 years. Attacks of diminishing severity and frequency were reported by the family to occur up to the age of 60 years. There were 14 clinically documented attacks of ataxia that occurred before the age of 15 years.

33. Dominant Adult Type of Periodic Vestibulocerebellar Ataxia (Farmer and Mustian, 1963)

Farmer and Mustian (1963) described 16 cases in 4 generations in a rural North Carolina kindred. Symptoms included recurrent attacks of vertigo, diplopia, and ataxia beginning in early adult life followed by progressive cerebellar ataxia in some individuals. The age of onset was from 23 to 42 years with attacks varying in duration from a few minutes to 2 months.

34. Dominant Type of Spastic Ataxia (Ferguson and Critchley, 1929)

Mahloudji (1963) reported a pedigree of 18 affected persons in 5 generations with a spastic ataxia resembling multiple sclerosis. Inconstant features included dysarthria and optic atrophy. The age of onset was from 35 to 45 years and the average duration from 5 to 10 years.

35. DOMINANT TYPE OF ATAXIA WITH PERSISTENT FASCICULATION (SINGH AND SHAM, 1964)

Singh and Sham (1964) described a kindred in India with 8 affected individuals in 3 generations. Clinical features included dysarthria, intention tremor, and persistent fasciculation, especially of the tongue, calf, and thigh muscles. There was generalized muscular weakness and atrophy resulting in inability to walk by the age of 24 years.

36. RECESSIVE TYPE OF ATAXIA–TELANGIECTASIA (KOREIN ET AL., 1961)

Tadjoedin and Fraser (1965) reported 6 new cases of a condition first described by Louis-Bar (1941) and found that the 60 cases reported in the literature were compatible with autosomal recessive inheritance. The condition is characterized by cerebellar ataxia with onset in infancy and inability to walk by the age of 10 years. Telangiectasia are seen on the bulbar conjunctiva and later on the skin. A deficiency of γ-globulin causes undue susceptibility to respiratory infection. The muscles may show evidence of denervation atrophy (Strich, 1966).

37. DOMINANT TYPE OF HUNTINGTON'S (1872) CHOREA WITH AMYOTROPHY

A single dominant gene with 100% penetrance has been found to cause the choreic movements and dementia associated with Huntington's chorea in which muscular wasting is rarely observed. Vessie (1932) traced about 1000 cases in 12 generations from 2 brothers from Suffolk, England. Reed and Chandler (1958) found the frequency of cases in lower Michigan to be 4.12×10^{-10}. The age of onset was from childhood to nearly 80 years, with most cases developing symptoms between 30 and 40 years.

38. RECESSIVE TYPE OF AMYOTROPHY SIMULATING REFSUM'S DISEASE AND HURLER'S DISEASE (SHY ET AL., 1967)

Shy *et al.* (1967) described a 21-year-old Negro female with progressive ptosis, external ophthalmoplegia, retinitis pigmentosa, ataxia, absent deep tendon reflexes, and an elevated cerebrospinal fluid protein.

39. Dominant Type of Amyotrophy with Pick's Disease (Pick, 1892)

By reexamination of a family with Pick's disease, Schenk (1959) increased the number known to be affected to 20 individuals occurring in 5 generations. Muscular wasting of unknown cause is a rare manifestation of this disorder.

40. Dominant Type of Amyotrophy with Alzheimer's Disease (Alzheimer, 1907)

Schottky (1932) reported cases of Alzheimer's disease, which usually occurs sporadically, in 1 family over 4 generations. Lowenberg and Waggoner (1934) reported affection of a father and 4 of his children. Wheelan and Race (1959) found a possible linkage between the gene of the MNS locus and the gene for Alzheimer's disease. Muscular wasting is a rare and inconstant finding in some cases.

41. Recessive Type of Amyotrophy in the Syndrome of Marinesco et al. (1931) and Sjögren (1950)

Sjögren (1950) described 2 kindreds with 12 affected individuals in which all 5 sets of parents had a first cousin relationship. Garland and Moorhouse (1953) reported 2 cases in which the parents of each case were either a first cousin and first cousin once removed or full second cousins. Features included cerebellar ataxia, congenital cataracts, and retarded somatic and mental development; distal muscular weakness was an inconstant feature.

42. Recessive Type of Infantile Amyotrophy with Neuroaxonal Dystrophy (Seitelberger, 1952)

Crome and Weller (1965) reported a brother and sister who died at 12 and 18 months, respectively. This degenerative encephalopathy is similar to Hallervorden–Spatz disease, with widespread focal swelling and degeneration of axons. Though muscular wasting may occur, it is as yet uncertain whether the lower motor neurons are involved.

43. Dominant Type of Agenesis of Cranial Nerve Nuclei (Möbius, 1888)

Van der Wiel (1957) observed 46 affected individuals in 6 generations with congenital paralysis of sixth and seventh cranial nerves in a kindred which was previously reported by Wilbrand and Saenger (1921). Fortanier and Speijer (1935) reported 15 cases occurring in 3 generations.

44. DOMINANT TYPE OF CONGENITAL ABSENCE OF PECTORAL MUSCLE WITH SYNDACTYLY (POLAND SYNDROME, 1841)

Clarkson (1962) observed that all reported cases of unilateral synbrachydactyly and ipsilateral aplasia of the sternal head of the pectoralis major muscle had been sporadic. It is still uncertain whether this disorder is ever genetically determined.

C. Disorders Principally Involving the Peripheral Nerves

45. DOMINANT TYPE OF PERONEAL MUSCULAR ATROPHY (CHARCOT AND MARIE, 1886; TOOTH, 1886)

Dyck *et al.* (1963) reported 29 cases in 4 generations and found the gene to be 100% penetrant if subclinical cases were identified by nerve conduction velocity measurement. Similar findings were found by Killian and Kloepfer (unpublished) from a study of over 100 cases in 4 kindreds of Southwestern Louisiana. Onset often was in childhood with foot-drop. Some individuals showing a 50% reduction in motor nerve conduction velocity were typically unaware of the affection and could not be detected clinically. Killian and Kloepfer identified two living homozygotes who simulated cases of the Dejerine–Sottas syndrome.

46. RECESSIVE TYPE OF PERONEAL ATROPHY (CHARCOT–MARIE–TOOTH DISEASE)

Allen (1939) described 8 cases in 6 families from Western North Carolina. Parents were consanguineous in 4 of the 6 families. Compared to other reported families of cases of peroneal atrophy, the symptoms in these cases were more severe.

47. DOMINANT X-CHROMOSOMAL TYPE OF PERONEAL ATROPHY (WORATZ, CITED BY BECKER, 1966b)

Woratz (cited by Becker, 1966b) reported 6-generation kindred in which 10 affected fathers had 15 affected daughters and 8 normal sons; 26 affected mothers had 23 affected sons, 21 affected daughters, and unaffected offspring of both sexes. Males were more severely affected than females.

48. RECESSIVE X-CHROMOSOMAL GENE SIMULATING
    CHARCOT–MARIE–TOOTH DISEASE AND FRIEDREICH'S ATAXIA
    (VAN BOGAERT AND MOREAU 1939)

Van Bogaert and Moreau (1939) described 3 males in 2 generations. It is possible that a larger kindred might have shown an autosomal dominant mode of transmission.

49. RECESSIVE X-CHROMOSOMAL TYPE OF PERONEAL ATROPHY WITH
    DEGENERATION OF OPTIC AND ACOUSTIC NERVES (ROSENBERG AND
    CHUTORIAN, 1967)

Rosenberg and Chutorian (1967) investigated 2 brothers and a son of a sister. The initial symptoms were delay in walking until the age of 2 years, hearing loss, lower leg atrophy observed at the age of 5 years, and marked difficulty in walking at the age of 15 years. Iwashita et al. (1970) reported affected male and female siblings from Korea with a similar disorder, which may, however, have been due to a different genetic entity.

50. DOMINANT TYPE OF HYPERTROPHIC INTERSTITIAL NEUROPATHY
    (DEJERINE AND SOTTAS, 1893)

Bedford and James (1956) reported 7 cases in 4 generations. Histological diagnosis was based on "onion bulb" formation in peripheral nerves. Croft and Wadia (1957) restudied a family reported by Russell and Garland (1930) and found 12 affected individuals over 5 generations. Austin (1956) reviewed the literature and found an age of onset from childhood to 20 years. The incidence was 1 in 1,000,000. Andermann et al. (1962) reported a deviant of Dejerine–Sottas syndrome in which 7 cases occurred over 4 generations and the affected persons showed nystagmus, distal muscular weakness, distal sensory changes, and pes cavus.

RELATIONSHIP BETWEEN CHARCOT–MARIE–TOOTH DISEASE AND DEJERINE–SOTTAS DISEASE. Dyck and Lambert (1968), in a series of careful studies, have attempted to clarify this confused situation in the light of their clinical, electrophysiological, and histological findings in a series of affected individuals and families. They identify first a benign hypertrophic demyelinating neuropathy of Charcot–Marie–Tooth type of dominant inheritance, and second a recessively inherited hypertrophic neuropathy of the Dejerine–Sottas type that is much more severe, of earlier onset, and usually leads to confinement to a wheelchair in the third decade or later. They also identify another variety of dominantly inherited Charcot–Marie–Tooth disease in which the neuropathy is due to axonal degeneration and not demyelination, and motor conduction velocity in the affected nerves is not severely reduced. A further rare type of Charcot–Marie–Tooth disease in which sensory abnormalities are absent appears to be a spinal muscular atrophy. They have seen several families in which features of Charcot–Marie–Tooth disease are combined with those of Friedreich's ataxia or hereditary spastic paraplegia.

51. DOMINANT TYPE OF NEUROTROPATHY WITH NEUROFIBROMATOSIS (VON RECKLINGHAUSEN, 1882)

From a Michigan study of all cases located, Crowe *et al.* (1956) found the frequency to be 1 in 2500 to 1 in 3300 and the ratio of old to new mutations was 1:1. Features included six or more café-au-lait spots 1.5 cm or larger in diameter, pendulous cutaneous and subcutaneous tumors, scoliosis, pseudoarthrosis of the tibia, pheochromocytoma, meningioma, glioma, acoustic neuroma, optic neuroma, mental retardation, and hypertension. Muscular wasting only occurs when motor nerves are compressed or distorted by one or more neurofibromata.

52. RECESSIVE TYPE OF HEREDOPATHIA ATACTICA POLYNEURITIFORMIS WITH HIGH SERUM CONCENTRATION OF PHYTANIC ACID (REFSUM, 1952)

Baker (1962) found evidence of recessive transmission in 15 reported families in which the parents were consanguineous. Features included ataxia, nystagmus, nerve deafness, atypical retinitis pigmentosa, hypertrophic peripheral neuropathy, high serum concentrations of phytanic acid (Klenk and Kahlke, 1963), abnormal electrocardiograms, and dry scaly skin. The onset age was usually at from 20 to 30 years. While finding evidence to support the proposal that patients accumulated phytanic acid due to a block at the initial α-oxidation step, Herndon *et al.* (1969) discovered a method to identify heterozygous carriers by using fibroblasts grown from skin biopsies.

53. DOMINANT TYPE OF HEREDOPATHIA ATACTICA POLYNEURITIFORMIS WITHOUT AN INCREASE IN SERUM CONCENTRATION OF PHYTANIC ACID (FURUKAWA ET AL., 1968)

Furukawa *et al.* (1968) reported a pedigree with 10 affected individuals over 3 generations. Features simulated Refsum's disease except that there was no increase in the concentration of phytanic acid in the serum.

54. Dominant Type of Recurrent Pressure Palsies of Peripheral Nerves (Staal et al., 1965)

Earl *et al.* (1964) reported 4 families. Staal *et al.* (1965) reported a Dutch family with transmission over 4 generations in which features included recurring transient unilateral peroneal nerve palsies, occurring especially after prolonged work in a kneeling position.

55. Dominant Type of Neuropathy with Interstitial Nephritis and Deafness (Alport, 1927)

Whalen *et al.* (1961) described 17 cases over 4 generations. Shaw and Glover (1961) reported 33 cases over 4 generations, with more oocytes passing the gene to gametes than expected. Males were more severely affected than females. Chiricosta *et al.* (1970) reported 18 affected individuals in 4 generations. Deafness was associated with renal foam cells in 30% of the cases. Other features included cataracts and sphcrophakia.

56. Dominant Type of Polyradiculocordonal Syndrome

Hicks and Camp (1922) reported a family with 10 affected persons. This family was restudied by Denny-Brown (1951), who found 11 affected individuals over 3 generations. The onset was typically between 15 and 36 years, with a painless foot ulcer developing from a corn which perforated and in some cases extended to the bone; there were shooting pains in the legs and sometimes in the arms; and deafness began at about 40 years. Autopsy findings included loss of ganglion cells in the sacral and lumbar dorsal roots. A number of families with distal muscular atrophy simulating that of Charcot–Marie–Tooth disease have been reported with associated features of sensory neuropathy similar to those described by Denny-Brown (1951).

57. Dominant Type of Portuguese Amyloid Neuropathy (Wohlwill, 1942; Andrade, 1952)

Heller *et al.* (1964) summarized the features of the Portuguese type of amyloid neuropathy as being lowered general health, gastrointestinal disturbances, premature impotence in males, disorders of pain sensation with distal paraesthesiae and renal involvement with pericollagenous amyloid. Males were more severely affected. The usual age of onset was from 20 to 30 years, the lower extremities were first affected and the duration of the illness was 7 to 10 years. Many cases occurred in north coastal areas of Portugal and in Portuguese descendants living in Brazil.

58. DOMINANT TYPE OF INDIANA AMYLOID NEUROPATHY (RUKAVINA ET AL., 1956)

In a clinical study of 29 cases of Swiss origin, Rukavina *et al.* (1956) found variable expressivity of peripheral neuropathy, skin changes, hepatic enlargement and dysfunction, eye changes, gastrointestinal disturbances, splenomegaly, and often a bilateral carpal tunnel syndrome. Schlesinger *et al.* (1962) reported 22 cases over 3 generations. Neuropathic signs occurred first in the upper extremities, usually in the 40s, and progressed to generalized neuropathy with a duration of about 20 years. Males were more severely affected. Van Allen *et al.* (1969) reported a kindred of Scottish–English–Irish origin with 15 affected individuals (8 verified) over 2 generations. The clinical features and age of onset overlapped those described both for the Indiana and the Portuguese (Andrade, 1952) types of neuropathy.

59. RECESSIVE TYPE OF A-α-LIPOPROTEINEMIC NEUROPATHY (TANGIER DISEASE)

This condition (Fredrickson, 1966) was first found on Tangier Island. Presenting features in children included characteristically large tonsils, enlarged liver, spleen, and lymph nodes; and hypocholesterolemia. The thymus was loaded with cholesterol esters. Heterozygotes (Kocen *et al.*, 1967) show low α-lipoprotein levels in the serum. Affected families have been reported in Missouri and Kentucky. Presenting features in adolescents include relapsing peripheral neuropathy.

60. RECESSIVE TYPE OF A-β-LIPOPROTEINEMIC NEUROPATHY (BASSEN AND KORNZWEIG, 1950)

Farquhar and Ways (1966) noted that 19 cases had been described. Using paper electrophoresis, they found an absence from the plasma of β-lipoprotein. There was progressive degeneration of posterolateral columns and cerebellar tracts, acanthocytosis, pigmentary degeneration of the retina, and malabsorption of lipids. Onset features before 5 years included a slow gain in stature and weight, steatorrhea, and abdominal distension. Loss of muscle strength, nystagmus, and progressive ataxia occurred in late childhood. Peripheral nerves showed extensive central and peripheral demyelination. Consanguinity was found in 7 out of 14 families. Salt *et al.* (1960) diagnosed the condition in a 17-month-old female and identified the heterozygous state in both parents and a paternal grandfather.

61. RECESSIVE ADULT TYPE OF NEUROPATHY WITH METACHROMATIC LEUKODYSTROPHY (VAN BOGAERT AND DEWULF, 1939)

Presenting symptoms after the age of 16 years were typically mental changes suggesting schizophrenia, with the later development of disorders of movement and posture. There was a greater sulfatide excess in the cerebral gray than in the white matter. As in the infantile variety, the peripheral nerves may also be affected.

62. RECESSIVE INFANTILE TYPE OF METACHROMATIC LEUKODYSTROPHY WITH DIFFUSE CEREBRAL SCLEROSIS (GREENFIELD, 1933)

Diagnosis of this condition may be confirmed by the absence of arylsulphatase A from the urine (Austin *et al.*, 1966), by slowing of motor nerve conduction velocity, and by metachromatic deposits demonstrated histologically in sural nerve biopsy specimens. Bass *et al.* (1970) reported 4 cases and 12 heterozygous carriers studied in Virginia. Five out of 6 heterozygotes had significantly lower levels of leukocyte arylsulfatase A activity. The age of onset was usually from 1 to 2 years, with death from pneumonia occurring usually between the ages of 3 and 6 years. Clinical features apparent early in the second year included an unsteady gait, increasing muscle weakness, areflexia, and hypotonia.

63. RECESSIVE TYPE OF GLOBOID CELL SCLEROSIS (KRABBE, 1958)

The finding of multinucleate "globoid cells" in the brain is the diagnostic feature of this form of diffuse cerebral sclerosis in which peripheral nerves may also be involved. The onset is usually at 4–5 months, and the condition is fatal in infancy. There are many reports of involvement of several siblings. Parents were first cousins in the family presented by van Gehuchten (1956).

64. RECESSIVE JUVENILE TYPE OF NEUROPATHY (SCHUTTA ET AL., 1966)

Schutta *et al.* (1966) reported 5 affected individuals in 3 sibships and cited from the literature 14 cases with a similar age of onset at from 4 to 10 years and a similar duration of the illness of from 1 to 11 years.

65. RECESSIVE TYPE OF PERIPHERAL NEUROPATHY INDUCED BY ISONIAZID

The method of Sunahara *et al.* (1961) made possible the identification of three genotypes related to slow and fast inactivation of isoniazid. Slow inactivators of isoniazid treated with this drug developed a peripheral neuropathy with onset symptoms of painful paraesthesiae in the hands and feet (Jones and Jones, 1953).

66. DOMINANT TYPE OF NEUROPATHY WITH PARAPROTEIN IN SERUM, CEREBROSPINAL FLUID, AND URINE (GIBBERD AND GAVRILESCU, 1966)

Gibberd and Gavrilescu (1966) described a pedigree with 4 affected individuals in 3 generations. Features included motor weakness and sensory impairment developing at about the age of 50 years. There was delayed motor nerve conduction velocity and histological evidence of demyelination with Schwann cell proliferation.

67. DOMINANT TYPE OF SWEDISH ACUTE INTERMITTENT NEUROPATHY WITH PORPHYRIA

Waldenstrom and Haeger-Aronsen 1963) reviewed the literature of about 600 cases in Sweden where attacks initiated by barbiturates included severe abdominal colic, constipation, and often an acute and severe sensorimotor polyneuropathy. Urine obtained during attacks became dark when left standing and contained excess porphyrins. The ratio of affected males to females was 2:3. The prevalence was 1.4 per 100,000 for the entire country and 1% for the Sapland area. Nonpenetrant carriers were identified.

68. DOMINANT TYPE OF SOUTH AFRICAN VARIEGATED NEUROPATHY WITH PORPHYRIA (DEAN, 1963)

Dean (1963) estimated that the 8000 affected individuals in South Africa were all descendants of certain Dutch settlers. All showed a constantly increased fecal excretion of protoporphyrin and coproporphyrin. Macalpine *et al.* (1968) described 16 presumably affected individuals noted over 10 generations in European royal families.

69. DOMINANT TYPE OF PORPHYRIA CUTANEA TARDA WITH NEUROPATHY

Ziprkowski *et al.* (1966) reported occurrence over 3 generations.

D. Disorders Believed to Affect Primarily the Myoneural Junction or Voluntary Muscle

70. RECESSIVE TYPE OF MYASTHENIC MYOPATHY (JOHNS ET AL., 1966)

These authors reported a sibship with 4 out of 8 siblings affected. The age of onset was in adolescence with evidence of proximal myopathy. The myasthenic reaction became more prominent after the disorder had existed for 10 years.

71. DOMINANT TYPE OF MYASTHENIA GRAVIS (NOYES, 1930)

Hermann (1966) described affection in a father and son. McQuillen (1966) reported affection in a father and three offspring. Electromyography suggested that there was a myopathy in addition to a disorder of the myoneural junction. The onset age was at 13–14 years.

72. RECESSIVE TYPE OF MYASTHENIA GRAVIS

From a review of the literature, Celesia (1965) found that affection of 2 or more cases was limited to sibships in 18 out of 22 studies. However, since no increased consanguinity was reported, it is entirely possible that further studies may show that a dominant type of inheritance is involved with frequent apenetrance in the heterozygous parent. It must also be stressed that the evidence that myasthenia gravis is genetically determined is very incomplete, as the great majority of cases are sporadic.

73. DOMINANT TYPE OF MYASTHENIA GRAVIS WITH HYPERTHYROIDISM (RENNIE, 1908; GREENBERG, 1964)

A. G. Engel (1961) found that most of the 93 cases reported in the literature were female. Hyperthyroidism was present in 79 of the 93 cases. Greenberg (1964) reviewed 18 reports of familial cases and reported affection in two sisters. Of all myasthenic patients, 6% develop hyperthyroidism yet only 1% of hyperthyroid patients develop myasthenia.

74. Recessive Type of Pseudocholinesterase Deficiency (Suxamethonium Paralysis)

Lehmann and Liddell (1966) described features of prolonged apnea after administration of the muscle relaxant suxamethonium in patients shown to have a low serum level of pseudocholinesterase. Three genotypes were identified. Two further alleles were a silent gene and a gene for fluoride inhibition. A nonallele was responsible for an electrophoretic variant. Rubinstein *et al.* (1970) reported 4 siblings with the silent gene.

75. Recessive X-Chromosomal Severe Type of Duchenne Muscular Dystrophy (Walton and Nattrass, 1954; Becker, 1962)

Usually the onset is before the age of 5 years with inability to walk by 12 years and death by 20 years. The heart muscle is constantly involved, and the female carriers may be identified by serum creatine kinase estimation and other methods (Thompson, 1969; Emery, 1969). The frequency of the gene for this type of muscular dystrophy seems to be the highest of all genes causing muscular dystrophy. The incidence has been estimated to be 159.6×10^{-6} and the mutation rate to be $5.4–7.2 \times 10^{-5}$ (Stevenson, 1953; Zellweger *et al.*, 1965). Walton and Gardner-Medwin (1969) have reviewed the recent literature.

76. Recessive X-Chromosomal Milder Type of Muscular Dystrophy (Becker, 1962, Ascending Type B)

Robert and Guibaud (1969) reviewed 8 sibships and the Becker type B. In a pedigre reported by Zellweger and Hanson (1967), the age of onset was from 5 to 15 years in 7 cases.

77. Recessive X-Chromosomal Benign Type of Muscular Dystrophy (Emery and Dreifuss, 1960)

Emery and Driefuss (1966) reported a kindred in Virginia of English descent with 8 affected males occurring over 3 generations. The onset was at 4–5 years and included flexion contractures of the elbows and shortening of the tendons of Achilles. There was myocardial involvement, absence of calf enlargement, and weakness of orbicularis oris and lower facial muscles.

78. RECESSIVE X-CHROMOSOMAL BENIGN TYPE OF
    MUSCULAR DYSTROPHY (MABRY ET AL., 1965)

Mabry *et al.* (1965) described a kindred in Kentucky with 9 affected
males over 2 generations. The onset features at 11–13 years included
myocardial involvement. Serum creatin phosphokinase and aldolase activity was normal. It is uncertain whether the three benign X-chromosomal disorders described here are different entities or variants of a
single genotype.

79. RECESSIVE GENE SIMULATING THE DUCHENNE TYPE OF
    MUSCULAR DYSTROPHY (KLOEPFER AND TALLEY, 1958)

Skyring and McKusick (1961) reported 9 cases in 2 sibships in which
all 4 parents were related. All clinical manifestations including the age
of onset and duration were similar to those of the Duchenne sex-linked
recessive type of muscular dystrophy. However, Penn *et al.* (1970) believe that the accuracy of diagnosis in many of the reported cases is
in doubt.

80. DOMINANT FACIOSCAPULOHUMERAL TYPE OF MUSCULAR
    DYSTROPHY (LANDOUZY AND DEJERINE, 1885)

Morton *et al.* (1963) reported a study based on all cases located
in Wisconsin. These authors, like Walton and Nattrass (1954), found
that the onset was usually in adolescence, mildly affected cases occurred,
and there was initial involvement of facial and scapulohumeral muscles.
The incidence rate was 38×10^{-5}.

81. RECESSIVE LIMB–GIRDLE TYPE OF MUSCULAR DYSTROPHY
    (ERB, 1891; VON LEYDEN, 1876: MÖBIUS, 1888)

Morton and Chung (1959) found the age of onset to range from
10 to 50 years, with a median of 30 years. There was no clinical distinction between recessive and sporadic types. Incidence was 385×10^{-5}.
Stevenson *et al.* reported a mutation rate of 3.1×10^{-5}.

82. SPORADIC TYPE OF LIMB–GIRDLE MUSCULAR DYSTROPHY
    (MORTON AND CHUNG, 1959)

Morton and Chung (1959) found 41% of all limb–girdle cases in
Wisconsin to be sporadic and indistinguishable clinically from the autosomal recessive types. The incidence rate was 268×10^{-5}, and the mutation rate was 3.6×10^{-5}.

83. Dominant Limb–Girdle Type of Muscular Dystrophy Limited to Females (Henson et al., 1967)

Henson *et al.* (1967) described an Indiana kindred with 8 affected individuals in 2 generations. The onset age was between 4 and 21 years. Many abortions were noted in the kindred.

84. Dominant Type of Distal Myopathy with Adult Onset (Welander, 1951)

Welander (1951) reported 249 cases in 72 kindreds. In 89% of the cases, small muscles of the hand were involved first. Onset ages were 20–77, with a mean of 47 years. Welander (1957) described a homozygote.

85. Dominant Type of Childhood Distal Myopathy (Biemond, 1955)

Biemond (1955) reported a kindred with 19 affected individuals in 5 generations. Age of onset was from 5 to 15 years. Features simulated the distal myopathy described by Welander (1951) except for early onset, initial weakness in both hands and feet simultaneously, and progression for as long as 60 years. Scrutiny of the histological evidence strongly suggests that this condition was a neuropathy akin to Charcot–Marie–Tooth disease and not a myopathy.

86. Dominant Ascending Distal Variety of Myopathy (Barnes, 1932)

Barnes (1932) reported a 6-generation pedigree with many affected individuals. Typically muscular hypertrophy was observed at about 9 years, pseudohypertrophy at about 21 years, and the terminal stages were characterized by distal atrophy at about 50 years, with inability to walk at about 80 years.

87–89. Dominant, X-Linked Recessive and Autosomal Recessive Atrophic Distal Types of Myopathy (Milhorat and Wolff, 1943)

Milhorat and Wolff (1943) reported 125 cases in 75 families. Presenting features included weakness in the lower extremities, pain and fatigue in the thighs and legs, enlarged calves, and muscle fiber replacement by fat in biopsy specimens. Onset ages ranged from 30 to 38 years in one family. All cases were classified as follows: 26 were sex-linked recessive in 4 families; 27 were autosomal recessive in 5 families; and 14 were autosomal dominant in 2 families. Almost certainly this early report included cases of several different nosological entities.

90. Dominant Proximal Myopathy (Schneiderman et al., 1969)

Schneiderman et al. (1969) found 18 cases over 5 generations in a kindred in which proximal limb muscles were mainly affected but facial muscles were spared. The onset was gradual at about 30 years, and progression was slow up to the age of 70–80 years with eventual inability to walk. Evidence was found for linkage with the Pelger–Huet anomaly with a recombination fraction of 0.26.

91. Dominant Type of Ocular Myopathy (Kiloh and Nevin, 1951)

Kiloh and Nevin (1951) reported 5 cases with an onset occurring typically between the ages of 20 and 25 years. Presenting features included ptosis, which usually progressed to complete ophthalmoplegia. Typically, the upper facial muscles were also weak, and in 25% of the cases, there was a variable degree of weakness and atrophy of the neck, trunk, and limb muscles. Penetrance was greatly reduced. This condition has been described in association with hereditary ataxia (see Walton and Gardner-Medwin, 1969).

92. Dominant Type of Oculopharyngeal Muscular Dystrophy (Victor et al., 1962; Bray et al., 1965; Barbeau, 1966)

Victor et al. (1962) described a pedigree with 9 persons affected over 3 generations. Of the various families reported, a number have been of French Canadian descent. Except for an average age of onset of 40 years and involvement of the pharyngeal muscles, the clinical features of this condition simulate those of the entity described by Kiloh and Nevin (1951).

93. Possibly Dominant Type of Ophthalmoplegia with Weakness of Facial, Pharyngeal, Trunk and Extremity Muscles (Kearns, 1964)

Kearns (1964) reported 9 unrelated cases of ophthalmoplegia in which there was also pigmentary degeneration of the retina, cardiomyopathy, deafness, small stature, and an increase in the cerebrospinal fluid protein.

94. Recessive Benign Congenital Type of Myopathy (Batten, 1910; J. W. A. Turner and Lees, 1962)

J. W. A. Turner and Lees (1962) described a 50-year followup of a pedigree with 6 out of 13 siblings affected. The first symptoms were those of diffuse infantile hyptonia with delay in walking, followed by many years of persistent weakness with little or no deterioration. The nosological status of this disorder and its relationship to congenital muscular dystrophy (see Walton and Gardner-Medwin, 1969) remain in doubt.

95. Recessive Type of Slowly Progressive Congenital Muscular Dystrophy Producing Arthrogryposis (Pearson and Fowler, 1963)

Pearson and Fowler (1963) found a brother and sister who had the atypical features of moderate contractures and muscular atrophy, widespread muscle weakness, and associated skeletal abnormalities.

96. Recessive Type of Rapidly Progressive Congenital Muscular Dystrophy Producing Arthrogryposis (DeLange, 1937)

DeLange (1937) reported 6 affected individuals in 2 second cousin sibships. Except for more rapid progression, these cases showed features similar to those of the begnign congenital myopathy described by Batten (1910) and J. W. A. Turner and Lees (1962).

97. Dominant Type of Nemaline Myopathy (Shy et al., 1963)

Affected individuals, who are usually "floppy infants," often show a high arched palate, skeletal deformities of varying types, thin underdeveloped musculature and diffuse muscular weakness. The muscle fibers are found to contain rod- or threadlike bodies made up of protein material similar to that contained in the normal Z band. The specificity of this pathological change is not in doubt. Asymptomatic relatives often showed minor signs of the condition. Spiro and Kennedy (1965) reported

5 cases over 3 generations and Lindsey *et al.* (1966) 3 cases over 3 generations.

98. Dominant Type of Nemaline Myopathy with Crystalline Intranuclear Inclusions (Jenis, et al., 1969)

Jenis *et al.* (1969) reported a case of nemaline myopathy that was indistinguishable from the type first described by Shy *et al.* (1963) except for the presence of crystalline intranuclear inclusions. Opinion is at present divided upon the question as to whether this is a new entity or an unusual manifestation of the entity described by Shy *et al.* (1963).

99. Dominant Congenital Type of Muscular Hypoplasia (Gibson, 1921; Krabbe, 1958)

The existence of this entity is also in doubt, because restudy of cases reported earlier (Ford, 1961) suggested to Hopkins *et al.* (1966) that the condition was nemaline myopathy.

100. Dominant Type of Congenital Central Core Disease (Shy and Magee, 1956)

Shy and Magee (1956) reported a kindred with 5 affected individuals over 3 generations. Onset features included delay in walking up to 5 years and diffuse muscular weakness. A number of subsequent reports of this condition have described additional features apart from the "central cores" within the muscle fibers. Dubowitz and Roy (1970) reported a pedigree with 4 cases occurring over 3 generations; 95% of the muscle fibers in a 29-year-old female were of histochemical type 1, and all had central cores, whereas 50% of the fibers in her 4-year-old son were of type 1 and relatively few had central cores.

101. Recessive Congenital Type of Megaconial Myopathy (Shy and Gonatas, 1964)

Shy and Gonatas (1964) described 2 affected female siblings at ages $1\frac{1}{2}$ and 8 years who showed diffuse muscular weakness, and a muscle biopsy showed large bizarre mitochondria in the fibers. A previous diagnosis of Werdnig–Hoffmann disease had been made in the sibling that died at $1\frac{1}{2}$ years.

102. Recessive Congenital Type of Pleoconial Myopathy (Shy et al., 1966)

Shy et al. (1966) reported 2 male siblings and a paternal male first cousin who showed features of proximal muscular weakness and wasting, prolonged episodes of flaccid paralysis, salt craving, and large numbers of mitochondria in biopsy specimens of voluntary muscles.

103. Dominant Congenital Type of Myopathy with Salt Craving and Numerous Large Mitochondria Aligned with Large Lipid Bodies (Sporo et al., 1970)

Spiro et al. (1970) described a 13-year-old boy who had a maternal aunt with a peculiar gait. In infancy he was a "floppy baby" with celiaclike diarrhea who first walked at 2 years of age. By the age of 13 years running was difficult.

104. Recessive Type of Myopathy with Mitochondrial Abnormalities

Coleman et al. (1967) described 2 cases with progressive proximal and subsequently distal muscle weakness at ages 5 and 10 years. The mitochondria were unusually large. Van Wijngaarden et al. (1967) described an onset at 4 years of age in a male patient whose I.Q. was 78 at the age of 11 years. There was hypertrophy of the calves and weak deltoids. The mitochondria were abnormal in shape, size, number, and location.

105. Dominant Congenital Type of Centronuclear Myopathy (Spiro et al., 1966)

Sher et al. (1967) reported an affected mother and 2 daughters. Typical features included failure to walk by the age of 2 years, ptosis, café-au-lait spots, hypertelorism, and marked generalized weakness of most muscles, including the temporales, masseters, and sternocleidomastoids. Central nuclei were observed in 80–98% of the muscle fibers, except in the mother who had a mixture of small centrally nucleated fibers and normal fibers. Coleman et al. (1968) reported a mother and a daughter with features similar to those in the cases reported by Sher et al. (1967). Kimoshita and Cadman (1968) reported an affected 6-year-old girl with a second cousin who was believed by the family to be similarly affected.

106. Dominant Type of Congenital Myotubular Myopathy
(Van Wijngaarden et al., 1969)

Van Wijngaarden *et al.* (1969) described a pedigree with 7 cases occurring over 3 generations in which 2 of the 4 female transmitters showed only histochemical manifestations. Features included infantile hypotonia, extraocular, facial and neck muscle involvement in survivors and areflexia; 2 out of 5 affected females had increased serum creatin phosphokinase levels; 85% of the muscle fibers were of small diameter with a peripheral rim of myofibrils and a central zone containing either a single nucleus or a collection of sarcoplasmic components. These authors suggested the possibility of X-chromosomal inheritance, but if the family reported by Sher *et al.* (1967) represents the same entity, the occurrence of affected daughters of a mother with partial manifestations of the disorder makes the mode of transmission most probably dominant with apenetrance or a greater range of expressivity in the female. Only mildly affected females would be expected to become mothers, and thus the pedigree might give appearances suggesting X-chromosomal transmission. Van Wijngaarden *et al.* (1969) believed their cases to be morphologically and histochemically different from those of Spiro *et al.* (1966).

107. Dominant Congenital Type of Myosclerosis
(Myodysplasia Fibrosa Multiplex, Lowenthal, 1954)

Lowenthal (1954) reported 4 affected siblings in one family and 4 affected over 3 generations in a second family. Features included symmetrical congenital contractures of the joints with possible sclerosis of both muscle and skin.

108. Dominant Type of Myopathy with Lipofibrocalcareous Anomaly (Mattioli-Foggia et al., 1963)

In two Italian families these authors reported 4 affected individuals. The nososlogical status of this disorder is obscure.

109. Dominant Congenital Type of Myopathy with Arachnodactyly (Marfan, 1896)

Myopathy of undetermined cause is a feature in only a fraction of individuals with this dominant gene which has a wide range of expressivity. Some individuals have no obvious clinical features apart from slenderness; others have only one cardinal sign, some two cardinal signs, and some all three cardinal signs. The cardinal signs are excessive extremity length with an abnormal metacarpal index, cardiovascular abnormalities, and subluxation of the lens.

110. Dominant Type of Dystrophia (Myotonia Atrophica, Steinert, 1909)

Hariga *et al.* (1969) reported two inbred kindreds with 18 affected individuals over 2 and 4 generations, respectively, with the typical wasting in temporal, neck, and distal limb muscles; myotonia; cataract; hypogonadism; frontal balding; and evidence of cardiac involvement. The literature is reviewed by Walton and Gardner-Medwin (1969).

111. Dominant Type of Myotonia Congenita (Thomsen, 1875)

Thomsen (1875) described his own family with 64 affected individuals occurring over 7 generations without a break, but the pedigree of Birt (1908), who was also affected personally, suggested incomplete penetrance in some generations. Presenting symptoms included a "strangled" cry and feeding difficulty in infancy, but symptoms were sometimes delayed until the second decade. Stiffness and impaired relaxation of limb and ocular movement were typically accentuated by rest and cold, but were relieved by exercise. Diffuse hypertrophy of muscles tended to persist throughout life, even though myotonia diminished progressively.

112. Recessive Type of Myotonia Congenita (Becker, 1966a)

Becker (1966a) distinguished the recessive type of myotonia from the dominant type (Thomsen, 1875) by an age of onset at 4–6 years, more severe symptoms, and a greater frequency in the population.

113. Recessive Type of Myotonia, Dwarfism, Diffuse Bone Disease, and Unusual Eye and Face Abnormality (Aberfeld et al., 1965)

Aberfeld *et al.* (1965) reported a male and a female sibling who showed in infancy talipes equinovarus, changes in facial features (pinched face) by the age of 2 years, an unsteady gait with stiff slightly flexed knees, hip pain by the age of 3 years, blepharophimosis, a high pitched voice, and pigeon-breast deformity. Mereu and Porter (1969)

and Aberfeld *et al.* (1970) each reported affected male and female siblings with clinical features and an age of onset similar to those of the
family reported in 1965.

114. Dominant Type of Paramyotonia Congenita (Eulenburg, 1886)

LaJoie (1961) reported a kindred with 72 affected individuals over
8 generations and Hudson (1963) reported a kindred with 16 affected
over 5 generations. Features included myotonia evident only on exposure
to cold and attacks of generalized muscular weakness beginning in childhood accompanied by a rise in the level of serum potassium; these
attacks sometimes disappeared in later life. The relationship of this condition to adynamia episodica hereditaria and to myotonic periodic paralysis (see below) is uncertain.

115. Recessive Type of Hypotonia in Infancy with Glucose 6-Phosphate Deficiency (von Gierke, 1929)

Features included diarrhea with steatorrhea (Crawford, 1946), pyrexia, delay in walking, muscle weakness, and hypoglycemic convulsions.
Williams *et al.* (1963) identified heterozygotes.

116. Recessive Type of Hypotonia and Muscle Weakness with Amylo-1,6-glucosidase Deficiency (Forbes, 1953)

Typically, the onset has been in young children with features which
were indistinguishable from those associated with glucose-6-phosphatase
deficiency. Genetic transmission has been autosomal recessive except
for the family reported by Oliver *et al.* (1961) in which a 50-year-old
father and 2 of his 4 offspring were affected.

117. Recessive Type of Hypotonia and Muscle Weakness with Amylo-1,4-glucodase (Acid Maltase) Deficiency (Pompe, 1932)

Nitowsky and Grunfeld (1967) reported a 5-month-old boy and identified the mother and a half sister as heterozygous carriers from cultured
fibroblast cells of the skin. In a second family these authors reported
3 offspring who died at 5 months, 4 months, and 3 weeks. They identified
as heterozygous carriers both the parents and a female sibling. Features
of the homozygote included cardiomegaly with heart failure, mental
retardation, large protruding tongue, diffuse muscular weakness, and
inability to stand until 24 months. Many affected children die in infancy;

those who survived in this family were unable to walk at 9–15 years. The prevalence in North Carolina from 1950 to 1963 was estimated to be 1 in 400,000 (Sidbury, 1965). Milder cases presenting with a myopathy in adult life have been reported by Hudgson *et al.* (1968) and others.

118. RECESSIVE TYPE OF PHOSPHORYLASE DEFICIENCY WITH PAIN AND MUSCLE WEAKNESS FROM EXERCISE (McARDLE, 1951)

Schmid and Hammaker (1961) restudied the family of McArdle (1951), which had 3 out of 13 offspring affected from first cousin parents. Features included pain, muscle cramps, and sometimes weakness that developed in all muscles after exercise but was relieved typically by rest. Fifty to 66% of cases reported occasional myoglobinuria after exercise. The onset age typically was early childhood. Dawson *et al.* (1968) reported 4 affected out of 5 siblings and suggested that brief painful cramps during exercise might be useful for the detection of the heterozygote.

119. RECESSIVE TYPE OF MUSCLE WEAKNESS WITH AMYLO-1,4–1,6-TRANS-GLUCOSIDASE DEFICIENCY (ANDERSON, 1956)

Anderson (1956) reported 2 male siblings. One died at the age of 7 months and the age of onset in the other was 10 months.

120. RECESSIVE TYPE OF MYOPATHY WITH MYOGLOBINURIA AND ABNORMAL GLYCOLYSIS (LARSSON ET AL., 1964)

Larsson *et al.* (1964) described 5 sibships with 1 to 4 individuals affected in each sibship (total of 14 affected). Features included fatigue, dyspnea, and palpitation followed by stiffness, tenderness, and muscular pain following exercise. Exacerbations sometimes were accompanied by myoglobinuria.

121. RECESSIVE TYPE OF MYOPATHY WITH MYOGLOBINURIA AND A POSSIBLE DEFECT OF LIPID METABOLISM (W. K. ENGEL ET AL., 1970)

W. K. Engel *et al.* (1970) reported concordant 18-year-old monozygotic twins with symptoms of intermittent muscle cramps and myoglobinuria appearing hours after exercise or induced by fasting on an isocaloric high fat, low carbohydrate diet.

122. RECESSIVE TYPE OF ABNORMAL LIPID STORAGE IN
     VOLUNTARY MUSCLE

Bradley *et al.* (1969) reported a young woman, product of a first cousin marriage, who had diffuse muscular weakness, and whose muscle biopsy showed an excessive accumulation of lipid in type 1 muscle fibers as well as abnormal mitochondria.

123. DOMINANT TYPE OF MYOPATHY WITH ABNORMAL GLYCOLYTIC
     BREAKDOWN OF PHOSPHOHEXOISOMERASE (SATOYOSHI
     AND KOWA, 1967)

Satoyoshi and Kowa (1967) described 5 cases over 3 generations, with symptoms of muscle pain, stiffness, and weakness occurring a few hours after moderately heavy exercise. The onset age was 35 years.

124. RECESSIVE TYPE OF MUSCLE WEAKNESS WITH
     PHOSPHOFRUCTOKINASE DEFICIENCY (LAYZER ET AL., 1967)

Tarui *et al.* (1965) reported 3 affected out of 5 offspring of first cousin parents. Features included marked weakness and stiffness of muscles after vigorous or prolonged exercise. The heterozygous carrier state was detected in the mother but not in the father. The onset was in childhood.

125. DOMINANT TYPE OF MUSCLE WEAKNESS WITH
     HYPOKALEMIC PERIODIC PARALYSIS

Cusins and Rooyen (1963) described a kindred with 8 affected individuals over 3 generations. A low serum potassium level accompanied attacks that which were precipitated by exposure to cold, exercise, alcohol, and sodium chloride. The frequency of attacks ranged from once in a lifetime to once a day. Potassium chloride relieved the attacks. The extensive literature is reviewed by McArdle (1969).

126. DOMINANT TYPE OF ADYNAMIA EPISODICA HEREDITARIA
     WITHOUT MYOTONIA OF THE EYELIDS (GAMSTORP, 1956)

Gamstorp (1956) reported 68 cases of paralysis with increased levels of serum potassium during attacks. Episodes typically developed when sitting after exercise during the day, but they could be provoked by administration of potassium. The frequency of attacks was usually once a week with a duration of about 1 hr. The incidence in Sweden was 2×10^{-6}. The recent literature is reviewed by McArdle (1969).

127. DOMINANT TYPE OF MUSCLE WEAKNESS WITH ADYNAMIA EPISODICA HEREDITARIA (MYOTONIC PERIODIC PARALYSIS, VAN'T HOFF, 1962)

Samaha (1965) reported a kindred with 28 affected individuals over 6 generations; these patients had myotonia of eyelids similar to that noted in the kindred reported by van't Hoff (1962) which contained 9 cases over 4 generations. Attacks of eyelid myotonia lasting 15–20 sec occurred both between and during attacks. Cases seen outside Sweden have typically been of this type.

128. DOMINANT TYPE OF NORMOKALEMIC PERIODIC PARALYSIS (POSKANZER AND KERR, 1961)

Poskanzer and Kerr (1961) described a pedigree with 21 affected cases in 5 generations. Features included attacks which lasted for days or even weeks and which were more severe than the ones occurring in the families described by Gamstorp (1956) or van't Hoff (1962). Attacks took place mainly at night. The serum potassium was normal during the attacks. There was a favorable response to the administration of sodium chloride.

129. DOMINANT TYPE OF GENERALIZED MYOSITIS OSSIFICANS

Tunte *et al.* (1967) reviewed 350 cases in the literature including 2 sets of concordant monozygotic twins. Transmission was over 2 generations in 5 families, and an additional 13 cases of deformity of the extremities were mentioned in other members of the families (1 parent in 7 cases). These authors presented 23 additional probands and their families. A parent was affected in 2 of the 23 probands. Features included hypogenitalism, infantilism, and anomalies of tooth position.

130. RECESSIVE TYPE OF MYOPATHY WITH HOMOCYSTINURIA (GERRITSEN ET AL., 1962; CARSON AND NEILL, 1962)

Features included mental retardation in 2 out of 3 cases. Ecotopia lentis was a constant feature in patients aged over 10 years. Skeletal features suggested the Marfan syndrome, and the muscles were slender and weak. Schimke *et al.* (1965) ascertained 20 cases while screening for a presumed Marfan syndrome. Spaeth and Barber (1967) found 91 cases in 38 institutions while screening urinary amino acid excretion in almost 10,000 mentally retarded patients. Prevalance in institutions for mental retardation was 0.02%. Five percent of the patients with dislocated lenses were found to be affected.

131. RECESSIVE TYPE OF PAROXYSMAL MYGLOBINURIA (KOREIN ET AL., 1959)

Korein *et al.* (1959) found in the literature 40 reports of this condition which was described first by Meyer-Betz (1910). Hed (1953) reported 3 affected male siblings in a sibship of 6 with an onset at from 15 to 22 years.

132. RECESSIVE TYPE OF POLYMYOPATHY IN WERNER'S DISEASE (WERNER, 1904)

Epstein *et al.* (1966) included 94 sibships from the literature in a gentic analysis showing an autosomal recessive pattern of inheritance. Consanguinity was increased in the parents. Features included premature graying, baldness, wizened and prematurely aged facies, short stature, diffuse muscular weakness, spheropoikiloderma, trophic ulcers of the legs, juvenile cataracts, hypogenitalism, and osteopetrosis.

133. RECESSIVE TYPE OF MYOPATHY AND PSEUDOHYPERTROPHY WITH MYXOEDEMA IN THE SYNDROME OF KOCHER (1892) AND DEBRE AND SEMELAIGNE (1935)

Cross *et al.* (1968) reported 2 affected sisters in an inbred Amish group.

134. DOMINANT TYPE OF CUSHING'S DISEASE WITH MYOPATHY (SIPPLE, 1961)

Pleasure *et al.* (1970) studied a pedigree with 5 affected individuals over 3 generations. Sarosi and Doe (1968) investigated a family with 3 out of 4 siblings affected. From a review of the literature, occasional associated features in this disorder, which is only rarely genetically determined, included parathyroid adenomas, pheochromocytoma, and medullary carcinoma of thyroid with amyloid stroma.

E. Disorders of Unknown Etiology and Pathogenesis in Which the Neuromuscular Apparatus May Be Involved

135. DOMINANT TYPE OF MYOKYMIA WITH HYPERHIDROSIS (ISAACS, 1961)

Grund (1938) described 2 affected brothers. Gamstorp and Wohlfahrt (1959) reported 3 unrelated cases with evidence suggesting dominant inheritance in one. Isaacs (1961) investigated an isolated case in a 12-year-old male and a second isolated case in a 53-year-old male; both showed continuous muscle activity in the electromyogram. The 3 cases of Mertens and Zschoeke (1965) described as examples of neuromyotonia were probably similar to the cases of Isaacs (Gardner-Medwin and Walton, 1969).

136. DOMINANT TYPE OF MUSCULAR CONTRACTURE AND MYOPATHY (HAUPTMANN AND THANNHAUSER, 1941)

Sood and Goyal (1969) reported 6 cases including a father and his offspring with an onset by the age of 2 years. Hauptmann and Thannhauser (1941) studied a French Canadian kindred with 9 cases over 3 generations. Features included an inability to bend the head forward, a slight "webbed neck," difficulty in bending the back and in extending the arms, and underdevelopment of limb–girdle muscles. This condition may well be a form of familial myosclerosis (Walton and Gardner-Medwin, 1969).

137. RECESSIVE TYPE OF KUSKOKWIM DISEASE WITH JOINT CONTRACTURES AND HYPERTROPHY OF ASSOCIATED MUSCLE GROUPS (PETAJAN ET AL., 1969)

Petajan *et al.* (1969) described 10 cases in 4 families from the region of Alaska drained by the Kuskokwim river. This region of 10,000–12,000 population had 17 verified cases in 7 families. Features included multiple joint contractures, especially at the knees and ankles, with hypertrophy of associated muscle groups.

138. SPORADIC LETHAL TYPE OF PRADER–WILLI SYNDROME (1963)

Johnsen *et al.* (1967) reported 7 isolated cases presenting extreme infantile hypotonia, followed by uncontrollable hyperphagia, mental retardation, plethoric obesity, and short stature with small hands and feet. Since cases do not reproduce, all cases appear to be due to new mutations. Zellweger and Schneider (1968) reported 2 new cases and presented excellent statistical findings based on an analysis of stigmata recorded for 93 cases reported in the literature. It is not certain whether the muscular hypotonia is purely central or whether there is peripheral neuromuscular dysfunction in this condition.

139. Dominant Type of Craniocarpotarsal Dysplasia (Whistling Face Syndrome, Weinstein and Gorlin, 1969)

Temtany (1966) described 3 kindreds with cases occurring over 3 generations. Weinstein and Gorlin (1969) reported a case in which there was difficulty in feeding because of a small mouth; blepharophimosis was noted at the age of 7; eyes were sunken, and there were flexion contractures of the thumbs.

VI. Summary and Conclusions

The basis for the selection of the neuromuscular and muscular genetic disorders listed in this chapter is explained, pertinent genetic terms are defined, and methods of assessing the risk of affection among relatives according to the type of inheritance are mentioned. Whenever available, information given for each disorder listed has included the pattern of inheritance, identifying names with references, and citation of one or two leading studies which give useful information as well as information about identifying clinical symptoms. Other information given, where possible, has included ages of onset, duration, incidence, penetrance, expressivity, mutation rate, and comments on carrier identification and linkage relationships.

Since the eventual prevention of the expression of genes responsible for neuromuscular and muscular disorders will probably be based upon very precise knowledge of gene action at the biochemical level, and since different genes influence different biochemical pathways, it is important that each entity be delineated as precisely as possible. It is to be expected that there will be much overlap in symptoms and signs among the various disorders, and many of those listed in this chapter may be artificially defined because of the lack of precision of current diagnostic techniques, but this fact should demand an even greater effort to identify the various mutant genes. There is no reason to expect less progress in the future than during the past 15 years in defining precisely each neuromuscular and muscular disorder. Every effort should be made to report new information based upon modern diagnostic methods in cases whose relatives have been reported previously. Such followup studies are invaluable in the delineation of genotypes.

REFERENCES

Aagenaes, O. (1959). *Acta Psychiat. Neurol. Scand.* **34,** 489–494.
Aberfeld, D. C., Hinterbuchner, L. P., and Schneider, M. (1965). *Brain* **88,** 313–322.
Aberfeld, D. C., Nambu, T., Vye, M. V., and Grob, D. (1970). *Neurology* **22,** 455–462.
Allen, W. (1939). *AMA Arch. Intern. Med.* **63,** 1123–1131.
Alport, C. A. (1927). *Brit. Med. J.* **1,** 504–506.
Alzheimer, A. (1907). *Allg. Z. Psychiat. Gerichtl. Med.* **64,** 146–148.
Andermann, F., Lloyd-Smith, D. L., Huntingdon, M., and Mathieson, G. (1962). *Neurology* 712–724.
Anderson, D. H. (1956). *Lab. Invest.* **5,** 11–20.
Andrade, C. (1952). *Brain* **75,** 408–427.
Aran, F. A. (1850). *Arch. Gen. Med.* **24,** 172–214.
Armstrong, R. M., Fogelson, H., and Silbergerg (1966). *Arch. Neurol.* **14,** 208–212.
Asano, N., Kizu, M., Yamada, T., Asano, N., and Kijima, C. (1960). *Jap. J. Hum. Genet.* **5,** 139–146.
Austin, J. H. (1956). *Medicine* (*Baltimore*) **35,** 187 237.
Austin, J. H., Armstrong, D., Shearer, L., and McAtee, D. (1966). *Arch. Neurol.* (*Chicago*) **14,** 259–269.
Baker, A. B., ed. (1962). "Heresopathia Atactica Polyneuritiformis," 2nd ed. Publ. 4, pp. 2286–2287. Harper (Hoeber), New York.
Barbeau, A. (1966). *In* "Progressive Muskeldystrophie, Myotonie, Myasthenie" (E. Kuhn, ed.), Springer, New York.
Barnes, S. (1932). *Brain* **55,** 1–46.
Bass, N. H., Witmer, E. J., and Dreifuss, F. E. (1970). *Neurology* **20,** 52–62.
Bassen, F. A., and Kornzweig, A. L. (1950). *Blood* **5,** 381–387.
Batten, F. E. (1910). *Quart. J. Med.* **3,** 313–328.
Becker, P. E. (1962). *Rev. Can. Biol.* **21,** 551–566.
Becker, P. E. (1966a). *In* "Progressive Muskelodystrophie, Myotonie, Myasthenie" (E. Kuhn, ed.), pp. 247–255. Springer, New York.
Becker, P. E. (1966b). *Humangenetik* **5,** Part 1, 427.
Bedford, P. D., and James, F. E. (1956). *J. Neurol., Neurosurg. Psychiat.* **19,** 46–51.
Bell, J., and Carmichael, E. A. (1939). *In* "The Treasury of Human Inheritance," Vol. 4, Part 3, pp. 169–172. Cambridge Univ. Press, London and New York.
Biemond, A. (1955). *Acta Psychiat. Neurol. Scand.* **30,** 25–38.
Birt, A. (1908). *Montreal Med. J.* **37,** 771–784.
Blumel, J., Evans, E. B., and Eggers, G. W. N. (1957). *J. Pediat.* **50,** 454–458.
Bonduelle, M., Bouygues, P., and Faveret, C. (1959). *Presse Med.* **67,** 1630–1633.

Bradley, W. G., Hudgson, P., Gardner-Medwin, D., and Walton, J. N. (1969). *Lancet* 1, 495–498.

Brandenberg, F. (1910). *Arch. Rass.-Gesellschaftsbiol.* 7, 290–305.

Brandt, S. (1949). *Amer. J. Dis. Child.* 78, 226–236.

Bray, G. M., Kaarsoo, M., and Ross, R. T. (1965). *Neurology.* 15, 678–684.

Brown, M. R. (1960). *N. Engl. J. Med.* 262, 1280–1282.

Carson, N. A. J., and Neill, D. W. (1962). *Arch. Dis. Childhood* 37, 505–513.

Celesia, G. G. (1965). *Arch. Neurol. (Chicago)* 12, 206–210.

Charcot, J. M., and Marie, P. (1886). *Rev. Med. (Paris)* 6, 97–138.

Chiricosta, A., Jindal, L. S., Meluzals, J., and Koch, B. (1970). *Can. Med. Ass. J.* 102, 396–401.

Clarkson, P. (1962). *Guy's Hosp. Rep.* 111, 335–346.

Coleman, R. F., Nienhius, A. W., Brown, W. J., Munsat, T. L., and Pearson, C. M. (1967). *J. Amer. Med. Ass.* 199, 624–630.

Coleman, R. F., Thompson, L. R., Nienhuis, A. W., Munsat, T. L., and Pearson, C. M. (1968). *Arch. Pathol.* 86, 365–376.

Crawford, T. (1946). *Quart. J. Med.* 15, 285–298.

Creutzfeldt, H. G. (1920). *Z. Gesamte Neurol. Psychiat.* 57, 1–18.

Croft, P. B., and Wadia, N. H. (1957). *Neurology* 7, 356–366.

Crome, L., and Weller, S. D. V. (1965). *Arch. Dis. Childhood* 40, 502–507.

Cross, H. E., Hollander, C. S., Rimoin, D. L., and McKusick, V. A. (1968). *Pediatrics* 41, 413–420.

Crowe, F. W., Schull, W. J., and Neel, J. V. (1956). "A Clinical, Pathological, and Genetic Study of Multiple Neurofibromatosis." Thomas, Springfield, Illinois.

Cusins, P. J., and Rooyen, R. J. (1963). *S. Afr. Med. J.* 37, 1180–1183.

Davidenkow, S. (1939). *Arch. Neurol. Psychiat.* 41, 694–701.

Davison, C. (1954). *J. Neuropathol. Exp. Neurol.* 13, 50–59.

Dawson, D. M., Spong, F. L., and Harrington, J. F. (1968). *Ann. Intern. Med.* 69, 229–235.

Dean, G. (1963). "The Porphyrias. A Story of Inheritance and Environment." Lippincott, Philadelphia, Pennsylvania.

Debre, R., and Semelaigne, G. (1935). *Amer. J. Dis. Child.* 50, 1351–1361.

Dejerine, J., and Sottas, J. (1893). *C. R. Soc. Biol.* 45, 63–96.

DeLange, C. (1937). *Acta Paediat (Stockholm)* 20, Suppl. 3, 1–51.

Denny-Brown, D. (1951). *J. Neurol., Neurosurg. Psychiat.* 14, 237–252.

Dick, A. P., and Stevenson, C. J. (1953). *Lancet* 1, 921–923.

Drachman, D. B., and Banker, B. Q. (1961). *Arch. Neurol. (Chicago)* 5, 77–93.

Dubowitz, V., and Roy, S. (1970). *Brain* 93, 133–146.

Duchenne, G. B. A. (1868). *Arch. Gen. Med.* 1, 5–25, 179–209, 305–321, 421–443, and 552–588.

Dyck, P. J., and Lambert, E. H. (1968). *Arch. Neurol. (Chicago)* 18, 603–618 and 619–625.

Dyck, P. J., Lambert, E. H., and Mulder, D. W. (1963). *Neurology* 13, 1–11.

Earl, C. J., Fullerton, P. M., Wakefield, G. S., and Schutta, H. S. (1964). *Quart. J. Med.* 33, 481–498.

Eldridge, R., Ryan, E., Rosario, J., and Brody, J. A. (1969). *Neurology* 19, 1029–1037.

Emery, A. E. H. (1969). *J. Neurol. Sci.* 8, 579–587.

Emery, A. E. H., and Dreifuss, F. E. (1966). *J. Neurol., Neurosurg. Psychiat.* 29, 338–342.

Engel, A. G. (1961). *Arch. Neurol. (Chicago)* 4, 663–674.

Engel, W. K., Kurland, L. T., and Klatzo, I. (1959). *Brain* **82**, 203–220.

Engel, W. K., Vick, N. A., Cleuck, C. J., and Levy, R. I. (1970). *N. Engl. J. Med.* **282**, 697–704.

Epstein, C. J., Martin, G. M., Schultz, A. L., and Motulsky, A. G. (1966). *Medicine* (*Baltimore*) **45**, 177–222.

Erb, W. (1891). *Deut. Z. Nervenheilk.* **1**, 13–94 and 173–261.

Espinosa, R. E., Okikiro, M. M., Mulder, D. W., and Sayre, G. P. (1962). *Neurology* **12**, 1–7.

Eulenburg, A. (1886). *Neurol. Zentralbl.* **5**, 265–272.

Farmer, T. W., and Mustian, V. M. (1963). *Arch. Neurol.* (*Chicago*) **8**, 471–480.

Farquhar, J. W., and Ways, P. (1966). *In* "The Metabolic Basis of Inherited Disease" (J. B. Stanbury *et al.*, eds.), 2nd ed., Chapter 24, pp. 509–522. McGraw-Hill (Blakiston), New York.

Ferguson, F. R., and Critchley, M. (1929). *Brain* **52**, 203–225.

Forbes, G. B. (1953). *J. Pediat.* **42**, 645–653.

Ford, F. R. (1966). "Diseases of the Nervous System in Infancy, Childhood and Adolescence," 5th ed., pp. 285–288. Thomas, Springfield, Illinois.

Fortanier, A. H., and Speijer, N. (1935). *Genetica* **17**, 471–486.

Fredrickson, D. S. (1966). *In* "The Metabolic Basis of Inherited Disease" (J. B. Stanbury, *et al.*, eds.), 2nd ed., pp. 486–508. McGraw-Hill (Blakiston), New York.

Friedreich, N. (1863). *Arch. Pathol. Anat. Physiol. Klin. Med.* **26**, 391–419 and 433–459; **27**, 1–26.

Furukawa, T., Takagi, A., Nakao, K., Sugita, H., and Tsukagoshi, H. (1968). *Neurology* **18**, 942–947.

Gamstorp, I. (1956). *Acta Paediat.* **45**, Suppl. 108, 1–126.

Gamstorp, I., and Wohlfahrt, G. (1959). *Acta Psychiat. Scand.* **34**, 181–194.

Gardner-Medwin, D., and Walton, J. N. (1969). *Lancet* **1**, 127–130.

Gardner-Medwin, D., Hudgson, P., and Walton, J. N. (1967). *J. Neurol. Sci.* **5**, 121–158.

Garland, H., and Moorhouse, D. (1953). *J. Neurol., Neurosurg. Psychiat.* **16**, 110–116.

Gerritsen, T., Vaughn, J. G., and Waisman, H. A. (1962). *Biochem. Biophys. Res. Commun.* **9**, 493–496.

Gibberd, F. B., and Gavrilescu, K. (1966). *Neurology* **16**, 130–134.

Gibson, A. (1921). *AMA Arch. Intern. Med.* **27**, 338–350.

Gilbert, G. J., McEntee, W. J., and Glaser, G. H. (1963). *Neurology* **13**, 365–372.

Gilman, S., and Horenstein, S. (1964). *Brain* **87**, 51–66.

Gomez, M. R., Clermont, V., and Bernstein, J. (1962). *Arch. Neurol.* (*Chicago*) **6**, 317 323.

Green, J. B. (1960). *Neurology* **10**, 960–962.

Greenberg, J. (1964). *Arch. Neurol.* (*Chicago*) **11**, 219–222.

Greenfield, J. G. (1933). *J. Neurol. Psychopathol.* **13**, 289–302.

Greenfield, J. G. (1954). "The Spino-cerebellar Degenerations." Blackwell, Oxford.

Grund, G. (1938). *Deut. Z. Nervenheilk.* **146**, 3–14.

Hariga, J., Mathys, E. T., Meckler, L., and Jonckheer, M. (1969). *Acta Neurol. Psychiat. Belg.* **69**, 215–233.

Hauptmann, A., and Thannhauser, S. J. (1941). *Archiv. Neurol. Psychiat.* **46**, 654–664.

Heck, A. F. (1964). *J. Neurol. Sci.* **1**, 226–255.

Hed, R. (1953). *AMA Arch. Intern. Med.* **92**, 825–832.

Heller, H., Sohar, E., and Gafni, J. (1964). *Pathol. Microbiol.* **27**, 833–840.

Henson, T. E., Muller, J., and Demyer, W. E. (1967). *Arch. Neurol.* (*Chicago*) **17**, 238–247.

Herndon, J. H., Steinberg, D., and Uhlendorf, B. W. (1969). *N. Engl. J. Med.* **281**, 1034–1038.

Herrman, C. (1966). *Neurology* **16**, 75–85.

Hicks, E. P., and Camp, M. B. (1922). *Lancet* **1**, 319–321.

Hill, W., and Sherman, H. (1968). *Arch. Neurol.* (*Chicago*) **18**, 350–357.

Hirano, A., Kurland, L. T., Krooth, R. S., and Lessell, S. (1961). *Brain* **84**, 642–661.

Hoffmann, J. (1893). *Deut. Z. Nervenheilk.* **3**, 427–470.

Hohmann, H. (1957). *Nervenarzt* **28**, 323–325.

Holmes, G. (1907). *Brain* **30**, 466–489.

Hopkins, I. J., Lindsey, J. R., and Ford, F. R. (1966). *Brain* **89**, 299–310.

Hudgson, P., Gardner-Medwin, D., Worsfold, M., Pennington, R. J. T., and Walton, J. N. (1968). *Brain* **91**, 435–462.

Hudson, A. J. (1963). *Brain* **86**, 811–826.

Hunt, J. R. (1917). *Brain* **40**, 58–148.

Hunt, J. R. (1921). *Brain* **44**, 490–538.

Huntington, G. (1872). *Med. Surg. Reporter* (*Phil.*) **26**, 317–321.

Isaacs, H. (1961). *J. Neurol., Neurosurg. Psychiat.* **24**, 319–325.

Iwashita, H., Inoue, N., Araki, S., and Kuroiwa, Y. (1970). *Arch. Neurol.* (*Chicago*) **22**, 357–364.

Jacob, A. (1921). *Z. Neurol.* **64**, 147–228.

Jenis, E. H., Lindquist, R. R., and Lister, R. C. (1969). *Arch. Neurol.* (*Chicago*) **20**, 281–287.

Johns, T. R., Dreifuss, F. E., Crowley, W. J., and Fakadej, A. V. (1966). *Neurology* **16**, 307.

Johnsen, S., Crawford, J. D., and Haessler, H. A. (1967). APS.

Johnston, A. W., and McKusick, V. A. (1962). *Amer. J. Hum. Genet.* **14**, 83–94.

Jones, W. A., and Jones, G. P. (1953). *Lancet* **1**, 1073–1074.

Kaeser, H. E. (1964). *Deut. Z. Nervenheilk.* **186**, 379–394.

Kearns, T. P. (1964). *Trans. Amer. Ophthalmol. Soc.* **63**, 559–625.

Kennedy, W. R., Alter, M., and Sung, J. H. (1968). *Neurology* **18**, 671–680.

Killian, J. M., and Kloepfer, H. W. Unpublished data.

Kiloh, L. G., and Nevin, S. (1951). *Brain* **74**, 115–143.

Kimoshita, M., and Cadman, T. E. (1968). *Arch. Neurol.* (*Chicago*) **18**, 265–271.

Klenk, E., and Kahlke, W. (1963). *Hoppe-Seyler's Z. Physiol. Chem.* **333**, 133–139.

Kloepfer, H. W., and Talley, C. (1958). *Ann. Hum. Genet.* **22**, 138–143.

Kocen, R. S., Lloyd, J. K., Lascelles, P. T., Fosbrooke, A. S., and Williams, D. (1967). *Lancet* **1**, 1341–1345.

Kocher, T. (1892). *Deut. Z. Chir.* **34**, 556–626.

Koerner, D. R. (1952). *Ann. Intern. Med.* **37**, 1204–1220.

Korein, J., Coddon, D. R., and Mowrey, F. W. (1959). *Neurology* **9**, 767–785.

Korein, J., Steinman, P. A., and Senz, E. H. (1961). *Arch. Neurol.* (*Chicago*) **4**, 272–280.

Krabbe, K. H. (1958). *Acta Psychiat. Neurol. Scand.* **33**, 94–102.

Kugelberg, E., and Welander, L. (1956). *Arch. Neurol. Psychiat.* **75**, 500–509.

Kurland, L. T. (1957). *Proc. Staff Meet. Mayo Clin.* **32**, 449–462.

Kurland, L. T., and Mulder, D. W. (1955). *Neurology* **5**, 249–268.

LaJoie, W. J. (1961). *Arch. Phys. Med. Rehabil.* **42**, 507–512.

Landouzy, L., and Dejerine, J. (1885). *Rev. Med.* (*Paris*) **5**, 253–366.

Larsson, L. E., Linderholm, H., Muller, R., Ringqvist, T., and Sornas, R. (1964). *J. Neurol., Neurosurg. Psychiat.* **27**, 361–380.

Layzer, R. B., Rowland, L. P., and Ranney, H. M. (1967). *Arch. Neurol.* (*Chicago*) **17**, 512–523.

Lehmann, H., and Liddell, J. (1966). *In* "The Metabolic Basis of Inherited Disease" (J. B. Stanbury *et al.*, eds.), pp. 1356–1369. McGraw-Hill (Blakiston), New York.

Lindsey, J. R., Hopkins, I. J., and Clark, D. B. (1966). *Bull. Johns Hopkins Hosp.* **119**, 378–406.

Londe, P. (1894). *Rev. Med.* (*Paris*) **14**, 212–254.

Louis-Bar, D. (1941). *Confin. Neurol.* **4**, 32–42.

Lovelace, R. E., Schotland, D. L., and DeNapoli, R. A. (1966). *Trans. Amer. Neurol. Ass.* **91**, 286–288.

Lowenberg, K., and Waggoner, R. W. (1934). *Arch. Neurol. Psychiat.* **31**, 737–754.

Lowenthal, A. (1954). *Acta Neurol. Psychiat. Belg.* **54**, 155–165.

Macalpine, I., Hunter, R., and Rimington, C. (1968). *Brit. Med. J.* **1**, 7–18.

McArdle, B. (1951). *Clin. Sci.* **10**, 13–36.

McArdle, B. (1969). *In* "Disorders of Voluntary Muscle" (J. N. Walton, ed.), 2nd ed., pp. 607–638. Churchill, London.

McKusick, V. A. (1968). "Mendelian Inheritance in Man," 2nd ed., p. 76. Johns Hopkins Press, Baltimore, Maryland.

McKusick, V. A. (1968b). "Mendelian Inheritance in Man," 2nd ed., p. 228. Johns Hopkins Press, Baltimore, Maryland.

McKusick, V. A. (1971). "Mendelian Inheritance in Man," 3rd ed. Johns Hopkins Press, Baltimore, Maryland.

McQuillen, M. P. (1966). *Brain* **89**, 121–132.

Mabry, C. C., Roeckel, I. E., Munich, R. L., and Robertson, D. (1965). *N. Engl. J. Med.* **273**, 1062–1070.

Mahloudji, M. (1963). *J. Neurol., Neurosurg. Psychiat.* **26**, 511–513.

Marfan, M. A. B. (1896). *Bull. Soc. Med. Paris* **13**, 220–226.

Marinesco, G. (1915). *C. R. Soc. Biol.* **78**, 481–483.

Marinesco, G., Draganesco, S., and Vasiliu, D. (1931). *Encephale* **26**, 97–109.

Mattioli-Foggia, C., Cunego, A., and Gragnani, W. (1963). *Proc. Int. Congr. Hum. Genet. 2nd, 1961* p. E82.

May, D. L., and White, H. H. (1968). *Arch. Neurol.* (*Chicago*) **19**, 331–338.

Menzel, P. (1891). *Arch. Psychiat. Nervenkr.* **22**, 160–190.

Mereu, T. R., and Porter, I. H. (1969). *Amer. J. Dis. Child.* **117**, 470–478.

Mertens, H. G., and Zschoeke, S. (1965). *Klin. Wochenschr.* **43**, 917–925.

Meyer-Betz, F. (1910). *Deut. Arch. Klin. Med.* **101**, 35–127.

Milhorat, A. T., and Wolff, H. G. (1943). *Arch. Neurol. Psychiat.* **49**, 641–654.

Möbius, P. J. (1888). *Muenchen. Med. Wochenschr.* **35**, 91–94 and 108–111.

Morton, N. E., and Chung, C. S. (1959). *Amer. J. Hum. Genet.* **11**, 360–379.

Morton, N. E., Chung, C. S., and Peters, H. A. (1963). *In* "Muscular Dystrophy in Man and Animals" (G. H. Bourne and N. Golarz, eds.), p. 327. Karger Basel.

Nitowsky, H. M., and Grunfeld, A. (1967). *J. Lab. Clin. Med.* **69**, 472–484.

Noyes, A. P. (1930). *R. I. Med. J.* **13**, 52–59.

Oliver, L., Schulman, M., and Larner, J. (1961). *Clin. Res.* **9**, 243.

Otto, A. W. (1841). "Monstrorum sexcentorum descriptio anatomica, Vratislaviae."

Pearson, C. M., and Fowler, W. G. (1963). *Brain* **86**, 75–88.

Penn, A. S., Lasak, R. P., and Rowland, L. P. (1970). *Neurology* **20**, 147–159.

Petajan, J. H., Momberger, G. L., Aase, J., and Wright, G. (1969). *J. Amer. Med. Ass.* **209**, 1481–1486.

Pick, A. (1892). *Prager Med. Wochenschr.* **17**, 165–167.

Pleasure, D. E., Walsh, G. O., and Engel, W. K. (1970). *Arch. Neurol. (Chicago)* **22**, 118–125.

Poland, A. (1841). *Guy's Hosp. Rep.* **6**, 191.

Pompe, J. C. (1932). *Ned. Tijdschr. Geneesk.* **76**, 304.

Poskanzer, D. C., and Kerr, D. N. S. (1961). *Amer. J. Med.* **31**, 328–342.

Prader, A., and Willi, H. (1963). *Verh. Kongr. Psych. Entw.-Stor. Kindes-Alt., 21st, 1961* Part 1, p. 353.

Reed, T. E., and Chandler, J. H. (1958). *Amer. J. Hum. Genet.* **10**, 201–225.

Refsum, S. (1952). *J. Nerv. Ment. Dis.* **116**, 1046–1050.

Refsum, S., and Skillicorn, S. A. (1954). *Neurology* **4**, 40–47.

Rennie, G. E. (1908). *Rev. Neurol. Psychiat., Praha* **6**, 229–233.

Research Group on Neuromuscular Diseases of The World Federation of Neurology. (1968). *J. Neurol. Sci.* **6**, 165–177.

Robert, J. M., and Guibaud, P. (1969). *Lyon Med.* **221**, 1485–1491.

Rosenberg, R. N., and Chutorian, A. (1967). *Neurology* **17**, 827–832.

Roussy, G., and Levy, G. (1934). *Rev. Neurol.* **62**, 763–773.

Rubinstein, H. M., Dietz, A. A., Hodges, L. K., Lubrano, T., and Czebotar, V. (1970). *J. Clin. Invest.* **49**, 479–486.

Rukavina, J. G., Block, W. D., Jackson, C. E., Falls, H. F., Carey, J. H., and Curtis, A. C. (1956). *Medicine (Baltimore)* **35**, 239–334.

Russell, W. R., and Garland, H. G. (1930). *Brain* **53**, 376–384.

Salt, H. B., Wolff, O. H., Lloyd, J. K., Fosbrooke, A. S., Cameron, A. H., and Hubble, D. V. (1960). *Lancet* **2**, 325–329.

Samaha, F. J. (1965). *Arch. Neurol. (Chicago)* **12**, 145–154.

Sarosi, G., and Doe, R. P. (1968). *Ann. Intern. Med.* **68**, 1305–1309.

Satoyoshi, E., and Kowa, H. (1967). *Arch. Neurol. (Chicago)* **17**, 248–256.

Schenk, V. W. D. (1959). *Ann. Hum. Genet.* **23**, 325–333.

Schimke, R. N., McKusick, V. A., Huang, T., and Pollack, A. D. (1965). *J. Am. J. Amer. Med. Ass.* **193**, 711–719.

Schlesinger, A. S., Duggins, V. A., and Masucci, E. F. (1962). *Brain* **85**, 357–370.

Schmid, R., and Hammaker, L. (1961). *N. Engl. J. Med.* **264**, 223–225.

Schneiderman, L. J., Sampson, W. I., Schoene, W. C., and Haydon, G. B. (1969). *Amer. J. Med.* **46**, 380–393.

Schottky, J. (1932). *Z. Neurol. Psychiat.* **140**, 333–397.

Schutta, H. S., Pratt, R. T. C. Metz, H., Evans, K. A., and Carter, C. O. (1966). *J. Med. Genet.* **3**, 86–91.

Schwarz, G. A., and Liu, C. N. (1956). *Arch. Neurol. Psychiat.* **75**, 144–162.

Seeligmuller, A. (1876). *Deut. Med. Wochenschr.* **2**, 185–186.

Seitelberger, F. (1952). *Z. Nervenheilk.* **5**, 213–219.

Shaw, R. F., and Glover, R. A. (1961). *Amer. J. Hum. Genet.* **13**, 89–97.

Sher, J. H., Rimalovski, A. B., Athanassiades, T. J., and Aronson, S. M. (1967). *Neurology* **17**, 727–742.

Shy, G. M., and Gonatas, N. K. (1964). *Science* **145**, 493–496.

Shy, G. M., and Magee, K. R. (1956). *Brain* **79**, 610–621.

Shy, G. M., Engel, W. K., Soniers, J. E., and Wanko, T. (1963). *Brain* **86**, 793–810.

Shy, G. M., Gonatas, N. K., and Perez, M. (1966). *Brain* **89**, 133–158.
Shy, G. M., Silberberg, D. H., Appel, S. H., Mishkin, M. M., and Godfrey, E. H. (1967). *Amer. J. Med.* **42**, 163–168.
Sidbury, J. B. (1965). *Progr. Med. Genet.* **4**, 32–58.
Silver, J. R. (1966). *Ann. Hum. Genet.* **30**, 69–75.
Singh, H., and Sham, R. (1964). *Brit. J. Clin. Pract.* **18**, 91–92.
Sipple, J. H. (1961). *Am. J. Med.* **31**, 163–166.
Sjögren, T. (1950). *Confin. Neurol.* **10**, 293–308.
Sjögren, T., and Larsson, T. (1957). *Acta Psychiat. Neurol. Scand., Suppl.* **113**, 1–112.
Skyring, A., and McKusick, V. A. (1961). *Amer. J. Med. Sci.* **242**, 534–547.
Smith, J. B., and Patel, A. (1965). *Neurology* **15**, 469–473.
Sood, S. C., and Goyal, B. G. (1969). *Indian J. Pediat.* **36**, 219–223.
Spaeth, G. L., and Barber, G. W. (1967). *Pediatrics* **40**, 586–589.
Spillane, J. D. (1940). *Brain* **63**, 275–290.
Spira, R. (1963). *Confin. Neurol.* **23**, 245–255.
Spiro, A. J., and Kennedy, C. (1965). *Arch. Neurol. (Chicago)* **13**, 155–159.
Spiro, A. J., Shy, G. M., and Gonatas, N. K. (1966). *Arch. Neurol. (Chicago)* **14**, 1–14.
Spiro, A. J., Princas, J. W., and Moore, C. L. (1970). *Arch. Neurol. (Chicago)* **22**, 259–269.
Staal, A., De Weerdt, C. J., and Went, L. N. (1965). *Neurology* **15**, 1008–1017.
Steinert, H. (1909). *Deut. Z. Nervenheilk.* **37**, 58–104.
Stevenson, A. C. (1953). *Ann. Eugen.* **18**, 50–93.
Stevenson, A. C., Cheeseman, E. A., and Huth, M. C. (1955). *Ann. Human Genet.* **19**, 165–173.
Strich, S. J. (1966). *J. Neurol., Neurosurg. Psychiat.* **29**, 487–499.
Strumpell, A. (1880). *Arch. Psychiat. Nervenkr.* **10**, 676–717.
Sunahara, S., Urano, M., and Ogawa, M. (1961). *Science* **134**, 1530–1531.
Sylvester, P. E. (1958). *Arch. Dis. Childhood* **33**, 217–221.
Tadjoedin, M. K., and Fraser, F. C. (1965). *Amer. J. Dis. Child.* **110**, 64–68.
Takikawa, K. (1953). *Jap. J. Hum. Genet.* **28**, 116.
Tarui, S., Okuno, G., Ikura, Y., Tanaka, T., Suda, M., and Nishikawa, M. (1965). *Biochem. Biophys. Res. Commun.* **19**, 517–523.
Temtany, S. A. (1966). Thesis, Johns Hopkins University, Baltimore, Maryland.
Thompson, W. H. S. (1969). *Clin. Chim. Acta* **26**, 207–221.
Thomsen, J. (1875). *Arch. Psychiat. Nervenkr.* **76**, 706.
Tizard, J. P. M. 1969). *In* "Disorders of Voluntary Muscle" (J. N. Walton, ed.), 2nd ed., pp. 579–606. Churchill, London.
Tooth, H. H. (1886). "The Peroneal Type of Progressive Muscular Atrophy." Lewis, London.
Tsukagoshi, H., Nakanishi, T., Kondo, K., and Tsubaki, T. (1965). *Arch. Neurol. (Chicago)* **12**, 597–603.
Tsukagoshi, H., Sugita, H., Furukawa, T., Tsubaki, T., and Ono, E. (1966). *Arch. Neurol. (Chicago)* **14**, 378–381.
Tunte, W., Becker, P. E., and Knoore, G. V. (1967). *Humangenetik* **4**, 320–351.
Turner, E. V., and Roberts, E. (1938). *J. Nerv. Ment. Dis.* **87**, 74–80.
Turner, J. W. A., and Lees, F. (1962). *Brain* **85**, 733–740.
Van Allen, M. W., Frolich, J. A., and Davis, J. R. (1969). *Neurology* **19**, 10–25.
van Bogaert, L. V., and Dewulf, A. (1939). *Arch. Neurol. Psychiat.* **42**, 1083–1097.

van Bogaert, L., and Moreau, M. (1939). *Encephale* **34**, 312–320.
van der Wiel, H. J. (1957). *Acta Genet. Statist. Med.* **7**, 348.
van Gehuchten, M. (1956). *Rev. Neurol.* **94**, 253–258.
van't Hoff, W. (1962). *Quart. J. Med.* **31**, 385–402.
van Wijngaarden, G. K., Bethlem, J., Meijer, A. E. F. H., Hulsmann, W. C., and Feltkamp, C. A. (1967). *Brain* **90**, 577–592.
van Wijngaarden, G. K., Fleury, P., Bethlem, J., and Meijer, A. E. F. H. (1969). *Neurology* **19**, 901–908.
Vessie, P. R. (1932). *J. Nerv. Ment. Dis.* **76**, 553–573.
Victor, M., Hayes, R., and Adams, R. D. (1962). *N. Engl. J. Med.* **267**, 1267–1272.
von Frischknecht, W., Bianchi, L., and Pilleri, G. (1960). *Helv. Paediat. Acta* **15**, 259–279.
von Gierke, E. (1929). *Beitr. Pathol. Anat. Allg. Pathol.* **82**, 497–513.
von Leyden, E. (1876). "Klinik der Ruckenmarks-Krankheiten," Vol. 2.
von Recklinghausen, F. D. (1882). "Uber die multiplen Fibrome der Haut und ire Beziehung zu den multiplen Neuromen." Hirchwald, Berlin.
Waldenstrom, J., and Haeger-Aronsen, B. (1963). *Brit. Med. J.* **2**, 272–276.
Walton, J. N., and Gardner-Medwin, D. (1969). *In* "Disorders of Voluntary Muscle" (J. N. Walton, ed.), 2nd ed., pp. 455–500. Churchill, London.
Walton, J. N., and Nattrass, F. J. (1954). *Brain* **77**, 169–231.
Weinstein, S., and Gorlin, R. J. (1969). *Amer. J. Dis. Child.* **117**, 427–433.
Welander, L. (1951). *Acta Med. Scand.* **141**, Suppl. 265, 1–124.
Welander, L. (1957). *Acta Genet. Statist. Med.* **7**, 321–325.
Welander, L. (1963). *Proc. Int. Congr. Hum. Genet., 2nd, 1961* Vol. 3, pp. 1629–1636.
Werdnig, G. (1891). *Arch. Psychiat. Nervenkr.* **22**, 437–470.
Werner, C. W. O. (1904). Doctoral Dissertation, Kiel University, Schmidt and Klaunig, Kiel.
Whalen, R. E., Huang, S., Peschel, E., and McIntosh, H. D. (1961). *Amer. J. Med.* **31**, 171–186.
Wheelan, L., and Race, R. R. (1959). *Ann. Hum. Genet.* **23**, 300–310.
Wilbrand, H., and Saenger, A. (1921). *Muenchen. Wiesbaden* **8**, 179.
Williams, H. E., Johnson, P. L., Fenster, F., Laster, L., and Field, J. B. (1963). *Metab., Clin. Exp.* **12**, 235–241.
Wohlfahrt, S. (1926). *Acta Med. Scand.* **63**, 195–219.
Wohlwill, F. (1942). *Amat. Lusit.* **1**, 373–391.
Worster-Drought, C., Greenfield, J. G., and McHenemy, W. H. (1940). *Brain* **63**, 237–254.
Yudell, A., Dyck, P. J., and Lambert, E. H. (1965). *Arch. Neurol. (Chicago)* **13**, 432–440.
Zellweger, H., and Hanson, J. W. (1967). *Arch. Intern. Med.* **120**, 525–535.
Zellweger, H., and Schneider, H. J. (1968). *Amer. J. Dis. Child.* **115**, 588–598.
Zellweger, H., Brown, I., McCormick, W. F., and Jun-bi Tu (1965). *Ann. Paediat* **205**, 413–437.
Ziprkowski, L., Krakowski, A., Crispin, M., and Szeinberg, A. (1966). *Isr. J. Med. Sci.* **2**, 338–243.

AUTHOR INDEX

Numbers in italics refer to the pages on which the complete references are listed.

SUBJECT INDEX

R

Rabbit
  ileum, ATP tachyphylaxis, 94
  muscle, vitamin E deficiency and
    actomyosin in, 176
    amino acids, 176
    chemical analysis of, 178
    creatine loss, 176–177
    nucleic acids, 176
    oxygen uptake, 175
    protein, 175–176
    therapy, 176
Rabies virus, myocardial involvement, 240
Rainey's corpuscles, 252
Rat
  diaphragm, ATPase in, 331
  muscle
    esterase activity, 303
    striated, 5-nucleotidase in, 327–328
  rectus abdominis, 5-nucleotidase in, 323
  vitamin E deficiency
    histopathology, 169
    muscle appearance, 163
    muscle lesions, 169
      inclusion bodies, 172
      nuclear membrane, 172
      ultrastructure, 172–173
    myometrium pigment, 197–198
    paralysis and, 161–163
"Receptor reserve," 112
Receptors, 56
  agonist/antagonist relationship, 106–108
  agonist combination, 58
  alkylating agents and, 110–111
  anesthetics, local and, 110
  blocking agents, 103–114
    competitive, 104–109
      agonist selectivity, 107
      criteria, 104–106
      theory, 104–105
    noncompetitive, 109–110
    selectivity, 104
  chemical modification, 113–114
  drug interactions, 57–58
  nicotinic, 113
  protectors, 111

sensitivity model, 114
spare, 111–113
Refsum's disease, 491–492
Reoviruses, 233–236
  infection
    antigen, 235
    immunofluorescence study, 235–236
    inoculation route and, 234
    myocardial damage, 234–235
Rheumatic heart disease, coxsackie virus
  role, 242
  experimental, 242–243
Ryanodine, 130

S

Sarcocystin, 257
Sarcocystis
  animals infected, 254–255
  birds and, 257
  cysts, 256–257
  development in muscle, 252–253
  diagnosis, 253
  humans and, 256, 257, 261
  incidence, 252, 254–255
  meat as food, and, 257
  morphology, 253–254
  pathogenicity, 256
  transmission mode, 253
Sarcocystis muris, 255
Sarcoidosis, 451
Schiff procedures, *see* Oxidative-Schiff
  procedures
Schiff technique, *see* Periodic acid-Schiff
  technique
Schistosomes, 283–284
  fish and, 284
Scleroderma, 390–391
Sclerosis
  amyotrophic lateral, 483
    recessive, 484
  cerebral, 498
  globoid cell, 498
Selenium, 178–186
  deficiency, 179
  nutrition role, 180
  toxic effects, 178–179
  vitamin E interrelationship, 179
Semliki Forest virus, 237–238
Skeletal muscle
  acetylcholine effect, 61

6
B 7
C 8
D 9
E 0
F 1
G 2
H 3
I 4
J 5